Intercellular Communication in Plants:
Studies on Plasmodesmata

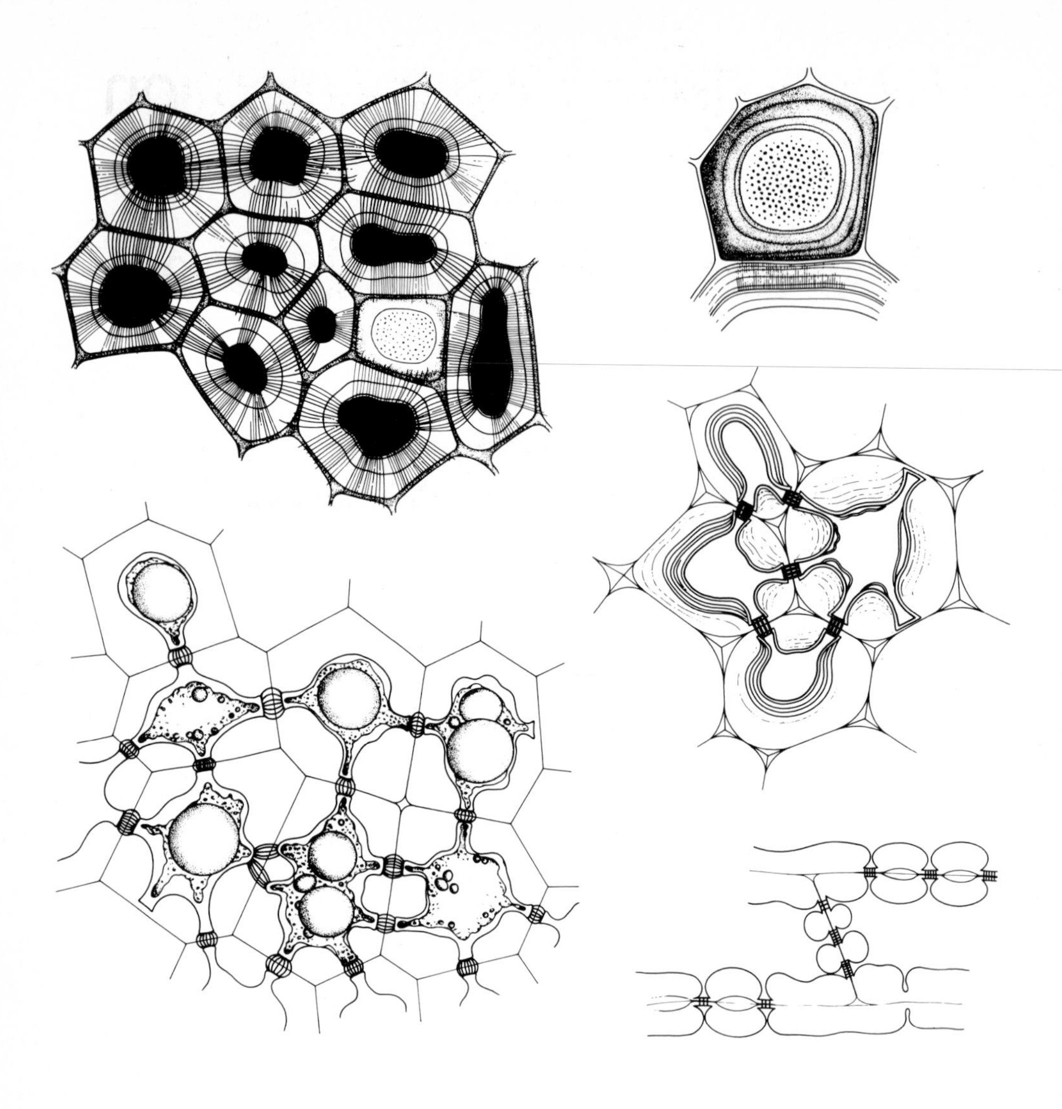

A Selection of Tangl's (1879) Drawings of *„Offene Communicationen"* in Endosperm Tissue

Top: *Strychnos nux-vomica "protoplasmic processes in the connecting canals... stained with iodine-potassium iodide";* left – thick section (x 390), right – showing the processes *"as points in profile"* (x 525).
Lower left: *Areca oleracea* stained with chlor-zinc iodide showing *"threads penetrating the middle lamella...between the ends of the pit canals"* (x 525).
Centre and lower right: *Phoenix dactylifera,* treated with chlorzinc iodide, cells viewed in transverse (x 590) and longitudinal (x 420) section.
All re-drawn and reduced x 0.85 from the original

Intercellular Communication in Plants: Studies on Plasmodesmata

Edited by
B. E. S. Gunning and A. W. Robards

With 90 Figures

Springer-Verlag Berlin Heidelberg New York 1976

Prof. B. E. S. Gunning, Research School of Biological Sciences,
Dept. of Developmental Biology, The Australian National University, Box 475, P. O.,
Canberra, A. C. T. 2601, Australia

Dr. A. W. Robards, Dept. of Biology, University of York,
Heslington, York, Y01 5DD, U.K.

ISBN 3-540-07570-4 Springer-Verlag Berlin Heidelberg New York
ISBN 0-387-07570-4 Springer-Verlag New York Heidelberg Berlin

Offsetprinting: Beltz Offsetdruck, Hemsbach/Bergstr. Bookbinding: Brühlsche Universitätsdruckerei, Gießen.

Preface

This Volume attempts to summarise and integrate a field of study in its entirety: the nature of plasmodesmata, and the part these intercellular connections play in the life of the plant. Except in the all-embracing early reviews of the pre-electron microscope era, there has been a tendency for the subject to be approached from disparate points of view: plant physiologists, developmental biologists, biophysicists, virologists and cytologists all contributing to the corpus of knowledge, but often without a full appreciation of each others' goals and problems, and sometimes misinterpreting each others' findings. In June 1975 a group of about 40 specialists in these various disciplines, all with a common interest in intercellular communication in plants, met for two days, presented papers, talked, argued, and in general pooled their knowledge. Out of a synthesis of manuscripts and discussions there has emerged, by an editorial process of elimination of unnecessary duplication and insertions to ensure completeness of coverage, the present book - not so much a straight record of a conference, as a Monograph based on the proceedings.

All of the Chapters are reviews and most include hitherto unpublished data or theoretical treatments. The work of reviewing some aspects of the subject has been unusual in that it has involved searching the literature for information which the authors of the papers may not have realised they had provided - a plasmodesma spotted in the corner of an electron micrograph, but not originally deemed worthy of comment, might in one of the present contexts represent a vital piece of a jigsaw. We cannot possibly claim to have uncovered all such clues, but we hope that our collation and redaction will provide not only a new starting point for those entering the field, but an illuminating review for those already in it, or who work in peripheral areas, perhaps unaware of the relevance of topics considered here.

In attempting to edit the transcripts of the open discussions we were presented with special problems, and have interpreted our duties very liberally in this area. Points raised in discussion that have been adequately dealt with in the text have been deleted; more often, we have taken the opportunity to expand upon original answers by asking the contributor to provide fuller details. We have also added relevant information, in reply to questions, where this was not presented at the meeting itself. In this way we hope to have covered many of the minor points that might otherwise have been neglected.

As regards nomenclature, the singular *plasmodesma* and plural *plasmodesmata* are by now widely used in the botanical literature and we follow the trend, avoiding variations such as *plasmodesm(s)* or *plasmadesma(ta)*. For the major compartments of the plant body we have followed Münch (1930), who introduced *symplast* and *apoplast*; his adjectives *symplastisch* and *apoplastisch* are translated as *symplastic* and *apoplastic*. The symplast may be considered as made up of *symplasm*, should this concept be required (as for *protoplast* and *protoplasm*), but all contributors agreed to avoid *symplasmic* and *symplasmatic* (both commonly used in the literature).

Système Internationale units have been used throughout, except that we have followed the Royal Society recommendations in continuing to employ familiar units such as the bar and the poise.

For those of us in the Department of Developmental Biology in the Australian National University Research School of Biological Sciences the production of this book has been very much a collaborative endeavour, and the Editors wish to thank their colleagues for all their help and forebearance. We also thank the contributors, without whom no comprehensive coverage could have been attempted. Dr. Ian Wardlaw, of the Division of Plant Industry, C.S.I.R.O., Canberra, deserves special mention for his skill and perception in leading the discussion at our plenary session: many of his contributions have been incorporated in the chapters and the open discussions that follow each of them, and it is a pleasure to record our gratitude to him. To the Director of the Research School, Professor Sir Rutherford Robertson, go our thanks for his encouragement and provision of the necessary financial assistance. Debbie Cooper, Marion Gunning, Joanne Hughes, Shirley Patroni, Hilary Payne, Val Rawlings, Ruth Robards and Sandy Smith have all contributed in innumerable and invaluable ways to the preparation of the text and the illustrations. We especially thank Gill Mayoh for masterminding and executing the exacting and arduous task of preparing the manuscript in 'camera-ready' form.

B.E.S.G., A.W.R.
September 1975

Contents

Discussants

Apart from authors, whose addresses appear on the title pages of their Chapters, the following contributed to the Open Discussions:

ASHFORD, A.	School of Biological Sciences, University of New South Wales, Kensington, N.S.W. 2033, Australia
BAIN, J.	Division of Food Research, C.S.I.R.O., Delhi Road, North Ryde, N.S.W. 2113, Australia
BOSTROM, T.	School of Biological Sciences, The University of Sydney, Sydney, N.S.W. 2006, Australia
BROWNING, A.	Department of Developmental Biology, Research School of Biological Sciences, The Australian National University, Box 475, P.O., Canberra City, A.C.T. 2601, Australia
COWAN, I.	Department of Environmental Biology, Research School of Biological Sciences, The Australian National University, Box 475, P.O., Canberra City, A.C.T. 2601, Australia
FINDLAY, G.	School of Biological Sciences, The Flinders University of South Australia, Bedford Park, S.A. 5042, Australia
GOODCHILD, D.	Division of Plant Industry, C.S.I.R.O., Canberra, A.C.T. 2601, Australia
GRESSEL, J.	Department of Plant Genetics, The Weizmann Institute of Science, Rehovot, Israel
HELMS, K.	Division of Plant Industry, C.S.I.R.O., Canberra, A.C.T. 2601, Australia
HUGHES, J.	Department of Developmental Biology, Research School of Biological Sciences, The Australian National University, Box 475, P.O., Canberra City, A.C.T. 2601, Australia
KING, R.	Division of Plant Industry, C.S.I.R.O., Canberra, A.C.T. 2601, Australia
KUO, J.	Department of Botany, University of Western Australia, Nedlands, W.A. 6009, Australia
LORIMER, G.	Department of Environmental Biology, Research School of Biological Sciences, The Australian National University, Box 475, P.O., Canberra City, A.C.T. 2601, Australia
LUMLEY, P.	Department of Botany, Monash University, Clayton, Vic. 3168, Australia

QUAIL, P. — Department of Developmental Biology, Research School of Biological Sciences, The Australian National University, Box 475, P.O., Canberra City, A.C.T. 2601, Australia

ROBERTSON, R. — Research School of Biological Sciences, The Australian National University, Box 475, P.O., Canberra City, A.C.T. 2601, Australia

WARDLAW, I. — Division of Plant Industry, C.S.I.R.O., Canberra, A.C.T. 2601, Australia

WATSON, L. — Taxonomy Unit, Research School of Biological Sciences, The Australian National University, Box 475, P.O., Canberra City, A.C.T. 2601, Australia

Pfeffer (1897), on Plasmodesmata

„Allgemeine physiologische Erwägungen lassen für Erreichung und Unterhaltung des harmonischen Zusammenwirkens, sagen wir kurz für die Reizverkettung, eine Continuität der lebendigen Substanz so nothwendig erscheinen, daß man dieselbe fordern müßte, wenn sie nicht schon entdeckt wäre. Damit ist aber wohl verträglich, daß die Verbindungen außerdem noch zum Stofftransport benutzt werden, ja daß sie in concreten Fällen diesem Zwecke vorwiegend oder gar ausschließlich dienstbar gemacht sind.“

“*General physiological considerations on the establishment and maintenance of correlative harmony, in short for the chain of stimulus transmission,* render a continuity of the living substance so essential that it would be necessary to propose it, even if it were not already discovered. *However, it is quite likely that the connections would also be utilised in the transport of substances and even, in particular cases, principally or solely for this purpose.*“
(Trans. D. J. Carr)

1

INTRODUCTION TO PLASMODESMATA

B.E.S. GUNNING

Department of Developmental Biology, Research School of Biological Sciences, The Australian National University, Box 475, P.O., Canberra City, A.C.T. 2601, Australia

1.1. INTRODUCTION TO PLASMODESMATA

Plasmodesmata are narrow strands of cytoplasm that connect neighbouring plant cells, penetrating through the intervening cell walls (Fig. 1.1.). The plant physiologist sees them as structures which elevate a plant from a mere collection of individual cells to an interconnected commune of living protoplasts. Cells and tissues that are remote from direct sources of nutrient can, it is thought, be nourished by diffusion or bulk flow through plasmodesmata, and materials can pass through them to and from the long-distance transport tissues of the vascular system. They also represent potential pathways for the passage of signals, either electrical or hormonal, that could integrate and regulate the activities of different parts of the commune. Their existence presents a challenge to the biophysicist: to compute in quantitative terms what their contribution to the functioning of the plant might be. They are objects of fascination to the cell biologist, tantalisingly close in their ultrastructural details to the limits of resolution of available methods of observation, and posing many problems in connection with their mode of development and the regulatory systems that govern their frequency and disposition and possibly, their functioning. These are the fields of research - none of them new, but not previously brought together for comprehensive and integrated study - with which this Volume is concerned.

Intercellular connections are not, of course, to be found in unicellular organisms. Even amongst multicellular plants, there are many simple types which lack plasmodesmata, mostly aquatic plants consisting of simple or branched filaments of cells, or of single or double layered two-dimensional expanses of cells. In such organisms, every cell is in contact with the surrounding water from which its nutrients are derived, and often the cells in the plant body are essentially identical copies, behaving as individuals in a colony, with little or no sign of differentiation or division of labour. We may conjecture that problems of nutrient supply or co-ordination may have limited the evolution of more complex plants than these, a pre-requisite for exploitation of the third dimension in the architecture of multicellular tissues being the emergence of transport systems which could cope with the nutritional requirements of internally-located cells, remote from direct external supplies , and ensure developmental regularity in organisms with differentiated cells. Judging by their taxonomic distribution amongst simple plants, it seems reasonable to suggest that the critical evolutionary breakthrough was the appearance of intercellular supply lines in the form of plasmodesmata.

Following the evolution of plasmodesmata, the plant body is, in essence, composed of two major compartments, for which the terms *apoplast* and *symplast* are convenient (Münch, 1930). The term apoplast refers to the non-living part of the plant body, mainly cell walls and intercellular spaces, all external to the plasmalemma. The term symplast refers to the interconnected protoplasts, all bounded by a continuous plasmalemma. There can be discontinuities in both of these major compartments of the plant, with, on the one hand, certain tracts of symplast isolated by a failure to develop cell-to-cell connections, and on the other, regions of apoplast sealed off by the formation of impermeable barriers.

The concepts of symplast and apoplast apply to animals just as to plants, but whereas in evolving their transport systems animals have been able to make full use of their apoplast compartment, elaborating from it complex vascular networks through which the intercellular fluids of blood, lymph, etc., are mechanically pumped to the vicinity of all cells, plants have not. Once the crucial evolutionary step of producing a cell wall had been taken by the progenitors of the plant kingdom, the development of transport systems, so necessary to support morphological developments, was constrained. Filling a large part of the apoplast with semi-rigid wall material precluded the use of large-scale mechanical pumping mechanisms, although the possibility of transport by diffusion through the apoplast remained, as did that of bulk flow of water and dissolved solutes from regions of absorption towards regions of evaporation. The advent of the xylem, a region of the apoplast derived from part of the symplast (which first becomes isolated and then dies) provided a long distance vascular conduit, but the major disadvantages of the apoplast as a vascular pathway in plants remained: that where and when it can carry a bulk flow, it is unidirectional (towards evaporating surfaces), while where and when it can carry diffusive fluxes in *any* direction (following concentration gradients), it is of low carrying capacity and adequate only over short distances. By contrast, plasmodesmata can be thought of as allowing for both diffusion and mass flow, in directions that can be specified by concentration gradients or by osmotic or hydrostatic forces generated within the symplast. The phloem system of the higher plant can be regarded as an evolutionary extension of the possibilities opened up by the existence of intercellular connections, for the nutrients carried in bulk, not just in one direction, but from any source to any sink, pass from one sieve element to the next through the enlarged plasmodesmata of the intervening sieve plates, bounded all the while by the plasmalemma.

In short, the major long distance pathways of solute transport in the animal body are apoplastic, whereas plants have generated a symplast, and used it to cope with problems of long distance transport as well as of short distance transport.

Much of what follows in this book aims to fill in the background to, and provide evidence for the glib statements just made about the part played by plasmodesmata in the life of the plant. It will become apparent that the established facts are few and the opportunities for research many. Among the uncertainties is the fundamental one of whether plasmodesmata can in fact function as routes for cell-to-cell transport - a question not fully resolved after 96 years of investigation. If we

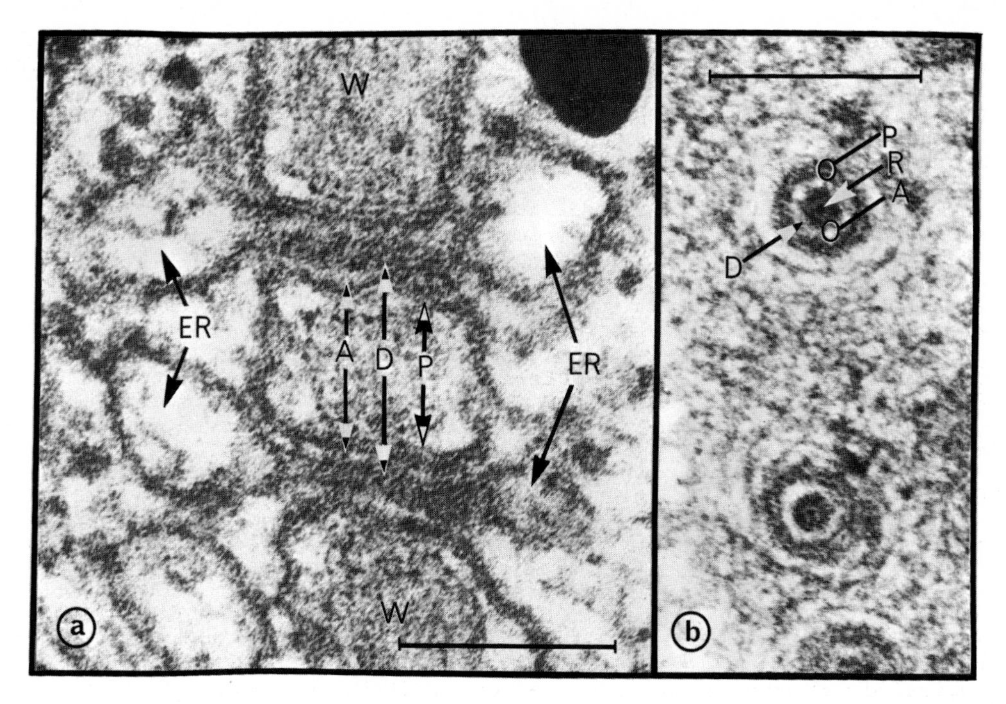

Fig. 1.1. Electron micrographs of plasmodesmata in longitudinal (a) and transverse (b) section.

In (a) a cell wall (W) passes vertically up the picture and the plasmodesmata pierce it at right angles. The plasmalemma (P) lines the plasmodesmatal canal, along the axis of which lies the desmotubule (D), here seen to be connected to the endoplasmic reticulum (ER).

In (b) the section is in the plane of the wall. The components of the plasmodesmata are: the plasmalemma (P), the desmotubule (D), with its central rod (R), and the cytoplasmic annulus (A).

Scale markers represent 100 nm, *Abutilon* nectary, distal wall of stalk cell

allow an affirmative to that question, there remains the question of what materials can be transported. To help answer these questions botanists can learn much from research upon other transport systems that are structurally comparable to plasmodesmata, and accordingly the remainder of this introduction will be devoted to a brief survey aimed at picking out some relevant points from the non-botanical literature.

Before examining any of their counterparts, we should know more about the nature of plasmodesmata themselves. A definition and a full description will be presented in Chapter 2, meanwhile it should be pointed out that in this book we are *not*, because of their anatomical or taxonomic restriction, considering under the heading of plasmodesmata the very large cytoplasmic channels that interconnect meiocytes in certain angiosperms (Heslop-Harrison, 1966a; Whelan, Haggis and Ford, 1974; Whelan, 1974), the pores of sieve plates (Esau, 1969), and the bewildering variety of pits and pores in red algae and fungi (Lee, 1971, and Chapter 3). We are concerned only with narrow, more or less cylindrical cytoplasmic channels (Fig. 1.1.), bounded by plasmalemma, and with an internal diameter in the range 30-60 nm, often carrying a hollow axial strand, the *desmotubule* (Robards, 1968a). The desmotubule is considered to be a derivative of the endoplasmic reticulum, and has external and internal diameters of 16-20 nm and 7-10 nm respectively; it may also contain an axial *central rod*. Thus, where a desmotubule is present, the complication arises that there are two, possibly independent, pathways for transport: one the desmotubule; the other the *cytoplasmic annulus* between the desmotubule and the inner face of the plasmalemma. Plasmodesmata are as long as the cell wall that they pierce is thick, and lengths in the range 0.1-1.0 μm are common. Their frequency varies, from less than one to more than fifteen per μm^2 of cell wall, implying that they occupy from much less than 1% of the wall surface to somewhat more than 1%, and that there can be as much, if not more, surface area of plasmalemma in the form of plasmodesmata than there is in the form of the ordinary bounding membrane of the cell. All of these features are treated fully later, but the above outline enables us to proceed and to assess the relevance of analogous systems now to be surveyed.

1.2. ANALOGUES OF PLASMODESMATA

Our main reason for looking at analogues of plasmodesmata is to glean evidence, however indirect, that structures of their size and shape might indeed function in transport, and to ask what sort of materials they might carry. Clearly, the subject of hydrodynamics is highly relevant, and it is surprising that with the exception of the seminal paper of Tyree (1970) on the symplast concept, this huge literature should have been so neglected. For instance, neither plasmodesmata nor their hydrodynamics receive any mention in a recent textbook on biorheology (including botanical aspects) (Scott-Blair and Spanner, 1974). Animal physiologists, by contrast, have been making full use of the following concepts.

1.2.1. The Hydrodynamics of Micropores: Theory

Much effort has been devoted to portraying the permeability to water and solutes of natural and artificial membranes in terms of the presence of *equivalent pores*, i.e. cylindrical channels piercing the membranes, of dimensions that could account for observed permeabilities and ultrafiltration properties. It is on this work that much of our understanding of the theory of chromatographic and other separation techniques rests. In most cases the equivalent pores are of molecular dimensions, and interactions between the permeating molecules and the walls of the pores dominate the flux characteristics. Such interactions have been assessed either by measuring reflection coefficients for the membranes and solutes, or by introducing correction terms into the classical equations that describe diffusion and convection through cylindrical channels (see review by House, 1974). The plasmodesmatal channels that we are concerned with are of larger radii than most of the equivalent pores in membranes, so that Fick's Law and Poiseuille's Law are more likely to provide accurate data on diffusive solute fluxes and convective flow rates respectively, but it is still useful to estimate the magnitude of the correction terms that are appropriate to plasmodesmata: their values tell us what is, and what is not, likely to be able to penetrate through the desmotubule and the cytoplasmic annulus. We need to examine both diffusion and convection, as there is evidence that both processes, and combinations of both, can participate in plasmodesmatal transport.

Let us first recall the laws of Fick and Poiseuille. The former deals with diffusion through a volume element, where the solute flux per unit area, J_s, is given by the free diffusion coefficient D, multiplied by the concentration gradient ($C_1 - C_2$ divided by length L), and the area fraction α that is available for diffusion i.e. :-

$$J_s = \frac{\alpha D\ (C_1 - C_2)}{L} \qquad \text{....(1)}$$

According to Poiseuille's Law, the volume flow J_v of a Newtonian fluid through a pipe (Fig. 1.2., central cylinder), is proportional to the fourth power of the radius R and to the pressure drop along the pipe ΔP, and is inversely proportional to the length L and to the viscosity η, i.e.:-

$$J_v = \frac{\pi R^4 \Delta P}{8 \eta L} \qquad \text{(divide by } \pi R^2 \text{ to obtain average velocity of flow)} \qquad \text{....(2)}$$

This, the Hagen-Poiseuille Law, named after the two who independently formulated it in 1839 and 1841, has to be modified to cover the more complex situation of flow through an annulus - which as already seen, is highly relevant to plasmodesmata. It becomes:-

$$J_v = \frac{\pi R_o^4 \Delta P}{8 \eta L} \left[(1-k^4) - \frac{(1-k^2)^2}{\ln(\frac{1}{k})} \right] \qquad \text{(divide by } \pi R_o^2 (1-k^2) \text{ to obtain average velocity} \qquad \text{....(3)}$$

where k (see Fig. 1.2., outer annulus) is the smaller radius of the annulus divided by the larger, R_o, so that when k is reduced to zero or becomes unity, i.e. the inner cylinder vanishes, we are left with the normal expression for flow through a cylindrical pipe[1].

The preceding two formulae underlie much of plasmodesmatal hydrodynamics, and the assumptions on which they rest should be recognised. First, it is assumed that steady state laminar flow occurs and that the liquid is incompressible. Laminar flow probably can occur in tubes of plasmodesmatal dimensions, the criterion here being the Reynolds number of the system, a dimensionless number equal to the fluid density times the average velocity of flow and the diameter, divided by the viscosity. Inserting reasonable values for the dimensions and the most rapid velocity of flow suggested so far to occur (Chapter 11), the conclusion is that for plasmodesmata the Reynolds number will be many orders of magnitude smaller than that (about 2100) at which turbulence starts. The diameter is so small that there should not even be any rippling.

A further assumption is that there are no 'end effects', that is, that the parabolic velocity profiles (Fig. 1.2.) build up quickly. An 'entrance length' for this to occur can be estimated as the product of the Reynolds number and 0.07 times the radius (Bird, Stewart and Lightfoot, 1960). On this basis a negligible correction factor is required, even for a plasmodesma in the thinnest of cell walls.

Other assumptions are less easy to apply to plasmodesmata. The flowing liquid should behave as a continuum, and there should be no 'slip' at the wall. Neither of these conditions is satisfied if the

---

[1] See also Chapter 5. Note that formula 12b in Tyree (1970) underestimates flow in an annulus by a factor of 2-3, and also, when the inner tube radius is reduced to zero, underestimates Poiseuille flow by a factor of 3. Tyree gives a correct formula in an erratum note published in Volume 27 (1971) of the Journal of Theoretical Biology.

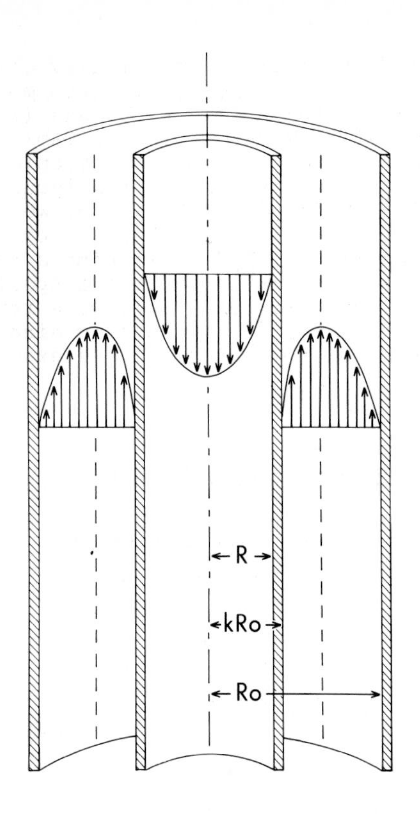

Fig. 1.2. Diagram illustrating laminar flow through a hollow cylinder (centre) and an annular pipe (outer compartment)

dimensions are such that, in the time taken for a molecule to pass through the tube, the random movements it undergoes as it diffuses bring about frequent interactions with the tube walls. In fact, when water passes along small tubes like plasmodesmata, the diffusive spread of the molecules within their transit times can be several hundred fold greater than the tube diameter (Taylor, 1953; Schindler and Iberall, 1973). What quantitative effects collisions, or even exchanges, between diffusing molecules and the irregularly disposed molecules of the wall might have is problematical, but it has been argued (Davis and Renkin, 1974) that their effects may be slight because the water molecules, which move in kinetic steps of about 0.2 nm every 10^{-11} sec or less, will interact many times with one another before they exchange energy with the wall of the tube. In other words, Poiseuille flow may still apply, except in tubes so narrow that the total diameter is of the same order as the molecular displacement per kinetic step.

Thus far the complication that molecules of various sizes may be present amongst the solvent molecules has not been raised. An immediate consequence is that small pores will automatically possess a capacity for ultra-filtration. Any molecule approaching the pore entrance in a zone less than half its own diameter from the wall of the pore will be impeded. The geometry is simply worked out, and for cylindrical and annular pores the cross sectional areas are effectively diminished by the factors given in Table 1.1., where a is the radius of the molecule in question, the pore has radius R and the annulus has radii R and kR (as in Fig. 1.2.).

Both for annuli and cylinders, these *steric hindrance to entry* correction factors are functions of a/R, and can be applied to diffusional fluxes into the plasmodesmata (or other pipes). For Poiseuille flow the correction is slightly less, as the edge hindrance actually constrains the solutes away from the non-flowing boundary and into more central regions where the flow rate is greater than the average (Fig.1. 2.). The appropriate factor for a cylinder is also given in Table 1.1. (Renkin, 1955). Considering either the desmotubule or the cytoplasmic annulus, and working out to what extent the entry of water molecules (a = 0.15 nm) and sucrose molecules (a = 0.44 nm) is hindered, we find (see Table 1.1.) that the gateway to a plasmodesma (either its cytoplasmic annulus or its desmotubule) has the potential of being a site where quite sensitive control over transport, at any rate transport by diffusion, could be imposed. We shall see later that in a wide range of plasmodesmata there are in fact constrictions where steric hindrance effects are likely to be much more dramatic than in the above examples.

TABLE 1.1.

CORRECTION FACTORS FOR TRANSPORT OF PARTICLES

	diffusion	convection
pipe	$\left(1-\frac{a}{R}\right)^2$ (a) 0.941 (b) 0.832	$2\left(1-\frac{a}{R}\right)^2-\left(1-\frac{a}{R}\right)^4$ (a) 0.9965 (b) 0.972
annulus	$\left[1-\frac{a}{R}\left(\frac{2}{1-k}\right)\right]$ (a) 0.9574 (b) 0.8752	?

Correction factors expressing the steric hindrance to the entry of particles (radius a) into a pipe (radius R) or an annular pipe (large and small radii R and kR), when passage of the particle is by diffusion and convection. Figures at (a) and (b) are examples representing the magnitude of the factors for molecules of (a) water (a = 0.15 nm) and (b) sucrose (a = 0.44 nm), for a pipe of radius 5 nm, and an annulus with R = 15 nm and k = 0.53

Once inside the pore, the passage of solute by both diffusion and convective flow is impeded by interaction with the walls. Wall correction factors, or *drag coefficients*, which seek to portray conditions within the pore as compared with those in a free solution, have been computed and used in a variety of biological and non-biological contexts (Paine and Scherr, 1975). Once again the ratio a/R is crucial: the closer it approaches to unity the greater the degree of ultrafiltration, and the smaller the *effective* area of the pore. Diffusive and convective solute fluxes require different treatment, but the differences are very slight up to a/R values of about 0.3. Derivation of the correction factors is exceedingly complex, but there are tables in Happel and Brenner (1965, page 320) and Paine and Scherr (1975); Verniori *et al.* (1973) present convenient graphs from which both the entry and wall corrections can be read off; Paine, Moore and Horowitz (1975) include graphs which combine all of the factors relating to diffusive transport, and from which the cross-sectional area of the pore that is effective for diffusion can be determined; and Renkin and Gilmore

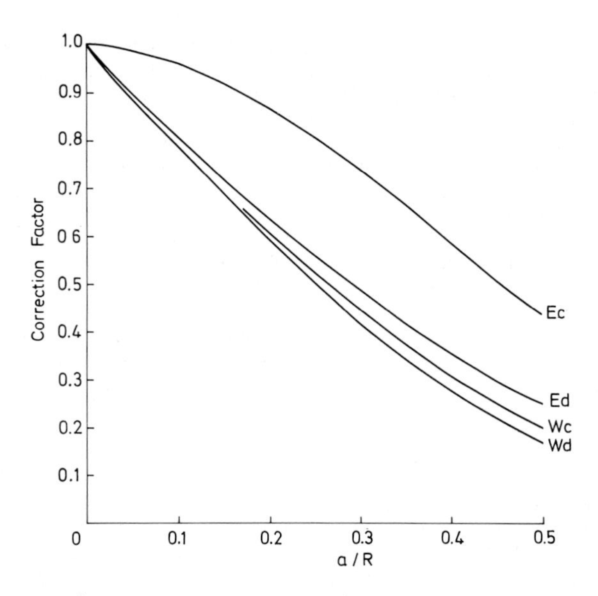

Fig. 1.3. Correction factors expressing the degree to which steric hindrance to entry (E) and interactions with walls (W) reduce the effective area for transport of particles (radius a) into and through pores (radius R), for both diffusive (d) and convective (c) movement. Note that the wall interaction factors (or *drag coefficients*) are often given as the reciprocals of the values graphed here

(1973) provide graphs not only for pores, but also for other geometries, including (of especial interest to botanists concerned with cell walls and p-protein in sieve tubes) fibre networks. Fig. 1.3. shows correction factors for both entry and wall interactions, and for both diffusion and convection, up to a/R = 0.5.

It is of interest to apply these tenets of hydrodynamic theory to the smallest of the tubes with which we are here concerned - the desmotubule (this, in fact, is all that we are able to do, for there is no comparable theory for transport through an annulus). As before, ignore any central rod, and assume a tube diameter of 10 nm. For purely diffusive transport of sucrose (a = 0.44 nm) in dilute solutions, the effective area is about 73% of the measured area. Considering only the convective transport, however, the flux is reduced but slightly, to about 96% of the uncorrected value. The relative contributions to total solute transport of diffusion and convection vary with the flow rate and the size of the solute. Obviously, with diffusion on its own there has to be a concentration drop from one end of the pore to the other, while with convection there is a pressure drop. Any ultrafiltration that occurs during convective flow leads to a concentration gradient that in turn generates a diffusive flux in the same direction as the convection. Verniori *et al.* (1973) have developed a comprehensive transport equation with terms that assess the contributions of diffusion, convection, and their interaction, in each case including the appropriate correction factors. Modified versions from which the extent of ultrafiltration can be estimated are given by Verniori *et al.* (1973) and Renkin and Gilmore (1973).

Application of hydrodynamic theory is all very well, but the theory was not developed with plasmodesmata in mind. We do not know, for instance, how much of the fluid inside plasmodesmata behaves as 'bulk' water (Cooke and Kuntz, 1974) as distinct from water molecules with their rotational and translational motions reduced through interaction with the macromolecules of the plasmalemma and desmotubule. It is probable that the walls of the tubules are permeable to water, and we do not know how this, and their molecular irregularities, might influence their rheological properties. Computation of the correction factors is based on the behaviour of single particles moving axisymmetrically through cylinders, and it is known that when numerous particles or molecules are present, the faster moving will collide with the slower, resulting in a tendency for migration towards the axis, and a change in the flux characteristics. Most importantly, we know nothing about the hydrodynamic behaviour of electrically charged particles in plasmodesmata. Tyree (1970) estimated that the effective diffusion coefficient of an electrolyte within a plasmodesma might be 33 times less than in free solution, the reduction being in part due to binding of some of the electrolyte to cytoplasmic components, and in part to electrostatic barriers in the pore.

Some of these problems are considered again later, or are attacked by different strategies. For the moment they loom so large that it is desirable to look briefly at some real, rather than theoretical, analogues of plasmodesmata, to try to find out whether, in fact, any of the foregoing is relevant.

1.2.2. The Hydrodynamics of Micropores: Practice

1.2.2.1. Artificial membranes and mica Dialysis tubing and cellulose membranes have been much studied, and their observed ultrafiltration characteristics are in reasonable conformity with theory (e.g. Renkin, 1955). Analysis is complicated, however, by the tortuosity of the channels in such material, resulting in the need to conceptualise equivalent pores. It is therefore something of a relief to know that the correction factors do predict very well the degree of ultrafiltration during diffusion through uniform straight pores drilled through mica by etching fission particle tracks, and covering the size range $R = 4.5$ up to 30 nm and $L = 3.5$ μm (Beck and Schulz, 1970). In reaching this conclusion a small protein, ribonuclease, with molecular radius 2.16 nm, was included in the tests, and it is noteworthy that for a desmotubule-sized pore diffusive ultrafiltration of ribonuclease would proceed as if the diffusion coefficient of the protein had been reduced about 20 fold.

1.2.2.2. Glomerular filtration Since the pioneering work of Pappenheimer (1953), most of the impetus to apply hydrodynamic theory to biological material has come from renal physiologists. Glomerular filtration from blood capillaries to renal tubules is completely opposite to cell-to-cell transport in plants, in that the pathway is entirely apoplastic, and geometrically constrained by the surrounding cells into narrow channels *outside* the cells. The channels themselves, nevertheless, or rather their equivalent pores, are remarkably similar to desmotubules, with radii from 3.6 to 5 nm, lengths up to about 0.3 μm, and carrying average flow rates of the order of 50 μm s^{-1}. Transport is driven both by diffusion and by hydrostatic pressures of 0.02-0.05 bar.

The theory conforms well with observations on flow rates and ultrafiltration properties for naturally-occurring molecules as well as for artificial probes such as polyvinylpyrrolidones, polyethylene glycols, and dextrans of selected molecular dimensions (see review by Renkin and Gilmore, 1973). For an equivalent pore of radius 5 nm, ultrafiltration for permeating molecules of 2.0, 2.5, and 3.0 nm radius amounts to 19%, 43% and 72% (% reductions from input concentrations). The largest of these molecular radii would correspond to a molecular weight of about 67,000.

Perhaps the most obvious conclusion for the student of plasmodesmata to draw from the huge amount of work on glomerular filtration is that there is no reason to doubt that plasmodesmata, and even their desmotubules, can be functional, and that whereas most cell proteins will be retained in their cells, small solutes can permeate through the symplast.

1.2.1.3. Nuclear envelope pores The two concentric membranes of the nuclear envelope are pierced by numerous pore complexes through which the nucleoplasm is considered to be in communication with the cytoplasm *via* diffusion. Numerous types of molecule have been used to probe the pores: ferritin, with radius about 5 nm, is just too large to penetrate; smaller molecules can pass, with greater ease as the size diminishes (Paine, Moore and Horowitz, 1975). Although electron microscopy of the pores would suggest that their lumen is larger, data on permeability of dextrans indicate a maximum patent channel of radius 4.5 nm. For amphibian oocytes the pore length is about 150 nm, so once again we are presented with a well-investigated analogue of a desmotubule. It is presumed that permeation of the tracers is by diffusion only, with the severe steric hindrance to entry and wall correction factors that this entails. The experimental data fit the equations describing diffusion as modified by the correction factors.

It is pointed out by Paine *et al.* (1975) that pore diameter sensitively controls the rate of transport. Thus a 1.0 nm change in pore radius would bring about a 2-3 fold change in the half-time for cytoplasm-nucleus mixing of a molecule the size of sucrose, a 10 fold change for a small protein (of radius 1.8 nm), and more than a 1000 fold change for a protein of radius greater than 3.1 nm. Clearly the possibility that cell-to-cell transport in plants could be regulated by modulations of plasmodesmatal dimensions must be borne in mind, and electron microscopists should be looking for ultrastructural equivalents of sphincters.

The other lesson that comes from observations on nuclear pores is that electron microscopy need not be a reliable guide to the patent diameter. The lumen looks wider than it actually is, a phenomenon that may or may not be relevant in the case of plasmodesmata.

1.2.3. Qualitatively Analogous Structures in Animals

There are channels between apoplastic compartments other than in kidney glomeruli. Recent work on capillary endothelial cells has revealed that they can be penetrated by narrow convoluted pores, possibly formed from catenations of vesicles. The diameter varies, but there are strictures of about 10 nm, and tracer molecules of 2 nm diameter can be carried (Simoniescu, Simoniescu and Palade, 1975).

Continuities within the animal equivalent of the symplast also exist. The morphology of the cell-to-cell connections is very different from those in plants, though there are striking superficial resemblances in the one animal tissue in which the apoplast becomes filled with an equivalent of the plant cell wall, i.e. bone: there the cell bodies of the osteocytes become separated, but remain in communication *via* elongated cytoplasmic strands that penetrate through the intervening matrix (Porter and Bonneville, 1968). Of greater relevance are the electrotonic junctions that can form between cells in a variety of animal tissues, both in the body and in tissue culture (Loewenstein, 1973; Bennett, 1973a,b).

The spread of an electric current that is applied within one cell to a neighbouring cell with less attenuation than would be expected if the plasmalemmas were intact implies that there is a specialised pathway for the electrical coupling. The phenomenon is seen in plants (Chapter 6), and has its counterpart in animals, where the 'gap junction', one of many types of cell-to-cell contact point (McNutt and Weinstein, 1973), is the most likely contender to be the electrotonic junction. We need not go into the structural details except to state that according to one model the individual intercytoplasmic connections are not much in excess of 1 nm in diameter (Bennett, 1973a,b). Many of these are grouped together in a hexagonal arrangement. Their penetrability varies from tissue to tissue and animal to animal, but in general small ions (K^+, Cl^-, $SO_4^=$, etc) can pass, as can sucrose (which would have an a/R ratio of 0.88), and a variety of dyes. With such diminutive channels it is not surprising that the cells can regulate the extent to which they are coupled, perhaps by conformational changes in the macromolecules of the junctions. Calcium ions and cyclic adenosine monophosphate are amongst the substances that can influence the communication system (Hax, van Venrooij and Vossenberg, 1974).

1.3. CONCLUSION

The remainder of this book will deal with plasmodesmata in detail. Meanwhile our excursion into the non-botanical literature has to some extent set the scene by drawing us to the optimistic conclusion that we need not have doubts about the efficacy of plasmodesmata as channels of transport solely on the ground of their minute dimensions. Equally small, or smaller, pores in other systems seem to be functional.

Tyree (1970) was the first to show that, small as plasmodesmata are, they still (in theory) provide a more efficient pathway than the alternative route from one plant cell to its neighbour, that is, by trans-plasmalemma transport into the cell wall, across the wall, and through the next plasmalemma. Several calculations to be presented later substantiate his conclusion. Instead of worrying about the small radii of the postulated conducting channels in plasmodesmata, we should perhaps be interpreting that same smallness as a necessity selected through evolution for its adaptive value. Thus plants appear to have evolved cell-to-cell lines of communication that are not too small to be functional, but are small enough to have ultrafiltration properties that permit the passage of low molecular weight materials, be they mineral solutes, organic nutrients, or hormonal messengers, while impeding or precluding the leakage of larger molecules such as enzyme proteins or ribosomal subunits. Further, within the confines of what is a very small structural entity, plasmodesmata would appear to possess two such potential pathways. Of these, the desmotubule, if it is indeed a derivative of the endoplasmic reticulum, is part of a closed, membrane-bound system which could in theory carry specially selected and sequestered solutes. It is of interest that the other pathway, the cytoplasmic annulus, is on the one hand much more prone to the hazards that might be associated with leakage of large cytoplasmic molecules, and on the other is often equipped with constrictions that would reduce this danger.

Thus far, the possible 'biological' properties of plasmodesmata have been neglected in favour of the 'physical'. Subsequent chapters will rectify the imbalance, but it would be wrong to conclude this introductory survey without noting that the plant might regulate cell-to-cell transport not only by relaxing or constricting the pathways, but also by inserting new plasmodesmata, or conversely, by occluding or breaking them. The placement of some sort of active-transport mechanism in plasmodesmata is another possibility. Additionally, the plant may use its plasmodesmata for purposes other than solute transport. Amongst those that have been mooted are roles in cell wall formation and breakdown, membrane flow from one cell to another, and the passage of electrical signals along the membranes.

1.4. OPEN DISCUSSION

The discussion opened with comments on the equations for flow through pipes. Concerning the viscosity term of the equations, WALKER pointed out that if there is a non-Newtonian material such as cytoplasm in the annulus, then the viscosity is not likely to be a constant: it will depend upon the shear rate, so that the velocity distribution across the pipe may not be as in Fig. 1.2. but more like that shown by Kamiya for slime moulds, where there is a comparatively stationary region near the wall of the pipe, a shear zone, and a wide central region where the velocity is much more uniform. He suggested that this would change the quantitative but probably not the qualitative conclusions regarding the transport capabilities of plasmodesmata. GUNNING accepted this caution, but noted that it should not apply to any transport that occurs in the lumen of the desmotubule, which is separated from the bulk cytoplasm by a membrane or a membrane derivative. Also, for the cytoplasmic annulus itself, it may be that ultrafiltration ensures that the fluid within the annulus is not simply a sample of bulk cytoplasm, but instead is a filtrate that behaves in a much more Newtonian fashion.

COWAN asked if the desmotubule moves along the axis of the plasmodesma. If it does then the boundary conditions that are assumed in the flow equations will be incorrect. GUNNING replied that it is simply not known whether the desmotubule moves or is static. It is assumed to be static, though one suggestion to the contrary is discussed in Chapter 11, where the possibility is raised that the special form of endoplasmic reticulum found in sieve elements is maintained by some form of membrane flow from the companion cells *via* the numerous plasmodesmata that interconnect the two cell types.

CARR then enquired about bulk flow through plasmodesmata, asking whether the presence of numerous particles affected the Reynolds number and the likelihood of generating non-laminar flow. The reply was that the Reynolds number for particle flow through plasmodesmata, as in the case of the Reynolds number for fluid flow, was far below the limit at which turbulence would be indicated. Where the particles are not crowded together, in other words as in a dilute solution, they would probably follow the lines of laminar flow. Where the particles are very crowded, as in a concentrated solution, then collisions tend to bounce the individual particles from the slower flowing laminae towards the axis generating a form of axial streaming.

GRESSEL was disturbed by the use of analogies with inert substances like mica: whilst mica may resemble to some extent fixed tissue, it is a poor substitute for living material where the membranes are flexible and mobile and can change their dimensions just as macromolecules can change their conformation. GUNNING agreed that the size range is such that control of transport by conformational changes in some of the plasmodesmatal components is a very real possibility. He said that there is at present no evidence for such phenomena, but that certain physiological observations pointed towards the possible existence of temporary closures or one-way valves for plasmodesmata.

The discussion closed with two fundamental questions: one from GRESSEL asking whether the hydrodynamic theory that had been described fits observations on transport through plasmodesmata; to which the reply was that so little is known about transport through plasmodesmata that we cannot tell whether the theory fits or not, although for viruses it seems that the theory most certainly does not fit, for large virus particles can move through plasmodesmata and yet, on the basis of ultrafiltration theory should be very strongly impeded. The other question was from BOSTROM who asked whether anybody had evidence that transport within the desmotubule does occur. GUNNING answered that his intuition told him somewhat teleologically that plants would not possess desmotubules unless they had some function, and then forecast that several suggestions for transport through the desmotubule would emerge in the subsequent papers, one of the tasks of the meeting being to evaluate these suggestions.

2

PLASMODESMATA IN HIGHER PLANTS

A.W. ROBARDS[1]

Department of Developmental Biology, Research School of Biological Sciences, The Australian National University, Box 475, P.O., Canberra City, A.C.T. 2601, Australia

2.1. INTRODUCTION

The first detailed description of "*offene Communicationen*", or protoplasmic connecting threads is usually attributed to Tangl (1879) although other workers subsequently contested his precedence. However, Tangl was the first botanist to write at length on these structures and his article stimulated a spate of publications which described such connections between cells from all parts of the plant kingdom (for fuller details refer to Meeuse, 1941b; 1957, and Chapter 14). In 1901 Strasburger used the term "*Plasmodesmen*" to describe the protoplasmic connections and, despite numerous other suggestions (see Meeuse, 1957), the word has survived the test of time and is now almost universally accepted (English - plasmodesma - Gk. *plasma*, form; *desma*, bond - Plu. plasmodesmata). Virtually all the early investigations involved treatment of cells to cause swelling of the wall so that plasmodesmata could be demonstrated by optical microscopy. This led to many criticisms and, as recently as 1964, Livingston reconsidered the "*nature of plasmodesmata in normal (living) plant tissue*", so highlighting the problem that has been with us for over 80 years. The advent of electron microscopy has done much to expand knowledge about the structure and variability of plasmodesmata; it has been of rather less value in helping us to understand the true nature and physiological function of these connections. In this Chapter I shall collate some of the information relating to the distribution, structure and possible physiological roles of plasmodesmata. The literature is extremely large and diffuse, and no attempt has been made to provide a comprehensive review.

---

[1]Written while on Sabbatical leave from the Department of Biology, University of York, Heslington, York, YO1 5DD, England (address for correspondence).

2.2. WHAT ARE PLASMODESMATA?

Most plant cells are joined to each other by *protoplasmic connecting threads* (see 2.3.). They are, therefore, closely linked, and together constitute something approaching a true syncytial structure. This, in itself, creates a fresh problem in defining what is a plasmodesmatal connection and what is merely an intercellular channel. Cytomyctic channels such as those described by Heslop-Harrison between angiosperm meiocytes (1966b) are not plasmodesmata, although they may each have been formed at the original site of a plasmodesma (for recent information on connections between angiosperm meiocytes, see Whelan, 1974; and Whelan, Haggis and Ford, 1974); similarly Weiling (1965) distinguishes between plasmodesmata and plasma channels in the pollen mother cells of *Lycopersicon* and *Cucurbita* on the basis of size (25.3 and 175 nm respectively) and structure (this contrasts with Ciobanu (1969) who refers to such channels as plasmodesmata, and has recorded the passage of organelles through them); Bisalputra and Stein (1966) examined the intercellular connections in *Volvox aureus* and concluded that they were "*cytoplasmic bridges*", not plasmodesmata (but see Dolzmann and Dolzmann (1964) for an opposing opinion). However, structures of intermediate size, which produce difficulties in definition are encountered. Meeuse (1957) attempted to solve the problem by a rather long definition: "*Plasmodesmata are continuous, tenuous threads of protoplasm connecting adjacent protoplasts through substances separating these protoplasts (generally speaking, plant cell walls, but also the mucilaginous intercellular substances in* Volvox *colonies, and, in the analogous case of animal tissues, substances such as bone, dentine and cartilage), thus establishing a correlated entity of interconnecting protoplasts. They are never so wide as to permit a real fusion of protoplasts or migration of protoplasmic inclusions, such as nuclei, plastids, microsomes, granula, under normal conditions, but may be converted into wider protoplasmic connections, or, conversely, arise in special cases by a reduction in diameter of wider protoplasmic strands connecting adjacent cells. They are situated in minute perforations of the cell wall, the plasmodesmoducts or plasmodesma channels (or pores).*"

Meeuse's comparison with animal intercellular connections is interesting because plasmodesmata have generally been considered exclusively a plant phenomenon. However, although the structure may be quite different, some junctions between animal cells pose many of the problems familiar to botanists (c.f. the 'cytodesma' of Komnick and Wohlfarth-Bottermann (1964), and the gap-junction model proposed by Bennett (1973a)).

Open cytoplasmic channels usually have no well defined core (desmotubule) but, unfortunately, nor do some plasmodesmata (see 2.4.), so a distinction cannot be made on this basis. For the moment it is probably both useful and sufficiently accurate to think of plasmodesmata as being small protoplasmic connections through which organelles and large macromolecules cannot normally pass (the special case of viruses is dealt with separately - Chapter 8), irrespective of whether or not a desmotubule is present. Juniper (in press) adopts a more liberal definition of plasmodesmata, finding it impossible to make a distinction on the basis of size alone. While far from satisfactory, it is impossible to produce a definition which will not exclude some of the structures which have been described in the literature and which must clearly be studied along with other plasmodesmata, although they may be distinctly different in structure (for example, the plasmodesmata found in *Bulbochaete* by Fraser and Gunning, 1969). One of the few features that appears common to all plasmodesmata (very few, possible, exceptions have been noted - see Chukhrii, 1971) is the continuity of the plasmalemma, through the plasmodesmatal canal, from cell to cell.

Despite almost 100 years of study by optical microscopy, the small size of plasmodesmata still needed electron microscopy for real advances in structural knowledge. It is interesting, however, to consider two relatively late papers which deal with plasmodesmata from an anatomical viewpoint. Both authors (Meeuse, 1957 and Livingston, 1964) refer to results from electron microscopy, available at the time that their papers were published. Even in 1964, Livingston considered the presence of plasmodesmata so poorly substantiated (despite an already growing literature from electron microscopy) that he felt it necessary to stress the particular point that his own study "*answers the criticism which has frequently been made in the past that these structures (plasmodesmata) are artifacts* ...".

Meeuse (1957) supported the suggestion of Strasburger (1882), that living cells in (higher) plants are connected by fine strands of protoplasm. The protoplasmic nature of plasmodesmata, assumed from their first discovery, and based mainly on ideas about their involvement in translocation, was not challenged seriously until 1930 when Jungers suggested that they are trapped nuclear spindle fibres[1]; however, Hume (1913) had earlier commented that "*it remains an open question whether they (plasmodesmata) arise from the spindle fibres and penetrate the wall* ab initio *or whether they may be formed subsequently*". (Hume was referring to plasmodesmata across a non-division wall). Meeuse (1957) collated the optical microscope evidence against the viewpoint of Jungers (see also Livingston, 1935, 1964). The main points he listed are: plasmodesmata have an affinity for protoplasmic stains; plasmolysis experiments either leave the plasmodesmata connected to the withdrawn protoplast, or may extract the threads from the wall (Strasburger, 1901; see also Burgess, 1971); enzymes move easily into plasmodesmata (Gardiner, 1897); plasmodesmata are only found where they can actually join protoplasts (e.g. not on outer walls); and plasmodesmata disappear from the walls of dying cells. Meeuse effectively summarizes the arguments and concludes that "*the accumulated evidence seems to be overwhelming and conclusive*" (that plasmodesmata are protoplasmic).

2.3. DISTRIBUTION AND FREQUENCY

The review by Meeuse (1941b) amply illustrates that optical microscopists had described plasmodesmata in most plant species, although many observations were contested. Electron microscopy has made the identification of plasmodesmata more certain but, at the same time, has imposed great problems in scanning sufficient material to obtain statistically useful results. However, it can be stated with certainty that plasmodesmata occur in angiosperms, gymnosperms, pteridophytes, bryophytes, and many algae. The distribution, frequency, and structure in fungi and blue-green algae are less well documented, although plasmodesmata are reported from both of these groups (see Chapter 3).

2.3.1. Distribution in Higher Plants

Statements that plasmodesmata occur between all living cells of higher plants (e.g. Meeuse, 1941b: Fahn, 1967) are now known to be

---

[1]Some of the earlier botanists distinguished spindle fibres from protoplasmic components - a separation that would not now be made. The nature of the traversing strand is discussed more fully in 2.4.2.

strictly false. However, it does seem that *all* cells in the early life of a plant are connected to each other (Schulz and Jensen, 1968a), and that the subsequent absence of plasmodesmata from particular walls is brought about by a pattern of loss - an area of investigation worthy of much fuller study. (A full account of the developmental consequences of isolation of cells from tissues is given in 13.4.).

The presence or absence of plasmodesmata, as well as their relative frequency between different cells, must relate to any functional role. This is frequently stated or implied in the literature, particularly where the performance demands very high symplastic fluxes between cells, or where cells would need to be isolated from their neighbours. As examples of the first case (high fluxes) the location and frequency of plasmodesmata seems particularly important in secretory glands (e.g. Cardale, 1971 - salt gland of *Aegiceras*; Hill and Hill, 1973 - salt gland of *Limonium*; Thomson and Liu, 1967 - salt gland of *Tamarix*; Ziegler and Lüttge, 1966, 1967 - salt gland of *Limonium*; [Lüttge, 1971, reviews plant glands - including the role of plasmodesmata]); in root nodule tissue (Pate, Gunning and Briarty, 1969; Sprent, 1972) and in the trigger hair of the Venus's fly-trap (Williams and Mozingo, 1971).

Lack of plasmodesmata between adjacent cells may be as significant as a high frequency, even if more difficult to prove! The relationships of reproductive and embryonic cells with each other have proved especially interesting, and, as a general rule, the structures of separate plant generations are *not* linked by plasmodesmata. Plasmodesmata are said to be absent between the zygote and surrounding cells of *Capsella* (Schulz and Jensen, 1968a,b), of *Quercus gambelii* (Singh and Mogensen, 1975), and of *Hordeum* (Norstog, 1972); also absent between the suspensor and embryo sac of *Capsella* (Schulz and Jensen, 1969), between the nucellus and megagametophyte of *Zea mays* (Diboll and Larson, 1966), between the nucellus and embryo sac and zygote and all other cells in *Myosurus* (Woodcock and Bell, 1968), between cells destined to form pro-embryoids in *Citrus* (Button, Kochba and Bornman, 1974), between adjacent pollen grains at later stages of development (Heslop-Harrison, 1964, 1966a), and from walls separating generative cells from each other and from surrounding tapetal cells (Ledbetter and Porter, 1970). A careful study of plasmodesmatal distribution during oogenesis in *Marchantia* has been made by Zinsmeister and Carothers (1974), who document the presence and absence of connections across different archegonial cell walls. These examples all refer to specific cases, but there can be little doubt that they are indicative of general categories of plasmodesmatal exclusion that further studies will reveal. (The significance of cell isolation in the development of alternating generations during reproduction is considered in 13.4.4.).

A more contentious area is the stomatal guard cell wall, where some authors have stated plasmodesmata to be absent (e.g. Allaway and Setterfield, 1972 - *Vicia* and *Allium*; Thomson and de Journett, 1970 - *Opuntia*; Brown and Johnson, 1962 - 16 species of grass), or 'infrequent' (Tucker, 1974 - Magnoliaceae), while other workers have reported their presence (Burgess and Fleming, 1973 - *Pisum*; Inamdar, Patel and Patel, 1973 - *Asclepiadaceae* [optical microscopy]; Pallas and Mollenhauer, 1972b - *Vicia* [!] and *Nicotiana*). Discrepancies between different observations - sometimes on the same species - might conceivably find an explanation in developmental changes, for example, the loss of plasmodesmata during maturation. If this is so, then it will be most important to determine, and cite, the age or state of development of the particular cells so that comparative observations can be made. (This point is discussed in detail in 13.4.2.).

Early statements relating to the inability to demonstrate plasmodesmata in certain walls may often be ascribed to the inadequacy of techniques. For example, Livingston (1935) cites the often quoted absence of plasmodesmata from cambial cells, but electron microscopy reveals their undoubted presence. Similarly, early reports that plasmodesmata are absent from tobacco callus tissue (Kassanis, Tinsley and Quak, 1958) have more recently been revised and plasmodesmata clearly shown (Kassanis, 1967; Spencer and Kimmins, 1969). This is in keeping with the increasing citations of plasmodesmata in different tissue culture and other *in vitro* systems (e.g. Fowke, Bech-Hansen, Constabel and Gamborg, 1974 - dividing cells of soybean; Haccius and Engel, 1968 - callus culture of *Cannabis* [optical microscopy]; Halperin and Jensen, 1967 - carrot suspension culture aggregates).

So far, in this account of distribution of plasmodesmata, I have confined the discussion to plasmodesmata in cell walls which would usually arise during the process of cell division. The formation and development of plasmodesmata are considered fully in Chapter 4, but at this point it is relevant to state that there is good reason to support the view that plasmodesmata can form across *non-division* walls.

To summarize, plasmodesmata are to be found in all higher plants that have been examined, pteridophytes, bryophytes, some algae and some fungi. Some mature cells may have no intercellular connections, while in others they are distributed in a manner which suggests a quantitative relationship with the fluxes of solutes in the symplast. Often different walls of the same cell have different plasmodesmatal frequencies (see the data for *Nicotiana*, *Zea mays* and *Hordeum* in Table 2.1.; and for vein phloem in Table 11.1.). Plasmodesmata are also found in walls which have not been formed during the normal process of cell division; they must therefore be thought of as dynamic structures, capable of secondary elaboration, and not simply as trapped structural remnants from cell plate formation.

2.3.2. Frequencies between Cells

In many cells, particularly when a secondary cell wall has been deposited, plasmodesmata are found grouped together: such groupings correspond to the primary pit fields and, later, to the pits. Much work remains to be done in determining whether the initial distribution of plasmodesmata across a wall is random, so that grouping occurs through selective loss, or whether the original pattern of distribution is already in a grouped form (see Chapter 4). This matter is important, not least because it means that calculations of plasmodesmatal frequencies must be made with great care to include samples from the whole intercellular face and not simply from the pits or pit fields. The range of plasmodesmatal frequencies is large (Table 2.1.). In presenting this information it was, however, a matter of greatest difficulty to select from the literature frequencies which can be regarded as statistically reliable. Counts made by optical microscopy are about an order of magnitude lower than those made by electron microscopy on the same cells (e.g. Krull, 1960, who compared her own data, obtained by electron microscopy, from *Viscum* with those obtained by Kuhla (1900) using optical microscopy on the same species). Even electron microscopy presents difficulties: plasmodesmata are either sectioned longitudinally, and frequencies determined on the basis of the estimated section thickness; or they are sectioned transversely, when it is an exacting task to make sure that the correct wall is being sectioned and that the areas being examined are representative of the whole interface.

Table 2.1. cites some plasmodesmatal dimensions and frequencies taken from the literature. These figures should be self-explanatory,

TABLE 2.1.

DIAMETER AND FREQUENCY OF PLASMODESMATA

(Data determined by electron microscopy except where stated)

1. ANGIOSPERMS AND GYMNOSPERMS

Species and Original Source	Cell Type	Diameter (nm)	Frequency (per μm^2)
Nicotiana tabacum	Various	500	
	Epidermis End walls		0.30-0.36
	Epidermis Side walls		0.18-0.25
Livingston, 1935	Outer cortex End walls		0.21-0.24
	Outer cortex Side walls		0.08-0.09
	Inner cortex End walls		0.09-0.13
	Inner cortex Side walls		0.09-0.11
	Basal septum, hair cells		0.14
	Data determined by optical microscopy from 'mature unswollen tissue'.		
	Callus cells	50-70 300-500*	-----
Spencer and Kimmins, 1969	*This larger diameter derived from optical microscopy; possibly a good example of the relative dimensions recorded from optical and electron microscopy.		
Pinus strobus	Cambium	< 200 (550)*	-----
Sequoia sempervirens	Xylem ray and wood parenchyma	200	-----
Lycopersicon esculentum	Parenchyma	200	-----
Livingston, 1964	Optical microscope measurements. *Diameter of central nodes.		
Lycopersicon esculentum and *Cucurbita maxima*	Pollen mother cells	25.3	-----
Weiling, 1965	Also plasma channels 175 nm diameter.		

Table 2.1 (continued)

Species and Original Source	Cell Type	Diameter (nm)	Frequency (per μm²)
Metasequoia glyptostroboides	Sieve areas	50-85 (max. 350)	-----
Kollmann and Schumacher, 1963	Median nodule 3.0 μm diameter		
	Phloem parenchyma - radial cell walls	50-60	30-40*
	Sieve cells - tapering radial end walls	50-500	7**
Kollmann and Schumacher, 1962	*Frequency in pit membrane <u>only</u> **Frequency in sieve area <u>only</u>		
Viscum album	Epidermal cells	----	0.02-0.04
	Cortical parenchyma	----	0.01-0.05
	transverse walls	----	0.14-0.38
Kuhla, 1900	Cambial radial walls	----	0.05-0.12
	tangential walls	----	0.12-0.19
	Pith parenchyma	----	0.04-0.22
	All figures rounded from three to two significant places - optical microscopy.		
Krull, 1960	Mature cortical cells of growing internodes	15*	0.6-2.4**
	Figures cited by Tyree, 1970, as *50-60 nm, and **2.0. The plasmodesmata studied by Krull are fine strands through tubes that frequently anastomose. She was unable to give detailed dimensions but referred to 15 nm diameter connections. Because she counted <u>all</u> channels cut transversely, and because she found the mean number of 'arms' per plasmodesma to be three, she divided the number of connections counted by three. The frequencies cited here include this downward adjustment.		
Cucurbita pepo	Root endodermal cells:		
	Inner tangential wall	----	6.2
Robards, 1975	Outer tangential wall	----	10.4
	Radial longitudinal wall	----	6.9

Table 2.1 (continued)

Species and Original Source	Cell Type	Diameter (nm)	Frequency (per μm²)
Phalaris canariensis López-Sáez *et al.*, 1966a	Root tip cells *Outer diameter	40-50*	-----
Pisum sativum Vian and Rougier, 1974	Parenchyma from root apex: thin (0.2 μm) walls thicker (several μm) Determined from sections of frozen material. The plasmodesmata are reported to have constrictions (necks) at each end	 60 40-50	-----
Pinus strobus Murmanis and Evert, 1967	Phloem parenchyma	30-50	-----
Murmanis and Sachs, 1969	Pit membranes of secondary xylem tracheids Plasmodesmatal diameter increases as cells age	30-50	-----
Pinus pinaster Carde, 1974	'Strasburger' cells of pine needle. There is a small constriction at each end of the plasmodesma. The frequency cited (*) applies to the pit field only. Over the whole wall, the frequency is much less (Carde - personal communication).	60	20-25*
Abutilon Gunning, this Volume	Distal wall of stalk cell of nectary hairs *Outer diameter (See Table 2.2.)	44*	12.5
Oenothera Brinckmann and Lüttge, 1974	Mesophyll cells *Assumptions of pore size made by A.W.R. **Frequency derived from my assumptions of pore diameter in conjunction with 0.38% open wall area cited by original authors	40 or 50*	3 or 2**
Dactylorchis fuchsii Heslop-Harrison, 1966b	Archesporial cells	28	>7

Table 2.1 (continued)

Species and Original Source	Cell Type		Diameter (nm)	Frequency (per μm^2)
Salsola kali	Wall between mesophyll and bundle sheath cells		44*	14**
Olesen, 1975	**Internal* diameter of pore **57 μm^{-2} within pit fields			
Zea mays	Transverse walls of cap cells			
Clowes and Juniper, 1968	Meristematic		25*	4.5
	Peripheral			0.81
	*At least during early stages of differentiation			
	Root cap cells			
Juniper and French, 1970 also Juniper and Barlow, 1969	Meristematic	Transverse	----	14.87
		Longitudinal	----	5.30
	Peripheral	Transverse	----	5.13
		Longitudinal	----	0.45
	Quiescent cell zone			
	Central		----	5.76
	Peripheral		----	1.52
Hordeum vulgare	End walls of young endodermal cells		46*	-----
Robards *et al.*, 1973; Robards, 1975 and unpublished results	Inner tangential wall of root endodermal cells (distance from root tip)	2 mm		1.2
		4 mm		0.7
		10 mm		0.7
		30-40 cm		0.7**
	Outer tangential wall of root endodermal cell		60-70*	0.4
	Radial wall of root endodermal cell			0.3
	*Outer diameter (see Table 2.2.) **Derived from sections cut in the plane of the cell wall (i.e. plasmodesmata sectioned transversely); in all other cases frequencies were determined from counts of plasmodesmata which had been sectioned longitudinally			
Helder and Boerma, 1969	Endodermis 5 mm from root tip		60-90*	5.0**
	*Measured in the mid-line of the wall. In the neck region the diameter of the pore is given as 30 nm **Derived and cited by Clarkson *et al.*, 1971 by making assumptions for area of endodermal wall			

Table 2.1 (continued)

Species and Original Source	Cell Type	Diameter (nm)	Frequency (per μm²)
Triticum aestivum Kuo *et al.*, 1974	Inner tangential wall of mestome sheath cells abutting phloem *The diameter is of the conducting pore, and *excludes* the plasmalemma **Derived by A.W.R. from the published figures - frequencies within pit field were within the range 25-38 μm^{-2}; the authors comment on sampling problems involved in such counts	50*	7.7**
Ledbetter and Porter, 1970	Stamen filament cells	40	-----
Allium cepa	Root meristem		
Strugger, 1957a		30-40*	6-7**
Strugger, 1957b		100-200	-----
	An interesting case, where the dimensions derived from electron microscopy (30-40 nm) have subsequently been 'corrected' to allow for presumed shrinkage. (*Cited by Tyree, 1970, as 80-100 nm). **The implication from Krull (1960-p.616) is that this frequency range applies to the pit-field only		
Scott *et al.*, 1956	Mature cortical cells of root *Calculated by Tyree, 1970, from micrographs of shadowed cell walls: a dubious method of obtaining plasmodesmatal dimensions, which will almost certainly be overestimated	80-120*	1.5*
Various	Various		
Buvat, 1957		25-50	-----
Buvat, 1963		20	-----
Buvat, 1969		<50	-----
Avena sativa Böhmer, 1958	Cortical cells of mature coleoptile *Derived and cited by Tyree, 1970	60-100*	3.6*
Tradescantia virginica Van Went *et al.*, 1975	Cross walls of staminal hairs *Outer diameter	45*	11

Table 2.1 (continued)

Species and Original Source	Cell Type	Diameter (nm)	Frequency (per μm²)
Tamarix aphylla	Walls between collecting and secreting cells of salt glands	80	17*
Thomson and Liu, 1967	Anticlinal walls of secretory cells	60	
	*Derived and cited by Tyree, 1970		
Tussilago farfara	Wall between companion cell and sieve element of minor vein (branched channels)		
Gunning *et al.*, 1974	Companion cell side	43	6
	Sieve element side	62	2
Tillandsia usneoides Dolzmann, 1965	Hair cell walls	50-70	-----
Utricularia monanthos	Transverse wall between pedestal and basal epidermal cell of trap hairs*	----	7
Fineran and Lee, 1975	Transverse wall between pedestal and terminal cells of hairs**	----	35
	*These plasmodesmata have four arms on the epidermal cell side but a single arm into the transfer cell (counted as single plasmodesmata) **Simple plasmodesmata		
Impatiens balsamina Jones, this Volume	Wall between giant cells induced by the root knot nematode *Meloidogyne incognita* *29 μm^{-2} within pit fields	50	2.7-7.6*
2. PTERIDOPHYTES			
Dryopteris filix-mas Burgess, 1971	Root meristem young primary walls	50-80*	140
	Pit fields in older primary walls	----	10-20
	*"Pore through the cell wall"		
Polypodium vulgare Fraser and Smith, 1974	Wall between basal and protonemal cells of gametophyte *Outer diameter, and containing a 22 nm desmotubule. The plasmodesma is within a 'wall canal' 110 nm wide	60*	-----

Table 2.1 (continued)

Species and Original Source	Cell Type	Diameter (nm)	Frequency (per μm^2)
Azolla filiculoides Robards, unpublished	Cortical cells of young root *Outer diameter	35*	-----
3. BRYOPHYTES			
Polytrichum commune Eschrich and Steiner, 1968	Leptoid end walls Parenchyma cell cross walls	---- ----	16-20 9-12
4. FUNGI			
Rhizopus sexualis *Gilbertella persicaria* Hawker *et al.*, 1966	Walls between suspensors and young gametangia	7.5-10	-----
Geotrichum candidum	Vegetative cells		
Wilsenach and Kessel, 1965		30-60	-----
Hashimoto *et al.*, 1964		20-70	-----
5. ALGAE			
Egregia menziesii *Fucus evanescens* Bisalputra, 1966	Vegetative cells	37.5	-----
Himanthalia lorea Berkaloff, 1963	Meristematic cells	50	-----
Laminaria spp. Ziegler and Ruck, 1967	Trumpet cell cross walls [1]	60	50-60
Laminaria groenlandica Schmitz and Srivastava, 1974a	Sieve elements[1]	60	42

[1]There is some difficulty in comparing the sizes of sieve plate pores and true plasmodesmata in these algae: the large diameters are generally of sieve plate pores, but have been included here for the sake of comparison, and because there is a range of cited dimensions

Table 2.1 (continued)

Species and Original Source	Cell Type	Diameter (nm)	Frequency (per μm^2)
Pelagophycus spp. Parker and Fu, 1965	Conducting elements	[300-800] [1]	4-6
Macrocystis pyrifera Parker and Philpott, 1961	Sieve tubes[1] *Ziegler (1963) considered that these pores were ordinary plasmodesmata rather than sieve plate pores. J. Parker (1964) discussed this interpretation. (See also Chapter 3).	30-50*	-----
Macrocystis pyrifera Ziegler and Ruck, 1967	Sieve tubes *This frequency is cited by Ziegler and Ruck (1967) as from Ziegler (1963), but such a frequency cannot be correct in combination with pores of the quoted diameter, and Ziegler (1963) does not explain the origin of the figure	[2,000-3,000] [1]	[1.0]*
Macrocystis pyrifera Parker and Huber, 1965	Sieve tubes *Calculated by A.W.R. from data cited	[2,400-6,000] [1]	0.008*
Bulbochaete hiloensis Fraser and Gunning, 1969; 1973	Filament cells *Outer diameter Pores at ends of plasmodesmata 8.5-15 nm diameter. The plasmodesmata are contained within 'wall canals' of about 120 nm diameter	75*	3.2
Oedogonium Pickett-Heaps, 1967	Vegetative cells	130*	-----
Coss and Pickett-Heaps, 1974	Antheridia *Outer diameter, calculated by Dr. H. Marchant from authors' micrographs	77*	-----
Ulothrix Floyd, Stewart and Mattox, 1971	Filamentous cells *Outer diameter, calculated by Dr. H. Marchant from authors' micrographs	28-36*	-----

[1]See footnote on previous page

Table 2.1 (continued)

Species and Original Source	Cell Type	Diameter (nm)	Frequency (per μm²)
Aphanochaete Stewart, Mattox and Floyd, 1973	Vegetative cells *Outer diameter, calculated by Dr. H. Marchant from authors' micrographs	20*	-----
Coleochaete scutata Marchant, unpublished	Vegetative cells	30-70	-----
Chara spp. Pickett-Heaps, 1967	Meristematic vegetative cells	48-64	-----
Chara spp. Marchant, unpublished	Internode/nodal	82-120	-----
Chara corallina Fischer *et al.*, 1974	Internodal/peripheral cell wall	100	12
	Internodal/central cell wall	118	14
Nitella translucens Spanswick and Costerton, 1967	Wall between mature internodal and nodal cells	55(71)*	2.5**
	Wall between mature nodal cells	52(68)*	4.5**
	*Figures in brackets are the *outer* diameters, obtained by adding 2x8 nm, for the thickness of the plasmalemma, to the diameter of the pore **Frequencies are much higher between immature cells: 6.8 μm^{-2} and 67 μm^{-2} for internodal/nodal, and nodal/nodal respectively. Frequencies calculated by A.W.R. from figures cited		

although it is not always clear exactly how the different authors have made their measurements. It is essential, if transport phenomena are to be considered, that frequencies and sizes of connecting channels are precisely determined and stated. From the data available, some general conclusions may be drawn. For example, the plasmodesmata of higher plants have an outer diameter of up to about 60 nm (external diameter of plasmalemma-lined tube). The frequencies range from less than 1.0 to more than 10.0 per square micrometre over whole wall surfaces (i.e. not within pit fields alone). The maximum frequency of face-packed plasmodesmata of 60 nm outer diameter would be almost 280 per μm^2. Figures of 180 (anticlinal walls of *Osmunda cinnamomum* shoot apex, calculated by Juniper (in press) from micrographs of Hicks and Steeves (1973)), and 140 (primary walls of *Dryopteris filix-mas* root meristem - Burgess (1971)) are very high frequencies indeed. The available data on these two ferns is limited and in view of the *relative* paucity of plasmodesmata in apical (e.g. Bowes, 1965; Havelange *et al.*, 1974) and intercalary (Leshem, 1973) meristems in angiosperms, further documentation is highly desirable (see also 4.5.). Although the frequencies per square micrometre do not appear very dramatic, even the smallest

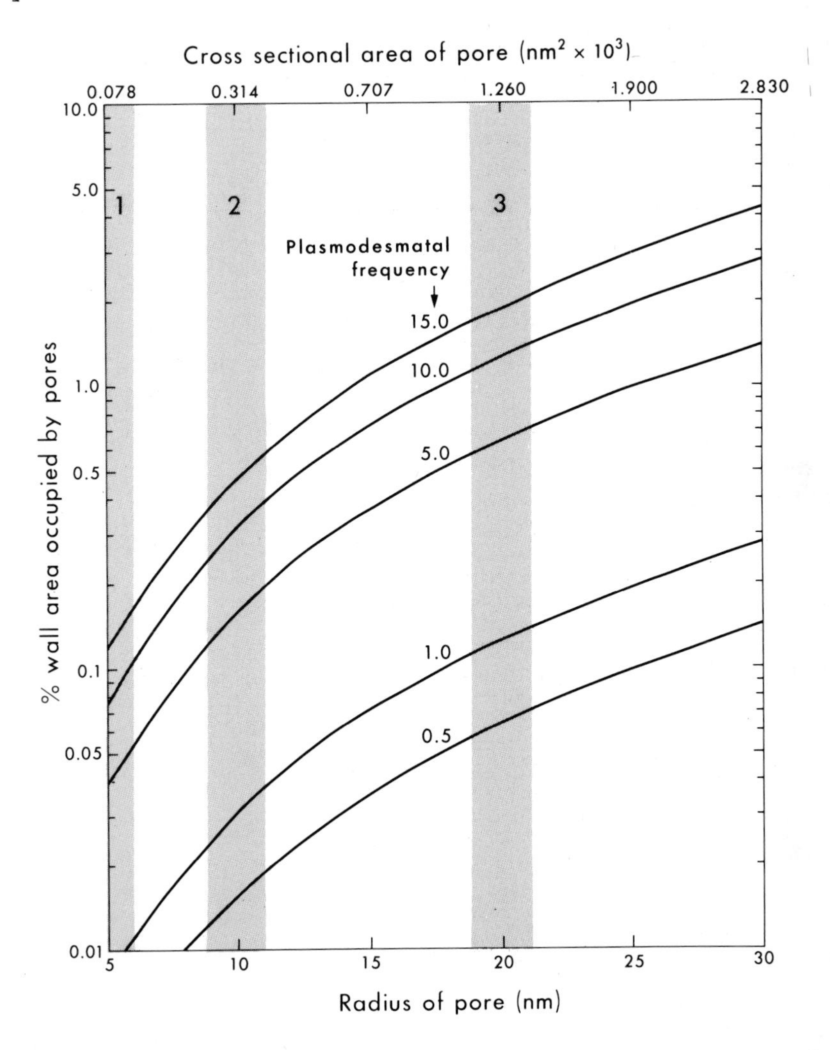

Fig. 2.1. Diagram relating the frequency and pore dimensions of plasmodesmata to the proportion of wall area occupied by pores. The shaded vertical columns show the wall area occupied at different frequencies for pore radii centred about: 1) 5 nm (e.g. the desmotubule); 2) 10 nm (e.g. the plasmalemma-lined pore); 3) 20 nm (e.g. a plasmalemma-lined pore without restrictions at the ends). It can be seen that the wall area occupied by pores is rarely likely to be much above 1%, and is usually much lower than this

meristematic cells will have between 1,000 and 10,000 connections with their neighbours (Clowes and Juniper, 1968).

Of more importance to the physiologist is the area available for transport from cell to cell. This parameter requires knowledge of the nature and size of the conducting pore - something that remains debatable. It is becoming increasingly popular to refer to the proportion of wall occupied by pores. That is, the total surface area of the presumptive conducting channels expressed as a proportion of the total wall area. Although no definitive figures can be cited, some limiting cases can be demonstrated (Fig. 2.1.). Using different pore sizes and frequencies it is seen that the open channels from cell to cell are unlikely ever to account for much more than 3% of the interface area and, in most cases, the proportion will be very much lower than this. (The figures cited apply to higher plants. In some algae - e.g. *Chara*, the plasmodesmatal area may occupy 5% or more of the wall between nodal and internodal cells - Table 9.1., see also Fischer, Dainty and Tyree, 1974). Finally, a neglected aspect of plasmodesmatal size and frequency is the amount of plasmalemma 'trapped' in plasmodesmata, as opposed to that which lines the cell surface. Some authors (e.g. Carde, 1974; Vian and Rougier, 1974) have commented on apparent ultrastructural differences between the plasmalemma within the plasmodesmata and that over the rest of the wall. In a wall 0.5 μm thick, having a plasmodesmatal frequency of 5 per μm^2 with a diameter of 60 nm, almost half as much plasmalemma is within the plasmodesmata as lines the rest of the wall; if the wall is 1.0 μm thick, then the amounts are approximately equal.

2.4. STRUCTURE

If anything meaningful is to be said about the possible functions of plasmodesmata, it is important that the details of fine structure should be well understood. It is also useful to have a uniform approach to the citation of dimensions and the terminology relating to these, sometimes complicated, structures. The cited dimensions vary greatly (Table 2.1.). From a structural viewpoint this might not be too significant. However, when measurements are related to theories of symplastic transport, as pointed out in Section 1.2.1., where it was shown that the carrying capacity of plasmodesmata can vary in relation to the *fourth* power of the radius, even small differences can have relatively profound effects. The system of nomenclature that I shall use is outlined in Fig. 2.2.

2.4.1. Optical Microscopy

The direct observation of plasmodesmata in untreated specimens by optical microscopy is restricted to relatively few cases: for example, the endosperms and cotyledons of seeds, such as those of *Strychnos nux-vomica* (Frontispiece) and *Aesculus hippocastanum*. Other than in such situations, plasmodesmata generally need to be stained, swollen and stained, or impregnated with a soluble salt of a metal (often silver) to render them visible. One of the simplest stains for plasmodesmata is iodine (Tangl, 1879), used either in simple solution, vapour form, or one of the often-cited complex forms such as aqueous or alcoholic Lugol solution (1% iodine in a 2% solution of potassium iodide). Applied following swelling agents such treatments show plasmodesmata, and this fact has been used to support the contention that plasmodesmata are protoplasmic in nature. A wide variety of other stains has been used to demonstrate the protoplasmic connections: aniline dyes, haematoxylin, methyl violet, etc.; no doubt individual users have each

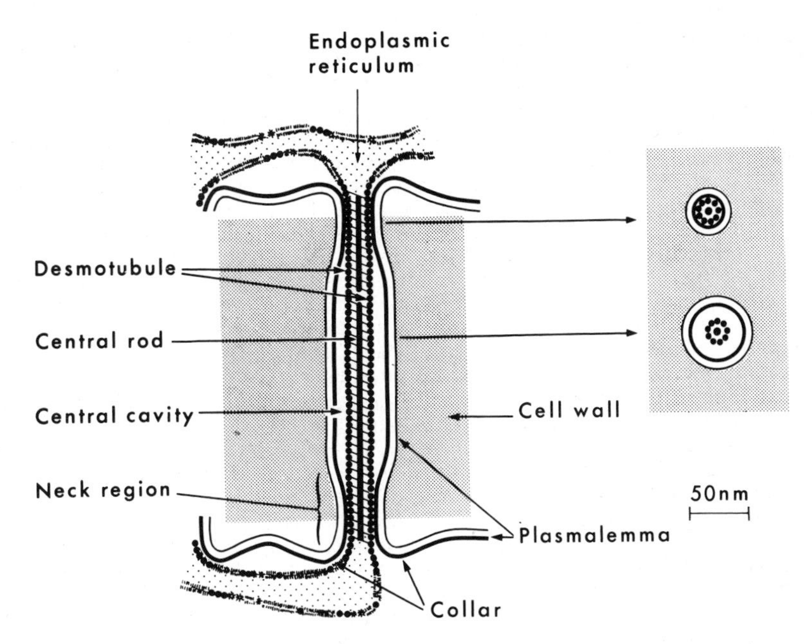

Fig. 2.2. The components of a simple plasmodesma. This diagram shows the various plasmodesmatal features seen in electron micrographs. It is not meant to imply any general uniformity of structure, nor specific features

found their particular recipe successful, but no clear picture emerges of any specificity. Swelling agents for demonstrating plasmodesmata have usually been sulphuric acid (followed, or accompanied, by iodine) or chlor-zinc-iodine (Schultz's solution). In either case, the cell wall framework swells and produces a blue or purplish colour while the plasmodesmata are yellow to brown. Part of the virtue of these methods appears to reside in the swelling of the wall, so allowing more space for the penetration and reaction of iodine molecules with the protoplasmic connection. Still more contrast may be obtained if sections, which have been stained with iodine after swelling, are further treated with a stain which reacts with the iodine: this is the basis of the pyoktanin method which uses methyl violet (see Meeuse, 1957 for further details). Even here, however, there appear to be no grounds for optimism relating to the specificity of the reaction, although several authors have suggested it (Jungers, 1930; Livingston, 1935). Impregnation of plasmodesmata with silver salts relies upon more rapid absorption and/or reduction of the reagent by the plasmodesmata than by other components - something that should possibly be kept well in mind when evaluating cytochemical precipitation methods involving silver salts (see Chapter 7). While having a useful place in the range of techniques for optical microscopy (e.g. Mühldorf, 1937), again, it may not be regarded as a specific reaction.

The conclusion to be drawn from techniques for optical microscopy is that, while a whole host of methods has been described (see Mühldorf, 1937; Meeuse, 1941b, 1957; Livingston, 1935, 1964), and each probably has its own usefulness, there is no single method that can be relied upon to demonstrate plasmodesmata specifically. Therefore, the usual precautions must be taken in analysing results from such methods: there is clearly a considerable risk of mis-identification, as well as the certainty that many are never seen at all by optical microscopy.

Optical microscopy showed that plasmodesmata comprised thin threads of protoplasm running through a pore in the cell wall. Beaded threads were ascribed to swelling artifacts (Mühldorf, 1937) or precipitation phenomena (Livingston, 1964). A thickening at the centre of a plasmodesma was thought by Meeuse (1957) to be another swelling artifact, but Livingston, from his examination of unswollen material, concluded that this 'node' is a constant morphological feature and not artifactual. Optical microscopy also demonstrated anastomosing plasmodesmata.

Although optical microscopy allowed real and important progress towards understanding the nature and distribution of plasmodesmata in plant cell walls (often confirmed subsequently by electron microscopy), it must be emphasised that, whatever the actual diameter of plasmodesmata, it always lies close to the limit of resolution of the optical microscope. Perhaps, therefore, it is not surprising to find that such measurements, when made from optical observations, are usually in the range of 0.1-0.5 μm (Table 2.1.). It has been suggested by some authors that the measured diameter is smaller than the *in vivo* condition (Münch, 1930; Livingston, 1935, 1964; Strugger, 1957a,b) because the techniques used to demonstrate plasmodesmata usually involve the use of wall swelling agents. Thus, Livingston (1935) suggested that mature, unswollen plasmodesmata of *Nicotiana* would have a diameter of 0.5 μm as opposed to the 0.2 μm measured after treatment with chemicals. Similarly Strugger (1957a) measured plasmodesmata to be 30-40 nm in diameter (electron microscopy), but he later corrected this to 100-200 nm (1957b), allowing for his estimate of shrinkage during processing. Although Strugger was using the electron microscope as his tool, his observations are clearly relevant to this section as they are among the earliest results from the electron microscope in this field of study, and his considerations were clearly influenced by the accumulated weight of evidence from optical microscopy. It appears to me probable that the optical microscope is a misleading instrument in this matter: plasmodesmata, as already stated, are usually demonstrated in tissue swollen in some harsh reagent, after which they appear as fine filaments close to the very limit of optical microscope resolution; under such circumstances, and with grouped plasmodesmata, one would expect diffraction phenomena to occur (particularly in the relatively thick sections usually studied) and, therefore, it would be unwise to regard these results as other than indicative of the uppermost limits of size, while not excluding much smaller structures.

2.4.2. Electron Microscopy

The electron microscope quickly confirmed the presence of plasmodesmata in material fixed either in osmium tetroxide (Buvat, 1957; Strugger, 1957a,b; Kollmann and Schumacher, 1962, 1963), or potassium permanganate (Porter and Machado, 1960a; Whaley, Mollenhauer and Leech, 1960; Esau, 1963). A detailed study by Scott, Hamner, Baker and Bowler (1956) of macerated, cleaned and shadowed onion root cell walls demonstrated the pores through the walls of pit-fields through which it was assumed that plasmodesmatal strands passed. Although the electron microscope established the morphological reality of plasmodesmata, it soon raised further problems largely attributable to the small size of the structures (or, at least, their components) even judged by electron microscope standards. In general, plasmodesmata seen in the electron microscope are about one tenth the size of those seen by optical microscopy. This led Clowes and Juniper (1968) to suggest that there may be two types of plasmodesma, a view that lacks much support. The discrepancies are largely related to the *apparently* greater size of plasmodesmata viewed by optical microscopy. This, in turn, may be contributed to by: diffraction effects of closely adjacent structures; branched plasmodesmata; and the possibility that, not only the plasmodesma, but also a restricted area of the wall around it, may stain differently from the remainder of the cell wall, and so increase the apparent size. (Both Taiz and Jones (1973) and Vian and Rougier (1974), cite electron microscopical evidence that would support this last possibility).

Examination of frozen-etched plasmodesmata with minimum pretreatment and no dehydration shows structures of general dimensions very similar to those obtained by conventional electron microscopy (own unpublished results); the same is true for plasmodesmata prepared by freeze-substitution (A.J. Browning, unpublished results). In ultrathin sections cut from frozen blocks of fixed plant tissue, plasmodesmata are of much the same size-range (Vian and Rougier, 1974) and, more significantly, so are those in frozen-cut sections of chemically untreated material (Roland, 1973; Vian and Rougier, 1974, Plate IId).

In the electron microscope the very small size of the plasmodesmatal canal, together with the included protoplasmic thread, has made it difficult to determine the true nature of the connection. From the very earliest micrographs of fixed and sectioned material it was noticed that the plasmodesmata were usually contacted at either end by a strand of endoplasmic reticulum. It therefore became a generalisation that the plasmodesma contains a strand of endoplasmic reticulum running through the wall from cell to cell (Wardrop, 1965; Frey-Wyssling and Mühlethaler, 1965; and many others). Esau (1963), with her usual care, avoids the unsupported view that endoplasmic reticulum necessarily traverses the wall: "....*endoplasmic reticulum is typically connected to the plasmodesmata; in fact, many investigators think that tubules of endoplasmic reticulum are structural elements of plasmodesmata*". This is a fair statement of fact, but there is little further *direct* evidence to clarify the nature of the connection. Porter and Machado (1960a) and Falk and Sitte (1963) were, like Esau, not prepared to accept direct endoplasmic reticulum continuity without reservations. It is ironic that, for many years, optical microscopists struggled with the inherent limitations to the resolution of their instruments in an effort to determine plasmodesmatal structure: then, as soon as detailed observations were made of thin-sectioned material in the electron microscope, it was seen that, once again, instrumental resolution (in association with specimen effects) imposes limitations in elucidating the crucial features of the connecting strands. Part of the resolution problem is outlined in Fig. 2.3. Further, the normal practice of presenting very slightly under-focused images may cause misleading phase and amplitude changes, which could affect image interpretation (Helder and Boerma, 1969; Robards, 1971). Chromatic loss due to the specimen itself will also adversely affect resolution. Still thinner sections can be cut, but it is then very difficult to obtain sufficiently high contrast in the structure of the desmotubule and endoplasmic reticulum to allow them to be seen and recorded at high resolution. The remarks above apply, in the main, to longitudinal sections of plasmodesmata. Transverse sections involve many of the same problems; they are, however, often easier to interpret as there is a greater possibility that the material (plasmalemma, desmotubule, etc.) is homogeneous throughout the thickness of the section (Fig. 2.3.).

Having mentioned these limitations to the ultrastructural study of plasmodesmata, I can now consider the results that are, in fact, available. The diversity of plasmodesmatal form is such that it will be considered separately later (Chapter 4). For the present I shall confine myself to *basic-type* plasmodesmata; these are simple plasmodesmata with no anastomosing desmotubules or complicated median nodules (Fig. 2.4.).

The ultrastructural demonstration of cytoplasmic continuity (Buvat, 1957; Strugger, 1957a) in cells fixed by osmium tetroxide was followed by studies in which potassium permanganate was used as the fixation agent, allowing membranes to be seen with clarity previously unobtainable. It was thus possible for Whaley, Mollenhauer and Leech (1960) to suggest the continuity of a strand of endoplasmic reticulum through the plasmodesmatal canal, and for Porter and Machado (1960a) to associate plasmodesmatal formation with cell plate deposition during mitotic telophase. In these and a large number of supporting papers, it

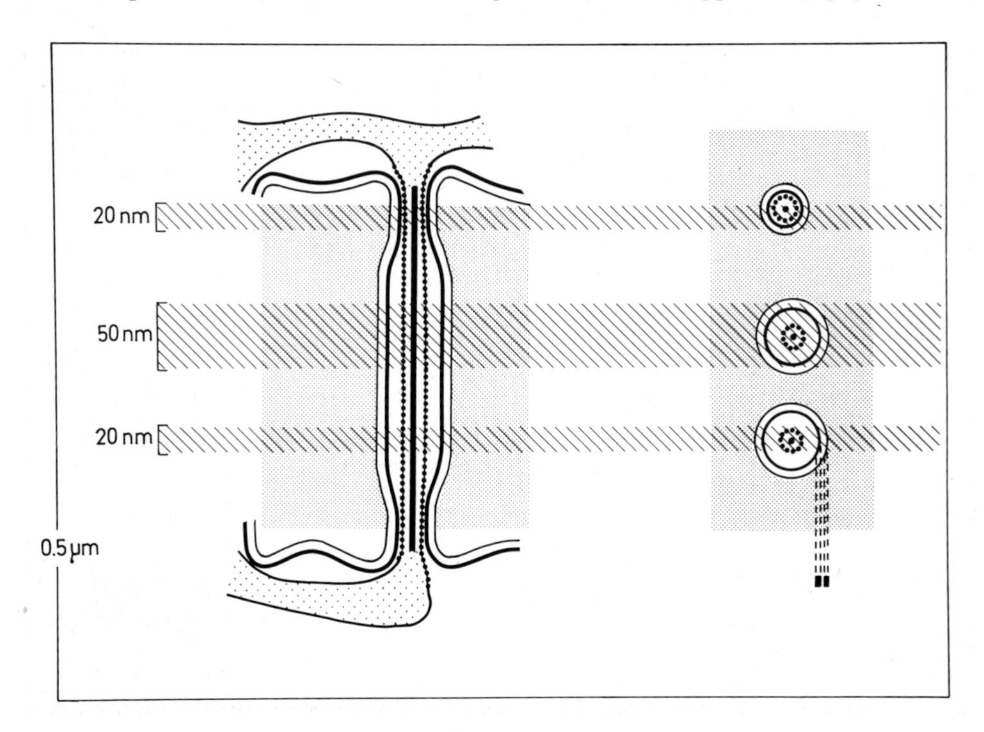

Fig. 2.3. Diagram illustrating the relationship of plasmodesmatal dimensions to section thickness. The height of the frame is equivalent to a section thickness of 0.5 µm (as, for example, might be used in high voltage electron microscopy): a complete plasmodesma could be contained vertically, while eight or more could be stacked horizontally upon one another within such a thickness. Resolution *within* the plasmodesma would be poor. Indeed, it is often impossible to see plasmodesmata orientated in the plane of a thick (0.5 µm - 1.0 µm) section viewed at high voltage (0.4 - 1.0 MeV) because the contrast of a thin component in a thick section is so low. A 'conventional' thin section (approx. 50 nm) includes the whole of a plasmodesma lying in the plane of a section, and gives reasonable resolution across parallel membranes. In the neck-region the image of such a section superimposes too much information for all details to be resolved. Even a *very* thin section (20 nm) encompasses half a plasmodesma cut vertically through the neck region. Longitudinal sections of plasmodesmata showing unit membrane structure of the plasmalemma with maximum clarity need to be of the order of 20 nm thick (lower right diagram)

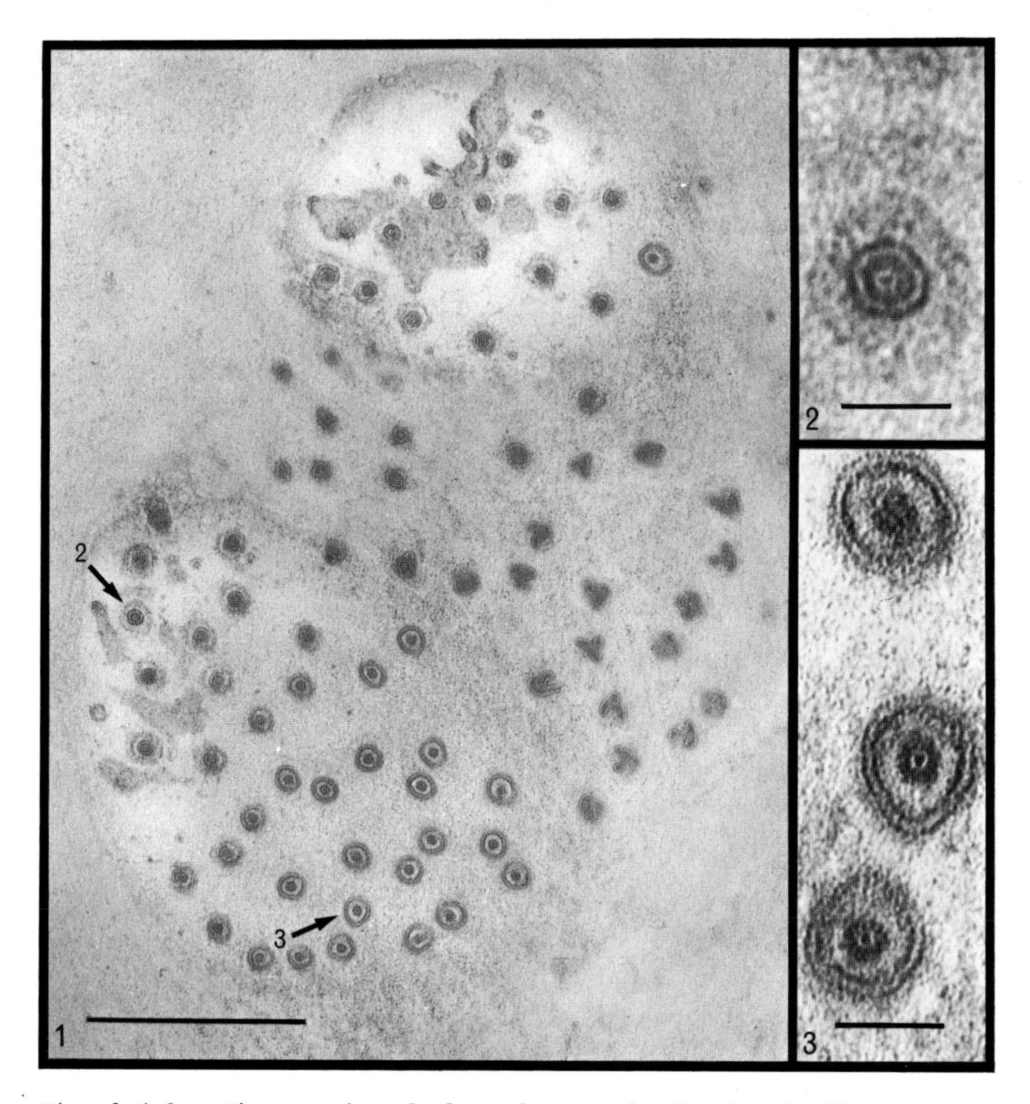

Fig. 2.4.1. Micrographs of plasmodesmata in the longitudinal radial walls of xylem ray cells of willow (*Salix fragilis*, L.) fixed in glutaraldehyde and osmium tetroxide. The section was slightly oblique to the plane of the wall so that plasmodesmata were cut at different levels. Those sectioned through the mid-line show a clear gap between the desmotubule and the plasmalemma (type indicated by '3'), whereas those cut through the neck region (Fig. 2.2.) have no such space ('2'). Such pictures provide the best evidence for a tight seal between plasmalemma and desmotubule in the neck region of some species (c.f. Fig. 1.1.). A desmotubule and central rod appear to be universally present. Observations on such plasmodesmata were used to construct the models illustrated in Figs. 2.6. and 2.7. These micrographs were first published in Robards (1968b). Scale markers = (1) 0.5 μm; (2) and (3) 50 nm

was demonstrated that profiles of endoplasmic reticulum become trapped as the vesicles converging upon the cell plate fuse with each other. Subsequently, López-Sáez, Giménez-Martín and Risueño (1966a), also using permanganate fixation, concluded that the plasmodesmatal canal is traversed by a strand of endoplasmic reticulum, tightly curved into a tubule, so that the inner opaque layer of the membrane appears as a central rod (Fig. 2.5.). López-Sáez *et al.* do not dwell on the possible function of plasmodesmata, merely stating that an *inter-tubular gap* (the gap between the plasmalemma and the endoplasmic reticulum strand [desmotubule] through the wall) would allow some connection between the hyaloplasm of adjacent cells whereas, if there is no continuous intertubular gap, "*the plasmodesm plays no physiological role in intercellular transport*". A very similar model has been proposed by Semenova and Tageeva (1972) who believe that the dark, axial, structure is a solid rod formed from a tightly constricted endoplasmic reticulum membrane tubule.

It is important to realise the implications of the statements by López-Sáez *et al.*, or Semenova and Tageeva: that communication, if it does occur, cannot take place through the supposed strand of endoplasmic reticulum (desmotubule), but rather through a possible *leaky seal* between the neck and the desmotubule (a point not clarified in these papers); if this channel, too, were closed, cytoplasmic continuity could not exist except in the form of a solid structure.

Improved methods of fixation for electron microscopy, such as the use of glutaraldehyde (Sabatini, Bensch and Barrnett, 1963), together with a basic dissatisfaction over the gap existing between structural and physiological approaches to the intercellular movement of ions and molecules, led me to carry out an investigation of plasmodesmatal ultrastructure using glutaraldehyde/osmium tetroxide fixed cells (Robards, 1968b). As I was working on differentiating xylem cells, and transport of solutes between thick-walled ray cells represents an area of particular interest, the plasmodesmata between ray cells were

---

Fig. 2.4.2. Plasmodesmata in the root of the water fern *Azolla filiculoides* (see also Fig. 4.4.). Fixation was in glutaraldehyde and osmium tetroxide. In the case of (2) and (3) tannic acid was added to the primary fixative solution. ▶

(1) Longitudinal sections of plasmodesmata through the outer tangential endodermal wall.

(2) Transverse sections of plasmodesmata fixed in a solution containing tannic acid.

(3) Detail of a plasmodesma from (2): the cavity between the desmotubule and plasmalemma is electron-opaque, due, presumably, to the negative staining effect of the tannic acid. The wall of the desmotubule appears particulate; there is a central rod. The same features are seen in (4), which was not treated with tannic acid; the desmotubule wall is less clearly contrasted. Note that these plasmodesmata appear narrower than those illustrated in Fig. 2.4.1. This contributes to the difficulty of showing continuity of the desmotubule through the wall (5). Such longitudinal sections indicate the relationship of the endoplasmic reticulum with the plasmodesmata (arrowed). In these rather narrow plasmodesmata there seems to be no neck, but simply a narrow space around the desmotubule throughout the pore (see Table 2.2. for dimensions). Scale markers = (1) 1 μm; (2) 0.5 μm; (3) and (4) 50 nm; (5) 0.5 μm

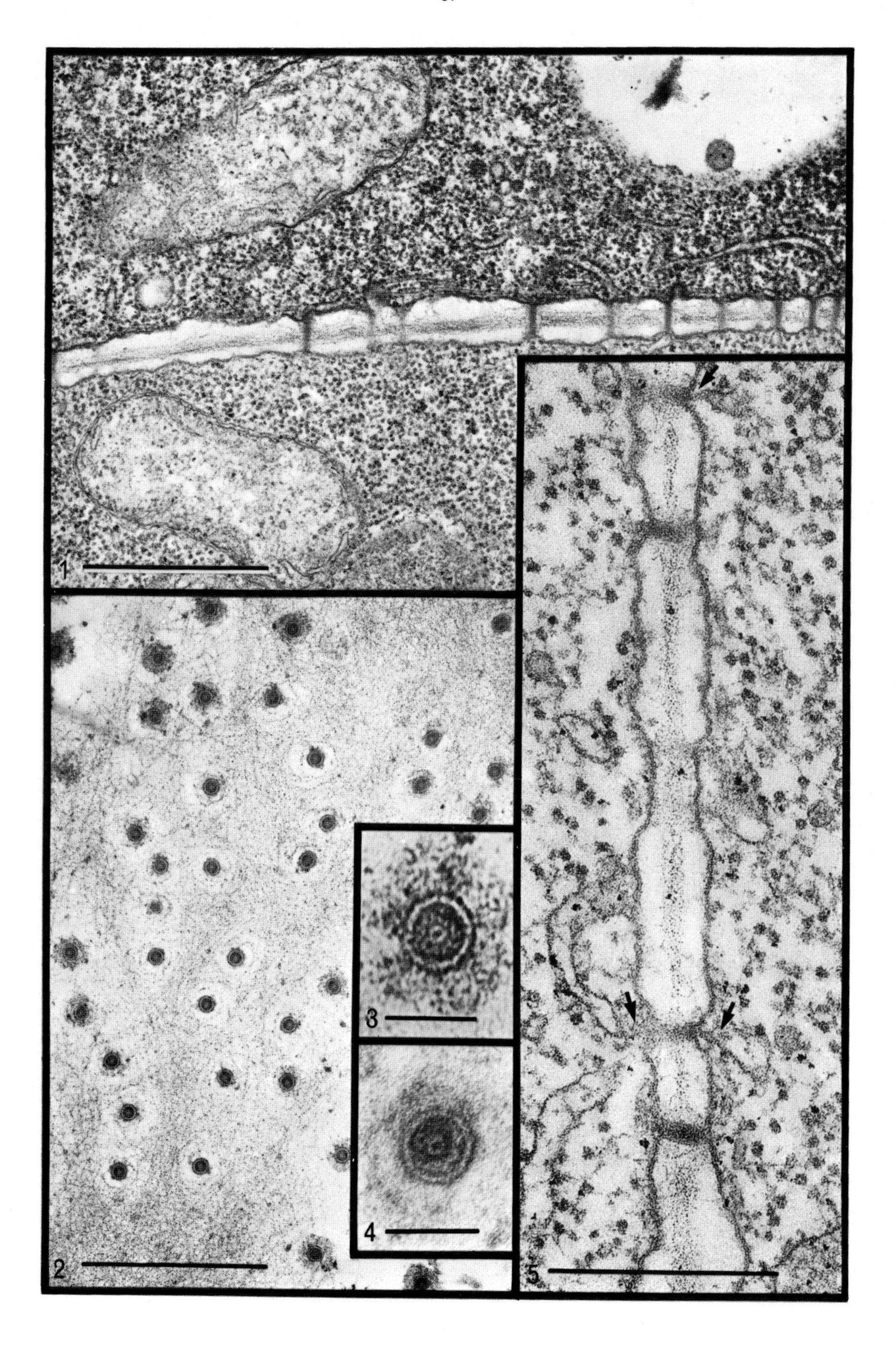
1
2
3
4
5

examined in both longitudinal and transverse sections. In general, my observations confirmed those of many other authors. The outer diameter of the plasmodesmata in the middle lamella region was about 60 nm; the core, in transverse section, comprised an opaque ring about 20 nm in diameter with a 4 nm thick wall; this tube contained a further opaque rod, roughly 4 nm in diameter. Transverse sections clearly demonstrated that the desmotubule[1] was closely encompassed by the plasmalemma in the collar region at either end of the pore (Fig. 2.4.1.). Thus, if the interpretation by López-Sáez *et al.* of their similar structures was correct, no intercellular translocation seemed possible. The wall of the 20 nm diameter tubule was strongly reminiscent of a microtubule dimension and structure, and this, together with the doubt that a lipoprotein bilayer could behave as suggested by López-Sáez *et al.* (particularly in curving about such a small radius, see Robertson, 1964, but also see 4.2.1.), led me to suggest that the desmotubule in fact had a structure similar to that of a microtubule. Rotational image reinforcement experiments suggested that the desmotubule had 11 subunits when seen in transverse section and, in the absence of other information to the contrary, I depicted the endoplasmic reticulum as abutting the desmotubule without being continuous with it (Fig. 2.6.). (Zee (1969), studying sieve plate pore initials, also used rotational reinforcement and concluded that the central core has a 14-fold subunit arrangement. He did not observe direct connections between the endoplasmic reticulum and sieve plate pore initials at this early stage of development. See 2.4.4.). This proposal, that the desmotubule has a microtubular structure, reflected earlier observations of a similar nature (e.g. Bajer, 1968a; Hepler and Jackson, 1968; O'Brien and Thimann, 1967a), as well as other comments in the literature concerning the possibility that an unmodified strand of endoplasmic reticulum passed through the canal. For example, Ledbetter and Porter (1970): "*If derived in the first instance from endoplasmic reticulum, it has come to be compressed by the narrowness of the pore into something no longer recognisable as endoplasmic reticulum membrane or tubule*".

---

[1]I first used the term *desmotubule* in 1968(a). The derivation from the Greek *desmos* (bond) seemed, and seems, appropriate and compatible with other terms such as plasmodesma and desmosome. My intention was that the word should refer unambiguously to a particular component of plasmodesmata, without pre-empting discussion about its nature.

Fig. 2.4.3. Plasmodesmata in the root of barley (*Hordeum vulgare*, L.).▶ Fixation was in glutaraldehyde and osmium tetroxide. In the case of (1) and (6), tannic acid was added to the primary fixative solution. (1) A pit field in the transverse end wall of a young endodermal cell showing the extremely frequent plasmodesmata. (2) Plasmodesmata through very young cortical cell walls - the plasmalemma-lined pores are relatively straight-sided, and endoplasmic reticulum appears to traverse the connection. (3) and (4) Plasmodesmata through young (State I) inner tangential endodermal walls showing different forms of dilation of the plasmalemma-lined cavity. (5) Transverse section of a plasmodesma through the wall of adjacent cortical cells: plasmodesmatal cavity, the desmotubule, and central rod can all be distinguished. (6) Detail from (1) showing the desmotubule negatively stained, presumably by tannic acid or its derivatives, which fills the space between desmotubule and plasmalemma. (7) Detail from (3). Scale markers = (1) 1 μm; (2) 0.25 μm; (3) and (4) 0.5 μm; (5), (6) and (7) 50 nm

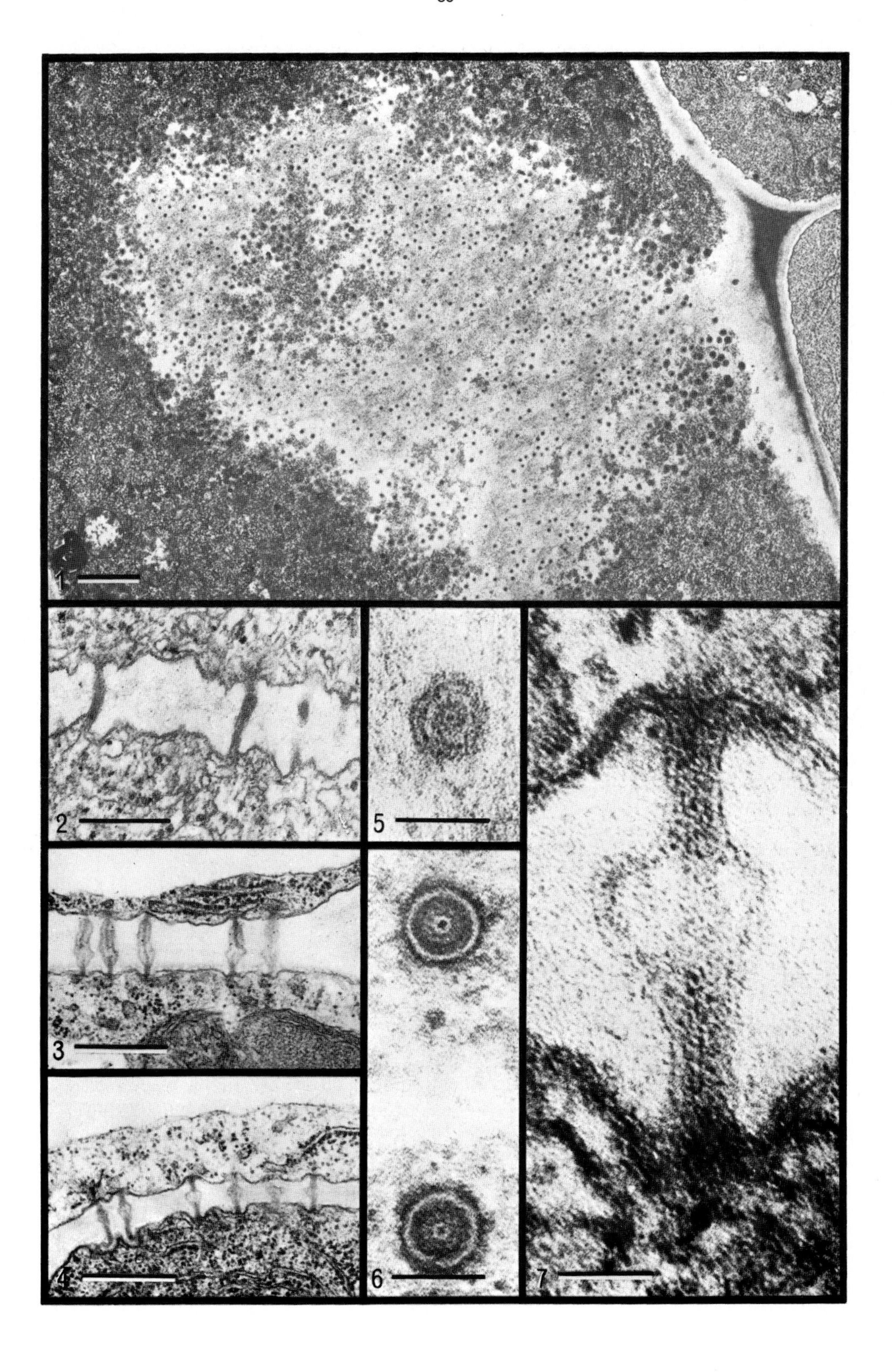
1
2
3
4
5
6
7

Later (Robards, 1971) I suggested a scheme whereby the continuity of endoplasmic reticulum and desmotubule can be accommodated while still retaining the essential structural features of the earlier proposal. This model is presented in Fig. 2.7.: it will be seen to comprise a direct continuity of endoplasmic reticulum from one cell to the next *via* the desmotubule, which is merely a modified form of membrane composed almost entirely of protein subunits in the form of a tubule, but *not* to be equated with a microtubule *sensu stricto*. The full evidence for proposing such a model is reviewed in the relevant paper. It is adequate for the present purposes to point out the salient features: that the cisternal cavity of the endoplasmic reticulum in one cell is continuous with that of its neighbour through the pore of the desmotubule; that the central rod seen in micrographs (Figs. 1.1., 2.2., 2.4.) is considered to be probably an artifact; and that there is no other open channel of communication through the plasmodesmatal pore (Robards, 1971). Such a model satisfies the probability that the desmotubule is in direct continuity with the endoplasmic reticulum, as suggested by so many authors; it explains why other workers have, in agreement with my own observations, remarked upon the similar appearance of the cross-section of a desmotubule to that of a microtubule; and it overcomes the objections to the desmotubule comprising a tubule of unmodified lipoprotein bilayer membrane. Molecular mechanisms whereby a lipoprotein bilayer could become modified to a largely proteinaceous tubule are discussed in 4.2.1.

The real nature of the central rod is a particularly difficult problem: I have provided one possible explanation for its artifactual presence on the basis of negative staining (Robards, 1971), but there is little evidence for or against such an argument. The belief of López-Sáez *et al*. (1966a), Semenova and Tageeva (1972), and others, that the central rod represents the inner lamina of a tightly furled endoplasmic reticulum unit membrane (or a derivative from it), should not be dismissed out of hand. However, such a conclusion effectively precludes any function of the desmotubule as an open channel of intercellular communication. If the central rod *is* present within a tubular connection from cell to cell, then it will have a profound effect upon the carrying capacity of this particular symplastic channel (Section 1.2.1.).

Criticisms of the protein subunit desmotubule have centred upon doubts that such a structure could branch/anastomose, or be in direct continuity with endoplasmic reticulum membranes (Wooding, 1968); the

---

Fig. 2.4.4. Thick (200 nm) sections of plasmodesmata through the inner ▶ tangential endodermal wall of barley (*Hordeum vulgare*, L.), fixed in glutaraldehyde and osmium tetroxide, viewed at 200 keV using a high voltage electron microscope (AEI EM7) at the British Steel Corporation Swinden Laboratories, Rotherham. Different tilt angles (indicated in the form X/Y on each micrograph) allow 3-dimensional arrangements to be studied (especially in combination with stereoscopic viewing). There are considerable problems in viewing plasmodesmata by this method: any section thickness greater than that of the structure being observed (in this case about 50-60 nm) will reduce clarity of that structure. Indeed, unless plasmodesmata are specifically heavily stained, they are invisible in 1.0 μm thick sections viewed at 500-1,000 keV (A.W.R. - unpublished observations). The micrographs presented here illustrate the manner in which tilt angle affects the apparent relationships between endoplasmic reticulum and plasmodesmata: check, for example, the points marked with arrows in the central (0/0) micrograph. Scale marker = 1.0 μm

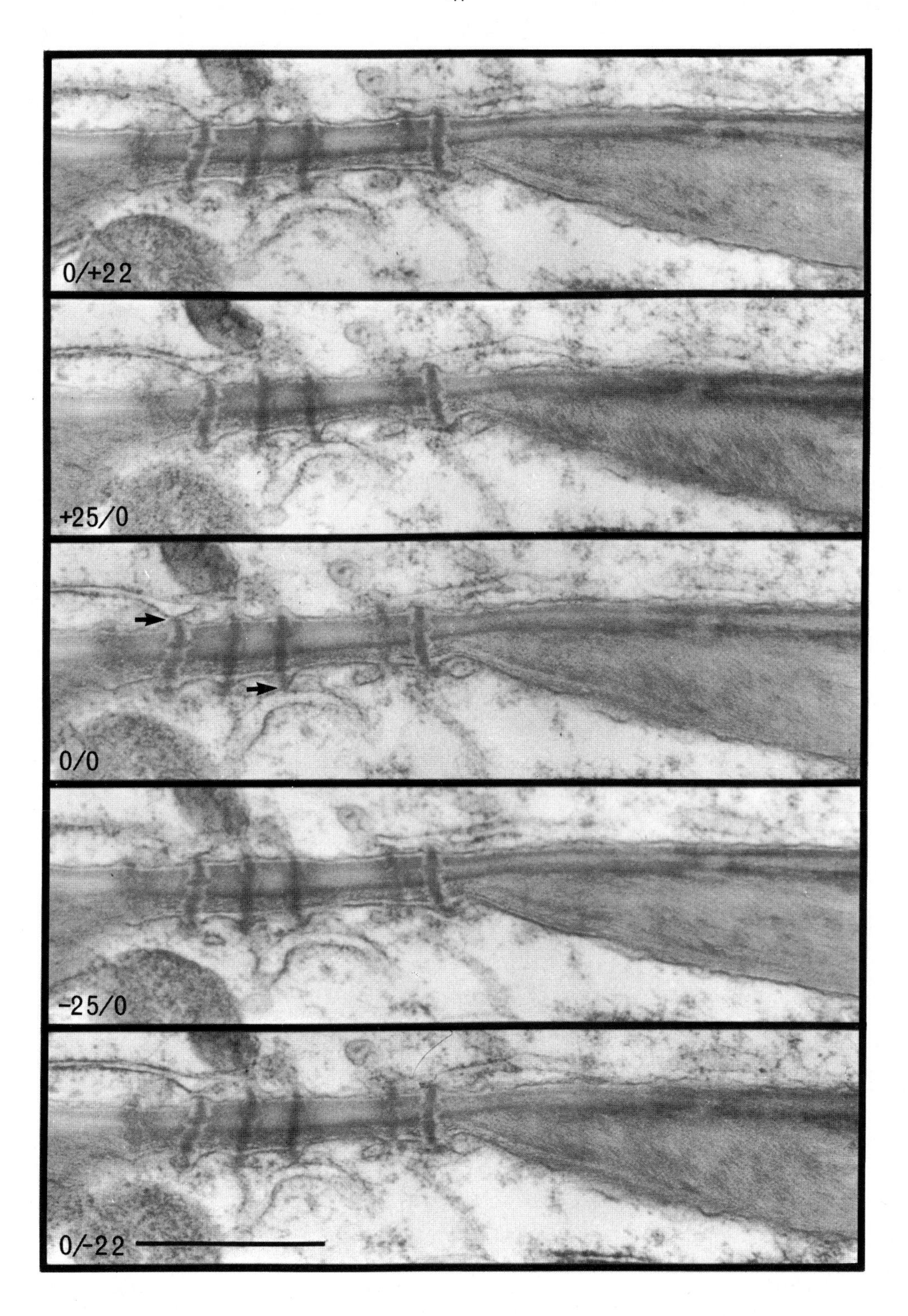
0/+22
+25/0
0/0
-25/0
0/-22

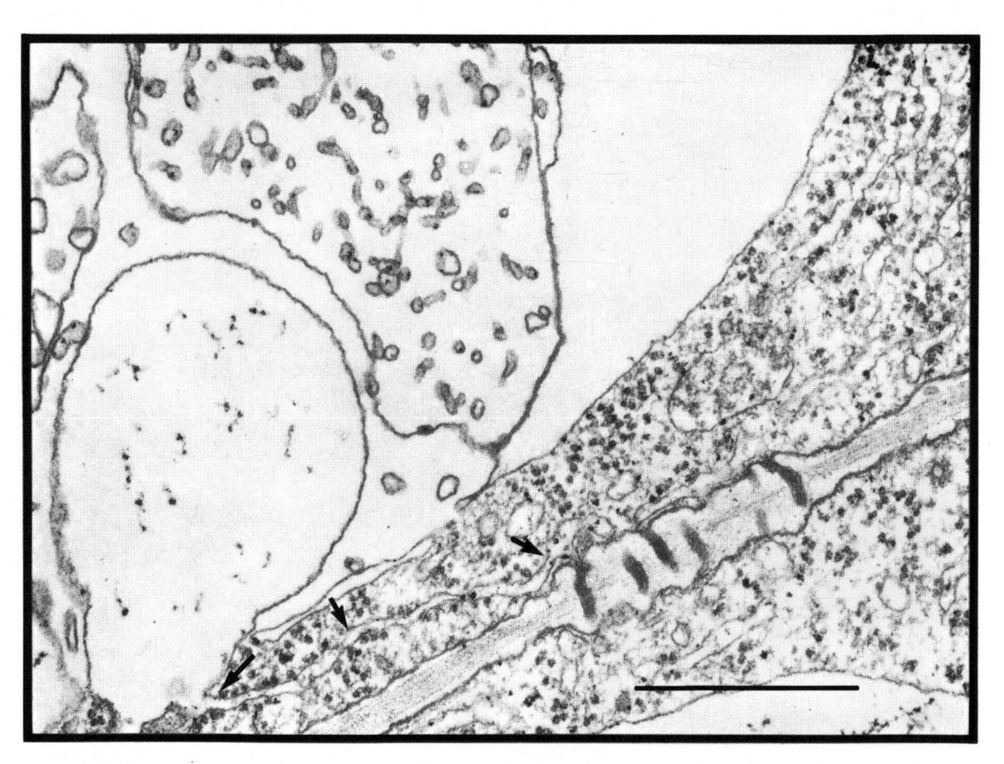

Fig. 2.4.5. Plasmodesmata traversing the wall of adjacent barley root cortical cells fixed in glutaraldehyde/formaldehyde followed by osmium tetroxide. There is clear continuity of what appears to be a strand of rough endoplasmic reticulum with a plasmodesma at one end, and a dilated vacuole-like cisterna at the other. If the endoplasmic reticulum is indeed an intercellular symplastic compartment (see Fig. 10.2B.), then arrangements of the type shown here might be expected. Scale marker = 0.5 μm

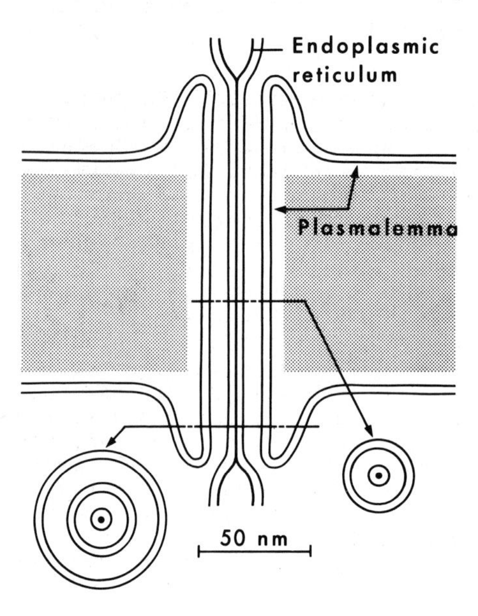

Fig. 2.5. Redrawn model of a plasmodesma derived by López-Sáez *et al.* (1966a) from observations on *Phalaris canariensis* after potassium permanganate fixation. The central spot (in transverse view) is considered by the authors to be the inner layer of the membrane of an endoplasmic reticulum tubule. An *inter-tubular gap* extends along the whole length of the plasmodesma between the supposed endoplasmic reticulum strand and the plasmalemma

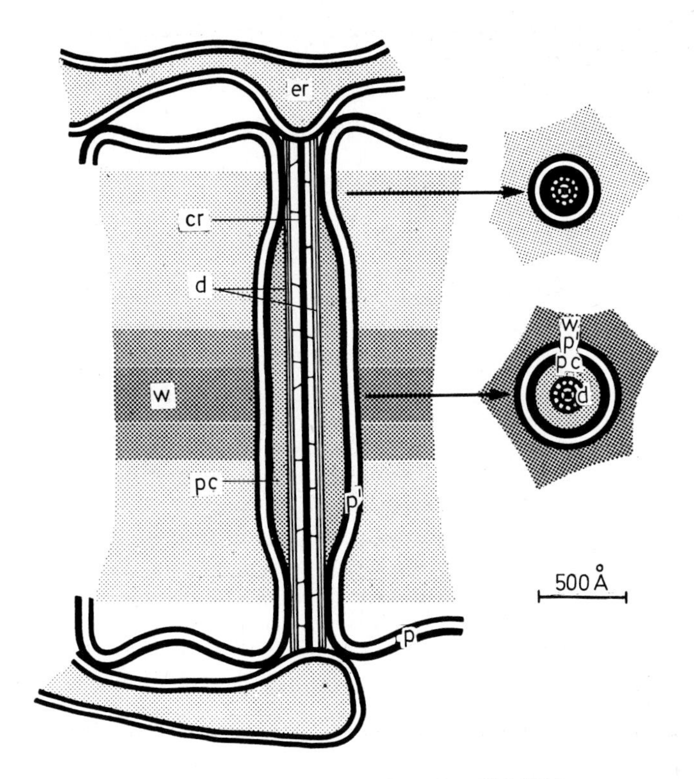

Fig. 2.6. Diagram reproduced from Robards (1968b) representing a simple plasmodesma viewed in longitudinal and transverse section. The diagram closely reflects the image as *seen* in the electron microscope. (This model was constructed from observations on plasmodesmata in the walls of xylem ray cells of *Salix fragilis*, but it is very similar to the situation in other plants - e.g. barley root plasmodesmata). *cr* - central rod; *d* - desmotubule; *er* - endoplasmic reticulum; *p* - plasmalemma; *p'* - plasmalemma through wall; *pc* - plasmodesmatal cavity (central cavity of Fig. 2.2.); *w* - cell wall

unlikely possibility that such a structure could stretch (Burgess, 1971); the fact that the desmotubule structure is largely preserved by permanganate fixation (Burgess, referring to the work of López-Sáez *et al.*); and the possibility that, whatever structure is finally seen, it represents a considerably shrunken representation of the original (Burgess, 1971). I have dealt with the problem of continuity and branching in a previous publication (Robards, 1971); suffice it to repeat here that even microtubules themselves are still at a relatively early stage of study and understanding, and that there are frequent new reports of them linking with each other or with membranes. Similarly, the extensibility of proteinaceous tubules is not known, and the effects of plasmolysis observed by Burgess (1971) would not themselves exclude the model proposed here. The ready, and spontaneous, formation of cylindrical structures from protein subunits in solution, ranging from viral particles to catalase, is another line of evidence that would be consistent with the desmotubule hypothesis. Preservation of the desmotubule (but not in such detailed structure) by permanganate fixation certainly does not preclude a protein subunit form: one has only to consider the case of flagellar fibrils to appreciate that, while many such structures are lost in permanganate fixation, others

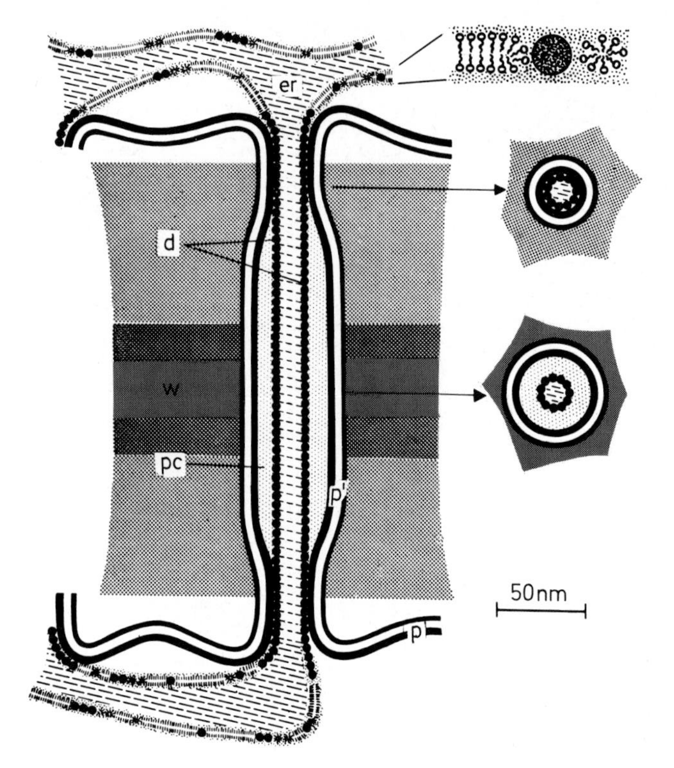

Fig. 2.7. Diagram reproduced from Robards (1971) showing an interpretative model derived from the previous illustration (Fig. 2.6.). Open continuity is shown between desmotubule and endoplasmic reticulum. The endoplasmic reticulum membrane was depicted as an expanded lipoprotein bilayer with included micelles (Lucy, 1964; and Glauert, 1968), although this is not an essential feature of the model. The reasons for deriving this model are given more fully in the text and in Robards (1968b; 1971). Labelling as in Fig. 2.6.

may still be preserved. The shrinkage problem is a severe potential artifact to interpret in the current context; it has already been touched upon in connection with plasmodesmatal size. Plasmodesmatal components, in common with other cell structures, may shrink during processing; how much, if at all, and what the effect will be, remains to be determined. Sections of frozen material, however, gratifyingly indicate plasmodesmata of much the same dimensions as from conventionally prepared material (Roland, 1973; Vian and Rougier, 1974) although the latter paper cites a rather narrow cavity (2-3 nm) within the desmotubule of chemically untreated plasmodesmata.

Burgess (1971) cites the presence of radiating spokes between the desmotubule and plasmalemma and considers that they may arise from processing damage. Similar spokes are evident in the earlier micrographs of Dolzmann (1965), which also show desmotubules appearing to open into the plasmodesmatal canal (and thus appear C-shaped in transverse section); similar observations have been made in my own laboratory, and they might quickly be dismissed as further processing artifacts if it were not for reports of similar structures in cytoplasmic microtubules (Cohen and Gottlieb, 1971).

Two recent reviews of plasmodesmatal ultrastructure have been provided. One is in Polish, by Wozny and Mlodzianowski (1973); in the other, Brighigna (1974) has studied the plasmodesmata of the water-absorbing scale of *Tillandsia* and, using arguments similar to those described above, arrives at the structural model depicted in Fig. 2.8., where the desmotubule is apparently sealed, and few unambiguous comments are made about its functional possibilities. An earlier review (in Russian - Sukhorukov and Plotnikova, 1965) considers the structure and function of both plasmodesmata and ectodesmata.

Some workers have stated that they find specific associations between paramural bodies (Marchant and Robards, 1968) and plasmodesmata (Kurkova, Vakhmistrov and Solovyev, 1974; Vakhmistrov, Kurkova and Solovyev, 1972), although whether this relationship is real or artifactual remains to be demonstrated. (See also the induction of paramural bodies near plasmodesmata by virus infection - 8.3.).

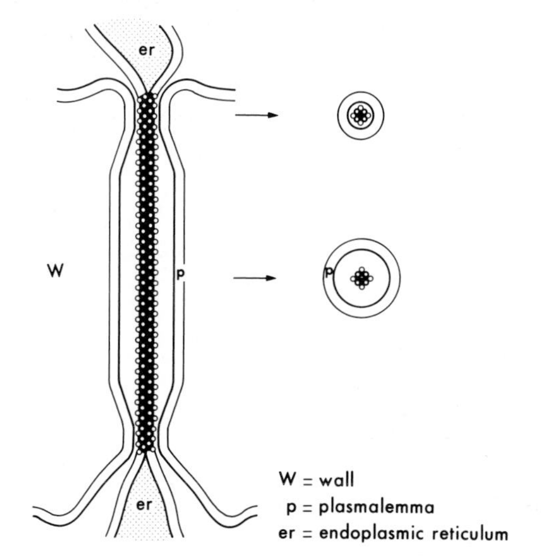

Fig. 2.8. Redrawn from Brighigna's (1974) interpretation of the plasmodesmata in the walls of the absorbing hairs of *Tillandsia usneoides*. *Two* globular components - of 7.0 and 4.5 nm diameter - are said to comprise the desmotubule. Functional capabilities *through* the desmotubule are not clear

One point is certain: *none* of the models described adequately depicts the structure of either a specific or a generalised plasmodesma. The model shown in Fig. 2.7. has at least two unresolved features critical to an understanding of its function: the presence or absence of a central rod through the desmotubule; and the 'tightness' of the neck on the desmotubule. The central rod is a common feature of plasmodesmata from widely different species (e.g. *Salix*, *Hordeum* and *Azolla* - Fig. 2.4.; *Abutilon* - Fig. 1.1.). Whether, as implied in Fig. 2.7., it is an artifact, remains to be determined. In *Hordeum* and in *Salix* (for example) the desmotubule is tightly invested by the plasmalemma at either end of the plasmodesma. If this seal is 'leak-proof', then the only opportunity for transport is through the desmotubule itself. If there is not a tight seal, then not only are two potential pathways opened up, but there exists the possibility for simultaneous bi-directional transport (Fig. 2.10.). Such plasmodesmata occur in *Abutilon* mature nectary stalk cells (Chapter 11), where there is a clear gap along the *whole* length of the plasmodesma between the plasmalemma and the desmotubule (Fig. 1.1.). There seems to be increasing evidence that plasmodesmata have a number of different basic structures or dimensions (contrast *Laminaria*; *Chara*; *Bulbochaete*;

Hordeum; Abutilon; Azolla, as examples of 'simple' plasmodesmata - see Table 2.1., and also Table 2.2. for detailed dimensions).

So far as the wall structure through which the plasmodesmatal pore runs is concerned, we have very little information. The early work of Scott *et al.* (1956) pointed to the primary pit fields as probable sites for the intercellular connections. Pit fields may give rise to pits traversed by plasmodesmata in the primary cell wall. It appears that most plasmodesmata traverse the walls of higher plants in these areas. The paper of Kollmann and Dorr (1969) implies that callose is found surrounding the plasmalemma of all plasmodesmata but this remains to be substantiated, both in higher plants and in algae. Callose has been found associated with algal plasmodesmata, although whether as an artifact or not remains contentious (Chapter 3); it is also found around phloem plasmodesmata; and, perhaps, around aging plasmodesmata in a moss (Eschrich and Steiner, 1968). Carde (1974) refers to a callose sheath, which he depicts as a cylinder of material in the neck region, around plasmodesmata, while Vian and Rougier (1974) and van Went *et al.* (1975) are among recent authors to comment on the different

TABLE 2.2.

DIMENSIONS OF PLASMODESMATA (nm)

	Outer diameter of plasmalemma	Inner diameter of plasmalemma	Outer diameter of desmotubule	Inner diameter of desmotubule	Central rod
Azolla young root cortical cells	35	25	16	7	3
Hordeum young (4 mm from tip) root endodermal cells	46	33	20	9	3
Hordeum older (120 mm from tip) root endodermal cells	60	44	20	10	4
Abutilon Distal cross wall of stalk cell of nectary hair (see Chapter 11)	44	29	16	10	3

All dimensions cited here are means of multiple measurements taken from micrographs of calibrated magnification. Measurements were made across the mid-line of plasmodesmata (not the neck region). The main feature contributing to the wider plasmodesmata of the older barley roots is the increased width of the gap between the plasmalemma and desmotubule.

appearance of the cell wall immediately surrounding plasmodesmata. Clowes and Juniper (1964) referred to a changed carbohydrate deposition around plasmodesmata. Taiz and Jones (1973) have demonstrated a component of the cell wall around barley aleurone plasmodesmata that is highly resistant to cellulase enzymes, so that 'wall tubes' are left after enzyme digestion of walls: these tubes enhance the apparent size of plasmodesmata seen in the light microscope. (See also *Abutilon* plasmodesmata, Chapter 11).

2.4.3. Variation in Structure

2.4.3.1. Gross structural differences From the preceding section it will be noted that even 'simple' plasmodesmata from different groups have different structures. More elaborate forms occur, some of which reflect development from the simple state and are considered in Chapter 4. Many plasmodesmata have some form of 'node' (median nodule) in the mid-line of the wall (Mittelknote - Krull, 1960; Cox, 1971) (Fig. 2.9.); othershave anastomosing arms (Fig. 2.9.), such types being frequently cited from phloem cells (e.g. Kollmann and Schumacher, 1962; Northcote and Wooding, 1966; Murmanis and Evert, 1967) as well as other sites (e.g. *Viscum* - Krull, 1960; *Tamarix* wood fibres - Fahn, 1967; secretory cells of *Nepenthes* - Clowes and Juniper, 1968; hairs of the trap of *Utricularia* - Fineran and Lee, 1975; Chapter 11); sometimes the median nodules are linked to each other by a large median cavity. The median nodule (Mittelknote) is usually seen as an enlargement of the plasmodesmatal cavity, together with a more complex 'knot'

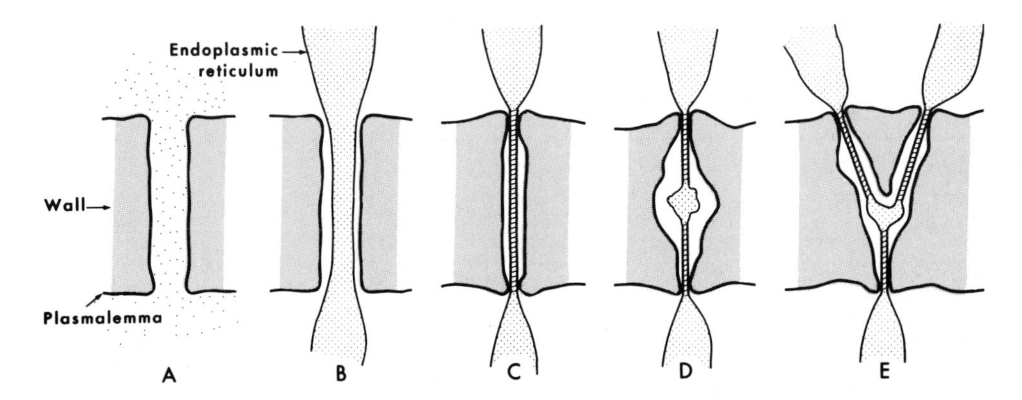

Fig. 2.9. Diagram illustrating plasmodesmatal variation.

A. Simple plasmalemma-lined pores with no desmotubule. A situation found mainly in the algae, but also occasionally reported from higher plants.

B. Loosely bound strand of endoplasmic reticulum. Such a profile is often seen during late stages of cell plate formation, but may persist in some cases. Two separate opportunities for symplastic transport are clearly available.

C. Tightly bound desmotubule. A constriction ('neck') appears to block any possible pathway between the plasmalemma and desmotubule.

D. A median nodule may form in the mid-line of the wall. Such regions often fuse to give a median cavity.

E. Desmotubules may anastomose, often with multiple connections on one side leading to a single channel on the other

structure of the desmotubule in the mid-line of the wall (Fig. 2.9D.). Median cavities arise when there is fusion of the central cavities of adjacent plasmodesmata, so forming large lacunae within the cell wall (Chapter 4).

Although there are few reports of plasmodesmata without desmotubules in higher plants (the micrographs of plasmodesmata through secretory cell walls of *Tamarix* published in Thomson and Liu, 1967 do not show clear desmotubules through all pores; also see Burgess, 1971 and Mueller, 1972), such 'open' pores are common in the algae, and increase the difficulty of relating the physiology of algal plasmodesmata to that of higher plants. In some cases (e.g. Dolzmann, 1965) the desmotubule may appear to be incomplete. Most of the variations described above have also been seen in varying degrees of elaboration in a relatively narrow sample of specimens studied in my laboratory, and it may thus be concluded that they are all of common occurrence. Indeed, the salient problem relating to structural variability is in determining whether the modified structures are indicative of the natural condition, or whether processing artifacts may lead to modifications of one type or another. In the case of plasmodesmata which possess a median nodule, as well as those which have anastomosing arms, there can be no doubt that this is how they exist in nature; these are relatively gross variations which we would not expect to be altered to any great extent by artifacts.

2.4.3.2. Minor structural variations In considering the nature of the desmotubule itself, the structure is closer to the realm of molecular dimensions, and it seems quite probable that artifacts could be readily induced. The complete absence of the desmotubule from a plasmodesma remains to be explained in acceptable functional terms (or *vice versa*!). An opening of the desmotubule into the central cavity (as in some of Dolzmann's plasmodesmata, 1965) need presumably have little effect upon translocatory function so long as the seal between the desmotubule and plasmalemma through the neck remains a good one.

The median nodule is apparently a secondary development from a 'simple' plasmodesma. It could be that this nodule is important in controlling translocatory fluxes or in completely stopping them; it may also be involved in enzyme-mediated processes initiated within the plasmodesmatal cavity. This idea is supported by the reports of enzyme activity associated with the plasmodesmata (Hall, 1969 - ATPase; Robards and Kidwai, 1969 - phosphatase; Ashford and Jacobsen, 1974a,b - phosphatases; also see Chapter 7), as well as the established hydrolytic activities during sieve plate pore formation. The anastomosing of desmotubules may occur with many tubules on either side linking into a relatively large central cavity or median nodule or by many desmotubules on one side fusing into relatively few, or only one, on the other (Fig. 2.9.). In such cases it is often found that the diameter of the single tubule is greater than that of the multiple ones (Wooding, 1968), although what significance this has for the molecular structure of desmotubules is unknown. Once again, one can only guess at the precise functional significance of such arrangements and speculate that they might be related to special requirements for the control of translocation.

2.4.4. Plasmodesmata in Phloem

Plasmodesmata in phloem tissue are of particular interest for their distribution: for example, as summarized by Shih and Currier (1969) in their contribution on cotton phloem: "*plasmodesmata connected parenchyma to parenchyma, parenchyma to companion cells, and companion cells to sieve elements. Their general absence between parenchyma cells and sieve elements points to a specific role of companion cells in sieve tube functioning*". This finding is in agreement with that of many other workers (e.g. Evert and Murmanis, 1965; Wooding and Northcote, 1965; Northcote and Wooding, 1968; Wark, 1965), with occasional reports to the contrary (Zee and Chambers, 1968 - *Pisum* primary phloem). Wooding and Northcote (1965) describe the common situation of complex plasmodesmata between the sieve elements and companion cells: from a single plasmodesma eight to fifteen arms develop on the companion cell side, but only a single tube, approximately twice the size of the others, is found on the sieve element side; similar reports have been widely published (e.g. Esau, 1973 - *Mimosa*). Behnke states that the desmotubule diameter of phloem plasmodesmata may be increased by up to six times compared to normal (cited by Kollmann and Dörr, 1969); as the radius of a narrow tube is an important parameter affecting flow-rate, such an observation is particularly germane. A median cavity is commonly associated with phloem plasmodesmata (Esau and Gill, 1973). Behnke and Paliwal (1973) have commented (in relation to *Gnetum*) that this may represent a phylogenetic stage prior to specialized connections between sieve tube members and companion cells. In *Isoetes*, Kruatrachue and Evert (1974) reported that the end walls of the sieve elements could have sieve plate pores and/or plasmodesmata, while the side walls have plasmodesmata only.

So far as the sieve plates are concerned, the plasmodesmata which initially traverse the wall appear to be quite normal. They may subsequently develop a median nodule, and from this general area hydrolysis of the wall occurs prior to the deposition of callose to form the completed sieve plate pore (e.g. Esau, Cheadle and Risley, 1962; Deshpande, 1974).

A detailed analysis of the structure of plasmodesmata (sieve plate pore initials) in differentiating sieve elements of *Vicia faba* was made by Zee (1969). The central core (desmotubule) is closely bounded by the plasmalemma, so that the inner, dark, layer of the latter appears to have fused with the central core. Zee's interpretation of the 25 nm diameter desmotubule substructure is similar to that of Robards (1968a, b), except that maximum rotational image reinforcement occurred where n = 14 instead of 11 (see 2.4.2.). In addition, Zee suggested the presence of seven further subunits at the inner margin of the desmotubular ring, these surrounding, and linking with, a central rod of about 15 nm diameter. Although such results are equivocal, and the image reinforcement technique requires most circumspect interpretation, they do strengthen the belief that the traversing strand of plasmodesmata, in widely different situations, is a structure with its own special organization, and not a simple tubule of unmodified endoplasmic reticulum membrane.

Phloem plasmodesmata, from their distribution and complexity, are clearly important in translocation and, presumably, especially in the loading and unloading of sieve elements *via* companion cells (Chapter 11). As in all other cases, however, this area requires intensive investigation to understand the mechanism of symplastic transport.

2.5. SOME CONCEPTS OF PLASMODESMATAL FUNCTION

Much of this Volume is concerned with the possible function of plasmodesmata in a variety of different situations. At the cost of a little duplication, it is useful briefly to mention some of the roles that have been envisaged, and some of the relevant evidence.

The very earliest papers on plasmodesmata assumed that they were involved in the translocation of material between cells. This point of view has been held more or less strongly with relatively few exceptions (see, e.g. Pfeffer, Vol. 1 p. 602, 1897). Where it has not been considered that translocation is the main function of plasmodesmata, then the transmission of stimuli has sometimes been suggested as an alternative possible function. Haberlandt (1914) summarized his ideas thus: "*the fact that protoplasmic connecting threads serve for the transmission of stimuli does not exclude the possibility that these structures may in certain cases be partly or entirely engaged in translocation of plastic materials*". This statement reflects the attribution of translocation as a subsidiary role in plasmodesmatal function and Haberlandt's views were probably coloured by the work of Pfeffer on the transmission of stimuli along plant parts (see Pfeffer, 1906). A similar sort of function is considered by Ledbetter and Porter (1970) who suggest that plasmodesmata may be important for the equilibration of membrane potentials and transfer of membrane-supported excitations; a possibility that is receiving increasing attention. The whole field of the application of electro-physiological techniques to the study of plasmodesmata is potentially a most rewarding one, and is well reviewed by Spanswick (1974; 1975), and is fully discussed in Chapter 6.

Most authors who deal with the question of function have speculated on a possible translocatory role of the intercellular connections. There is no *direct* proof that these structures serve such a purpose, and Clowes and Juniper (1968) state that "*we cannot assign any role to the plasmodesmata with any confidence*". Spanswick (1974) aptly cites a recent plant physiology textbook "...*the extent and importance of the role of plasmodesmata in intercellular transport is not known*" (Greulach, 1973); and Ziegler in a short article on "*What do we know about the function of plasmodesmata in transcellular transport?*" (1974) concludes "...*our knowledge of the function of plasmodesmata in transcellular transport is at present very poor and circumstantial*".

The main problem in attributing translocatory function to plasmodesmata is this: young cell walls, although they may contain plasmodesmata, are usually freely permeable and constitute no severe barrier to the passage of ions and small molecules, *provided that* such solutes can move easily across the plasmalemma: plasmodesmata at these sites are not, therefore, obviously obligatory for intercellular translocation of such solutes. (Quantitative treatments of plasmodesmatal transport *versus* trans-plasmalemma transport are given by Tyree (1970) and in Chapter 5.) Movement of materials from cell to cell across *thick* intervening walls could be assumed to take place *via* the plasmodesmata if the walls were impermeable, but it is difficult to show that they are; if it can be shown that *thick-walled* cells can communicate through their plasmodesmata, then it seems possible that the same channels serve for transport between younger cells also. Consequently, many experiments investigating plasmodesmatal function have been concerned with showing that these structures can be the only route for translocation and, secondly, that they are actually capable of allowing fluxes at high enough rates to account for experimental data (see, for example, Chapters 10 and 11).

Before the recent experimental approaches to the study of plasmodesmata are considered in later Chapters it is important to dwell briefly on the less direct, and often more circumstantial, evidence. While none of this is conclusive, when put together it makes a strong enough case to justify the devotion of further energies to the investigation of this problem. I have already made it clear that the distribution of plasmodesmata reflects the anticipated capacity for the passage of materials between cells. Some cells, where other evidence suggests the need for complete isolation, have no plasmodesmata at all; others have numerous plasmodesmata in the walls across which high flow rates would be expected; plasmodesmata with branching desmotubules are often found at very active sites of solute movement (such as sieve tubes/companion cells; secretory cells in the *Nepenthes* pitcher; salt glands; etc.) and, in extreme cases, may even show desmotubules enlarged to as much as six times their normal diameter. The stamen filament cells of wheat elongate at 2.0-3.5 mm min^{-1} and, according to Ledbetter and Porter (1970) are "*doubtless supplied by the abundant plasmodesmata*". Plasmodesmata are also present between cells with very thick walls which are considered to be impermeable (e.g. xylem ray cells; thick-walled endodermal cells) or to have barriers to apoplastic transport (e.g. the Casparian strip of young endodermal cells).

Just as there are features which favour acceptance of the idea that plasmodesmata are sites of intercellular communication, so also there are aspects which are difficult to reconcile with such a role. I mention only two: one is that neighbouring interconnected cells can exhibit morphological and/or biochemical differentiation one from another, hence plasmodesmata (or the membranes bounding the endoplasmic reticulum cisternae if this comprises a symplastic compartment) must be selective (as they commonly separate cells of quite different kinds [e.g. tannin cells and tannin-free cells, Esau, 1965; non-articulated laticifers and parenchyma cells]); the other difficulty is that, individually or together, plasmodesmata need to be capable of sustaining bidirectional fluxes (see Chapters 10 and 11). In the first case, even though the *manifestation* of a genetic dissimilarity (e.g. tannin) may be produced in, or sequestered into, a compartment of one cell, but not another, the fact remains that the plasmodesmata presumably allow a degree of intercellular continuity without at the same time jeopardising the individual genetic identity of adjacent cells. The second case arises from the consideration that, while we may envisage a unidirectional volume flow through some plasmodesmata where high flux rates occur, it is difficult to equate this with the certain necessity for at least some transport in the opposite direction. The possibility that plasmodesmata may be polarized to work in one direction only, and that different plasmodesmata through the same wall or barrier allow fluxes in different directions, is one that would be extremely difficult to test (see Kuo, O'Brien and Canny, 1974; also Chapter 12). It would, however, provide a simple means for controlling intercellular flow. A different concept is that single plasmodesmata may be able to carry fluxes in two opposing directions simultaneously through two different symplastic compartments (Fig. 2.10.). O'Brien and Thimann (1967a) suggested that the endoplasmic reticulum may act as a pathway for the intracellular transport of auxin, and that polarity of transport may in some way be associated with the evident connections of the endoplasmic reticulum to the complex plasmodesmata. Another alternative is that some molecules and ions would move through symplastic pathways while others would cross apoplastic ones - at least in part.

It seems to me that the current weight of *circumstantial* evidence is heavily inclined towards the view that plasmodesmata are functional in symplastic transport. Assuming that cells connect with each other through the plasmodesmata, what sort of connection exists? The structure has been discussed; we now look at the problem from a functional viewpoint. It must be stated at once that most physiologists have contemplated intercellular connections as small, but open, channels linking cytoplasm with cytoplasm (Fig. 10.2A.) - a situation apparently common in algae (see Chapter 3); this is different from the concept of a sealed strand of endoplasmic reticulum running from cell to cell.

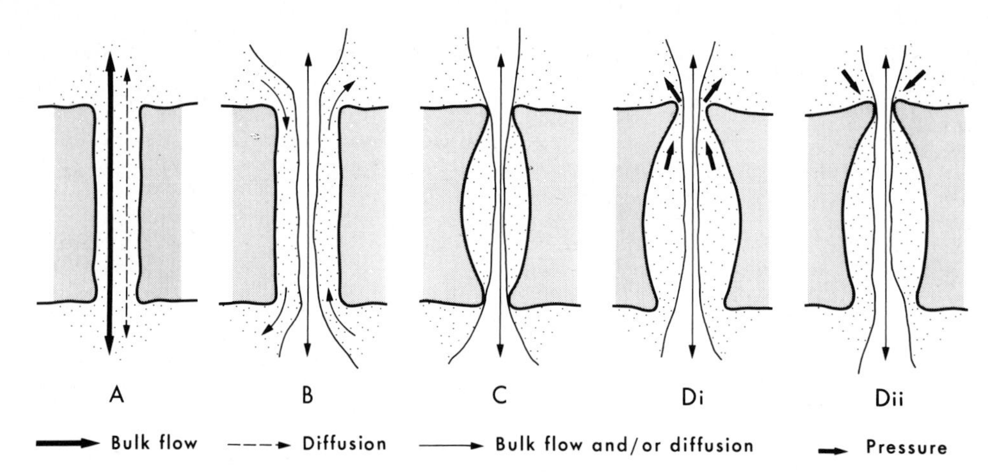

Fig. 2.10. Some theoretical and hypothetical possibilities for plasmodesmatal function.

A. An open tube, without desmotubule. Bulk flow or diffusion could occur in either direction so long as the rate of bulk flow in one direction was not so high as to reduce diffusion against it to negligible proportions. This possibility - bulk flow and/or diffusion - applies to all pathways depicted here, so long as the conducting pore is not too long, in which case the diffusive flux would be greatly reduced (see 1.2.1.).

B. An open tube containing a desmotubule, so allowing two possible pathways: through the desmotubule from endoplasmic reticulum cisterna to endoplasmic reticulum cisterna; or through the cytoplasmic annulus around the desmotubule (the *Abutilon* situation - Fig. 1.1.).

C. A desmotubule tightly sealed in the tube through the wall. The only apparent channel of intercellular communication appears to be through the desmotubule (the *Salix* and *Hordeum* situation - Fig. 2.4.).

D. A hypothetical concept for the operation of plasmodesmata as valves. If the plasmodesmata were permanently, or temporarily, asymmetrical - as shown - greater pressure from the cell on one side could keep the cytoplasmic annulus open (Di); alternatively, a reversal of pressure could close the cytoplasmic annulus, so leaving the desmotubule as the only conducting channel (Dii)

Functionally, the differences are important. In the former case the plasmodesmata may serve as no severe barrier for anything but the largest molecules (although open, but 'necked' plasmodesmata as found in *Bulbochaete* could modify this); in the latter, not only is space more restricted but the channel of communication is, at least in part, from the cavity of the endoplasmic reticulum in one cell to that of the next. A theoretical (Chapter 11) possibility is a *two* compartment connection, with an endoplasmic reticulum tubule passing through an *open* plasmodesmatal canal (Fig. 1.1., 2.10B.; see also Bräutigam and Müller (1975b - Fig. 12.), who propose a similar model). Buvat (1969) encourages the open channel view (although he has always considered that endoplasmic reticulum runs from cell to cell) by putting his concept of the 'plasmodia': the wall being a skeletal support for an internal medium which permeates the entire plant and through which circulate metabolites, water, ions and the products of cellular activity. Sutcliffe (1962) assumed that plasmodesmata function in the intercellular transport of ions, and "*small vacuoles may move through the protoplasmic connections and could act as vehicles in salt transport*", although we now know that this would be impossible through normal plasmodesmata. Brouwer (1965) and Minchin and Baker (1970) are among the authors who have considered models depicting plasmodesmata merely as open tubes through cell walls, but the evidence is clear that such models are much too simplistic and can only apply to a minority of cases in higher plants. Gamalei (1973) reports having followed the movement of osmiophilic droplets along plasmodesmata between ray parenchyma cells of spruce, and notes that the droplets are often larger than the diameter of the plasmodesma; a situation similar to that sometimes found in the intercellular movement of viruses. Gorin (1969) considers that enzymes can move through plasmodesmata - even of dead cells. Unfortunately, neither of these two papers provide the critical data that would allow an impartial and independent evaluation of the conclusions which are, therefore, still very much *sub judice*. Although *direct* experimental evidence is still required, there appear to be good reasons for believing that plasmodesmata act as pathways for symplastic transport and electrotonic coupling of cells. That is not to say that all plasmodesmata are functional in this way. There also seems to be abundant evidence showing that by secondary modification and/or occlusion, the function of plasmodesmata can be changed.

ACKNOWLEDGEMENT

I am grateful to the Agricultural Research Council and Science Research Council for financial assistance in support of some of the work described in this Chapter.

2.6. OPEN DISCUSSION

Most of the discussion centred around questions of fixation, staining and possible artifacts.

What was it that the early light microscopists saw? asked CARR, noting that plasmodesmata are below the theoretical limit of resolution. Did some diffraction phenomenon make them visible, or were they only seen when several were superimposed or side by side, or was it possible that it was sleeves of special wall material that made the plasmodesmata apparent? (see 2.4.2.). ASHFORD confirmed that wall sleeves do occur (see 4.9.) and that while some of the stains the early workers had used react with protoplasm, others are indeed stains for cell wall constituents. HUGHES and GUNNING described their efforts to repeat some of Tangl's work, by embedding date endosperm in both glycol methacrylate and Spurr's resin and observing sections using a Zeiss Photomicroscope III; with luck and dedication they hoped one day to be able to approach the quality of Tangl's 1879 drawings (see Frontispiece). Electron microscopy did at least confirm that the structures that are visible by light microscopy are indeed plasmodesmata!

GUNNING then made an extended comment about errors that arise when attempts are made to count plasmodesmata using micrographs showing transversely sectioned cell walls, i.e. with plasmodesmata sectioned longitudinally. The answer that is obtained depends very much on the thickness of the section and the diameter of the plasmodesmata in question (see also Spanswick and Costerton, 1967). The section includes within it parts of plasmodesmata as well as (if it is thick enough) whole plasmodesmata, and if *all* are counted then the sample volume is in fact a volume that is greater than that represented by the thickness of the section multiplied by area (or length of cell wall). Limits for the errors can be set, thus if R is the radius (μm) of the plasmodesma (outermost dimension), T the thickness of the section (μm) and F the frequency of the plasmodesmata per square micrometre:

(i) if *any* portion of a plasmodesma, however small, that is included within the section is detectable and is therefore counted then:

$$F = \text{count per } \mu\text{m of wall length}/(T + 2R)$$

(ii) if *at least* half a radius of a plasmodesma has to be present in the section for that plasmodesma to be detected and counted then:

$$F = \text{count per } \mu\text{m of wall length}/(T + R)$$

The true situation is likely to fall between these limits, with a plasmodesma being detectable when about one quarter of its radius lies within the section, in which case:

$$F = \text{count per } \mu\text{m of wall length}/(T + 1.5R)$$

Using the latter approximation to estimate F for the distal wall of the stalk cell of *Abutilon* nectary trichomes, and assuming a section thickness of 40 nm (this not measured accurately but merely a guess based on the resolution obtained in high magnification views such as Fig. 1.1.), the answer is F = 11.6, well within the standard error of direct counts based on face views of the wall in question, which gave F = 12.6, as described in 11.3. Had the correction for plasmodesmatal dimensions not been used (i.e. the 1.5R value), an erroneous answer of 30.7 would have emerged from the views of transversely sectioned walls.

For the above correction factor to be accurate, another condition must be met. The longitudinal axis of the plasmodesmata must not be at an angle to the plane of the section. If there is an angular deviation, the plasmodesmata will be sectioned obliquely. Because of this, parts of plasmodesmata that would not, with accurate orientation, be sectioned are in fact included, thus again increasing the apparent volume of the sample. The maximum increase in effective volume is given by ($\sec\theta + L/T \tan\theta$) where θ is the angle by which the plasmodesmatal axis deviates from the plane of the section and L is the length of the plasmodesmata. The effect is well illustrated in the micrographs of Hicks and Steeves (1973) which have been used to obtain such unusually high counts (see Table 2.1. and 2.3.2.), and where if L is taken as 0.2 μm and T as 0.05 μm, a θ of as little as 15° gives an effective doubling of the sample volume. Since the tilt is fairly clearly >15°, the plasmodesmatal frequency, while admittedly high, is probably not as high as has been suggested.

GIBBS, aware of long-term changes in plasmodesmatal frequency (Chapter 4), wondered whether short-term, even day-to-day, changes might occur, thereby contributing further to errors in estimating numbers per unit area. Possible, but no evidence as yet, was the verdict on this suggestion.

ROBARDS then took up the problem of estimating the cross-sectional dimensions of plasmodesmata from thin sections. If the sides of the pore and desmotubule are parallel (e.g. Fig. 1.1.), then the mean of a reasonable number of measurements will provide a good estimate of the various parameters. However, the plasmalemma-lined pore and sometimes even the desmotubule may vary in diameter through a single plasmodesma (see Fig. 2.4.3.). If this is the case, then citation of a mean from multiple measurements will have neither structural nor physiological significance.

A useful procedure under such circumstances is to record separately dimensions falling within specific ranges from all plasmodesmata under observation; if the section thickness can be reasonably estimated (T_S), and if total wall thickness (T_W) can also be chosen (e.g. from transverse sections of cell walls), then the number of section thicknesses through a wall (N_t) will be approximately known. The number of ranges of dimensions measured (R_x) should be $<N_T$. It is then possible to obtain a rough idea of the dimensions *and* shape of the plasmodesma as follows (this assumes: (a) all plasmodesmata being studied are similar, and (b) there is a continuous variation of size in one direction from mid-line to extremity of the pore).

Let us assume that the inner radius (R_i) of the plasmalemma-lined pore is the parameter of interest. The wall thickness (T_W) = 0.5 μm (500 nm); section thickness (T_S) = 50 nm; therefore N_T = 10. 100 measurements (ΣN) are made: they range between R_i = 25 nm and R_i = 10 nm. Therefore the measurements are each allocated to size classes at 1.5 nm intervals between 10 and 25 nm. They distribute as:

	a	b	c	d	e	f	g	h	i	j
R_i	10-11.5	11.5-13	13-14.5	14.5-16	16-17.5	17.5-19	19-20.5	20.5-22	22-23.5	23.5-25
N	12	11	9	5	6	7	3	15	15	17

Assuming also that the plasmodesmata are symmetrical about the mid-line, then the mean length of plasmalemma-lined tube of radius $R_{i(a-j)}$, over

$$\text{one half of a plasmodesma,} = \frac{N_{(a-j)} \times T_W}{\Sigma N}$$

$$\text{e.g. for } R_{i(a)}, \quad 0.5\ \frac{12 \times 500}{100} = 30 \text{ nm}$$

Similarly, (b) = 27.5, (c) = 22.5, (d) = 12.5, (e) = 15.0, (f) = 17.5, (g) = 7.5, (h) = 37.5, (i) = 37.5, (j) = 42.5. These values then represent the length over which the radial dimension is at each given size class, and realistic estimates of maxima and minima can be determined. Longitudinal sections are, however, still required in order to check on the appearance and number of constrictions, swellings, etc.

BAIN asked which is the best fixative to use for observing plasmodesmatal structure. No special recommendations can be made, said ROBARDS, although systematic studies are needed to check on dimensions. Various observations were offered from amongst those present - that glutaraldehyde prefixation gives very similar results to para-formaldehyde-glutaraldehyde mixtures (when the same cell wall is compared (HUGHES)); that freeze-substituted plasmodesmata are very like chemically-fixed ones (BROWNING); that fixatives incorporating tannic acid enhance certain aspects of ultrastructural detail (ROBARDS, Fig. 2.4.). It was also noted that Roland (1973) and Vian and Rougier (1974) have published a micrograph[1] of a cryosection of *unfixed* tissue showing plasmodesmata containing desmotubules; and that the cryosections of *fixed* material

[1] The same micrograph appears in both papers.

illustrated by Vian and Rougier (1974) also show desmotubules (though very narrow ones), neck constrictions, and tripartite plasmalemma. ROBARDS confirmed for QUAIL that the plasmalemma of the plasmodesma does stain with the phosphotungstic acid-chromic acid 'Roland' procedure (see Roland and Vian, 1971), and also with the periodic acid-thiocarbohydrazide-silver proteinate Thiéry procedure (Vian and Roland, 1972) and GUNNING reported a personal communication from Dr. M.E. McCully stating that it also stains like the rest of the plasmalemma when periodic acid - thiosemicarbazide - osmium vapour is used.

MARCHANT asked whether low temperature treatment or treatment with drugs such as colchicine, which disrupt microtubules, have any effect on the desmotubule. ROBARDS commented that, while both colchicine and vinblastine sulphate had been used, no clear effect on plasmodesmatal structure had been established. He added the caution that the effect might be quite different within the confined space of the plasmodesmatal canal, likening the situation to flagellar protein tubules which are relatively resistant to such disruptive agents.

There was considerable debate between the ultrastructuralists and the biophysicists about the structural complexity of plasmodesmata as compared with the relative simplicity of the models that have thus far been used. Are the complications really necessary? GOODWIN stressed that he had not seen a good longitudinal section showing a clear opening through a desmotubule. ROBARDS explained the technical difficulties of cutting sections sufficiently thin to reveal any such detail. He replied to GOODCHILD that stereo-pairs probably would not overcome the difficulty. Since good longitudinal views cannot be obtained, both the lumen of the desmotubule and the central rod remain of questionable reality. There was a pause while GIBBS, who had suggested that the central rod should be re-named a 'desmoquark', was ejected, whereupon GOODCHILD announced his desire to play devil's advocate. He argued that glutaraldehyde, when applied to a streaming cell, can bring about very dramatic changes in the cytoplasm and organelles as seen in the light microscope: hence what reason is there to be sure that the fixative, which might well move through the plasmodesmata as it penetrates pieces of tissue, does not modify or create the central rod or even the desmotubule? ROBARDS reiterated that the desmotubule appears after various fixation procedures - osmium tetroxide, aldehyde-osmium, and even permanganate - and best of all, it appears in unfixed cryosectioned material. It is not present in all plasmodesmata (see text), but all in all there is no more reason to suppose it to be an artifact than in the case of the endoplasmic reticulum in general. ROBARDS pointed to the widespread occurrence of images illustrating a central rod, but agreed that very little is known about it, or whether it is a real structure at all. A case could reasonably be made that it is an artifact of staining (Robards, 1971), but there is at present little evidence to support such an idea. A detailed study of this feature is needed.

VAN STEVENINCK pointed out that several French botanists had tried the technique of Golgi, impregnating tissues for long periods with osmium tetroxide or permanganate solutions at elevated temperatures. Cisternae of endoplasmic reticulum and dictyosomes seem to fill up with electron-opaque deposits, and claims of continuity from cell to cell had been made. The claims were perhaps somewhat premature, for example Benbadis, Lasselain and Deysson (1973) include one micrograph showing deposits in cisternae on either side of a cell plate, but not traversing it, and Chardard (1973) and Poux (1973) show impregnated cisternae on either side of cell walls, but the plasmodesmata between them are very much less electron-opaque. The technique had not, however, been applied specifically to reveal plasmodesmatal ultrastructure, and should be tried with this in mind. GUNNING added that the impregnation method, particularly in conjunction with high voltage electron microscopy, had shown that networks of extremely fine tubes are common between the endoplasmic reticulum and the forming faces of dictyosomes, the point being that the desmotubule is no longer an isolated, and hence especially suspect, example of a narrow tube in connection with conventional membranes.

Two experimental approaches to the controversy emerged. Measurements of the electrical resistance of cell junctions were, said GUNNING, acceptable to all as a means of assaying symplastic continuity. Would it not be possible to monitor the conductivity while glutaraldehyde was applied, thereby checking on whether the fixative introduces any marked changes as it cross-links the constituents to produce the structures that the electron microscopist sees. It was agreed that this experiment should be attempted. The other approach stemmed from the data on plasmodesmatal frequency that had been cited by ROBARDS. GRESSEL asked what happened to the vast numbers of plasmodesmata when cells are broken, as during biochemical isolation procedures? JONES thought that the plasmodesmata would pinch off and stay within the wall, quoting the work of Burgess (1971) on plasmolysed cells in support. GUNNING too had seen isolated plasmodesmata, complete with desmotubules, in walls between severely plasmolysed protoplasts. There was general optimism that it should be possible to prepare cell walls and to isolate plasmodesmata from them; further details of ultrastructure might then be open to study.

3

PLASMODESMATA IN ALGAE AND FUNGI

HARVEY J. MARCHANT

Department of Developmental Biology, Research School of Biological Sciences, The Australian National University, Box 475, P.O., Canberra City, A.C.T. 2601, Australia

3.1. INTRODUCTION

As well as occurring throughout the higher plants, plasmodesmata have been described from numerous algae and some fungi. While there is limited structural variation in plasmodesmata of higher plants, great diversity of cellular interconnections exists among the thallophytes and it is often difficult to make a clear distinction between true plasmodesmata and other protoplasmic continuities. In other words, plasmodesmata are just one of the very many types of cytoplasmic junction found among these organisms.

In this Chapter I will not attempt to define plasmodesmata in structural terms, but will discuss the various types of cytoplasmic interconnections between cells, concentrating particularly on those which most closely resemble the plasmodesmata of higher plants. In doing so I hope to be able to present algal and fungal plasmodesmata as specialized, yet far from the only form of cytoplasmic continuity. As well as reviewing the structural variations in cytoplasmic connections of algae and fungi I will briefly mention what is known of their functions. I have confined my attention, almost entirely, to the more recent work; concentrating especially on those studies made with the electron microscope. Readers interested in the great wealth of light microscopical information on cellular interconnections of algae are advised to consult Fritsch (1935, 1945) and other reviews indicated in the various sections of this Chapter.

3.2. CYANOPHYTA (BLUE-GREEN ALGAE)

Whether blue-green algae, as procaryotes, should even be considered in the Chapter is open to conjecture, but a brief discussion of their cellular interconnections provides a useful comparison with eucaryotic organisms. The distinctly filamentous blue-green algae (the Hormogonales or Oscillatoriales) are divided into two orders, the Nostocales

pseudo-parenchymatous thalli. Pit connections are conspicuous in the septa of most Floridophyceae. They also occur in the advanced members of the Bangiophyceae, but (as in parallel cases in the Chlorophyceae described later) are absent from some of the simpler multicellular genera e.g. *Bangia* (Dixon, 1963; Lee, 1971). To my knowledge there are no reports of true plasmodesmata being found in the red algae, although Mangenot (1924) and Dawes *et al.* (1961) use the term. Dixon (1973) considers that recent electron microscopy has rendered the term 'pit connection' most inappropriate to describe the septal structures. The 'pit' in fact contains a lenticular membrane-bound plug which separates rather than connects the cytoplasm of adjacent cells.

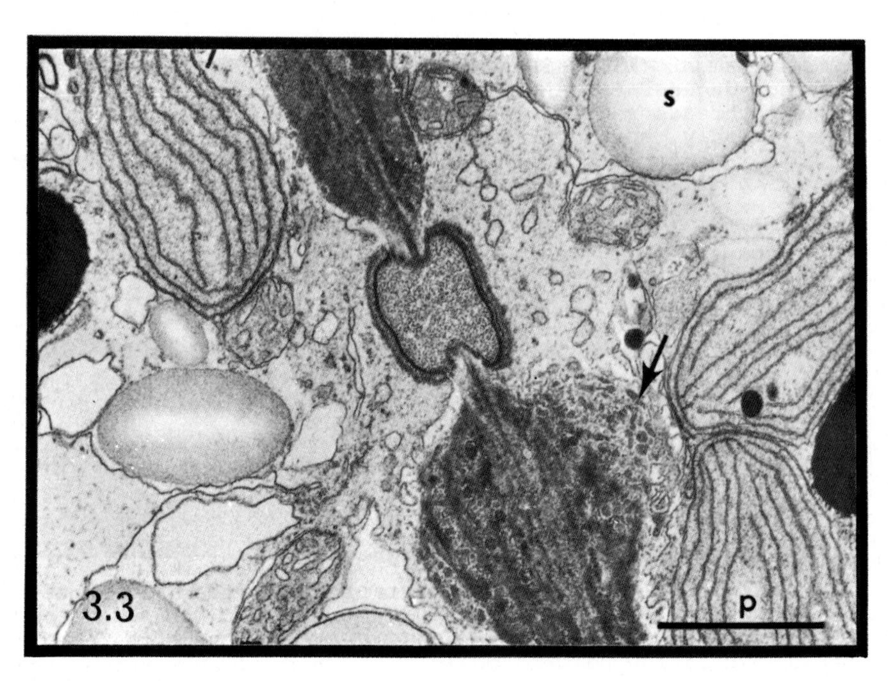

Fig. 3.3. Longitudinal section through a septal plug of a gametophytic filament of the red alga *Pseudogloiophloea*. Note the amorphous plug caps, the granular material condensing on the septum (arrow), the floridean starch granules (s) and plastids (p). Scale marker = 1 μm. Micrograph by courtesy of Dr. J. Ramus from Ramus 1969b

While Bisalputra *et al.* (1967) and Hawkins (1972) concluded that the plasmalemmas of adjacent cells are separated by the septal plug, Ramus (1969a,b), Lee (1971) and Duckett *et al.* (1974) showed the plasmalemma to be continuous between cells and that the plug itself and associated plug caps are enclosed by separate membranes. The plugs vary in diameter from 0.2-40 μm (Dixon,1973) and, at least in *Griffithsia*, consist of an acidic polysaccharide-protein complex in which the polysaccharide is neither cellulose nor pectin (Ramus, 1971). While the plugs are highly resistant to acids, they are readily solubilized in dilute alkali. Septation and the development of septal plugs in *Pseudogloiophloea* has been investigated in detail by Ramus (1969a,b). Plugs develop as depositions upon membranes occupying the aperture left by incomplete annular ingrowth of the septum. Their final size, shape and composition can vary widely within a given plant, depending on the nature of the cells (Ramus, 1969b; Bouck, 1962;

Tripodi, 1971; Hawkins, 1972; Duckett *et al.*, 1974). As with true plasmodesmata (see Chapter 4) secondary pits can be formed (see Bouck, 1962; Hawkins, 1972).

Dixon (1973) discusses the light microscopical observations on the dissolution of septal plugs that allows intercellular transfer of cytoplasm during carposporogensis and both sporangial and vegetative regeneration. Ramus (1971) had suggested that such fusions could be facilitated if the plugs were susceptible to enzymatic digestion and removal, but the findings of Duckett *et al.* (1974) indicate a different mechanism in *Nemalion*, where, during formation of the fusion cell in the carposporophyte, septal plugs that have been removed survive free in the cytoplasm. Pits and their plugs persist until dehiscence of spermatia (Kugrens and West, 1972a), or where percurrent development occurs, spermatial release may be through apertures left when plugs are removed (Duckett *et al.*, 1974). The pit connection between stalk cells and tetraspore mother cells has been observed to persist until, following meiotic cleavage, wall development around the proximal tetraspore severs it (Kugrens and West, 1972b). The morphogenetic programme governing pit and plug formation can be suppressed, for instance no pits develop in the walls that arise after tetraspore meiosis (Kugrens and West, 1972b). The red algae provide a most remarkable situation not reported from any other photosynthetic eucaryotes: amongst the Bangiaceae it has been found that one generation (the *Conchocelis* phase) develops pits while the alternate generation (the *Porphyra* thallus) does not (Lee, 1971). Considering that the septal complex of Basidiomycetous fungi can break down to allow nuclear migration (Geisy and Day, 1965), it is not implausible that a similar process could take place in the red algae. As during heterokaryon formation in fungi, red algae can also dissolve their cell wall to establish cell fusions, (Waaland and Cleland, 1974 and Chapter 13) but this is a process quite independent of septal plugs and pores.

Bouck (1962), when presenting the first ultrastructural study of sections of red algal pits commented *"it is clear that there are no other connections or plasmodesmata between the cells of the thallus, so protoplasmic continuity must be either maintained through the pit or not at all"*. Despite the plugging of septal pits, translocation (presumed to be symplastic), of ^{14}C leucine has been demonstrated in *Delesseria* and *Cystoclonium*, at rates in excess of 16 $\mu m\ s^{-1}$ over distances of about 100 mm, possibly along files of cells that are so long that the observed rate represents only 2 cells min^{-1} (Hartmann and Eschrich, 1969). Translocation between two different algae can also occur. ^{14}C labelled metabolites move from *Gracilaria* to the colourless parasitic rhodophyte *Holmsella* (Evans *et al.*, 1973) and Citharel (1972) reported the movement of ^{14}C glutamic acid from the brown alga *Ascophyllum* to an epiphytic rhodophyte *Polysiphonia*, where the host-epiphyte interface displays the interesting feature that some of the rhizoids of *Polysiphonia* have very thin localized areas of wall where cytoplasm is in closer contact with the host than elsewhere. When the rhizoids are pulled away, cytoplasm often leaks from these points. Attempts to demonstrate, with the light microscope, connections between host and epiphyte were unsuccessful (Rawlence and Taylor, 1970). Further, it should be stressed that Evans *et al.* (1973) were unable to confirm the presence of secondary pit connections with the electron microscope between *Gracilaria* and *Holmsella* as reported by Sturch (1924). Neither were secondary pit connections found between *Rhodymenia* and its red algal parasite *Halosacciolax* by Edelstein (1972). Thus it would appear that these total or partial parasites do not develop secondary cytoplasmic interconnections with their hosts. How metabolites cross the membranes between the algae is not known, but increased permeability of host membranes can be induced by certain parasitic fungi (Wheeler and Hanchey, 1968), by the angiosperm parasite *Cuscuta* (Wolswinkel, 1974b) and in numerous examples of algal-invertebrate symbioses (see review by Smith, 1974).

Various authors have drawn attention to the resemblance between Rhodophycean pit connections and the septal complexes found in Basidiomycetes and Ascomycetes and have suggested that the similarity indicates phylogenetic relationships between these apparently diverse groups of organisms (Denison and Carroll, 1966; Lee, 1971; cf. Saville, 1955). In particular, septation of *Pseudogloiophloea* (Ramus, 1969a) resembles that process in the Ascomycete *Ascodesmus*. Also, the membrane-bound bodies within the septal pores of some Ascomycetes Brenner and Carroll, 1968; Carroll, 1967; Kreger-van Rij and Veenhuis, 1969a,b) are not unlike Rhodophycean septal plugs. In summary, septal plugs of red algae apparently interrupt rather than interconnect the cytoplasm of neighbouring cells, yet translocation through the pores occupied by the plugs has been demonstrated. Dissolution or removal of the plugs can occur to facilitate intercellular transfer of cytoplasm.

3.4. PHAEOPHYTA (BROWN ALGAE)

For the first time in this Chapter, we now meet higher plant-like plasmodesmata.

3.4.1. Filamentous and Pseudoparenchymatous Thalli

There is a dearth of relevant information on filamentous and pseudoparenchymatous brown algae, but in view of the presence of plasmodesmata in heterotrichous green algae (see next section), it would be surprising if such connections were not present. In fact they have been seen in members of the order that is regarded (Fritsch, 1945) as the least advanced in the Phaeophyceae:- *Ectocarpus* sporelings and young sporophytes (Oliveira and Bisalputra, 1973), *Eudesme* (Cole, 1969), and *Leathesia* (Cole and Lin, 1968), all members of the Ectocarpales. The arrangement of branches in some of the pseudoparenchymatous types can be compact, though perhaps not so precise as in the red algae; however no evidence of formation of plasmodesmata between unrelated cells has so far been revealed.

3.4.2. Parenchymatous Thalli

Plasmodesmata have been seen in the structurally simpler orders, for example in the Dictyosiphonales (*Phaeostrophion*, Bourne and Cole, 1968), Scytosiphonales (*Petalonia*, Cole, 1970; Cole and Lin, 1970) and Dictyotales (*Dictyota*, Dawes *et al.*, 1961; Evans and Holligan, 1972a,b; *Zonaria*, Dawes *et al.*, 1961; Neushul and Liddle, 1968), but most of the observations centre upon the Fucales and, particularly, the Laminariales.

The large brown algae *Macrocystis*, *Nereocystis*, *Pelagophycus* and *Laminaria*, members of the Laminariales, have evolved a highly differentiated system for the transport of metabolites over long distances. Since the discovery by Reinke (1876) of sieve tubes in these algae, confusion has developed over the terminology of the conducting cells of the Laminariales (Esau, 1969; Schmitz and Srivastava, 1974a). Schmitz and Srivastava refer to individual conducting cells in *Laminaria groenlandica* as sieve elements and longitudinal files of these cells as sieve tubes, but point out that the cells in question do not rigorously meet the definitions of either sieve elements or sieve cells *sensu* Esau (1969). Schmitz and Srivastava use 'sieve element' to include all conducting elements that had been referred to previously as 'trumpet hyphae','sieve hyphae', 'conducting filaments' etc. which had been considered different to the 'true sieve tubes' of the larger Laminarians, *Macrocystis* and *Nereocystis*. In the view of these authors, there is phylogenetic specialization of the sieve elements among the Laminariales, ranging from those of *Laminaria* with numerous small pores to those with large but sparse pores as in *Macrocystis*.

Parker and Huber (1965) found filaments of cells, which they called trumpet hyphae, arising in the cortical parenchyma of *Macrocystis*, passing among the sieve tubes and entering the medulla. As these cells make no connection with the sieve tubes it is contended that they do not contribute appreciably to the metabolic and consequent functions of the sieve tubes in *Macrocystis*. The trumpet hyphae of *Macrocystis* are therefore not homologous with sieve tubes and probably do not function similarly; it is fortuitous that their trumpet shape resembles that of the genuine sieve elements (*sensu* Schmitz and Srivastava, 1974a) of *Laminaria*. Van Went and Tammes (1973) present some evidence that the trumpet shape of the sieve elements of *Laminaria* is an artefact produced during chemical fixation or cutting, but Schmitz and Srivastava (1974a) do not agree, and consider that the sieve elements become trumpet shaped by passive stretching to more than ten times their original length as the thallus grows.

Ultrastructural studies by Parker and Philpott (1961), Ziegler (1963), J. Parker (1964), Ziegler and Ruck (1967) and van Went and Tammes (1972) have confirmed the existence of a range of structures considered to be equivalent to sieve plates between sieve elements of the Laminariales. Ziegler's finding of much larger pores than Parker and Philpott in

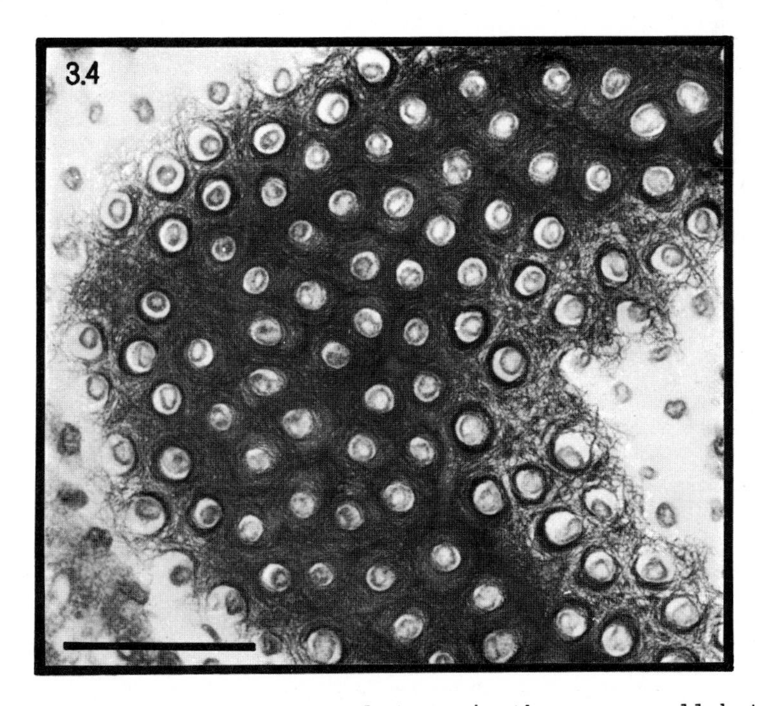

Fig. 3.4. Closely packed plasmodesmata in the cross wall between sieve elements of the brown alga *Laminaria groenlandica*. Scale marker = 0.5 μm. Unpublished micrograph by courtesy of Dr. K. Schmitz

Macrocystis led him to believe that what these latter authors had in fact observed were typical plasmodesmata between cells that were not sieve elements. While J. Parker (1964), in answering Ziegler (1963), refutes the claim that Parker and Philpott were not looking at sieve elements he does concede that their material may have suffered damage during processing for electron microscopy. Parker and Huber's (1965) measurements of the sieve plate pores of the same species of *Macrocystis* are in agreement with those of Ziegler. Ziegler and Ruck report that sieve plates of *Laminaria digitata* contain 20,000-30,000 pores of 50-100 nm diameter occupying one sixth of the wall area. Corresponding figures provided by Schmitz and Srivastava (1974a) for *Laminaria groenlandica* are:- 40,000 pores in the total plate (34.5 μm diameter), each pore being 60±2 nm in diameter (with a somewhat shrunken plasmalemma canal *within* the diameter), at a packing density of 42 μm^{-2}. Some of their micrographs show desmotubules and constricted cytoplasmic annuli. By contrast, *Macrocystis* has only 86-244 pores of 2.4-6 μm diameter (Ziegler, 1963; Parker and Huber, 1965). These latter authors stress that sieve elements of the Laminariales have sieve plates only in their end walls.

The mature sieve elements of *Laminaria, Nereocystis* and *Pelagophycus* are nucleate (Smith, 1939; Parker and Fu, 1965; Schmitz and Srivastava, 1974a) while those of *Macrocystis* lack nuclei (Parker and Huber, 1965). No equivalent of the P-protein of higher plant sieve elements has been seen. As the sieve elements mature their cytoplasm becomes rather disorganized and vesiculate while massive amounts of callose may be deposited on the sieve plates as well as along the longitudinal walls (B. Parker, 1964), and may eventually occlude the cells (Parker and Huber, 1965). Van Went *et al.* (1973b) claim to have reduced the extent of callose deposition by low temperature processing although Schmitz and Srivastava (1974a) doubt their conclusion. These latter authors consider that the electron-lucent area between the plasmalemma and the rim of the sieve pore is not callose, as thought by Ziegler and Ruck (1967) and van Went *et al.* (1973b), but *"an artifact caused by shrinkage of the wall during dehydration"*. Parker and Huber also point out that while the sieve tubes of the Laminarians have much in common with sieve cells of higher plants, their higher complement of organelles points to a greater metabolic independence, reflecting their lack of adjacent companion cells.

Schmitz and Srivastava (1974a) have produced the only evidence, to my knowledge, of secondary development of plasmodesmata between neighbouring cells of an alga. In other organisms plasmodesmata develop at the time of cell division presumably by the trapping of endoplasmic reticulum or by incomplete septation leaving strands of plasmalemma interconnecting the daughter cells (see Chapter 4). Schmitz and Srivastava illustrate recently formed cross walls penetrated by few plasodesmata and walls of older sieve elements containing a high density of plasmodesmata. They believe that the *"pores are formed by enzymatic digestion of parts of the wall and ....the plasmalemma and endoplasmic reticulum are involved in this process"*.

Considerable attention has been given to the transport of metabolites in the Laminariales (Parker, 1963, 1965, 1966; Nicholson and Briggs, 1972; Hellebust and Haug, 1972; Lüning *et al.*, 1971; Schmitz *et al.*, 1972; Steinbiss and Schmitz, 1973). Perhaps the most significant findings in the context of this discussion are the demonstrations that the sieve elements are in fact the pathways of translocation (Parker, 1965; Steinbiss and Schmitz, 1973) and that the sieve tube exudate is much lower in sodium content than sea water, but rich in D-mannitol, amino acids, protein and potassium (Parker, 1966), thus in certain of these respects resembling the phloem exudate of higher plants. Schmitz and Srivastava (1974b) report high concentrations (up to 1100 g m^{-3}) of ATP in the sieve element sap and have shown rapid incorporation of ^{32}P into ATP in the sap suggesting that considerable amounts of ATP are needed to transport metabolites through the sieve tubes.

The distribution of plasmodesmata has not been thoroughly mapped in any member of the Laminariales or Fucales. Parker and Fu (1965) consider them to be anatomically widespread in *Macrocystis* and van Went *et al.* (1973a) record symplastic interconnection between cells of the medulla and cortex of *Laminaria*. Cytoplasmic continuities in *Fucus* tissues were first demonstrated by Hick (1885) and subsequently by McCully (1968) and Bisalputra (1966), who also found them in *Egregia*. Berkaloff (1963) adds another genus, *Himanthalea*, to the rather short list of Fucalean plants that have been examined. As in *Dictyota* and *Endarachme* (Dawes *et al.*, 1961; Evans and Holligan, 1972a,b), plasmodesmata in *Fucus* meristoderm seem to be arranged in pits (above references), whereas in the elongated filament cells of the midrib they are more uniformly distributed across the end walls, being absent from the side walls except probably at the rare points of contact between cells in this highly mucilaginous tissue (Fulcher and McCully, 1971). In none of these observations is there any compelling evidence that the endoplasmic reticulum traverses the plasmodesmata, but the quality of fixation of these refractory plants is so poor that the point cannot be considered settled.

This account of plasmodesmata in the Phaeophyceae can be concluded with two observations that are of recurring interest in this Volume. Fulcher and McCully (1971) find that, when *Fucus* cells are damaged, the plasmodesmata remain in the wall bounding the underlying undamaged cell. They break off on the 'wounded' side of the wall, and later they become isolated by deposition of fibrillar polysaccharide on the 'healthy' side. The second observation is not dissimilar: in *Zonaria* (Dictyotales) the numerous plasmodesmata which lead to the egg during its development become occluded (though they persist) by a thick layer of 'mucilage' lying between their distal extremities and the egg cell plasmalemma (Neushul and Liddle, 1968). The theme of symplastic isolation of reproductive cells, so self evident in most algae, where the cells usually disperse, is taken up again in the next section and in Chapter 13.

3.5. CHLOROPHYTA (GREEN ALGAE)

Where cellular interconnections are found in the green algae they generally resemble the plasmodesmata of higher plants. The only notable and well studied cytoplasmic interconnections that differ markedly from higher plant plasmodesmata are the cytoplasmic bridges of *Volvox*. Pocock (1933), Bisalputra and Stein (1966), Dolzmann and Dolzmann (1964) and Deason *et al.* (1969) consider that the intercellular cytoplasmic bridges of *Volvox* result from incomplete cytokinesis. While initially they are of sufficient diameter (0.25-1 µm) to accommodate the passage of mitochondria and other organelles, they become much smaller in diameter (0.1-0.3 µm) as the cells move apart after inversion. The only organelles then found in the stretched bridges are endoplasmic reticulum and ribosomes, although Ikushima and Maruyama (1968) found a layered, perforated medial body in a species of *Volvox* probably *V. globator*. Neither they nor Pickett-Heaps (1970) and Bisalputra and Stein (1966) found such a structure in other species. The variability of both the number and size of the cytoplasmic bridges have been used as taxonomic criteria. Whether or not these bridges

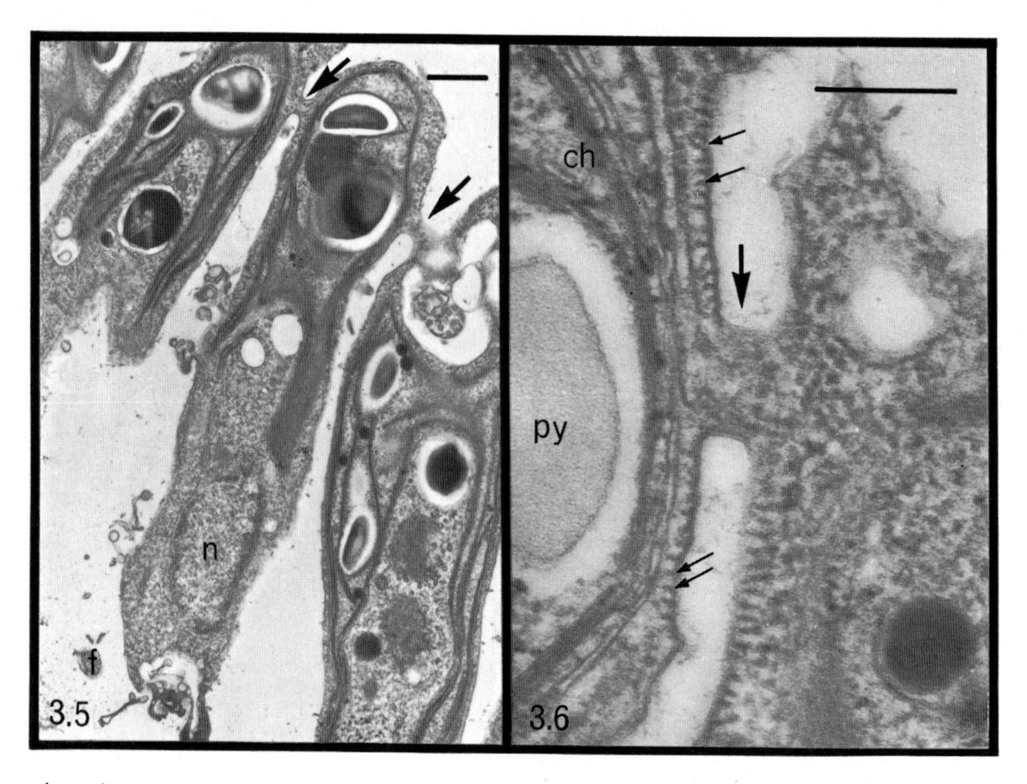

Fig. 3.5. Cells of an inverting colony of *Volvox tertius*, interconnected by cytoplasmic bridges (arrows). The bridges predominantly occur at the opposite end of the cell from that occupied by the nucleus (n) and from where the flagella (f) arise. Scale marker = 1 μm

Fig. 3.6. Cytoplasmic bridge of *V. tertius* (large arrow), flanked by striations (small arrows) on the plasmalemma. A chloroplast (ch) containing a pyrenoid (p) is nearby. Scale marker = 0.5 μm. Unpublished micrographs by courtesy of Dr. J.D. Pickett-Heaps

persist between cells throughout the life of the coenobia is also variable. Pickett-Heaps (1970) reports that the bridges break down soon after inversion in *V. tertius*, while in other species (e.g. *V. globator* and *V. aureus*) the bridges remain interconnecting cells in mature coenobia. According to Smith (1944) and Starr (1970) the only species of *Volvox* without cytoplasmic bridges are those belonging to the taxonomic section Merrillosphaera. During sexual reproduction in *Volvox aureus*, Deason *et al.* (1969) found cytoplasmic bridges between cells within developing sperm packets persisting until shortly before the sperm were released. Cytoplasmic bridges have also been reported between young cells in other members of the Volvocales, *Gonium* and *Pandorina* (Bock, 1926). Although tenuous cytoplasmic strands may occur between cells of *Gonium* there is no microscopical evidence of actual cytoplasmic continuity (Stein, 1965).

The function of cytoplasmic bridges of *Volvox* is enigmatic. Analyses of the patterns of differentiation in *Eudorina* (Gerisch, 1959, Goldstein, 1967), suggest that the distribution of somatic and reproductive cells is not determined by intercellular communication so much as by asymmetric divisions during colony formation. It has long been assumed that the bridges might somehow coordinate flagellar beat; for instance, Dolzmann and Dolzmann (1964) point out the close association between rough endoplasmic reticulum and the eyespot and suggest a possible involvement of the cytoplasmic bridges in coordinating the phototactic response of *Volvox aureus*. This is not likely if mature coenobia of some species, which swim normally, lack the bridges. Further, if the *direction* of flagellar beat was coordinated by the orientation of flagella themselves, as appears to be the case in *Eudorina* (Marchant, unpublished data), the flagellar beat need not be synchronized as a metachronal rhythm to produce the characteristic rolling, swimming motion of the coenobia. There is however one stage in the life cycle of *Volvox* where coordination of the cells must be of extreme importance and that is during inversion, when the young, hollow spherical colony turns itself inside out. Not only could the bridges provide intercellular communication to coordinate their complex movements but the connections are also at ideal sites on each cell to act as a hinge. Pickett-Heaps (1970) reported finding circular, or possibly spiral, striations on the plasmalemma around the cytoplasmic bridges. In another alga, *Sorastrum*, striations on the plasmalemma encircle cells at sites of constrictions (Marchant, 1974). One may speculate that contraction of the plasmalemma in the region of the cytoplasmic bridges is part of the cellular mechanism of inversion in *Volvox*.

Plasmodesmata are only known to occur in those green algae which utilize a cell plate for cytokinesis, never having been found in those chlorophytes whose cells are partitioned at cytokinesis by a centripetally growing furrow, such as *Ulothrix zonata*, *Klebsormidium*, *Microspora*, *Ulva* and *Microthamnion*. The presence of plasmodesmata and a cell plate in cytokinesis had been used to support the idea that those green algae with these characters, particularly the Ulotrichales and Chaetophorales, most likely exemplify the progenitors of the higher plants (Floyd, Stewart and Mattox, 1971; Stewart, Mattox and Floyd, 1973). However, on the basis of cytokinetic structures, (Pickett-Heaps, 1972; Pickett-Heaps and Marchant, 1972), the morphology of motile cells (Marchant *et al.*, 1973) and some biochemical criteria (Frederick *et al.*, 1973), it has become apparent that there are two major groups within the green algae; those in which the interzonal spindle keeps the daughter nuclei widely separated during cytokinesis as distinct from those algae in which the interzonal spindle collapses after telephase (Stewart *et al.*, 1974). The latter group contains most of the Ulotrichalean and Chaetophoralean algae, but it is the former group, with persistent interzonal spindles, where the algae now thought to exemplify the progenitors of the land plants are found. Hence these structures can no longer be regarded as indicators of close affinity with the higher plants.

Unlike the situation in higher plants and some fungi, the plasmodesmata of green algae are only found between daughter cells and, to my knowledge, never develop secondarily between adjacent cells derived from different parents. It is interesting to note that *Spirogyra*, an alga that uses an ingrowing furrow to initiate cytokinesis and a small central cell plate to complete septation, lacks plasmodesmata completely (Fowke and Pickett-Heaps, 1969). Various fungi, which either use an ingrowing cleavage furrow or fusion of vesicles for septation (see next section) i.e. the two cytokinetic mechanisms employed by *Spirogyra*, possess plasmodesmata. It would appear then that the ability to form plasmodesmata is not so much a consequence of a particular cytokinetic mechanism but rather depends upon whether an organism has the genetic capability for the expression of plasmodesmata.

Plasmodesmata of the Ulotrichales and Chaetophorales are often indistinguishable from their counterparts in higher plants; their size

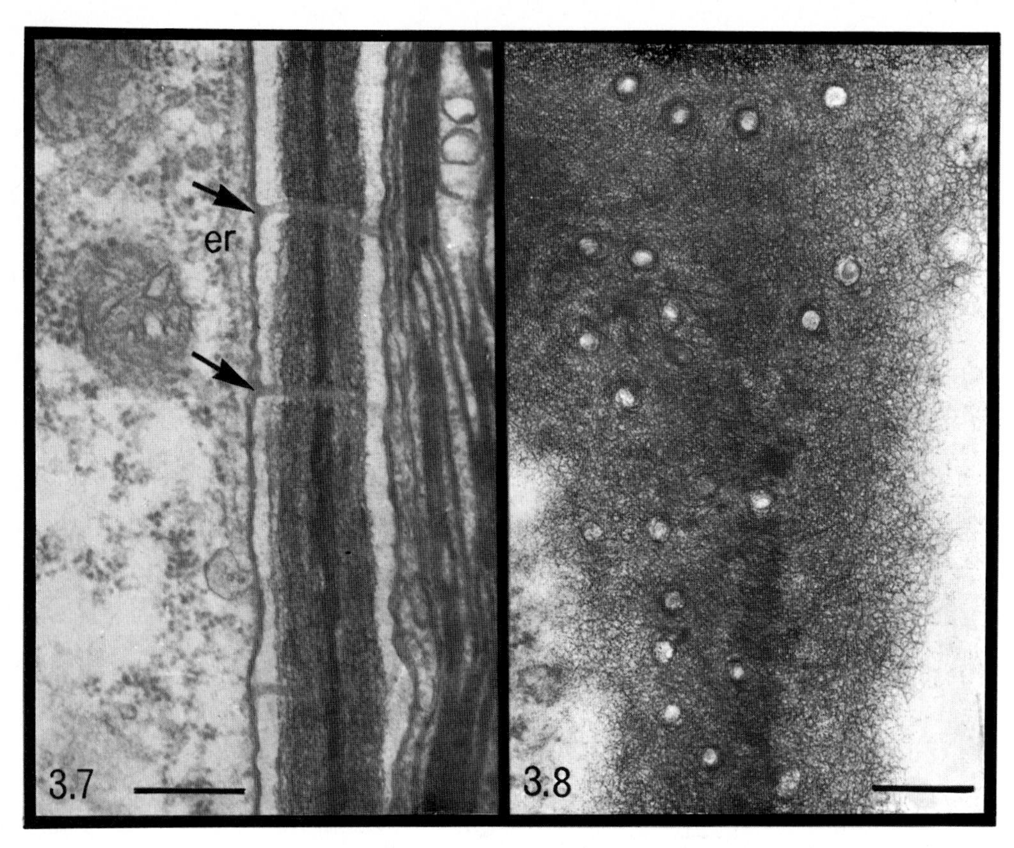

Fig. 3.7. Longitudinally sectioned plasmodesmata (arrows) within a wall of *Coleochaete scutata*. No internal structure is evident. Note the endoplasmic reticulum (er) adjacent to the plasmalemma, which has pulled away from the cell wall. Scale marker = 0.25 µm

Fig. 3.8. Transversely sectioned plasmodesmata of *C. scutata*. Plasmalemma can be seen lining, or contracted within, the plasmodesmata which apparently lack any internal structure. Scale marker = 0.25 µm

is much the same, they are lined with plasmalemma and some have an electron dense central core (Floyd, Stewart and Mattox, 1971; Robards, 1968b). The extensive studies of Stewart, Mattox and Floyd (1973) have revealed plasmodesmata in the following Ulotrichalean and Chaetophoralean genera, *Ulothrix* (but not *U. zonata)*, *Uronema*, *Schizomeris*, *Stigeoclonium*, *Chaetophora*, *Draparnaldia*, *Aphanochaete* and *Trentepohlia*. Other algae with plasmodesmata are *Coleochaete* (Marchant and Pickett-Heaps, 1973), *Fritschiella* (McBride, 1970) and *Ctenocladus* (Blinn and Morrison, 1974; c.f. Stewart, Mattox and Floyd, 1973). The presence or absence of the central core or desmotubule in the plasmodesmata of green algae is not well documented. Floyd, Stewart and Mattox (1971) illustrate desmotubules from *Ulothrix* and *Stigeoclonium*, and in *Aphanochaete* they show the desmotubule apparently connected to the endoplasmic reticulum (Stewart, Mattox and Floyd, 1973). While these authors state that "*electron-dense cores are seen in the plasmodesmata of nearly complete cell plates*" they make no mention of finding desmotubules between old cells. My own studies have not revealed desmotubules in the plasmodesmata between old cells of *Coleochaete* or *Aphanochaete*.

Unlike most other green algae in which the plasmodesmata are apparently random over the transverse walls, the plasmodesmata of *Trentepohlia* are localized in a central region of each septum, giving the appearance of a pit field (Brand, 1902; Stewart, Mattox and Floyd, 1973) reminiscent of the vascular plants. *Ctenocladus* reportedly has a single plasmodesma located in the centre of each septum (Blinn and Morrison, 1974). Among the siphonaceous green algae, which only rarely produce septa, *Codium* has been reported to possess plasmodesmata (Küster, 1933). Küster's drawings of these 'plasmodesmata' reveal that they would be better described as central septal pores and unfortunately, there is no electron microscopical confirmation of their existence.

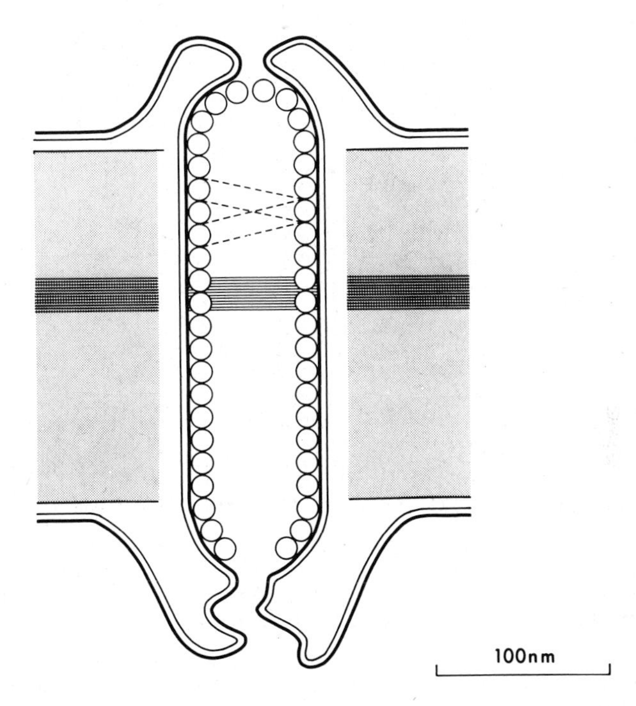

Fig. 3.9. Diagram of a plasmodesma in the green alga *Bulbochaete hiloensis*. Helically arranged particles line the lumen of the plasmodesma and the plasmalemma is domed over the ends, and constricts the orifices. Diagram from Fraser and Gunning (1969)

The order Oedogoniales contain only three genera: *Oedogonium*, *Oedocladium* and *Bulbochaete*. The plasmodesmata of the two genera that have been studied ultrastructurally, *Oedogonium* and *Bulbochaete*, are quite unlike those found in any other organisms. Fraser and Gunning's (1969) detailed study of these structures in *Bulbochaete* revealed that the plasmodesmata which interconnect the cells of this alga lack a desmotubule but that the lumen of the plasmodesmata is lined by particles which are apparently helically arranged. At each end of a plasmodesma the plasmalemma is raised in an annular dome constricting the orifices, and the suggestion is made that this constriction may act to regulate the movement of material between cells (see Chapter 1). Similar plasmodesmata to those of *Bulbochaete* have been found in *Oedogonium* by Pickett-Heaps (1971) and Coss and Pickett-Heaps (1974) although in this latter alga the constrictions at each end of the plasmodesmata are not conspicuous. During zoosporogenesis in *Bulbochaete* (Retallack and Butler, 1970) and *Oedogonium* (Pickett-Heaps, 1971) the plasmodesmata are plugged by the vesicle or hyaline layer which encloses the developing zoospore. The former authors suggest that this disruption of intercellular communication *"may allow the incipient zoospore to assume and develop its own polarity and peculiar morphological characteristics"*. Oogamous sexual reproduction is well developed in the Oedogoniales. During spermatogenesis in the nannandrous *Bulbochaete* investigated by Retallack and Butler (1973) and the macrandrous *Oedogonium* (Coss and Pickett-Heaps, 1974) plasmodesmata interconnect the spermatozoid mother cells, but these interconnections are severed by development of the vesicle which surrounds the spermatozoids. It is not clear at what stage the antheridial cells of *Oedogonium* are isolated from those of the parent. Similarly, it is not revealed exactly by Retallack and Butler at which stage the oogonium is isolated from the rest of the vegetative plant. They do report however that plasmodesmata interconnect the suffultory cells with the secondary oogonium and that this cell contracts basally during its maturation to become the oogonium.

Of the various algal plasmodesmata, those of the Characeae have probably been the most extensively studied. In *Nitella* and *Chara* there are two types of plasmodesmatal interconnections. The plasmodesmata traversing the walls between nodal cells are relatively simple plasmalemma-lined tubes of uniform diameter. However, interconnections between nodal and internodal cells and between nodes and lateral branches are not uniform in diameter but are dilated, giving rise to median sinuses between cells. In the meristematic regions the sinuses are simple, but in mature nodes they coalesce to produce large ana-

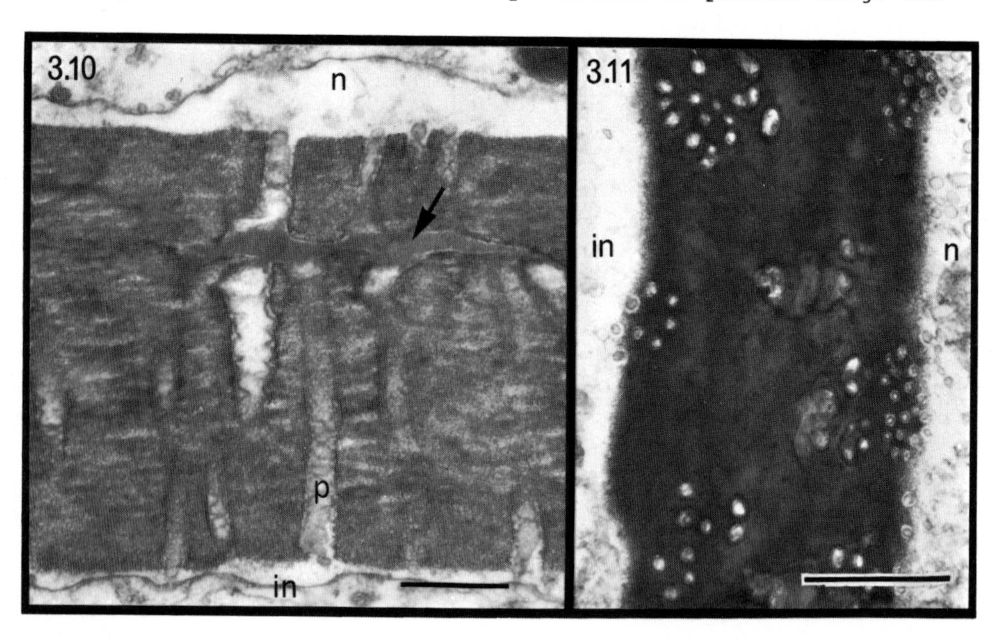

Fig. 3.10. Plasmodesmata (P) between young internodal (in) and nodal (n) cells of *Chara corallina* showing the plugging of the sinus within the wall by amorphous material (arrow). Scale marker = 0.5 μm

Fig. 3.11. Tangential section of a wall between a nodal (n) and internodal (in) cell of *C. corallina* revealing plasmodesmata, which lack internal structures, arranged in pit fields. Scale marker = 1 μm

stomosing cavities. Despite this growth of sinuses in the wall between mature nodal and internodal cells, the diameter of the openings at both ends of plasmodesmata remain constant (Spanswick and Costerton, 1967). Pickett-Heaps (1967) found that plasmodesmata between meristematic cells of *Chara* generally contained some central structure (desmotubule). The observation of central structures in *Chara* has been confirmed by Fischer *et al.* (1974). A desmotubule has also been seen in the plasmodesmata of *Nitella* (Spanswick and Costerton, 1967) but these authors stress that it was not a common finding in their material. Plasmodesmatal connections persist during spermatogenesis in *Chara* until shrinkage of the sperm stretches and ultimately severs them. When young, however, they are unusually wide (85 nm in diameter in Fig. 23. of Pickett-Heaps, 1968) and the continuity of the nuclear envelope and endoplasmic reticulum through them is especially clear cut - indeed the endoplasmic reticulum is apparently not at all modified to a desmotubule.

Fischer *et al.* (1974) draw attention to apparent plugging of the sinuses in plasmodesmata of mature node-internode junctions with electron-opaque material. They rightly conclude that it is not possible to determine from micrographs alone whether it is a result of a wounding reaction induced by fixation. Very recent evidence obtained from *Chara* by Fischer and MacAlister (1975) reveals that the sinuses are not plugged when processed for electron microscopy by freeze substitution rather than by the conventional aldehyde-osmium fixation. This is an important point in relation to quantitative aspects of electrotonic coupling across the nodes, as discussed further in Chapters 6 and 9.

3.6. FUNGI

Among the fungi there is an amazing variety of cellular interconnections. These pores can be conveniently subdivided into three broad classes; (1) the complex pore of the Basidiomycetes, (2) simple pores, including Ascomycete septal pores and (3) plasmodesmata.

The dolipore septum found in the Basidiomycetes is the most elaborate fungal cytoplasmic bridge. In this structure the central pore is surrounded by an annular swelling of the septum and over the pore are dome-shaped membranous vesicles which are continuous with the endoplasmic reticulum. There is considerable variation on this basic structure (Bracker, 1967). The taxonomic value of the type of pore in the fungal septum has been seriously questioned by Kreger-van Rij and Veenhuis (1971). The dolipore type of septum has been regarded as the Basidiomycete type and the simple pore as the Ascomycete type (Moore and McAlear, 1962). However, that the ascomycetous yeast *Endomycopsis* has a dolipore septum (Kreger-van Rij and Veenhuis, 1969a) and that some basidiomycetous yeasts have simple pores (Bracker, 1967) contradicts this rule. Geisy and Day (1965) report the breakdown of complex pores to simple pores through which nuclei migrate in heterokaryotic mycelia of *Coprinus lagopus*.

Septa of Ascomycetes usually only contain a single central pore varying in diameter from about 0.05-0.5 μm (Bracker, 1967). Unless plugged, these pores allow the passage of streaming cytoplasm and also nuclear migration. Trinci and Collinge's (1973) data reveal that there are considerable strain differences in the percentage of plugged septa at different distances from the periphery of spreading mycelia of *Neurospora*. Their results suggest that the hyphal density is related to

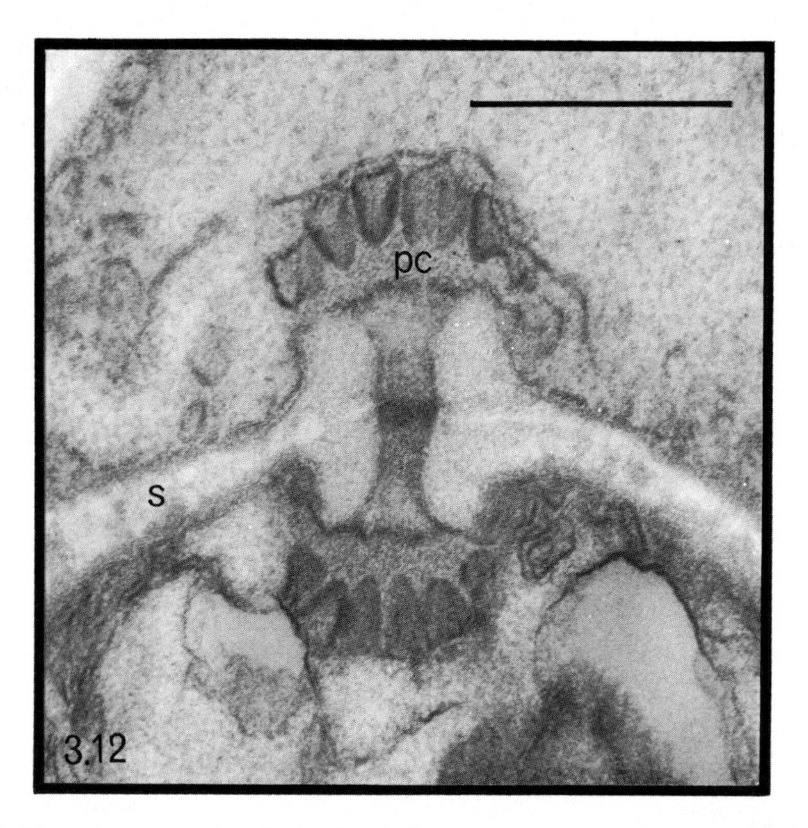

Fig. 3.12. Septum (s) of the basidiomycetous yeast *Leucosporidium capsuligenum* showing the dolipore structure. The pore cap (pc) is an elaboration of endoplasmic reticulum. Scale marker = 0.5 μm. Unpublished micrograph by courtesy of Drs. N.J.W. Kreger van Rij and M. Veenhuis

the degree of septal plugging. Dense colonies have their septal pores plugged closer to the periphery of the mycelium than do sparse colonies.

Multiperforate septa occur in the Ascomycete *Fusarium* (Reichle and Alexander, 1965) and Ascomycetous components of lichens (Wetmore, 1973). Wetmore concludes that the function of the extra pores is probably to facilitate movement of nutrients and water between cells and points out that extra pores, more often found in lichens, may be correlated with the metabolite relationship between the algal and fungal components of the lichen. The diameter of the pores in multiperforate and monoperforate septa is about the same.

The way in which multiperforate septa develop is unclear. Whether pores develop in the ingrowing septum or arise secondarily following complete septation is not known. One might suspect the latter, knowing that fungal cell walls may be dissolved at specific sites, for example, in clamp connections and hyphal anastomoses. Elaborations of the basic Ascomycete pore which would presumably restrict the passage of organelles, but not small molecules, are known to occur at the base of asci (Bracker, 1967; Carroll, 1967). A simple septum with a single central rimmed pore has also been found in the Phycomycetes, *Harpella* (Reichle and Lichtwardt, 1972) and *Linderina* (Young, 1969).

The first suggestion of genuine plasmodesmata between fungal cells came from Hashimoto *et al.* (1964) following an examination of shadow cast cell walls of *Geotrichum candidum*. These authors reported that each septum contained 20-50 pores and that the diameter of the pores was 20-70 nm. Wilsenach and Kessel (1965) and Kirk and Sinclair (1966) confirmed the existence of cytoplasmic connections between cells of *G. candidum* by electron microscopy of sectioned material. Unlike Hashimoto *et al.*, Wilsenach and Kessel concluded that the pores were arranged in groups and not irregularly distributed. Septation in *G. candidum* has been investigated by Hashimoto *et al.* (1973) and Steele and Fraser (1973). Both groups of workers found plasmodesmata in immature or partially formed septa although Schnepf (1974) has suggested that the elaborate arrangement of spokes and pores described by Hashimoto *et al.* from shadow cast isolated immature septa were in fact decoration on valves of a centric diatom contaminating their preparation of fungal septa.

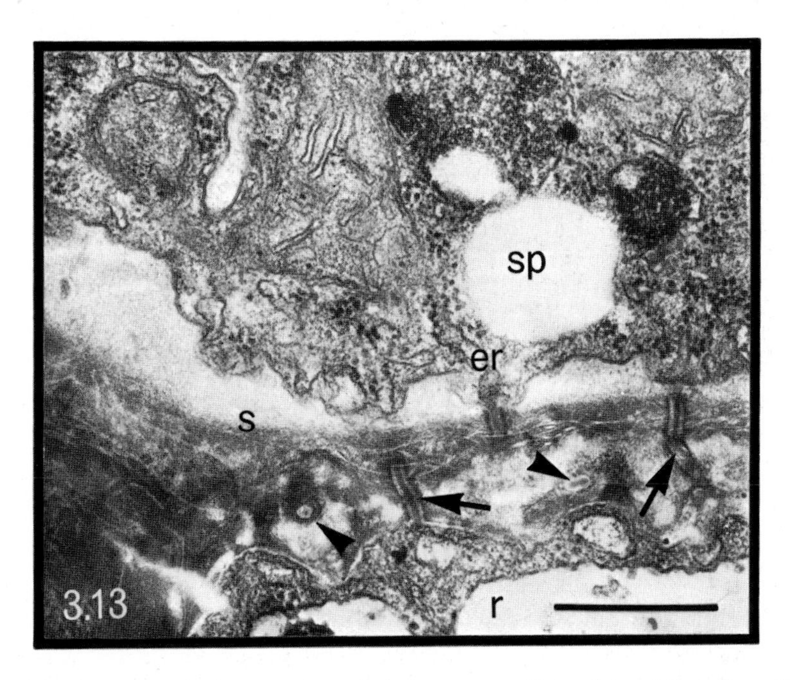

Fig. 3.13. Part of the septum (s) between the rhizoid (r) and sporangium (sp) of *Entophlyctis* showing both longitudinally (arrows) and transversely (arrowheads) sectioned plasmodesmata. The plasmodesmata contain a solid desmotubule with which the endoplasmic reticulum (er) appears continuous. Scale marker = 0.5 μm. Micrograph by courtesy of Dr. M.J. Powell, from Powell (1974)

Other fungi in which plasmodesmata have been found are *Endomycopsis* (Takada *et al.*, 1965), *Gilbertella* (Hawker *et al.*, 1966), *Rhizopus* (Hawker *et al.*, 1966; Hawker and Gooday, 1967) and the chytrids *Entophlyctis* and *Rhizophydium* (Powell, 1974). Unlike all the other fungi that have been investigated, where plasmodesmata interconnect nucleate cells, the connections found in *Entophlyctis* traverse a septum between the multinucleate sporangium and an *anucleate* rhizoid. As *Entophlyctis* produces its septum early in the development of the thallus, before zoosporangial growth is complete, Powell considers that the plasmodesmata may facilitate the transport of absorbed nutrients from rhizoids to the sporangium. One may speculate that the plasmodesmata may also pass nuclear information in the opposite direction, from the nucleate sporangium to the anucleate rhizoid.

Tubules interconnecting the endoplasmic reticulum of adjacent cells *via* fungal plasmodesmata have only been reported from some species of *Endomyces* while in other species of this genus the plasmodesmata contain a dense core (Kreger-van Rij and Veenhuis, 1972) as was also found in *Entophlyctis* (Powell, 1974) and *Geotrichum* (Steele and Fraser,1973). The dense cores are unlike the desmotubules of higher plants (Robards, 1968b) in that they appear solid. According to Kreger-Van Rij and Veenhuis (1972), the distribution of plasmodesmata in septa of various species of *Endomyces* is quite variable. Interestingly, in *E. magnesii*, *E. overtensis* and *E. tetrasperma* the plasmodesmata are arranged more or less evenly spaced in a circle near the edge of the septum, with very few inside the circle.

As discussed in the preceding section of the Chapter, the plasmodesmata of green algae are only known to occur between cells which at cytokinesis become partitioned by a cell plate formed by the fusion of vesicles. There are no examples, to my knowledge, of algal plasmodesmata occurring between cells separated by a transverse wall which develops by annular furrowing, though other cytoplasmic interconnections, for example, cytoplasmic bridges in *Volvox* and septal pores of red algae,develop between cells separated in this way. Algal cells may be partitioned by furrowing or by a cell plate (Pickett-Heaps, 1972) while cytokinesis amongst the fungi is achieved by annular ingrowth of septa. The wall material of the developing septum is added in one of two ways; either by progressive constriction by the ingrowing septum (Bracker and Butler, 1963; Kreger-Van Rij and Veenhuis, 1971; Young, 1969) or by the centripetal fusion of vesicles aligned in the plane of the developing septum (Hawker and Gooday, 1967). This latter method, in some respects, is not unlike the development of cell plates in higher plants and green algae. The most conspicuous differences between fusion of vesicles to form septa in fungi and cells plates in algae and higher plants is that the growth of the latter is generally centrifugal while being centripetal in the fungi. How plasmodesmata arise in septa growing by annular infurrowing, unless by dissolution of part of the cross wall following its completion, is difficult to visualize. Yet, Steele and Fraser (1973) illustrate plasmodesmata in developing septa of *Geotrichum*. If the septa grow by the fusion of vesicles, trapped organelles, such as the endoplasmic reticulum seen by Kreger-Van Rij and Veenhuis (1972) in incomplete septa, may lead to the formation of plasmodesmata. While the development of the more complex pores in fungal septa is reasonably well documented, the origin of fungal plasmodesmata is as yet far from clear.

Growth of fungal hyphae is restricted to their tips and as fungi can grow for considerable distances over nutrient-free surfaces, metabolites must be translocated in hyphae to the growing tips. Trans-

---

Fig. 3.14. and 3.15. Longitudinally and transversely sectioned plasmodesmata within a septum (s) of *Endomyces geotrichum*, the perfect stage of the imperfect fungus *Geotrichum candidum*. Note the endoplasmic reticulum (er) associated with the desmotubule. While the desmotubules of some fungi appear solid (e.g. *Entophlyctis* Fig. 3.13.), desmotubules of *E. geotrichum* (arrows, Fig. 3.15.) apparently have some substructure. Scale markers = 0.1 μm. Micrographs by courtesy of Dr. C.E. Bracker, paper in preparation ▶

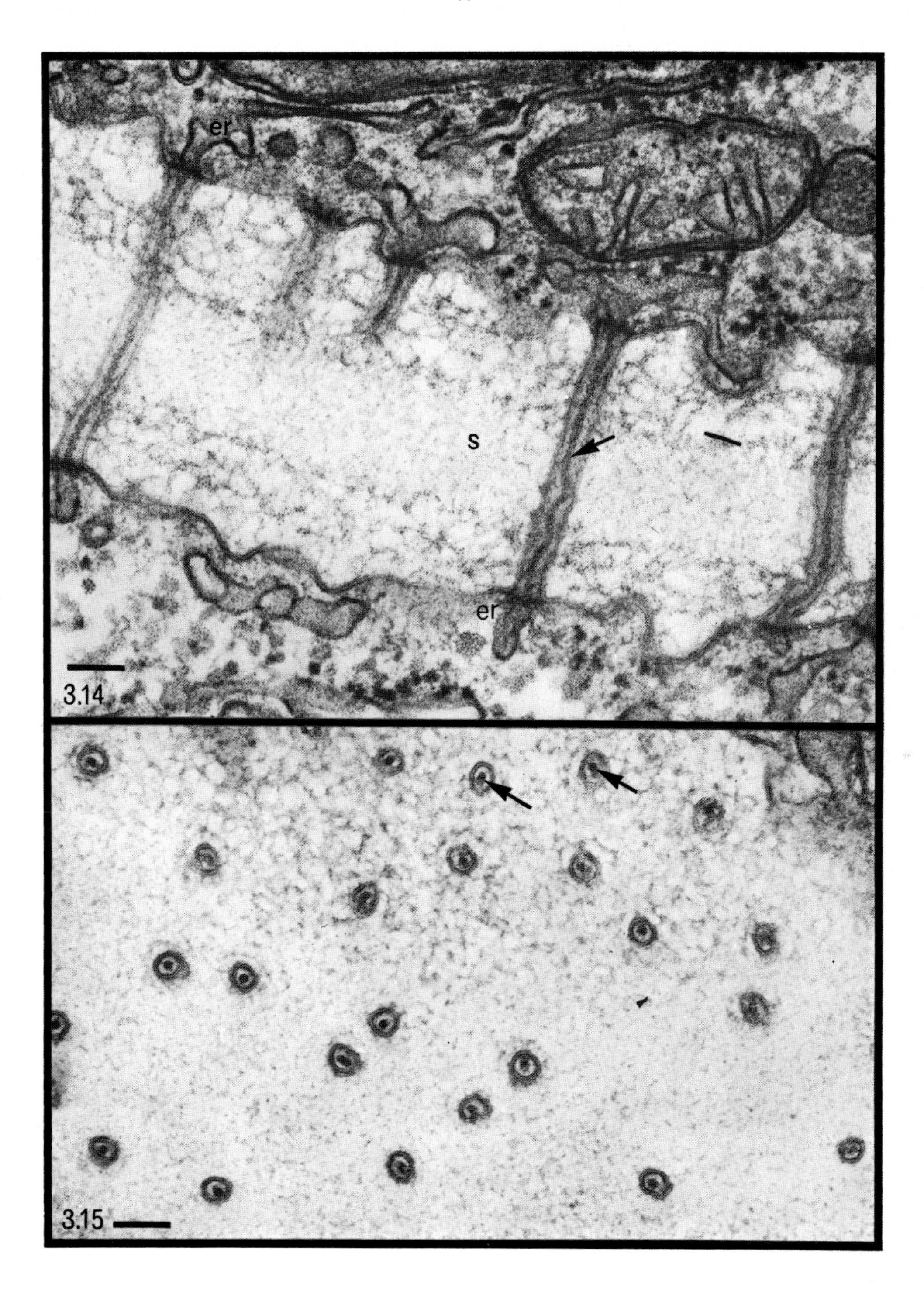
er
s
er
3.14
3.15

location within fungal hyphae has recently been reviewed by Jennings *et al.* (1974). They conclude that there are four ways in which material may move within hyphae: by diffusion, by cytoplasmic streaming, by nuclear migration and by osmotic flow. The latter is bulk flow of nutrients driven by evaporative water loss from certain sites of the mycelium. Greatest resistance to bulk flow along a hypha will obviously be presented by the septal pores and where present, plasmodesmata.

3.7. CONCLUSIONS

Unlike higher plants, the principal way in which metabolites can be translocated between cells of aquatic algae and fungi is *via* a symplastic pathway, as apoplastic transport is clearly impossible without losses into the surrounding medium. Evidence reviewed in this Chapter leaves little doubt that algal and fungal plasmodesmata and other cytoplasmic interconnections are involved in symplastic transport of metabolites as in higher plants (Tyree, 1970) but other roles of cytoplasmic interconnections in algae and fungi are not well understood. In differentiated organisms where there is division of labour between cells it is obviously necessary to transport nutrients to non-photosynthetic cells e.g. hair cells of *Bulbochaete* (Fraser and Gunning, 1973), rhizoids, and to and from specialized synthetic sites, e.g. cyanophycean heterocysts (Wolk, 1968; Stewart *et al.*, 1969; Wolk and Wojciuck, 1971). It is significant that all the groups of algae that have evolved complex thalli containing specialized cells or groups of cells, or which have localized meristematic regions, have also evolved some form of cytoplasmic interconnection. Clearly, there is no apparent need to translocate photosynthate or other solutes between cells that are capable of independent existence. The variety of intercellular connections that is to be found among the algae and fungi strongly suggests that there is considerable selective advantage for cellular continuity and that these intercellular bridges have diverse evolutionary origins.

To me, it seems most unlikely that transport of nutrients *per se* is the sole function of plasmodesmata in algae and fungi. The non-random distribution of heterocysts in blue-green algal filaments (Fogg, 1949; Wilcox *et al.*, 1973) and hair cells in thalli of *Coleochaete* (Korn, 1969) require that there is some form of intercellular communication on which the gradients that establish positional effects are based. Plasmodesmata and the other cytoplasmic bridges of algae and fungi may provide those elements of communication which determine patterns of an organism's growth and differentiation - an hypothesis discussed further in Chapter 13.

3.8. OPEN DISCUSSION

It seemed extraordinary to GUNNING that plasmodesmata in the fungi should be so similar in their ultrastructure to those of green plants. Correspondence between those of green algae and higher plants is less puzzling, as evolutionary links can be postulated. *Either* the selection pressures which have shaped plasmodesmata are very uniform as between fungi and green plants, *or* the sporadic occurrence of plasmodesmata in a few fungi (amongst the taxonomically much more widespread septal pores of various sorts) is a consequence of a different evolutionary origin. Is there, for instance, any evidence that those fungi that possess genuine plasmodesmata might have an evolutionary origin from algae by loss of plastids? First, MARCHANT responded, although the plasmodesmata of fungi closely resemble those of algae and land plants there are some subtle structural differences between them. For instance fungal "desmotubules" (when they occur) are not generally tubular but look more like a solid rod. Also the constrictions commonly found at either end of higher plant plasmodesmata are apparently lacking in the fungi. That there are few detailed studies on fungal plasmodesmata makes generalization risky. Secondly, regarding the possible origin of plasmodesmata-containing fungi from algae by loss of their plastids MARCHANT considered that while such a possibility could not be ruled out, the diverse taxonomic distribution of the fungi in question makes a polyphyletic origin for plasmodesmata seem more likely. Support for this concept would, said GUNNING, be gained if plasmodesmata were to be found in some of those algal groups whose evolution parallels that of the green algae: what, for instance, of the heterotrichous members of the Heterotrichales in the Xanthophyceae and the Chrysotrichales in the Chrysophyceae? These need to be examined, agreed MARCHANT; there are no reports of plasmodesmata in *Tribonema* (Heterotrichales), but in this particular case, just as in the plasmodesma-less *Microspora* in the green algae, the cell walls consist of H-pieces and fragmentation of the filaments is common. Even without evidence from these groups, however, the presence of plasmodesmata in brown algae is another good indicator of a polyphyletic origin.

BOSTROM expressed his interest in the fact that in, for example, an alga like *Chara*, although plasmodesmata are formed during cell plate formation, they have no internal structure at all; therefore isn't it possible that plasmodesmata can be formed by something other than endoplasmic reticulum trapping? MARCHANT pointed out that, in the meristematic region of *Chara*, some internal structures had been reported by Pickett-Heaps (1967) and Fischer *et al.* (1974) in plasmodesmata. However, he stressed that, among those algae which utilize a phycoplast for cytokinesis, it was necessary to think of other methods of plasmodesmatal formation than trapping of cisternae of endoplasmic reticulum. He cautioned that the quality of fixation of giant algae was not yet sufficiently good to be sure about the persistence of desmotubules from cell plates in the meristematic zone to walls in mature nodes.

WALKER made the point that in *Chara* all adjacent cells are different, and that this alga is a good example to show that intercellular connections do not prevent cells (in this case multinucleate) from embarking on their own developmental pathways. CARR said that intercellular gradients of *morphogenetic* information must occur, from blue-green algae upwards, but he reiterated the idea that there must nevertheless be some way of stopping the free transmission of *genetic* information from cell to cell. GIBBS, however, warned that we should be open minded on that point, as viruses which contain genetic information can move symplastically. MARCHANT added that in the case of the chytrid *Entophlyctis* where a septum separates the nucleate sporangium from the anucleate rhizoid, it would be interesting to know whether nuclear information passes *via* the plasmodesmata to the rhizoid.

There was a number of questions concerning the confusing terminology of brown algal 'sieve tubes', 'trumpet hyphae', etc. (see text). LUMLEY asked whether there are any plasmodesmata between trumpet hyphae (etc.) and the neighbouring cells. The reply was that there are not: although there are transverse connections between longitudinal files of sieve elements in *Laminaria*; side wall plasmodesmata do not, it seems, appear until the state of evolution represented by the leptoids of bryophytes.

WARDLAW asked whether, in those complex brown algae with a well developed system for translocation, there was evidence of movement of inorganic ions, particularly phosphate, in the sieve elements as there is in the phloem of higher plants. MARCHANT replied that this particular question had been tackled by a number of workers, some of whom had found that the speed of translocation in these algae did not exceed the rate of simple diffusion while others had shown that the midrib is a preferred pathway for long-distance transport. The most recent work on this problem was that of Schmitz and Srivastava (1975), who, in double labelling experiments with ^{32}P phosphate and ^{14}C labelled assimilates, showed no movement of ^{32}P while the assimilate moved at velocities of up to 0.25 m h^{-1}. MARCHANT thought that this was a somewhat surprising result considering that Schmitz and Srivastava had previously shown that ^{32}P was rapidly incorporated into sugar phosphates and nucleotides by the sieve tube sap of *Macrocystis* and that apparently this organically bound phosphate did not move.

Since plasmodesmata have been found in the fungi, and since the haustoria of some parasitic angiosperms can become symplastically connected to their hosts, GOODCHILD wondered whether any connections had been found in fungal haustoria. MARCHANT said no, and then added that in fungi transport from host to parasite seemed to be enhanced by enlarging the surface area of the haustorial interface, possibly along with parasite-induced solute leakage from the host, rather than by symplastic means.

CARR showed pictures by Genkel and Bakanova (1966), who suggested that filamentous algae which (they claim) possess (Cladophorales) or do not possess (Zygnemales and Ulotrichales) plasmodesmata provide suitable material for testing whether or not the Hechtian strands of cytoplasm resulting from plasmolysis bear any relationship to plasmodesmata. MARCHANT replied that *Ulothrix*, but not *U. zonata*, does in fact have plasmodesmata, and that he was not aware of any electron microscopical evidence for plasmodesmata in *Cladophora* or *Chaetomorpha*; it was also pointed out that Hechtian strands can connect a plasmolysed protoplast to walls which undoubtedly contain no plasmodesmata, e.g. the exterior walls of onion epidermal cells, or of the fern gametophytes studied by Smith (1972).

4

THE ORIGIN AND DEVELOPMENT OF PLASMODESMATA

M.G.K. JONES

Department of Developmental Biology, Research School of Biological Sciences, The Australian National University, Box 475, P.O., Canberra City, A.C.T. 2601, Australia

4.1. INTRODUCTION

Current ideas on the structure, occurrence and function of plasmodesmata in higher plants are outlined in Chapter 2. In this Chapter further aspects of the formation, distribution and modification of plasmodesmata will be considered. 'Primary formation' will be used to refer to the formation of plasmodesmata at the time of cytokinesis, when a new wall is laid down, and 'secondary formation', to refer to formation of plasmodesmata in cell walls which have already formed.

4.2. PRIMARY FORMATION OF PLASMODESMATA

4.2.1. Formation of Plasmodesmata during Cytokinesis that is Coupled to Nuclear Division

The formation of plasmodesmata is a normal part of the process of cytokinesis. Tangl (1879), Russow (1883) and Gardiner (1900, 1907), thought that plasmodesmata might be trapped spindle fibres, but later workers (see Chapters 2 and 14) decided that they were 'protoplasmic'. With the advent of the electron microscope, the similarity of the desmotubule within a plasmodesma to a cytoplasmic microtubule led to the suggestion that the desmotubule is a nuclear or phragmoplast microtubule which becomes trapped and embedded within the developing cell plate during cytokinesis, and subsequently remains there (Allen and Bowen, 1966; O'Brien and Thimann, 1967a; Robards, 1968b; Juniper and Barlow, 1969). A number of reasons for abandoning the idea soon became apparent. During cytokinesis, microtubules of one daughter cell are rarely continuous through the cell plate to the other daughter cell (although they can be) (Newcomb, 1969). Plasmodesmata with desmotubules occur amongst those algae in which cytokinesis is achieved *via* a phycoplast stage, with microtubules aligned *in the plane of* the forming plate (Chapter 3). The desmotubule may stretch if cells are plasmolysed (Burgess, 1971); it may depart markedly from tubular form (O'Brien and Thimann, 1967a; Wooding, 1968) and it is preserved by potassium permanganate fixation (López-Sáez, Giménez-Martín and Risueño, 1966a). These are properties not expected of cytoplasmic microtubules. In addition, the oft-proposed continuity between the desmotubule and endoplasmic reticulum (Buvat, 1957; Frey-Wyssling and Mühlethaler, 1965; Hepler and Newcomb, 1967) is inconsistent with the desmotubule being identical to a cytoplasmic microtubule. Further evidence, including the formation of plasmodesmata with desmotubules in developing walls where microtubules are absent (4.2.2.) and in plasmodesmata which have formed secondarily through a previously unperforated wall (4.3.), completely rule out the theory that the desmotubule is a trapped cytoplasmic microtubule. The theory has, however, served to emphasise the regular nature of the desmotubule in some circumstances, and its subunit structure.

The alternative to the above theory, which is now more generally accepted, is that plasmodesmata form at sites where fusion of cell plate vesicles is prevented by the presence of strands of endoplasmic reticulum (Buvat and Puissant, 1958; Porter and Machado, 1960a; Kollmann and Schumacher, 1962; Frey-Wyssling, López-Sáez and Mühlethaler, 1964; López-Sáez, Giménez-Martín and Risueño, 1966a; Hepler and Newcomb, 1967; Burgess, 1971). Micrographs showing that endoplasmic reticulum cisternae do indeed traverse the cell plate when it is forming by vesicle fusion are convincing (e.g. Fig. 4.1.A,B,C; Hepler and Newcomb, 1967; Hepler and Jackson, 1968; Evert and Deshpande, 1970; Burgess, 1971; Gunning and Steer, 1975) although it must be admitted that most of the work on cytokinesis has been directed towards a consideration of the origin of cell plate vesicles and the role of microtubules, and comments on the formation of plasmodesmata have usually been asides. Robards (1971) suggested that the structural changes from endoplasmic reticulum membrane to desmotubule are brought about directly or indirectly by forces exerted when the membrane becomes trapped in the cell plate. This concept has been mentioned in Chapter 2, and the consequent implications for the nature of the symplast are a recurring

---

Fig. 4.1. Early stages of the formation of plasmodesmata in the root tip of *Phaseolus vulgaris*. Micrographs kindly supplied by P.K. Hepler and E.H. Newcomb (Figs. 4.1B. and C. from Hepler and Newcomb, 1967) with permission from Academic Press ▶

Fig. 4.1A. Section normal to the plane of a forming cell plate (CP). Elements of endoplasmic reticulum (ER) traverse the fusing vesicular material, and microtubules (M) are also present. Bar represents 0.25 μm

Fig. 4.1B. Section similar to Fig. 4.1A. but a slightly later stage of cell plate formation. More extensive fusion of cell plate (CP) material has occurred, and elements of endoplasmic reticulum (ER) are becoming constricted by the fusing vesicular aggregates. Part of a nucleus (N) is also evident. Bar represents 0.25 μm

Fig. 4.1C. Face view of a cell plate (CP). Of particular interest are the plasmalemma-delimited annuli of cytoplasm within each of which the circular profile of a constricted element of endoplasmic reticulum can be seen (unlabelled arrows). It is presumed that the tubules of endoplasmic reticulum are trapped within the coalescing cell plate and give rise to plasmodesmata. At this stage it is evident that the diameter of the tubules of endoplasmic reticulum, which will become the desmotubules of the plasmodesmata, is smaller than that of the cytoplasmic microtubules (M). Bar represents 0.25 μm

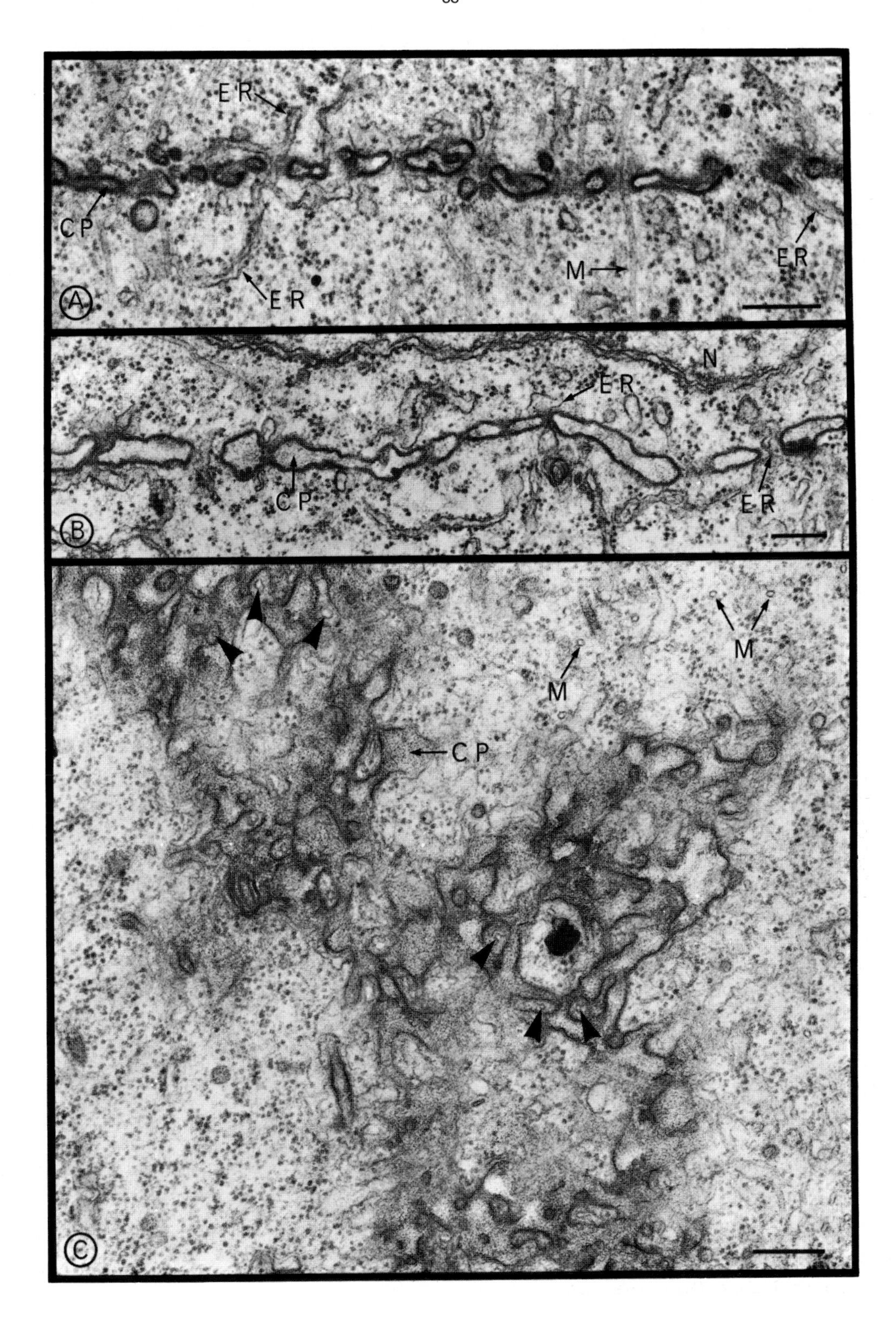
ER
CP
M
ER
ER
A
N
ER
CP
ER
B
M
M
CP
C

theme in this Volume. Two further points arise. The plasmalemma forms from fusion of the limiting membranes of the cell plate vesicles, and lines the plasmodesma. Cell plate vesicles may originate from Golgi bodies or the endoplasmic reticulum (Whaley and Mollenhauer, 1963; Frey-Wyssling, López-Sáez and Mühlethaler, 1964; Hepler and Newcomb, 1967; Hepler and Jackson, 1968; Roberts and Northcote, 1970; Bajer, 1968a; Benbadis, Lasselain and Deysson, 1974). Why, then, do the endoplasmic reticulum membrane and the plasmalemma not fuse, and what kinds of forces act on the endoplasmic reticulum cylinder in the forming cell plate to cause its modification?

When two membranes approach each other closely, Van der Waal's forces will tend to attract them and electrostatic forces to repel them (Poste and Allison, 1971, 1973). Factors favouring membrane fusion are close contact, increased fluidity and a decrease of orderliness of the membranes: the converse will reduce the possibility of fusion. There is also some indication that membranes of a similar nature (e.g. Golgi vesicles and plasmalemma) are more likely to fuse than dissimilar membranes. Thus the different chemical compositions of the endoplasmic reticulum and plasmalemma (Keenan and Morré, 1970) would seem to be a prerequisite to prevent fusion between the vesicles which coalesce to form the cell plate, and the trapped endoplasmic reticulum. It has been suggested that the plasmalemma which lines the plasmodesmata is different from that elsewhere in the cell (Carde, 1974; Vian and Rougier, 1974), but if so the modification is not so obvious as that of the desmotubule. In developing plasmodesmata, therefore, the endoplasmic reticulum and the plasmalemma may be too ordered and of such composition that forces of repulsion are great enough to prevent fusion.

In considering why the cylinder of endoplasmic reticulum trapped in the forming cell plate is greatly modified, and yet the plasmalemma surrounding it, which is presumably acted on by similar forces, is less so, both steric and electrostatic effects must be considered.

Although the phospholipid molecules which constitute a membrane do not have rigid structures, they exhibit preferred shapes and tend to oppose forces of tension or compression which distort them (Israelachvili and Mitchell, 1975). As the curvature of the tubule of endoplasmic reticulum increases in tightness when it is trapped in the cell plate, the shapes of certain phospholipid and other molecules will suit either the inner or outer layers of the tightly curved bilayer. Those molecules that are unsuited by shape may diffuse laterally along the membrane, absenting themselves from the tightly curved regions. There is also evidence to suggest that in any curved bilayer membrane charged lipids may be expected to distribute themselves asymmetrically on the inner and outer faces. Further, as the radius of curvature decreases, the asymmetry of charge distribution will increase (Israelachvili, 1973). It therefore appears reasonable to suggest that as the curvature of the endoplasmic reticulum tubule increases and its environment becomes more unusual the sum of these effects is to cause lateral diffusion of membrane components, both protein and lipid, leaving a more concentrated and structured region of membrane proteins in the form of the desmotubule. Estimates for the rates of lateral diffusion of membrane components vary (e.g. Edidin and Fambrough, 1973; Lee, Birdsall and Metcalfe, 1973). The size of proteins, and factors such as the presence or absence of calcium ions or phase separation to form domains, makes estimation of the rate of lateral diffusion of proteins difficult: it is probably one or two orders of magnitude less than that for lipids. Taking values for lateral movement of lipids within a membrane at about 1.0 $\mu m\ s^{-1}$, and for proteins at about 0.1 $\mu m\ s^{-1}$ then for a plasmodesma of length 0.1 μm it may only take 0.1 to 1.0 second for lateral diffusion and molecular rearrangement to occur when a tubule of endoplasmic reticulum is converted into a desmotubule. This may explain why intermediate stages between trapping of a cylinder of endoplasmic reticulum and its conversion to a desmotubule have yet to be reported. The width of an endoplasmic reticulum membrane, about 6-8 nm, is normally less than that of the plasmalemma (Grove, Bracker and Morré, 1968), but is approximately the same width as the wall of the desmotubule. Analogous situations where similar membrane modification occurs are at nerve synapses and gap junctions (Singer, 1974), though in these cases the closely applied membranes are flat, not cylindrical as for plasmodesmata. In particular, a single protein species predominates in the gap junction of liver membranes (Goodenough and Stoekenius, 1972), whereas the proteins of the rest of the membranes are extremely heterogeneous. Thus there appears to be no reason either by nature of its regularity or radius of curvature why the desmotubule should not derive from a cylinder of endoplasmic reticulum.

Two factors which may help to cause the modifications to the endoplasmic reticulum cylinder do not apply to the plasmalemma which lines the plasmodesmata. First, it is not curved into such a tight cylinder as the desmotubule, and second the cell wall which bounds it on one side may have a stabilising influence. Thus it is not in such an abnormal environment as a trapped cisterna of endoplasmic reticulum. It is not really surprising that the endoplasmic reticulum and plasma membrane may respond differently to the various forces, since Sheetz and Singer (1974) provide evidence that even the two halves of an asymmetric bilayer membrane may respond differently to various perturbations while remaining coupled ('bilayer couple hypothesis').

It should be noted, however, that desmotubules have been reported in plasmodesmata formed by secondary perforation of intact walls (see 4.3.). In such cases the desmotubule must have been formed by a different series of events. This is added evidence that the desmotubule is not formed simply as a result of constriction of the endoplasmic reticulum by the cell plate.

As discussed above, it is possible to interpret some aspects of the structure of plasmodesmata in terms of what is known about membrane substructure. What remains obscure, however, is the nature of the mechanisms which control the siting of plasmodesmata, and the forces which determine the final size of the membranous canals. In most micrographs of developing cell plates, fenestrae, which are relatively large compared with plasmodesmatal dimensions, are illustrated, and it is surmised that these narrow progressively until plasmodesmatal dimensions are reached. We do not know whether the gradual closure is caused by expansion of the wall polysaccharides until the forces which are created are balanced by opposing forces such as the resistance to further compression by the plasmalemma and the desmotubule. Nor do we know whether the wall that surrounds the plasmodesma is especially modified to prevent enlargement or further encroachment on the pore. Views of shadow cast pit areas clearly show that microfibrils are laid down tangentially to parts of the plasmodesmatal pores in a manner which most probably provides a stabilising effect (Frey-Wyssling and Müller, 1957; Veen, 1970).

It would be of great interest to know whether, in a cell in which plasmodesmata normally form during cytokinesis, the formation of the cell plate and its plasmodesmata are obligatorily coupled. Many anti-mitotic treatments act at stages which are too early to be useful in answering the question, e.g. colchicine, hexachloro-cyclohexane and high pressure all prevent cell plate formation (Risueño, Giménez-Martín and López-Sáez, 1968), as do caffeine and aminophylline (López-Sáez, Risueño and Giménez-Martín, 1966b; Paul and Goff, 1973).

However, treatment of root tips with caffeic acid or calcium deficiency (Paul and Goff, 1973), digitonin (Underbrink and Olah, 1968; Olah and Hanzely, 1973), or with vinblastine, acenaphthene and naphthalene (Mesquita, 1967; Mesquita and Mangenot, 1967; Mesquita, 1970) allows the formation, to varying extents, of cell plates which are highly distorted. Micrographs by Mesquita show that neither of the latter three substances prevent the appearance of plasmodesmata, and this may also apply to digitonin (Olah and Hanzely, 1973), even though the wall that they cross may be non-functional, and perhaps even floating freely in the cytoplasm of the affected cell. Cell plates that develop in the presence of cytochalasin appear to be less distorted, but are disoriented, and Fig. 21. of Palevitz and Hepler (1974) suggests that plasmodesmata may be present. Bajer (1968a) illustrates a forming cell plate after treatment with chloral hydrate, and states that cytokinesis is not significantly affected, and that some aspects of cell plate formation are especially clear. Further investigations along these lines might reveal much about the morphogenesis of plasmodesmata.

4.2.2. Formation of Plasmodesmata during Cell Wall Formation that is Independent of Nuclear Division

A coenocytic (free) nuclear endosperm stage occurs soon after fertilisation of the egg sac of many angiosperm and gymnosperm species. The free nuclear endosperm subsequently becomes divided into cells by the infurrowing of walls - a process that is independent of mitosis. In *Helianthus annuus*, Golgi bodies and vesicles predominate at the ends of the growing endosperm walls (Newcomb, 1973). In *Stellaria media* (chickweed) Golgi bodies, vesicles and endoplasmic reticulum are abundant near forming walls at the micropylar end of the embryo sac but only Golgi bodies towards the chalazal area. Microtubules and microfilaments are absent (Newcomb and Fowke, 1973). Similarly in wheat free nuclear endosperm, wall-like processes furrow into the vacuole and new walls are laid down along lines marked out by ingrowths of the plasmalemma into the cell cytoplasm (Buttrose, 1963; Mares, Norstag and Stone, 1975). However, care should be taken when deciding the ontogeny of plasmodesmata in cellular endosperm walls, as subsequent mitosis and cytokinesis may occur in some cells (Newcomb and Fowke, 1973). Newcomb (1973) comments that the freely growing endosperm walls do have a few plasmodesmata; D.J. Mares, C. Jeffrey, K. Norstag and B.A. Stone (personal communication) and Pierre (1970) have noted plasmodesmata in such walls respectively in wheat endosperm and similar walls formed during the cellularisation of the female prothallus in pines. It would be interesting to know more about the formation of plasmodesmata in these walls and for example if they occur in walls in *Stellaria* which are reported to grow without the participation of endoplasmic reticulum.

Plasmodesmata of apparently normal form have been reported in few fungal genera (see 3.6.), and the observations of Powell (1974) are especially relevant. She examined the formation of the septum between the multinucleate sporangium and *anucleate* rhizoid of a chytrid phycomycete, *Entophlyctis*. As in other fungi, this septum forms by centripetal accumulation of wall material. Vesicles, but not microtubules, are present at the inwardly growing edge. Plasmodesmata (see Fig. 3. 13.) are present within the mature septum; they contain desmotubules (20-40 nm diameter) apparently in continuity with endoplasmic reticulum. Here then, are plasmodesmata quite clearly formed independently of nuclear division and of microtubules.

Some further variations should be mentioned before concluding the sections on the formation of plasmodesmata during cytokinesis. We have seen that plasmodesmata can be formed whether or not the cytokinesis is coupled to nucleokinesis, whether or not microtubules are present, and whether or not the microtubules (if present) are parallel to or perpendicular to the newly-forming wall. In most cases desmotubules arise at the time of plasmodesmatal formation, but the examples of the fungus *Geotrichum* (Steele and Fraser, 1973) and the alga *Bulbochaete* (Fraser and Gunning, 1969) show that for an ingrowing septum and a phycoplast type of cell plate, respectively, desmotubule formation is not obligatory. The final variation is that the morphogenetic programme for the formation of plasmodesmata can be initiated but then negated, if not entirely suppressed, as in cell divisions that produce reproductive cells (see 13.4.4.).

4.3. SECONDARY FORMATION OF PLASMODESMATA

The controversy as to whether plasmodesmata form solely during cytokinesis, or can also form in completed walls after cytokinesis, arose almost as soon as plasmodesmata were described. The historical aspects of this problem have been documented in Chapters 13 and 14. There is now a growing body of evidence, albeit rather fragmentary, that post-cytokinetic or secondary formation of plasmodesmata can occur. The ontogeny of plasmodesmata in some situations where secondary formation of plasmodesmata might be expected to occur, such as in the radial walls of cambial cells or between cells on either side of a successful graft junction await elucidation. The different situations where some evidence for the secondary formation of plasmodesmata exists are given below.

4.3.1. Fusion of Like Cells

At certain times during development, cells of the same plant which were not in contact when formed may fuse (postgenital fusion). When a tylose develops, the protoplast of a xylem parenchyma cell bulges out through a pit between lignified thickenings like a hernia, and the nucleus may migrate into the tylose. Many tyloses may arise from different parenchyma cells, and occlude the vessel lumen. After the tyloses stop growing, thickened walls form around them and the presence of simple pits between adjacent tyloses suggests (but it is not yet proved) that plasmodesmata form between the tyloses (Esau, 1948b). A factor which complicates this interpretation is that a tylose may divide after it has developed. A more convincing example, described by Czaninski (1974), is that plasmodesmata can form between tyloses and vascular parenchyma cells across the pit regions between the lignified secondary thickenings. The half plasmodesmata that she illustrates are connected *via* extensive median cavities which run laterally in the centre of the wall for several micrometres.

Postgenital fusion occurs as a normal event in several families of flowering plants. Boeke (1971, 1973a,b) has studied the fusion of carpels which form the false septum in the gynoecium of *Capsella bursa-pastoris*. The septum is formed from two outgrowths of the carpels, which become appressed to each other and fuse at their margins in the gynoecial cavity. Subsequent location of the place of fusion is facilitated since during development most of the septal mesophyll breaks down to leave only the fused cells and some adjoining cells. Before fusion, both carpels possess distinct cuticles. On fusing, the two cuticles are at first evident, but then they merge to form a single layer which gradually breaks down, starting where the epidermal cells are in closest contact. This is one of the few examples where removal of cuticular material may occur. In the walls between fused epidermal cells, plasmodesmata are found where cuticular material has been removed before secondary wall thickening has occurred. Plasmodesmata which run from one epidermal cell to the lumen between that cell and the opposite epidermal cell are interpreted by Boeke as being obliquely sectioned. The presence of plasmodesmata between fused cells is restricted to a short time period; during subsequent wall thickening nearly all the plasmodesmata disappear. Similarly, plasmodesmata were recorded between fused epidermal cells of carpels in the pod of *Lathyrus vernus*. However, no plasmodesmata were observed in the fused marginal cells of the carpels in the gynoecium of *Trifolium repens*, but they were infrequent in other walls as well.

4.3.2. Fusion of Unlike Cells

In some situations where cells that originate from different species, and subsequently come in close contact with one another (interspecific postgenital contact), plasmodesmata can form across the interspecific cell walls. Bennett (1944), in a study of virus transmission by the parasitic plant dodder (*Cuscuta subinclusa*), growing on *Nicotiana tabacum* and *Nicotiana glauca*, reported that after staining sections with gentian violet, strands which were interpreted as plasmodesmata were present between parasite and host. Where the host and parasite wall were in close contact, strands passing through the outermost cell wall of the parasite met similar strands traversing the wall of the adjacent host. Dörr (1968a) and Kollmann and Dörr (1969) studied the intracellular *searching* hyphae of *Cuscuta odorata*, and confirmed that apparently normal plasmodesmata develop between the parasitic and host cells. The period over which complete plasmodesmata are present is limited, since away from the growing hyphal tip deposition of wall material, presumably by the host cells, appears to block off the plasmodesmata and leave only half plasmodesmata on the hyphal side of the interspecific wall. Transfer cell-like wall ingrowths, but apparently no plasmodesmata, occur where the *feeding* hyphae contact host sieve elements.

A similar instance is documented by Tainter (1971). The hyphae of *Arceuthobium pusillum* (eastern dwarf mistletoe) parasitic on *Picea mariana* (black spruce) penetrate to the phloem parenchyma cells of the needle traces of the host. Here, in localised regions, plasmodesmata occur between host and parasite cells, whose walls appeared to have fused together. It is suggested that carbohydrates enter the parasite *via* these plasmodesmata (see 13.9. for further discussion of the literature on parasitic plants).

The non-division walls separating the cells of the periclinal chimaeras *Cytisus adami* and *Solanum tubingese* provide further examples of close interspecific cell contact. In the first case, a layer of epidermal cells originating from *Cytisus purpureus* overlies tissues of *Laburnum vulgare*; in the second, the epidermal cells are from *Lycopersicon esculentum* overlying tissues of *Solanum nigrum*. Following work by Buder (1911), Hume (1913) examined the interspecific junctions by optical microscopy and claimed that connecting threads occur in the interspecific walls. Burgess (1972) extended the work on *Cytisus adami* by electron microscopy and found pits in the interspecific wall, which frequently contain plasmodesmata-like structures. Direct continuity of these across the wall was not observed. However the half plasmodesmata, which occur on both sides of the wall are apparently joined by an indirect bridge between their ends. Burgess suggested that half plasmodesmata are initiated on each side of the wall, and that the plasmodesmata initiated from one cell can only penetrate up to half way across the wall, so that, to complete a connection, half plasmodesmata must be initiated on opposing regions of the interspecific wall. It is interesting that the structure of the half plasmodesma is substantially the same as that of a normal plasmodesma, including the desmotubule.

4.3.3. Perforation of Existing Walls

Plasmodesmata are reported to reform after sliding growth between cells (Livingston, 1964), and are described as becoming "*an increasing feature*" of the latex vessel wall between a young vessel and neighbouring cells (see Dickenson, 1964; and 13.5. for an account of the growth and connections formed by non-articulated laticifers). Apart from these reports there are at present two good examples of *de novo* formation of plasmodesmata between like cells, that is formation which occurs after the completion of the wall in question. The first is in the brown alga *Laminaria groenlandica* (Schmitz and Srivastava, 1974a) and the second in nematode-induced giant cells in higher plants (Jones and Dropkin, in press; Jones, unpublished).

During the development of sieve elements in the intercalary growing region of *Laminaria groenlandica*, the recently formed cross walls of the innermost cortical cells have but few plasmodesmata-like pores, but sieve areas form later on in these walls. At several places the plasmalemma makes close contact with the wall, and at these sites (presumptive pore sites) the wall stains much more lightly than elsewhere. Cisternae of endoplasmic reticulum lie parallel and close to the plasmalemma at this stage, and subsequently complete pores are found. Schmitz and Srivastava (1974a) believe that the pores, which are similar in diameter to higher plant plasmodesmata, are formed by enzymatic digestion, and that the plasmalemma and endoplasmic reticulum are involved in the process. Older sieve areas have many more connecting pores between elements than there are presumptive pore sites of younger cross walls. The final sieve pore frequency reaches an average number of 42 pores μm^{-2}, giving a total of 40,000 connections occupying 12% of the sieve area. These observations contrast with the development of sieve plates in higher plants, where the pores are thought to be elaborated from the pre-existing plasmodesmatal connections.

Root-knot nematodes (*Meloidogyne* spp) induce the formation of giant cells in host plant roots. Each successful nematode induces about six giant cells to form, and these reach their maximum dimensions of up to 500 μm long and 200 μm in diameter about 3 weeks after induction (Jones and Northcote, 1972). It is not clear whether some wall breakdown occurs during the early stages of their initiation, but 3 days after induction no wall fragments are present and subsequent enlargement is by repeated synchronous mitoses without cytokinesis. The inner walls have been examined by scanning electron microscopy after removal of the cytoplasmic contents of the cells (Jones and Dropkin, in press). Although small pits (diameter 0.5 to 1.5 μm) containing perforations through the wall are present in all walls of the giant cell initials 3 days after induction, pits in the walls between giant cells and normal cells are soon lost (Fig. 4.2.). In contrast, pits in the walls between neighbouring giant cells increase in size up to about 12 μm 20 days after induction. An increase in the number of perforations within the pits also occurs. Parallel studies by transmission electron microscopy have shown that the perforations through the walls seen after cytoplasmic digestion represent sites where plasmodesmata occur (Fig. 4.2.). The frequency of plasmodesmata in the pit fields is 29±5. The number of plasmodesmata per μm^2 of wall has been estimated and values are given in Table 4.1. Between days 3 and 33 after induction, individual giant cells increase in maximum length up to two times and in maximum diameter up to seven times. Cell division occurs outside the giant cells and accommodates for the increase in

TABLE 4.1.

THE FREQUENCY OF PLASMODESMATA IN THE WALLS SEPARATING ADJACENT GIANT CELLS

Number of days after induction	Frequency of plasmodesmata per μm^2 of wall
3	3.0
6	4.7
10	6.4
15	4.0
22	7.7
33	4.2

Fig. 4.2A-E. Scanning electron micrographs of giant cells induced in ▶ the roots of balsam (*Impatiens balsamina*) by the root-knot nematode *Meloidogyne incognita*. The roots were fixed in glutaraldehyde/acrolein, postfixed in osmium, split with a sharp razor blade and the cytoplasmic contents digested out by treatment with periodic acid and potassium hydroxide (Jones and Dropkin, in press) to reveal the walls

Fig. 4.2A. A giant cell initial 3 days after induction. Note the small pits in the walls (arrows). Bar represents 20 μm

Fig. 4.2B. Pits (arrows) in wall similar to those in micrograph A. They are regions where plasmodesmata occur, and in some cases the wall in these pits has been removed by the digestion used to remove the cytoplasm. Bar represents 2 μm

Fig. 4.2C. Giant cells (GC) in root 10 days after induction. The giant cells are surrounded by a gall, and an unusually large abnormal xylem element (X) has formed next to the giant cells. Bar represents 200 μm

Fig. 4.2D. The cell wall separating two giant cells 17 days after induction. The pit fields (arrows) have a much greater diameter than at day 3. The wall is much thinner in the pits than elsewhere, and deposition of wall materials has been greatly reduced. Between the pits wall ingrowths (I) typical of transfer cells have also been deposited. Ribs of thicker wall material traverse some of the pits. Bar represents 20 μm

Fig. 4.2E. Higher magnification of pit fields shown in D. Perforations through the wall are evident (arrows), although their diameter is considerably greater than that of the plasmodesmata from which they were probably derived. Bar represents 4 μm

Fig. 4.2F. Transmission micrograph of pit field like those in E and F in a wall between two giant cells. Numerous plasmodesmata (PD) traverse the walls in these areas. The thickened portion of the wall was probably part of a rib of wall material which traversed the pit field. Bar represents 2 μm

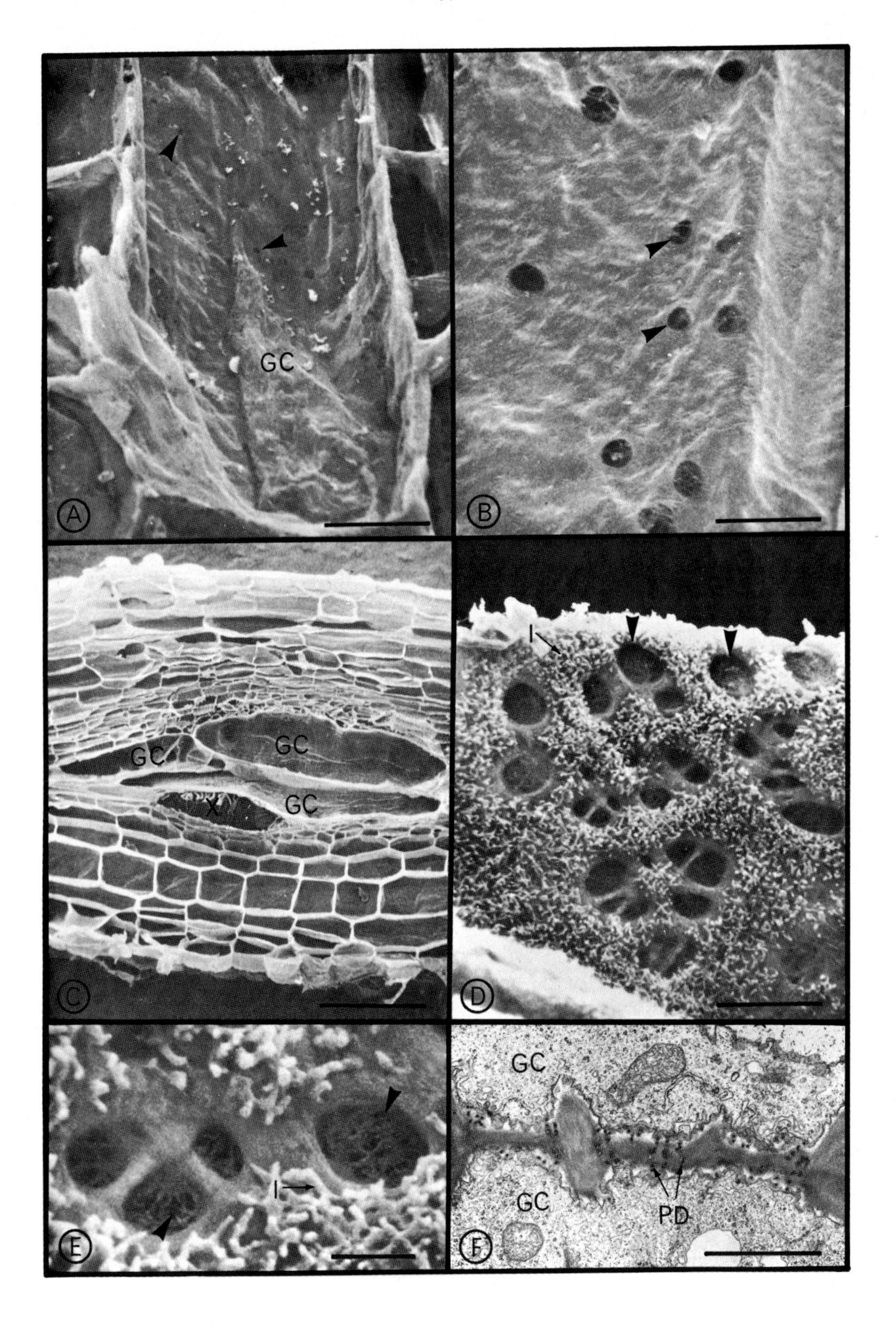
A
GC
B
C
GC
GC
GC
X
D
I
E
I
F
GC
GC
PD

size of the giant cells; the walls surrounding giant cells thus undergo a dramatic extension. Bearing in mind that the overall shape of giant cells is variable, it is possible to make a rough estimate of the total number of plasmodesmata in the walls separating the cells on day 3 and 33 after induction. Assuming wall areas of 225x21 μm^2 and 350x100 μm^2 and values of plasmodesmatal frequency from Table 4.1., total plasmodesmatal numbers on days 3 and 33 are 14,175 and 147,000 respectively, an increase of 10.4 times. Thus there is little doubt that formation of new plasmodesmata has occurred.

4.3.4. Branched Plasmodesmata

Branched plasmodesmata typically occur between companion cells and sieve elements, as well as at other sites (2.4.3.1.). There is no evidence to suggest that the branches form during cell plate deposition, so that the branching, which occurs only on the companion cell side of a plasmodesma, is a secondary modification of a simple plasmodesma (see Fig. 11.2. and 11.2.3.3.). However, since new connections must be made through the wall from the companion cell side to the median nodule, the formation of these tubules may be considered to be another example of secondary formation of plasmodesmata. The problems of understanding how this modification occurs are less than those for the examples of secondary formation of plasmodesmata cited earlier, since the simple plasmodesma already marks the site where the branches will occur.

The lateral division of plasmodesmata as wall extension proceeds has been suggested as a mechanism to account for the existence of median sinuses from which a number of plasmodesmata branch to the cells on both sides (Krull, 1960; Murmanis and Evert, 1967). This would also constitute another type of secondary plasmodesmatal formation, but a more probable explanation is that dissolution of the central part of the wall has occurred to join existing plasmodesmata. A distinction should be made between branched plasmodesmata of the companion cell-sieve element type and those which are branched on both sides of the wall (e.g. compare Figs. 9. and 10. in Gunning, Pate and Briarty, 1968). Some of the latter may form by lateral coalescence of two or more plasmodesmata.

4.3.5. Mechanism of Secondary Formation of Plasmodesmata

It is difficult to identify a possible intermediate stage in secondary formation of plasmodesmata. When partial plasmodesmata are observed, it is impossible without resorting to serial sectioning or high voltage electron microscopy to know whether they cross the wall indirectly, or are indeed stages of formation or occlusion of the structures. In cases where the majority of plasmodesmata are formed during cytokinesis, the identification of possible secondarily formed plasmodesmata is even more difficult. Intermediate stages in the secondary formation of plasmodesmata may have been observed in *Cytisus adami* and in *Laminaria groenlandica* (Burgess, 1972; Schmitz and Srivastava, 1974a).

Since there are so few documented cases of secondary formation of plasmodesmata, it is inevitable that suggested mechanisms will be speculative. However, there is a number of requirements to be met for such an event. To perforate an otherwise intact wall, wall degrading enzymes are needed. Possibly only pectinases and hemicellulases are required and the cellulose microfibrils may be separated mechanically. If not, cellulases are also necessary. These enzymes must be strictly localised. Possibly they are present within the wall and require only activation, alternatively they may be directed to the site from the cytoplasm. The mechanism by which a site for secondary perforation of the wall is selected by the cell(s), and how the wall digesting enzymes are localised there is unknown. Presumably the control is mediated in some way by the plasmalemma and the observations of Schmitz and Srivastava (1974a) on *Laminaria groenlandica* are consistent with this. The kind of sequence that would be required for secondary formation of plasmodesmata is shown in Fig. 4.3. There is no doubt, however, that higher plant cells can synthesise the required enzymes, and can closely control their sites of action. Situations where wall degradation occurs as part of the normal developmental sequence include: formation of sieve plate pores, the development of xylem vessels, hydrolysis of non-lignified secondary walls in xylem elements, leaf abscission and endosperm digestion in germinating grains.

It is not clear whether hydrolysis must be initiated in cells on both sides of a wall (Fig. 4.3A.), or whether once initiated on one side a perforation may be completed from that side (Fig. 4.3B.) Burgess's (1972) observations on the interspecific wall of *Cytisus adami* suggest that a half-plasmodesma, once initiated, stops at or before the middle of the wall. Thus co-ordination between the cells on both sides of the wall will be necessary to complete the connection, and since blind pits occur in the walls this requirement may not always be met. An explanation for the observations of half plasmodesmata in interspecific walls is that a cell may possess only those enzymes capable of perforating its own wall, and not that of another species. This suggestion is compatible with the sealing off of the plasmodesmata through the walls of parasitic hyphae by deposition of wall by the host (Kollmann and Dörr, 1969). Where secondary formation of plasmodesmata occurs between cells of the same plant, the connections appear to be more direct than in *Cytisus adami* (Boeke, 1971, Jones, unpublished), although in the fused carpels Boeke (1971) shows half plasmodesmata which stop at the cuticular layer. Thus between like cells perforation of the wall may occur from one side only. However, co-ordination between two cells, as occurs when secondary thickenings in xylem cells form (Hepler and Fosket, 1971), might be expected to be more controlled between like cells than across an interspecific wall. Hydrolysis of the middle of the wall to form a cavity might be a method of completing the connection between two slightly misaligned half plasmodesmata (Fig. 4.3A.).

Many of the pit membranes in the interspecific wall of *Cytisus adami* contain no structures resembling plasmodesmata, but where they are present they include a desmotubule. Desmotubules are also present in plasmodesmata between parasitic haustoria and host cells (Tainter, 1971) and in walls separating nematode induced giant cells (Jones unpublished). In these cases, therefore, desmotubules can form without the trapping of endoplasmic reticulum during wall formation.

The plasmodesmatal frequency in the walls between giant cells stays relatively constant, or indeed increases slightly, between days 3 and 15 after induction when wall extension is most rapid. From this, and the lack of observations of possible intermediate stages in secondary plasmodesmatal formation, it seems that the process must take place quickly and be well controlled.

It is noteworthy that median cavities are present in plasmodesmata formed secondarily through walls brought together by fusion (Czaninski, 1974; Kollman and Dörr, 1969) as well as in normally formed plasmodesmata (section 4.6.). In the former case two secondary walls will have become fused and the mid line of demarcation will contain no middle lamella, whereas in normal walls there will be a middle lamella at the region where most of the hydrolysis giving rise to the median cavities occurs.

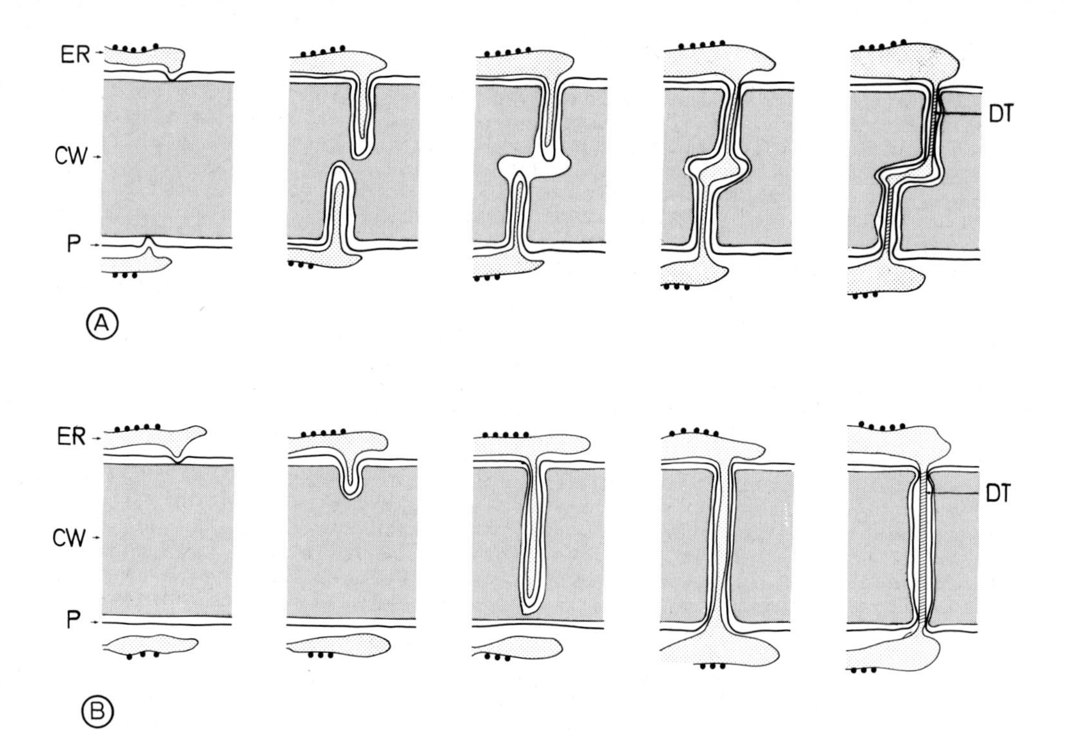

Fig. 4.3. Hypothetical schemes to illustrate the kind of processes which may occur during secondary formation of plasmodesmata CW = cell wall DT = desmotubule ER = endoplasmic reticulum P = plasmàlemma

Fig. 4.3A. Half plasmodesmata initiated from the cells on both sides of the wall. The sites of initial wall degradation are located at the plasmalemma where it is more closely appressed to the wall than elsewhere. Wall degrading enzymes are localised here, either bound to the outer surface of the plasmalemma or released into the wall. These enzymes may be supplied *via* the endoplasmic reticulum which overlies the region and subsequently follows the invagination of the plasmalemma. Progressive perforation of the wall from both sides results in the formation of half plasmodesmata, and if these are misaligned the fusion of the membranes of the two half plasmodesmata may be accomplished after the formation of a median cavity within the wall. The desmotubule is depicted as forming after perforation and fusion has been completed, but the endoplasmic reticulum may be modified to the desmotubular form as soon as the endoplasmic reticulum membrane follows the invagination of the plasmalemma

Fig. 4.3B. A similar scheme to that shown in A, except that perforation is initiated from the cell on only one side of the wall and is completed from that side. The need for co-operation in sites of initiation and half plasmodesmata alignment shown in A is thus diminished to a requirement for eventual membrane recognition and fusion

4.4. DISTRIBUTION OF PLASMODESMATA

A survey of some of the situations where plasmodesmata occur, and their frequencies, is given in Chapter 2. Plasmodesmata are rarely distributed symmetrically about a cell. Usually transverse walls in axial structures have a higher frequency of plasmodesmata than longitudinal walls (Juniper and Barlow, 1969; Cutter and Hung, 1972; Phillips and Torrey, 1974a,b). Juniper and Barlow (1969) have studied plasmodesmatal distribution in the maize root tip in some detail. Only in the quiescent centre, in which the plane of division is random, is the plasmodesmatal frequency the same in transverse and longitudinal walls. The difference in frequency per unit area on transverse and longitudinal walls increases from a factor of about 3 to one of about 11 moving from the columella cap initials to the central root cap and cap periphery. A similar but more protracted alteration in frequency probably occurs in the files of cells running from the meristematic cells to the root stele and cortex *via* the zone of elongation, since the ratio of length to breadth is much greater in these cells than it is in the elongating cells of the cap. Juniper and Barlow (1969) stress that the potential for cell communication in such a file of cells is much greater between the cells in the file than between the cells of adjacent files. When cells differentiate, characteristically there is a decrease in plasmodesmatal number per unit area between cells that are increasing in volume. In this sense cells become progressively more independent of their neighbours.

In rapidly growing tissues, such as the maize root tip, more complicated plasmodesmatal types with median sinuses and branches are not usually found. They normally occur in older tissues. Different types of plasmodesmata may occur on different walls of the same cell, as in companion cells. Apart from in the most recently formed walls, plasmodesmata are usually grouped together in pit fields where the walls remain thinner than elsewhere. For example, in the needles of *Abies balsamea* (balsam fir) Chabot and Chabot (1975) illustrate isolated plasmodesmata in newly formed walls of mesophyll cells, but they are confined to pits in mature mesophyll cells. The processes which lead to the formation of these pit fields are discussed in the next section. There are walls, however, such as cross-walls in the nectaries of *Abutilon* (see Chapter 11), the cross walls of *Tradescantia* stamen hairs (Roelofsen, 1959), walls in conducting cells in brown algae (Schmitz and Srivastava, 1974a) or mosses (Hébant, 1972), in which a relatively high number and uniform distribution of plasmodesmata is maintained. The regularity and repeatability of pit field distribution within particular walls is striking, and this factor has been used, after counting the numbers of plasmodesmata within measured pit field areas, to estimate the total number of plasmodesmata per unit area between particular cell types (Clarkson, Robards and Sanderson, 1971; Robards, Jackson, Clarkson and Sanderson, 1973; Kuo, O'Brien and Canny, 1974). Estimates of the frequency of plasmodesmata per μm^2 within pit fields range from 10-20 (Burgess, 1971), 27 (Clarkson *et al.*, 1971), 23-36 (in walls between giant cells, Jones, unpublished), 25-38 (Kuo *et al.*, 1974), 54-60 (Olesen, 1975). In a detailed study of pit field distribution in the mestome sheath cells which surround the longitudinal veins of wheat leaves, Kuo *et al.* (1974) found that pit distribution in the inner tangential walls of the cells is remarkably symmetrical about the saggital plane of a vascular bundle. Pits are most common in those cells adjacent to the metaphloem, and absent adjacent to a tracheary element. Where one sheath cell abuts both a phloem parenchyma and a tracheary element, pits are abundant in the sheath cell wall adjacent to the parenchyma cell and absent next to the xylem element. Pit field sizes also vary between sheath cells in different sized vascular bundles, from $3.0 \pm 0.3\ \mu m^2$ to $5.4 \pm 0.4\ \mu m^2$. Although pit fields occur between cells of the mestome sheath and both xylem parenchyma cells and phloem tissues in some bundles, there is no symplastic connection between xylem parenchyma and any phloem cell. This would allow a separation of water fluxes from the xylem to transpiring surfaces and sugar fluxes *via* the mestome sheath cells to the phloem (see Chapter 11). Kuo *et al.* (1974) remark that the pattern of pit fields develops before the mature functions are assumed, indicating that the sheath cells can recognise their neighbours and pre-programme the required connections between them. Whether the pre-programming occurs as early as cytokinesis has not been determined. There are other examples where plasmodesmatal distribution is not pre-programmed but rather is under the control of feedback systems that operate during development. Included in this category are secondarily formed plasmodesmata between fused cells (Boeke, 1971) and those between adjacent nematode-induced giant cells (Jones and Dropkin, in press).

The use of the term 'pit' to describe the regions between the secondarily thickened walls of differentiating xylem elements is confusing, since they are seemingly not sites where many plasmodesmata cross the wall. Indeed, plasmodesmatal connections between differentiating xylem elements and adjacent cells are rare, and the thickenings are deposited over or between the plasmodesmata that are present with about equal frequency (O'Brien, 1970). There seems to be no obvious connection between the distribution of plasmodesmata in differentiating xylem elements and the patterns of secondary thickenings. Unless special provisions, such as are seen in reproductive tissues, are provided, once a cell becomes isolated from other living cells and hence from symplastic sources of nutrient it will die. The 'stretched' ray cells in some gymnosperms provides an example (Ziegler, 1964). Part of the programmed differentiation of xylem elements must include the severing of plasmodesmatal links with living cells, and this process may start before deposition of wall thickenings.

4.5. CONTROL OF FREQUENCY AND DISTRIBUTION OF PLASMODESMATA

With the evidence given elsewhere in this Volume, it is clear that the frequency and distribution of plasmodesmata control the degree of symplastic communication between cells. It is therefore important to consider how these parameters are regulated.

The plane of the cell division is the first factor to affect plasmodesmatal distribution. Cytokinesis in a plane perpendicular to the preceding division, or an asymmetric division, three of which occur during the formation of stomata (Pickett-Heaps and Northcote, 1966b) are obvious examples. The plane of division is determined before mitosis, possibly at the time the preprophase band of microtubules form (Pickett-Heaps and Northcote, 1966a,b; Burgess and Northcote, 1967), in response to appropriate morphogenetic gradients within the tissue.

Plasmodesmata normally form when the cell plate develops. Control of their frequency and distribution may therefore be exerted at this stage. Two alternative mechanisms to explain subsequent plasmodesmatal distribution exist. Either plasmodesmata in the newly formed walls are relatively uniformly distributed at a high frequency, or they are localised into distinct regions, primary pits, within the forming cell plate. Glancing sections of newly formed walls suggest that the former alternative occurs, although adequate information on this point is lacking. Burgess (1971) recorded 140 plasmodesmata per μm^2 in the root meristem of *Dryopteris filix-mas*, and from micrographs of recently divided cells in the shoot apex of *Osmunda cinnamomea* (Hicks and Steeves, 1973), Juniper (in press) estimated 180 and 80 plasmodesmata per μm^2 were present in anticlinal and periclinal walls of superficial promeristem cells. Although Burgess (1971) describes the distribution of the plasmodesmata as random, whilst it is indeed uniform the plasmodesmata are arranged in rows. These may be straight, curved or whorled, and are composed of from 3 up to about 20 plasmodesmata. Similar patterns are seen elsewhere (Fig. 4.4., *Azolla* root meristem cell). The trapping of cisternae of endoplasmic reticulum within the forming cell plate need not therefore be random, and control of location of the cisternae near the growing cell plate would affect frequency and distribution of plasmodesmata within it. It would be interesting to visualise by serial sectioning the patterns of endoplasmic reticulum membranes underlying the rows of plasmodesmata. Another type of cell plate that would repay examination in this regard is that which is formed in guard mother cells in developing stomatal complexes as there is evidence that plasmodesmata are not laid down in that part of the wall destined to become the stomatal pore, though they do form elsewhere in the same wall (Ziegler, Shmueli and Lange, 1974).

The pattern of plasmodesmatal connections can be further modified by selective loss and by area dilution as a result of wall extension.

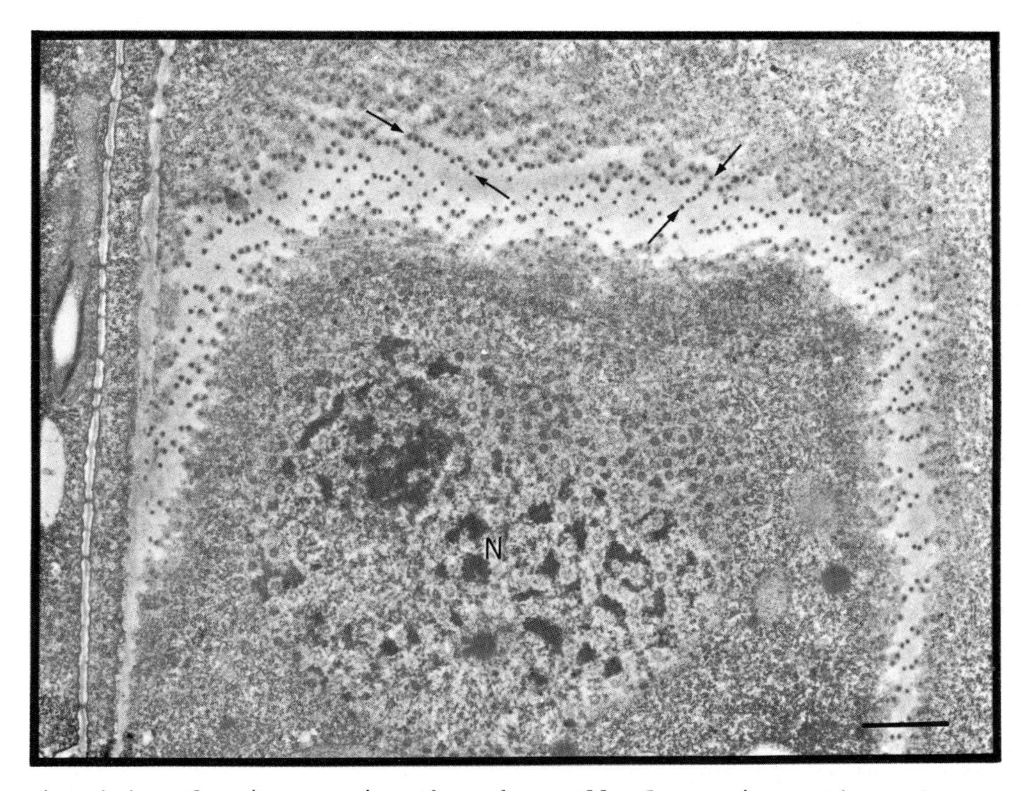

Fig. 4.4. Glancing section through a wall of a meristematic cell in an *Azolla* root tip. The plasmodesmata are relatively uniformly distributed within the wall, but are arranged in distinct rows and patterns, as indicated by the pairs of unlabelled arrows. The cell nucleus (N) is also evident. Micrograph reproduced from Gunning and Steer (1975), by permission. Bar represents 1 μm

Selective loss may occur through occlusion of plasmodesmata by deposition of wall polysaccharides. For example, secondary wall thickenings in developing xylem cells overlie plasmodesmata (Cronshaw and Bouck, 1965; O'Brien and Thimann, 1967b; Srivastava and Singh, 1972a); callose is deposited over the plasmodesmata of pollen mother cells during meiosis, and in subsequent divisions which give rise to microspores no plasmodesmata are present in the newly formed callose walls. Thus all connections between the diploid maternal tissues and the haploid microspores are severed (Heslop-Harrison, 1966a). Other examples of isolation of reproductive cells by occluding layers of wall material are seen in the algae (Chapter 3). Differentiated sieve elements have few or no plasmodesmata except those connecting to the companion cells (Chapter 11) and it would be of interest to know whether in juvenile stages there is a greater variety of cell-to-cell connections. There is good evidence that plasmodesmata, present in young stages, are lost from wall junctions in developing stomatal complexes (Kaufman, Petering, Yocum and Baic, 1970a; Srivastava and Singh, 1972b; Singh and Srivastava, 1973; Ziegler, Shmueli and Lange, 1974). It is also not unusual to see plasmodesmata embedded in walls in which both their ends are apparently covered by wall material. Plasmodesmata in cells of the pigment strand in the developing wheat seed are apparently progressively occluded by electron dense 'adcrusting material' (Zee and O'Brien, 1970). Sliding growth of one cell relative to its neighbour may sever plasmodesmata (Livingston, 1964), and during the germination of seeds of *Sinapis alba* plasmodesmata between adjacent myrosin cells are reportedly ruptured (Werker and Vaughan, 1974).

As cells expand, the plasmodesmatal frequency must fall assuming no secondary formation takes place. The frequency per unit area of plasmodesmata in transverse walls of the developing root cap of *Zea mays* is estimated at 4.5 per μm^2 in the meristematic cells and 0.81 per μm^2 in the peripheral cells of the cap apex (Clowes and Juniper, 1968). These figures were judged to be consistent with the dilution of the number of plasmodesmata per unit area which might result solely from wall expansion during transformation from the former to the latter cell type. Robards *et al.* (1973) comment similarly, that plasmodesmatal frequency in the inner tangential wall of the endodermis falls from 1.194 to 0.5 per μm^2 from just behind the meristem to 10 mm behind the barley root tip, and that this diminution is simply accounted for by cell expansion. Dilution of plasmodesmata by wall elongation is a reasonable explanation for the difference in frequencies between lateral and longitudinal walls of files of cells in root tips (Juniper and Barlow, 1969).

There is a different quantitative discrepancy which is less easily explained. The plasmodesmatal frequencies in newly formed walls given by Burgess (1971) and Juniper (in press), based on Hicks and Steeves, (1973), are nearly 10 times greater than those quoted for meristematic cells fixed some time after division, or even for pit fields (Table 2.1.). This might suggest that there is a considerable and rapid loss of plasmodesmata immediately after cell plate formation, and that the selective nature of this loss results in the formation of the localised pit areas. It is important that the unusually high counts reported by Burgess (1971) and Juniper (in press) (Hicks and Steeves, 1973) are checked. The published micrographs are few and it is not clear how representative they are (see Chapter 2).

Setterfield and Bayley (1957) showed that deposition of wall material occurs uniformly over the wall, and it seems unlikely that pit areas form as a result of differential extension of the walls between plasmodesmata that initially were relatively uniformly distributed. However, once formed, pit fields do not increase in number or significantly in size within an expanding wall, although this is not always the case (see section 4.3.3.). Apparently the microfibrils within the pit fields cohere so strongly that they do not slip past one another during the main growth phase (Roelofsen, 1965). Thus the distance between pit fields increases during extension growth. Since the thickness of the wall in pit fields is less than that elsewhere a limitation of secondary deposition of wall polysaccharides must occur (Frey-Wyssling and Müller, 1957; Jones and Dropkin, 1975). Ribs of wall material, in which further polysaccharide deposition does occur, are frequently observed to cross pit fields (Setterfield and Bayley, 1958; Fig. 4.2D.E. and F.). The fibrous nature of the wall between plasmodesmata in pit fields suggests that specifically it is the matrix polysaccharide deposition which is reduced (Burgess, 1971). The microfibrils in pit fields are well illustrated in shadowed wall preparations, where the microfibrils run tangentially to the plasmodesmatal perforations (Frey-Wyssling and Müller, 1957) and in some cases even appear to bend around the periphery of the pores (Roelofsen, 1959). Pit fields are frequently oval in shape with their long axes perpendicular to the long axis of the cell (Roelofsen, 1965; Veen, 1970). This is precisely the opposite orientation to that which would be expected if the shape of the pit fields reflected the extension of the wall in which they are located.

Secondary formation of plasmodesmata (section 4.5.) and branching of existing plasmodesmata could also contribute to alter initial patterns of distribution. It may well be that future research will show that secondary formation is a more common and widespread occurrence than heretofore suspected.

The final way by which cell communication *via* plasmodesmata may be controlled is by modification of existing plasmodesmata. Some of these types of modifications are considered further in the next section.

4.6. STRUCTURAL MODIFICATIONS OF PLASMODESMATA

The plasmodesmata that are formed at cytokinesis are simple, i.e. straight unbranched tubes lined by plasmalemma, and containing a desmotubule. If they are not lost or occluded in some way, they may remain without modification. Alternatively their morphology may become altered. The least obvious variations are changes in constriction of the neck region at either end of a plasmodesma, thus altering the separation between the plasmalemma and desmotubule. In the bundle sheath cells of *Aristida ascensionis* (Fig. 12.3E.) this type of constriction is particularly clear (Laetsch, 1971). Conversely, there is evidence that initially constricted plasmodesmata in the distal wall of the stalk cell of *Abutilon* nectary trichomes lose their constrictions at approximately the time when nectar secretion starts (Hughes and Gunning, personal communication). The diameter of the plasmodesma may also increase as a result of etching the wall away. It is difficult to rule out fixation artefacts, but certain micrographs give the impression that the presence or absence of desmotubules can vary even within a single wall (Pickett-Heaps, 1967, Fig. 4.; Schmitz and Srivastava, 1974a; Burgess, 1971). Another modification, which may clearly be related to the function of the plasmodesmata, occurs in bundle sheath cells of grasses (O'Brien and Carr, 1970; Laetsch, 1971): suberized lamellae occur in all the walls of mestome sheath cells of wheat and oat leaves, and numerous plasmodesmata connect these cells with those of the parenchyma sheath and vascular parenchyma cells (see Chapter 11). The suberized lamellae appear to constrict each plasmodesma so that the plasmalemma is pressed tightly against the desmotubule to give the plasmodesmata a distinct waist. Plasmodesmata are similarly modified in the parenchyma sheath of maize and sugarcane leaves. In the sugarcane leaves, the suberin layer is thicker in the pit fields where it is traversed by numerous plasmodesmata, and the constriction of the plasmodesmata from a diameter of 65 nm in the secondary wall to form a cylinder of about 32.5 nm diameter where they cross the layer is even more marked (see Fig. 12.3.). The desmotubule is observed clearly in this region (Laetsch, 1971). A similar modification occurs in endodermal cells of barley and maize roots, which also possess suberized lamellae (Robards, Jackson, Clarkson and Sanderson, 1973; Karas and McCully, 1973).

The next most obvious variation is the development of median nodules or Mittelknöten (Krull, 1960; O'Brien and Thimann, 1967a), in which a cavity develops within the middle lamella region of the wall. The extent of cavity development varies, and is seemingly greater in older tissues, and in anatomical situations where high fluxes through the plasmodesmata might be expected. More extensive lateral development of the cavities may lead to anastomosis of adjacent plasmodesmata in which two or more half-plasmodesmata extend from a central cavity to both sides of the wall. Relatively small median nodules are illustrated by Helder and Boerma (1969) in the walls of endodermal cells of young barley seedlings; they noted two main types, one with a spherical nodule and the other spindle shaped. They discuss the possibility that different fixation procedures may have caused these differences. More elaborate structures with branching from a central cavity are illustrated by Krull (1960) and O'Brien and Thimann (1967a); Murmanis and Evert (1967) show a particularly striking example of a cavity stretching for almost 6 μm in a ray parenchyma cell of *Pinus strobus*, and connections between adjacent plasmodesmata within the wall are well illustrated in cells of the halophyte *Salicornia* (Hess, Hansen and Weber, 1975). They also occur characteristically between files of xylem parenchyma transfer cells in stem nodes of *Helianthemum* (Jones, unpublished observations). It is reasonable to expect fusion of median cavities of plasmodesmata in pit fields in which the frequency of plasmodesmata is high. In plasmodesmatal arms the desmotubules exhibit their usual appearance, but they may return to a structure that is more normal for a membrane within the central cavities. Sometimes there is a proliferation of membranous material within the cavities, and it is difficult to see what connections there are within the cavities. The question of whether each desmotubule remains connected only through the original plasmodesma, or whether continuity occurs between the desmotubule membranes within a cavity to allow equilibration of desmotubule contents between the plasmodesmata is unanswerable at present. In some micrographs of the sieve element to companion cell (or albuminous cell) type of branched plasmodesmata the desmotubule clearly branches at the junction of the arms (Wooding, 1968).

There is evidence that the wall surrounding plasmodesmata may be modified. Callose may be deposited on the sieve tube side of plasmodesmata that connect with companion cells (Northcote and Wooding, 1968; Shih and Currier, 1969; Esau, 1973). Taiz and Jones (1973) found β-1,3-glucanase and 'Onozuka' cellulase resistant wall tubes surrounding plasmodesmata of the barley aleurone layer. It is thought that the wall tubes, 120 nm in diameter, are a modified wall layer secreted uniformly across the plasmodesmatal plasmalemma. The effect of the wall tube is to form a physical barrier between the plasmalemma and the cell wall, which might protect the plasmodesmata when the wall matrix is autolysed in response to secretion of gibberellic acid. Wall tubes with axial plasmodesmata actually grow into the cytoplasm in Chinese cabbage leaves infected by cauliflower mosaic virus (Bassi, Favali and Conti, 1974). Phillips and Torrey (1974a,b) also note that the wall surrounding plasmodesmata in cells of the root apex of *Convolvulus* stain less intensely in the electron microscope than the rest of the wall.

The formation of sieve pores has been the subject of many papers (Esau, Cheadle and Risley, 1962; Kollmann and Schumacher, 1962; Northcote and Wooding, 1968; Zee, 1969; Burr and Evert, 1973; Evert, Bornman, Butler and Gilliland, 1973; Esau and Gill, 1973; Evert and Eichhorn, 1974; Deshpande, 1974). The process is an extreme example of modification of plasmodesmata. A plasmodesma originally delineates the site where the pore will form, and before the sieve element develops, the plasmodesmata in longitudinal and lateral element walls are similar to those in parenchyma cell walls (Deshpande, 1974). First of all the pore site is established by formation of a pair of collar-like areas immediately surrounding each pre-pore plasmodesma, but separated in the middle lamella region. As the walls thicken callose is deposited in or on the wall in the collar areas, and this seems to prevent further deposition of normal wall polysaccharides over these areas. Significantly, profiles of endoplasmic reticulum overlie the callose pads at this stage. The next phase involves removal of middle lamella material and formation of a median cavity. The cavity frequently contains a proliferation of membranous material. The removal of the callose collars, starting from the central cavity, results in a widening of the original plasmodesmatal tubule throughout its length. In *Welwitschia* the median cavities of the developing sieve pores may merge and remain as such when the sieve pores are mature, and may contain membranous materials and even starch grains (Evert, Bornman, Butler and Gilliland, 1973). Moving further down the plant kingdom the degree to which plasmodesmata are modified during sieve plate development diminishes progressively.

A further situation where wall breakdown appears to begin at pit fields is in lateral root primordia in *Zea mays*. Karas and McCully (1973) report that the walls of pericycle cells which have divided become thinner and lose electron density, and reticulate regions of the wall around the pits are frequent. Complete wall perforations are developed secondarily between axial parenchyma cells in *Pinus strobus* (Murmanis and Evert, 1967), in development of certain laticifers (Karling, 1929) and in syncytia induced by cyst-nematodes in host plant roots (Jones and Dropkin, in preparation).

4.7. MECHANISMS OF STRUCTURAL MODIFICATION

There is no information available on how most of the modifications of plasmodesmata described above occur. Only where there is removal of wall material is there some relevant data. It applies also to problems raised in discussing the secondary formation of plasmodesmata (4.3.5.).

It seems reasonable to postulate that the types of enzyme activity needed to modify plasmodesmata should be present in them. Histochemical techniques have in fact indicated that ATPase activity may be detected at the plasmalemma and in plasmodesmata in meristematic cells (Hall, 1969), in branched plasmodesmata between companion cells and sieve elements (Gilder and Cronshaw, 1973a and b); and phosphatase in developing ray cells (Robards and Kidwai, 1969). Ashford and Jacobsen (1974a,b) have similarly found acid phosphatase and esterase (Ashford, personal communication) in plasmodesmata in the aleurone cell layer of imbibed barley grains (see also Chapter 7).

In the latter case there is a correlation with a wall degrading function. Fulcher, O'Brien and Lee (1972) showed that in the aleurone cells of wheat a phenol (probably ferulic acid)-carbohydrate complex exists, possibly joined by an ester linkage. The resistant cell wall component found around plasmodesmata of barley aleurone cells (Taiz and Jones, 1973) is probably composed of a similar complex (Ashford and Jacobsen, 1974a). The function of the plasmodesmatal esterase may thus be to cleave the complex and hence to allow or to initiate further wall digestion. Taiz and Jones (1970) thought that the barley aleurone cell wall contains a β-1,3-glucan which is hydrolysed by a β-1,3-glucanase. It is now known that the wall does not contain a β-1,3-glucan, but the major polysaccharides are an arabinoxylan (85%) and cellulose (8%) (McNeil, Albersheim, Taiz and Jones, 1975). Nevertheless it is significant that the gibberellic acid-induced wall degradation appears to be initiated around plasmodesmata (Jones, 1972) and that the localisation of the enzymes stained by Ashford and Jacobsen (1974a) corresponds exactly with these degraded wall regions. Thus there is reasonably good evidence that plasmodesmata can in this case secrete wall degrading enzymes.

The types and quantities of enzymes secreted will therefore control which parts of the walls surrounding plasmodesmata are modified, and to what extent. Median nodules or cavities might form as a result of secretion of pectinases, whereas enlargement of sieve plate pores might require secretion of pectinases, hemicellulases, β-1,3-glucanase and possibly also cellulases. The secretion of such enzymes could be synchronous or sequential.

The next questions are, where do such enzymes originate, and how do they reach the plasmodesmata? From results of pulse chase autoradiography Chen and Jones (1974) conclude that hydrolases in gibberellic acid treated barley aleurone cells are synthesised in the endoplasmic reticulum, but then move to the cytoplasm and are released from the cells without the participation of a membrane bound vesicle. However, Gibson and Paleg (1972) and Vigil and Ruddat (1973) suggest that the hydrolytic enzymes are secreted *via* lysosomal-like vesicles. In situations where general wall hydrolysis occurs, passage of enzyme across the plasmalemma, including that in plasmodesmata, could still give rise to patterned hydrolysis in the wall, in that the surface area of plasmalemma within the plasmodesmata may be very extensive (see 2.3.2.). In situations where only the wall surrounding the plasmodesmata is etched away, it is more understandable if it is postulated that the enzyme(s) are synthesised by ribosomes on the endoplasmic reticulum, passed into the lumen and transported directly to the plasmodesmata *via* the cisternae which are presumed to connect with the desmotubules. How the enzymes might pass out into the space between the desmotubule and the plasmalemma and thence into the wall is problematical.

4.8. CONCLUSIONS

It is manifest that the plant exerts control over the formation, distribution and structure of plasmodesmata. We can in many cases describe the end product generated by the control systems, but very little is known about the operational aspects. In that variations in the distribution and structure of plasmodesmata may be presumed to be of great functional importance there is a clear need to pursue work in this area. It is hoped that the material reviewed in this Chapter may aid in specifying some of the outstanding problems.

4.9. OPEN DISCUSSION

CARR provided details of early work on two other examples of secondary formation of plasmodesmata. One, discussed by Strasburger (1901) concerns the shoot apex, where the periclinal wall underlying the dermatogen undergoes enormous expansion without being augmented by new cell plates. Since there are frequent plasmodesmata in this wall secondary formation is indicated. The other example concerns non-articulated laticifers. Schmalhausen (1877) showed that these ducts grow intrusively in *Euphorbia*, and in 1891 Chauveaud described how the laticifer system of Euphorbiaceae, Urticaceae, Apocynaceae and Asclepiadaceae originates by intrusive growth and branching starting from a few cells initiated in the embryo. Kienitz-Gerloff (1891) observed plasmodesmata between parenchyma and latex ducts in *Euphorbia pulcherrima* but held their secondary origin to be improbable. Meyer (1896b) made the same observation for *Nerium oleander*, as did Strasburger, who added that such connections can scarcely be other than secondarily formed. Much later Meeuse (1941a) confirmed the existence of secondary plasmodesmata between laticifers and parenchyma in *Euphorbia* embryos. Such plasmodesmata may in fact be of considerable economic significanance. According to Daniel (1927) and Funk (1929) non-articulated laticifers do not fuse across a graft junction and Cramer (1930) makes the same claim for *Hevea brasiliensis* (para rubber), a crop for which grafting is now an important technique. When grafted, the foliage of one partner does have influences upon the quality of the latex formed in the trunk by the other (Ng, Rubber Research Institute of Malaya, pers. comm.). If the laticifers do indeed fail to join across the graft, it would seem that the 'influences' that affect the quality of the rubber harvest must pass *via* symplastic connections established in other parts of the junction, and thence *via* the plasmodesmata leading into the latex ducts.

ROBARDS added that the radial walls in stem cambium are similar to the inner periclinal wall of the shoot apex dermatogen; they grow predominantly by stretching of existing walls, yet plasmodesmata occur in the walls of mature derivatives. JONES remarked that divisions would occasionally occur in this plane, but ROBARDS felt that these would not be sufficiently frequent to account for the numbers of plasmodesmata actually observed; there is, he said, another puzzle, for the radial walls become very thick in winter, when they do not have many plasmodesmata, yet in the active growing period plasmodesmata are relatively abundant.

SMITH asked what signals might govern the formation of secondary plasmodesmata. Could they form in response to local ionic gradients? JONES knew of no evidence for this, but agreed that one might guess that some type of local gradient would be involved. The idea is consistent with the sorts of ion movements which might be expected to occur between adjacent nematode-induced giant cells. He quoted other evidence for collaboration between adjacent cells in wall metabolism, for example in the back-to-back formation of xylem thickenings.

WATSON asked if half plasmodesmata ever form in the outer wall of epidermal cells. The reply cautioned against confusing plasmodesmata with ectodesmata, and mentioned a report by Clowes and Juniper (1968) that remains of plasmodesmata can sometimes be found in the outer walls of root cap cells as they are sloughed off, having been surrounded by other cells at an earlier stage of development. Reference was also made to the work of Dolzmann (1964) illustrating the secondary closure of plasmodesmata during the formation of intercellular spaces in *Viscum*. Further, Juniper (1963), dealing with root tip cells of barley and maize, had proposed that protoplasmic connections are formed by the fusion of chains of 20 nm diameter vesicles which are possibly derived from the Golgi apparatus. He believed that these connections link up with the 'cytoplasmic endoplasmic reticulum', but are subsequently plugged by carbohydrate deposition so that endoplasmic reticular continuities are not found through the plasmodesmata of older cells. JONES then confirmed, in response to GOODCHILD's question, that Juniper and Barlow's data show conservation of plasmodesmata during cell enlargement in the root cap.

VAN STEVENINCK questioned whether it was possible to think of cell walls being digested away very locally, as during secondary formation of plasmodesmata. Yes, said JONES, it does happen - for example, in the enlargement of sieve plate pores, and the formation of median cavities in compound plasmodesmata. As detailed in the text, wall digesting enzymes are seen in certain cases to emanate from plasmodesmata - the best evidence coming from aleurone layers. GUNNING asked whether the more precise examples of local digestion imply that the relevant enzymes are attached to the plasmalemma of the plasmodesma - otherwise they would, as in aleurone, diffuse and digest their way throughout the wall. This might be so, agreed JONES, but there is evidence that the cell wall surrounding plasmodesmata can be specialised - in aleurone, for instance, a sleeve that resists digestion remains around the plasmodesmata when protoplasts are being prepared (Taiz and Jones, 1973); and there is a number of reports of differential staining of the peri-plasmodesmatal wall, including the recent work on cryo-cut sections by Vian and Rougier (1974). In several types of virus infection this peri-plasmodesmatal sleeve appears to grow preferentially (see Chapter 8).

The question of what happens to plasmodesmata during secondary thickening was raised by GOODCHILD. In many cases they survive, was the answer, but there are also many examples where they are lost: occlusion occurs in, for instance, the pigment strand of wheat seeds (see text) and in certain algae deposition of layers of 'mucilage' covers the plasmodesmata (see Chapter 3). CARR referred to the paper by Jungers (1930), who had probably seen plasmodesmata trapped in walls that had become greatly thickened, but interpreted them as non-protoplasmic striations of some sort. ROBARDS added that occlusion occurs in xylem development - a rather special situation where, since the cell dies, there is no 'need' for protection of the plasmodesmata, and lignified thickenings commonly are deposited on top of them. However, where cells continue to live and function after secondary thickening (e.g. xylem parenchyma and root endodermal cells) the secondary wall material is not deposited over pit fields. KUO contributed the point that lignification of sieve tubes (as in wheat leaves) does not occlude plasmodesmata (Kuo and O'Brien, 1974a). GUNNING described a different type of occlusion which can sometimes be seen in meristematic tissue, where fusion of a cell plate with the existing walls of a dividing cell may occur on top of plasmodesmata. In reply to ROBARD's comment that secondary wall material is not deposited over pit fields in living cells, JONES raised the point that there must normally be some wall deposition in pit fields since the walls are thicker than primary walls, and in many cases, such as plasmodesmata between sieve elements and companion cells, the wall is actually thicker just where each plasmodesma crosses the wall than elsewhere within the pit field.

Finally, CARR gave other examples where postgenital fusion occurs, and suggested that electron microscopy should be applied to see whether symplastic continuity develops. One was from Küster (1916) who described how, under the influence of a nematode infection, *Melilotus alba* leaflets fold in the primordial stage, the left and right adaxial surfaces fusing to produce a tubular structure. (GUNNING said that treatments with tri-iodobenzoic acid often give rise to tubular vegetative and floral parts: they too might repay examination). The other examples were from normal processes of flower development - the fusion that occurs between the stigmas of Asclepiadaceae (which, after flowering, separate again), or between the anthers of members of the Campanulales (Baum, 1948b). Other families where postgenital fusion occurs in carpel development are listed by Baum (1948a).

5

PHYSICO-CHEMICAL ASSESSMENT OF PLASMODESMATAL TRANSPORT

W.P. ANDERSON

Bioenergetics Unit, Research School of Biological Sciences, Australian National University, Canberra, A.C.T. 2601, Australia

5.1. INTRODUCTION

Although microscopists have been aware of the existence of cytoplasmic connections between plant cells for a very long time (see Chapters 2 and 14), it is only relatively recently that transport through plasmodesmata has been examined quantitatively. The treatment by Tyree (1970) is now generally taken as the basis for assessment of plasmodesmatal transport and it seems appropriate to introduce this Chapter by reviewing his formulation of the problem.

5.2. IRREVERSIBLE THERMODYNAMIC FORMULATION

Fig. 5.1. shows the composite junction between two plant cells; as is indicated there are two possible parallel pathways for movement from cell to cell: (i) through the plasmalemma of the first cell, across the cell wall and through the plasmalemma of the second cell; and (ii) through the plasmodesmata. Henceforth I shall refer to these pathways as the membrane pathway and the plasmodesmatal pathway, respectively. Flow of matter (water, small solutes, large macromolecules and possibly viruses) may in principle occur in both these pathways, although at (vastly) differing rates. In order to compute the flow rates, or fluxes, we need to know two quantities, the permeability coefficients (or more generally the phenomenological coefficients) for the pathway, and secondly the driving forces aausing the flow.

According to the formalism of irreversible thermodynamics (De Groot and Mazur, 1962) the flux J_i of any component i of a multi-component system, will be given by linear addition of all possible forces ($X_1..X_n$) acting on it, multiplied by the appropriate coefficients ($L_1..L_n$):

$$J_i = L_{i1}X_1 + L_{i2}X_2 \quad .... \; L_{in}X_n$$

For systems in which the chief flows are water and small solutes, the application derived by Kedem and Katchalsky (1958) is commonly used:

$$J_v = L_p \ (\Delta P - \sigma RT\Delta C_s)$$

$$J_s = \overline{C}_s \ (1-\sigma)J_v + \omega_s RT\Delta C_s$$

where J_v is the volume flux across a membrane, J_s is the solute flux, ΔP is the hydrostatic pressure difference, ΔC_s is the solute concentration difference, L_p is the hydraulic conductivity, σ is the reflection coefficient, ω_s is the solute permeability coefficient, and $\overline{C}$ is the mean solute concentration in the membrane phase. R is the gas constant and T is the absolute temperature.

The reflection coefficient σ takes up the cross-coefficients of solute and water flows and relates them to the perfection of the membrane as a semi-permeable barrier. In fact, it can be shown (Kedem and Katchalsky, 1958) that

$$\sigma = - \frac{\Delta P}{RT\Delta C_s}$$

For an ideal semi-permeable membrane, $\sigma = 1$ while a membrane completely non-selective between solutes and solvent has $\sigma = 0$.

5.3. TREATMENT BY TYREE

Tyree (1970) used this formalism to derive two sets of equations to describe solute and water flow in the two pathways across the plant cell-to-cell junction (Fig. 5.1.).

$$\left. \begin{aligned} J_v &= \alpha L_p(\Delta P-\sigma RT\Delta C_s) \\ J_s &= (1-\sigma)J_v\overline{C}_s + \omega_s RT\Delta C_s \end{aligned} \right\} \quad \ldots.(1)$$

and

$$\left. \begin{aligned} J'_v &= \alpha' L'_p(\Delta P-\sigma' RT\Delta C_s) \\ J'_s &= (1-\sigma')J'_v\overline{C}'_s + \omega'_s RT\Delta C_s \end{aligned} \right\} \quad \ldots.(2)$$

where the first set denotes the flows through the membrane pathway and the second through the plasmodesmatal pathway, and the only new parameters are α, the area-fraction of the membrane pathway (of the order of 0.99 and α', the area fraction of the plasmodesmatal pathway (of the order of 0.01, see Table 2.1.). Note that this formulation reco-

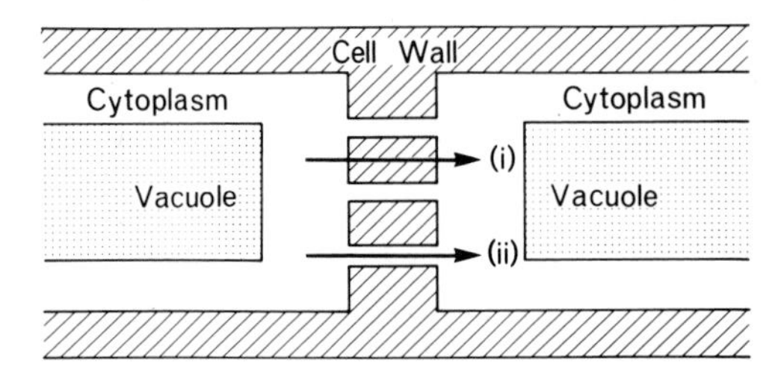

Fig. 5.1. Schematic representation of a plant cell-to-cell junction, showing the two possible transport pathways described in the text

gnises that the overall differences in pressure and concentration are identical by either pathway; the simplifying, but somewhat unrealistic, assumption is that the cell cytoplasm is a homogeneous thermodynamic phase, isobaric and of uniform concentration.

Strictly speaking, only non-electrolyte solutes are described in sets (1) and (2), which, again, strictly only apply to a two component system of water and a single solute. However, on this basis Tyree (1970), who recognised these restrictions, proceeded to show that the plasmodesmatal pathway is much more conductive than the membrane pathway, for all small solutes, but in certain species with membranes of high water permeability, a significant fraction of the cell to cell water movement may be conducted by the membrane pathway.

The argument essentially revolves around the values to be given to the various phenomenological coefficients that are dependent on the physical dimensions of the plasmodesmata, which are matters of continuing concern and controversy for the anatomists (Chapter 2). Let us briefly set out these coefficients:

(i) The *reflection coefficient*, σ, may be taken as 1 in the *membrane pathway* for all solutes likely to be present (experimental values range from about 0.9 upward); in the *plasmodesmatal pathway*, σ is zero because there is little likelihood of solute exclusion from the plasmodesmatal canals, except in the case of large colloidal particles (where the formalism is inappropriate anyhow) or electrolytes, if the plasmodesmata contained high densities of fixed electric charges (see also 1.2.1.).

(ii) The *hydraulic conductivity*, L_p, of the *membrane pathway* is taken from experimental values, obtained by plasmolysis, transcellular osmosis or measurements of self-diffusion of water on a number of plant cells. Typical values range from 5×10^{-8} m s^{-1} bar^{-1} in the *Characeae*, to 2×10^{-11} m s^{-1} bar^{-1} in *Zea mays* root cortical cells. (These are 0.5 times the literature values because there are two membranes in series.)

The hydraulic conductivity, $\alpha' L_p'$, of the *plasmodesmatal pathway* depends very strongly on plasmodesmatal geometry and frequency, and on the viscosity of the fluid filling the pores. For an array of right cylindrical pores, each containing a right cylindrical occlusion (Fig. 5.2.) the expression is

$$\alpha' L_p' = \frac{N\pi}{8L\eta}\left[r_1^4 - r_2^4 - \frac{(r_1^2 - r_2^2)^2}{\log_e r_1/r_2}\right]$$

(an alternative to equation (3) in 1.2.1.)

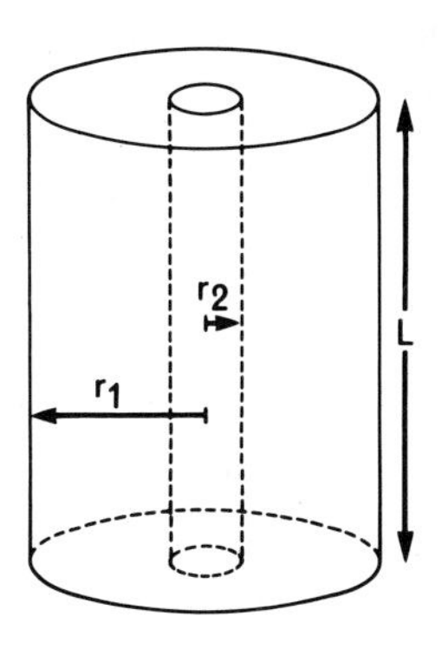

Fig. 5.2. Diagram of the plasmodesmatal pore geometry used by Tyree (1970). The parameters indicated are described in the text

Using this, Tyree (1970) calculated $\alpha' L_p'$ for a range of plasmodesmatal frequencies N from 1.5 to 4 μm^{-2}; a range of values of outer pore radius, r_1, from 30 nm to 60 nm; viscosities, η, from 0.5 to 2.0 poise; the pore inner radius r_2 was held at 3 nm and the pore length, L, at 500 nm (see 10.5.1.2. for comments on these values). The calculated values of $\alpha' L_p'$ ranged from 10^{-4} to 2 x 10^{-5} m s^{-1} bar^{-1}, *at least two orders of magnitude higher than the* L_p *values for the membrane pathway.*

(iii) The *solute permeability*, ω_s, for the *membrane pathway* is available from experimental measurements for a large variety of solutes in many different plant cells. Values range from 10^{-5} mole s^{-1} m^{-2} bar^{-1} for water to 10^{-8} mole s^{-1} m^{-2} bar^{-1} for ions. (Again these are 0.5 times the literature values because there are two membranes in series). For the plasmodesmatal pathway, $\alpha'\omega_s'$ is computed for the same pore geometry as before, a right cylindrical pore with a central occlusion.

$$\alpha'\omega_s' = \frac{N\pi(r_1^2 - r_2^2)D_s'}{RTL}$$

where D_s' is the diffusion coefficient of the solute in the pore fluid. Tyree's arguments about the values of D_s' will not be reproduced here, except to say that the colligative properties of cytoplasm will reduce D_s' by no more than the factor 3, of the diffusion coefficient of the solute in aqueous solution, D_s. However, there is direct evidence (Spanswick and Costerton, 1967) from electrical measurements of the resistance to the flow of ions that carry current through the nodes of *Nitella translucens*, that the apparent diffusion coefficient in the plasmodesmatal pores is 1/330 of the value in aqueous solution. Tyree assumed a factor 10 of this 330 might be taken up by compartmentation of the ions into organelles (which would reduce the concentration of mobile ions in the cytoplasm below the measured value) and therefore suggests $D_s' = 1/33\ D_s$ for ions. The additional restriction on ions may be due to fixed charges in the pore. On the basis just outlined, Tyree (1970) found values of $\alpha'\omega_s'$ in the range 7 x 10^{-4} to 2.7 x 10^{-4} mole m^{-2} s^{-1} bar^{-1}, *which are up to several orders of magnitude larger than equivalent solute permeabilities for the membrane pathway.*

Using these various parameter values, Tyree finally calculates the fluxes which might be conducted through a symplast of 10 cells in file (e.g. radially across a root cortex). His assumption is that the flux is rate-controlled by plasmodesmatal transit, i.e. that solutes are distributed in the cytoplasm of each cell by a combination of diffusion and cyclosis sufficiently rapidly not to rate limit the overall transport. This point will be taken up in the next section. However, on this basis, Tyree (1970) shows that the observed xylem exudation salt fluxes from excised roots of *Allium cepa* could be maintained across the cortical symplast by a salt concentration difference of only 0.1 mole m^{-3} (0.1 mmolar) at each cell junction (but see also 10.5.1.2.).

5.4. REASSESSMENT OF SOLUTE PERMEABILITY

The implicit assumption in what I have just written about the effective solute permeability of the plasmodesmatal pathway, $\alpha'\omega_s'$, is that transit across the plasmodesmata is the rate-controlling process for solute movement from cell to cell. Put another way, it is assumed that a combination of diffusion and cyclosis causes sufficiently rapid solute re-distribution that we may take the cytoplasm of each cell as being a well-stirred, homogeneous phase for small solutes. In fact, this assumption was quite explicit in the original treatment (Tyree, 1970) and was justified by comparison of the presumed rate of cyclosis with the much slower mean diffusional drift velocity of solutes in the plasmodesmata. As Tyree himself now realises (Tyree, Fischer and Dainty, 1974) this argument is in error, and as Walker (see Chapter 9) first pointed out, a better comparison is between the solute flow (mole s^{-1}) carried by cytoplasmic cyclosis, and the solute flow through the plasmodesmata.

On this basis, it becomes apparent that the transfer of radio-labelled solute across the nodes of members of the *Characeae* is a process likely to be rate-controlled by cyclosis, while in the higher plants where rates of cyclosis are not well documented, the position is quite open. Tyree, Fischer and Dainty (1974) take the most favourable set of circumstances for the example of symplastic movement across *Zea mays* root cortex, and justify the original assumption, that plasmodesmatal transit is the rate-controlling process. However, the general view of anatomists is that the 'circumferential' (i.e. around the circular cross section of the cylindrical cells) cyclosis assumed by Tyree *et al.* (1974), is much less likely than longitudinal cyclosis, along the long axis of the cylindrical root cortical cells. Under such circumstances, it becomes even more likely that cyclosis is the rate-controlling event (see later). This idea is contrary to the view currently held by root physiologists (e.g. Anderson, 1975a,b).

After hearing Dr. N.A. Walker's discussion of nodal transport in *Chara corallina* at a meeting in Melbourne in May 1975, I decided to extend the treatment he produced, by including diffusional transfer of solute within the cyclosis stream (Fig. 5.3.). The mathematics of this situation is set out in the Appendix and numerical values are given in Tables 5.1. and 5.2. Two types of cell-to-cell transport are discussed, which I have termed *co-current* and *counter-current* (see Fig. 5.4.).

In the case of co-current transport, the concentrations in the flowing cytoplasms decrease-increase exponentially as the flows pass along the junction, as does their difference at any point, while their sum remains constant (see Fig. 5.5.). The total plasmodesmatal transport will be given by:

$$ZJ' = \int_0^Z \alpha' \omega_s' RT(C_1 - C_2)dz$$

where Z is the total length of the junction in the direction of flow. Values may be found in Table 5.1. It might be noted that plasmodesmatal transport is a function of cyclosis rate v, and increases as v increases. In this sense, v is always a rate control on plasmodesmatal transport.

For counter-current transport, the concentrations in the flowing cytoplasms decrease-increase linearly, as does their sum at any point. Their difference remains constant at all points along the junction (see Fig. 5.6.). The total plasmodesmatal transport is given by:

$$ZJ' = \alpha' \omega_s' RTZ(C_1 - C_2)$$

which is again a function of v, because $(C_1 - C_2)$ is a function of v (see Table 5.2.). Note that at the slowest cyclosis rate (v=2.77 μm s^{-1})

over the longest junction (200 μm) and with the highest solute plasmodesmatal permeability, ($\alpha'\omega_g' = 10^{-3}$ mole m^{-2} s^{-1} bar^{-1}), the effective concentration difference across the plasmodesmata is only 36% of the difference in entrant concentrations of the afferent and efferent streams.

Tyree's (1970) treatment completely omits these considerations, which must be seen as a serious shortcoming. His basic assumption of perfect mixing can, on the present analysis, be thought of as the special case when v is infinite. Furthermore, the recent retraction of perfect mixing for *Chara corallina* nodal transport (Tyree *et al.*, 1974) shows little appreciation of the real effects of streaming velocity on the concentration profiles at the plasmodesmata, such as are set out here.

The three obvious assumptions incorporated in the treatment given in the Appendix are:

(i) The plasmodesmata in the common cell wall form a perfect array, so that we may take the permeabilities as constants over the area under consideration;

(ii) There is zero volume transfer across the junction (v is constant in both cells).

(iii) The pores are open and the possibility of continuity of endoplasmic reticulum is ignored.

Should any of these assumptions prove untrue the conclusions would have to be modified. The effect of the first is hard to quantify; it basically implies that we should use a stochastic procedure rather than continuous mathematics. The assumption that there is zero volume transfer from cell to cell minimises the effect of cyclosis rate on symplastic transport. If solute transport from the high concentration to the low were coupled to volume flow in the same direction, plasmodesmatal transport would increase, hence enhancing the possibility that cyclosis is an effective rate control on overall symplastic transport.

5.5. PLASMODESMATAL ELECTRICAL RESISTANCES

It is a long established fact (Dainty, 1962) that the electrical conductance of Characean membranes is about 300 times the permeability measured by radio-labelled ions; this difference has long been thought to be due to the effect of H^+ permeability, inaccessible to tracer experiments. In contrast, the electrical conductance through the nodal plasmodesmata of *Nitella translucens* (Spanswick and Costerton, 1967) is at least 10 times *smaller* than the $^{36}Cl^-$ tracer permeability for nodal transport in the very similar *Chara corallina* (Tyree *et al.*, 1974). This discrepancy is puzzling; it seems certain that cytoplasmic electrical conductivity is not affected by cyclosis rates (at least for realistic streaming velocities), while apparent tracer plasmodesmatal permeability certainly is (see Tables 5.1. and 5.2.). These factors combine to aggravate the discrepancy, and the whole problem seems worth early and exhaustive analysis.

5.6. SUMMARY

After a review of the formalism best suited to discussion of plasmodesmatal transport, a novel description of solute transfer from cell to cell as affected by plasmodesmatal permeability and rates of cyclosis is given.

Two systems are described, with cytoplasm streaming co-currently and counter-currently on either side of the cell-to-cell junction. In the first, concentration differences from cytoplasm to cytoplasm are shown to vary exponentially along the length of the common cell junction; in the latter, the concentration difference is constant.

In both cases, plasmodesmatal transport is a function of cyclosis rate. The assumption of perfectly mixed cytoplasm (Tyree, 1970) is seen to be a special case ($v = \infty$), but seems to be a reasonable approximation for counter-current junctions.

Finally the electrical resistance anomaly is put up as an intriguing puzzle, which must soon be thoroughly investigated.

5.7. APPENDIX

Consider a cell-to-cell junction to be a rectangular section in the (x,z) plane, of unit length in the y-direction (see Fig. 5.3.). Cells are usually right circular cylinders, or perhaps more realistically regular hexagonal cylinders, but because the cytoplasm (the region of flow in the following analysis) occupies only the outer 5% of the hexagon diameter, the above approximation to a rectangular section is good.

Matter must be conserved, so that we may add the flows across any element of volume, $2\delta z$ in length, unit thickness in the y-direction, and kx in width, where k is given by:

$$k = \frac{\text{Area of junction}}{\text{Area of cyclosis stream}}$$

For a 'typical' higher plant cell, with 1 μm thick cytoplasm, $k = 10^6$.

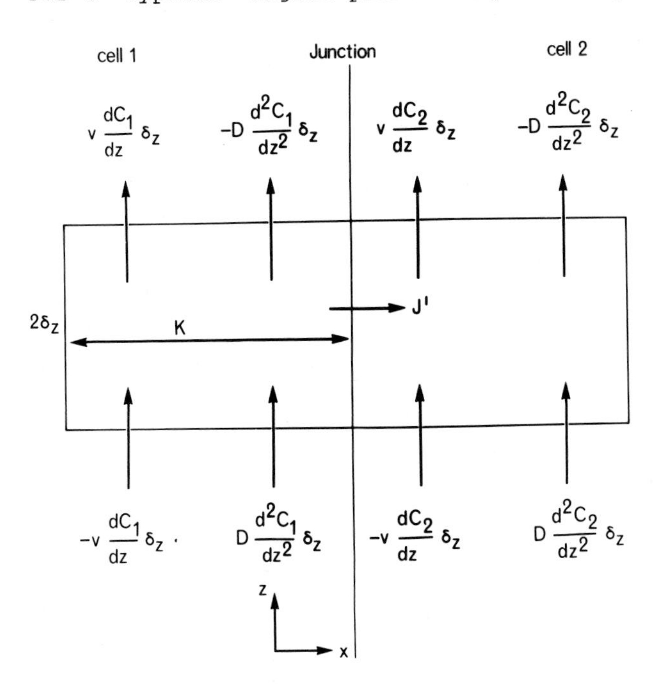

Fig. 5.3. Sketch of the (x,z) plane of the cell-to-cell junction, showing the various component fluxes of solute transfer to and from an element of volume of dimensions k in the x direction, unity in the y direction and $2\delta z$ in the z direction. See Appendix for mathematic details

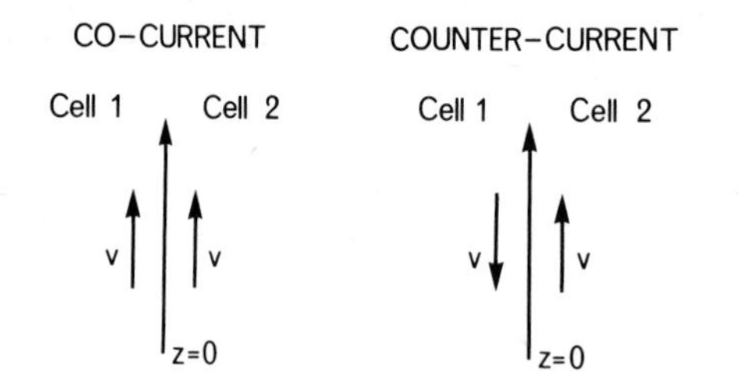

Fig. 5.4. A diagram to explain the terminology of co- and counter-current junctions

CASE I: CO-CURRENT TRANSPORT

For the case of co-current flow (Fig. 5.4.), we have for mass conservation:

$$-kJ' + D\frac{d^2C_1}{dz^2} - v\frac{dC_1}{dz} = 0 \qquad \text{....(1A)}$$

$$+kJ' + D\frac{d^2C_2}{dz^2} - v\frac{dC_2}{dz} = 0 \qquad \text{....(2A)}$$

Now we introduce two new variables,

$$F = C_1 + C_2$$

$$G = C_1 - C_2$$

and then add, then subtract (1A) and (2A) to obtain

$$DF'' - vF' = 0 \qquad \text{....(3A)}$$

$$DG'' - vG' - 2PG = 0 \qquad \text{....(4A)}$$

where we have written $2kJ' = 2k\alpha'\omega_g'RT(C_1-C_2) = 2PG$

Integrating (3A) once,

$$F'-vF/D = h$$

and again,

$$F = g + j\exp vz/D + hz$$

where h, g and j are integration constants.

Physically we require that F remains finite as z tends to infinity;

therefore $j = 0$, $h = 0$

$$F = C_1 + C_2 = g$$

when $z = 0$, $F = C^* + 0 = g$ ....(5A)

The simplifying assumption here is that the entrant concentration of the afferent stream (C_1) is C^*, and the entrant concentration of the efferent stream is zero. Otherwise, C^* is the difference in entrant concentrations of afferent and efferent streams, and because this translation is linear, the analysis proceeds as given here, with C^* =

$(C_1{}^o - C_1{}^o)$ with $C_1{}^o$ the entrant, afferent concentration, and $C_2{}^o$ the entrant, efferent concentration.

Therefore, quite generally, equation (5A) tells us that the sum of the concentrations in the two cytoplasmic streams at any point along the junction is a constant and equal to the difference in entrant concentrations (Fig. 5.5.).

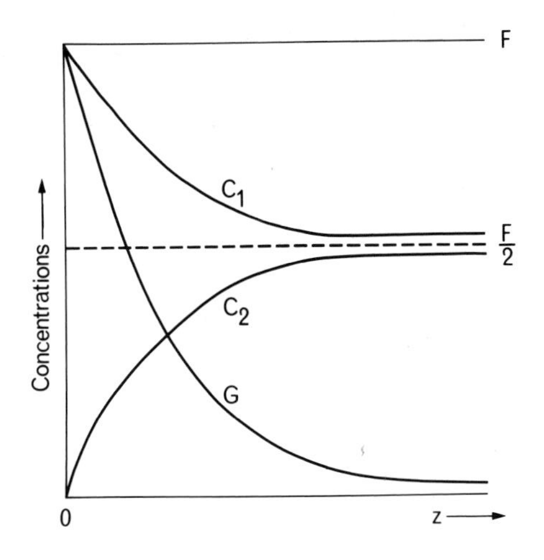

Fig. 5.5. A graph of the concentration profiles within the streaming cytoplasm of cells 1 and 2 along the length of junction, and of the derived functions $F = C_1 + C_2$ and $G = C_1 - C_2$, for co-current flow

Integration of equation (4A) is more interesting; it is second order, constant coefficients, and therefore gives

$$G = K_1 \; exp \left\{ b + \sqrt{b^2 + f^2} \right\} z + K_2 \; exp \left\{ b - \sqrt{b^2 + f^2} \right\} z$$

with $b = v/2D$; $f^2 = 2K\alpha'\omega_s' RT/D$.

Again G remains finite, as $z \rightarrow \infty$; $K_1 = 0$.

Further, when $z = 0$, $G = C^*$ (see earlier);

$$\text{thus } G = C^* \; exp \left\{ b - \sqrt{b^2 + f^2} \right\} z \qquad \text{....(6A)}$$

Finally, if we assume the flux of solute is driven by a concentration gradient (we must because in this analysis there is zero volume transfer from cell to cell), then the plasmodesmatal transport is

$$ZJ' = \int_0^Z \alpha' \omega_s' RT . G dz$$

Briefly consider the effect of variation in v on the total plasmodesmatal flux. If we write G is $f(v)$, then,

$$ZJ' = \left[\frac{\alpha' \omega_s' RT}{-f(v)} \; exp \left\{ f(v) . z \right\} \right]_0^Z$$

$$\text{or} \quad ZJ' = \frac{\alpha' \omega_s' RT}{f(v)} \left[1 - exp(-f(v) . Z) \right]$$

Now $|f(v)|$ decreases as v increases, and because $[1 - exp(-f(v).Z)]$ has boundaries from 0 to 1, then ZJ' increases as v increases.

CASE II: COUNTER-CURRENT TRANSPORT

For the case of counter-current flow (Fig. 5.4.) we have for mass conservation:

$$-kJ' + D\frac{d^2C_1}{dz} - v\frac{dC_1}{dz} = 0 \qquad \text{....(7A)}$$

$$+kJ' + D\frac{d^2C_2}{dz^2} + v\frac{dC_2}{dz} = 0 \qquad \text{....(8A)}$$

Using again the transformation

$$F = C_1 + C_2$$

$$G = C_1 - C_2$$

we may write for equations (7A) and (8A), on addition and then subtraction;

$$DF'' - vG' = 0 \qquad \text{....(9A)}$$

$$DG'' - vF' - 2PG = 0 \qquad \text{....(10A)}$$

where again $2kJ' = 2k\alpha'\omega_g' RTG = 2PG$

Integrating (9A) yields:

$$DF' - vG = \Omega = \text{constant} \qquad \text{....(11A)}$$

Now substituting (11A) into (10A);

$$DG'' - (v^2/D + 2P)G = v\Omega/D$$

or $$G'' - \beta^2 G = \lambda \qquad \text{....(12A)}$$

where $\beta^2 = (v^2/D^2 + 2P/D)$; $\lambda = v\Omega/D^2$

The solution to (12A) is

$$G = K_1 exp\beta z + (K_2 exp{-\beta z}) - \lambda/\beta^2$$

Again, we must require physically that G remains finite, therefore, $K_1 = 0$.

Consider now the concentrations when $z = 0$; call the emergent concentration $C_1{}^o$, and if the entrant concentration of the efferent stream is zero, then

$$G_o = C_1{}^o = K_2 - \lambda/\beta^2$$

When $z = Z$, $G_Z = C^* - C_2{}^Z = (K_2 exp{-\beta Z}) - \lambda/\beta^2$

where $C_2{}^Z$ is the emergent concentration of the efferent stream.

Because of the conservation requirements,

$$C_2{}^Z = C^* - C_1{}^o$$

thus $G_Z = C_1{}^o = G_o = -\lambda/\beta^2$ ....(13A)

The concentration difference at any point is therefore a constant, and is equal to the emergent concentration of the afferent stream (Fig. 5.6.).

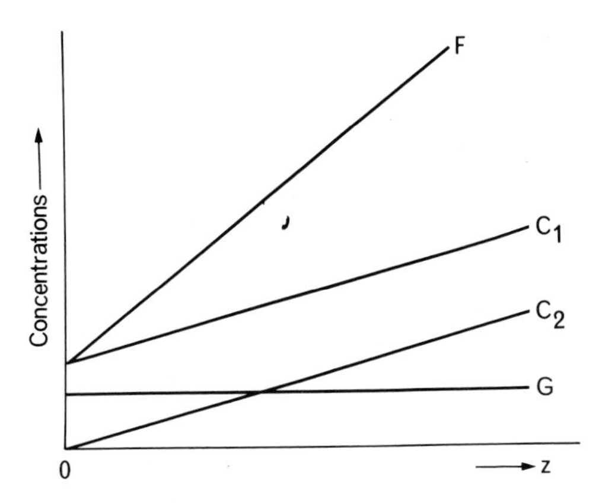

Fig. 5.6. As in Fig. 5.5. but for the counter-current case

Substitute (13A) into (11A) and integrate to give,

$$F = (\Omega/D - v\lambda/\beta^2 D)z + \rho$$

Applying again the boundary conditions given above, we find

$$F = Z\frac{2(C^* - C_1{}^o)z}{Z} + C_1{}^o$$

or $$C_1 = Z\frac{(C^* - C_1{}^o)z}{Z} + C_1{}^o$$

Finally the transport across the plasmodesmata will be

$$ZJ' = Z\alpha'\omega_s'RTC_1{}^o \qquad \text{....(14A)}$$

and it will also equal $v(C^* - C_1{}^o)$.

Hence $$C_1{}^o = \frac{vC^*}{v + Z\alpha'\omega_s'RT} \qquad \text{....(15A)}$$

It is easy to show from (15A), that C_1 increases (i.e. total plasmodesmatal transport increases) as v increases, to a maximum $C_1{}^o = C^*$ when $v = \infty$.

$$\frac{1}{C_1{}^o} = \frac{1}{C^*} + \frac{Z\alpha'\omega_s'RT}{C^*v}$$

5.8. OPEN DISCUSSION

WALKER pointed out that osmotic flow of water is not possible if plasmodesmata are open and their reflection coefficient is zero, yet much of the present theory of water transport in cells and tissues is

TABLE 5.1.

TRANSPORT ACROSS A 'CO-CURRENT' JUNCTION

For solute permeability $\alpha'\omega_s' = 10^{-4}$ mole m^{-2} s^{-1} bar^{-1}:-

v cm h^{-1} (μm s^{-1})	z (μm)	0	1	2	5	10	15	20	100
1	G	1.000	0.933	0.872	0.710	0.504	0.358	0.254	0.001
(2.8)	ZJ'x10^4	0	2.4	4.6	10.3	17.7	22.9	26.6	35.5
2	G	1.000	0.935	0.874	0.715	0.511	0.365	0.261	0.001
(5.6)	ZJ'x10^4	0	2.4	4.6	10.4	17.8	23.1	26.9	36.4
5	G	1.000	0.939	0.881	0.729	0.531	0.387	0.282	0.001
(13.9)	ZJ'x10^4	0	2.3	4.6	10.4	18.1	23.6	27.6	38.5
10	G	1.000	0.944	0.892	0.751	0.564	0.423	0.318	0.003
(27.8)	ZJ'x10^4	0	2.4	4.6	10.6	18.6	24.6	29.1	42.7
20	G	1.000	0.954	0.910	0.789	0.623	0.491	0.388	0.009
(55.6)	ZJ'x10^4	0	2.4	4.6	10.9	19.4	26.2	31.5	51.0

For solute permeability $\alpha'\omega_s' = 10^{-3}$ mole m^{-2} s^{-1} bar^{-1}:-

v cm h^{-1} (μm s^{-1})	z (μm)	0	1	2	5	10	15	20	100
1	G	1.000	0.803	0.645	0.334	0.110	0.037	0.012	0.000
(2.8)	ZJ'x10^3	0	2.1	3.9	7.4	9.9	10.7	11.0	11.1
2	G	1.000	0.804	0.646	0.336	0.113	0.038	0.013	0.000
(5.6)	ZJ'x10^3	0	2.1	3.9	7.4	10.0	10.8	11.1	11.3
5	G	1.000	0.807	0.652	0.343	0.118	0.040	0.014	0.000
(13.9)	ZJ'x10^3	0	2.2	4.0	7.5	10.1	10.9	11.2	11.4
10	G	1.000	0.813	0.660	0.355	0.126	0.045	0.016	0.000
(27.8)	ZJ'x10^3	0	2.2	4.0	7.6	10.3	11.3	11.6	11.8
20	G	1.000	0.823	0.677	0.377	0.142	0.054	0.020	0.000
(55.6)	ZJ'x10^3	0	2.2	4.0	7.8	10.7	11.8	12.3	12.5

Values of concentration difference, G (molar units m^{-3}), and plasmodesmatal transport, ZJ' (molar units m^{-2} s^{-1}), for entrant, afferent concentration 1 molar unit, and entrant, efferent concentration 0, for a junction of length z with co-current streaming at velocity v. Values for transport at different positions across a given junction can be read off from the table (unlike the counter-current junction)

TABLE 5.2.

TRANSPORT ACROSS A 'COUNTER-CURRENT' JUNCTION

For solute permeability $\alpha'\omega_s' = 10^{-4}$ mole m^{-2} s^{-1} bar^{-1}:-

v $cm\ h^{-1}$ ($\mu m\ s^{-1}$)	Z (μm)	10	20	50	100	200
1	G	0.9914	0.9827	0.9580	0.9193	0.8507
(2.8)	$ZJ' \times 10^4$	24.2	48.0	116.9	224	416
2	G	0.9960	0.9913	0.9789	0.9580	0.9193
(5.6)	$ZJ' \times 10^4$	24.3	48.4	119.5	234	448
5	G	0.9984	0.9965	0.9913	0.9827	0.9661
(13.9)	$ZJ' \times 10^4$	24.4	48.6	121	240	472
10	G	0.9992	0.9982	0.9956	0.9913	0.9827
(27.8)	$ZJ' \times 10^4$	244	48.7	121	242	480
20	G	0.9996	0.9991	0.9978	0.9956	0.9913
(55.6)	$ZJ' \times 10^4$	24.4	48.8	123	243	484

For solute permeability $\alpha'\omega_s' = 10^{-3}$ mole m^{-2} s^{-1} bar^{-1}:-

v $cm\ h^{-1}$ ($\mu m\ s^{-1}$)	Z (μm)	10	20	50	100	200
1	G	0.9193	0.8507	0.6950	0.5326	0.3629
(2.8)	$ZJ' \times 10^3$	22.4	41.5	84.8	130	177
2	G	0.9580	0.9193	0.8201	0.6950	0.5326
(5.6)	$ZJ' \times 10^3$	23.4	44.9	100	170	260
5	G	0.9827	0.9661	0.9193	0.8507	0.7401
(13.9)	$ZJ' \times 10^3$	24.0	47.1	111	208	361
10	G	0.9913	0.9827	0.9580	0.9193	0.8507
(27.8)	$ZJ' \times 10^3$	24.2	48.0	117	224	415
20	G	0.9956	0.9913	0.9785	0.9580	0.9193
(55.6)	$ZJ' \times 10^3$	24.3	48.4	119	234	448

Values of concentration difference, G (molar units m^{-3}), and plasmodesmatal transport, ZJ' (molar units m^{-2} s^{-1}), for entrant, afferent concentration 1 molar unit, and entrant, efferent concentration 0, for a junction of length Z with counter-current streaming at velocity v. Unlike the co-current junctions, every counter-current junction is unique and the transport characteristics at low Z values do *not* represent transport at different positions across a larger junction

coleoptile parenchyma and stem apices are in symplasts. The cells in the stem apices of pumpkin are particularly well linked, as a current injection to a cell on one side of the apex causes a strong membrane polarisation in a cell on the opposite side (Fig. 6.1.). Cells in the single tissue type examined where there is no organized growth, callus, show no evidence of symplastic linkage.

TABLE 6.3.

ELECTRICAL COUPLING OF PLANT CELLS*

Species	Coupling	Reference
Chara braunii	High	Sibaoka (1966)
Chara australis	High	Skierczyńska (1968)
Nitella translucens	High	Spanswick and Costerton (1967)
Elodea canadensis leaf	0.20±0.04**	Spanswick (1972)
" " "	0.72***	" "
Avena coleoptile	0.13±0.02	" "
" "	Nil	Goldsmith *et al.* (1972)
Zea root	0.24±0.05	Spanswick (1972)
Oenothera leaf	High	Brinckmann and Lüttge (1974)
Elodea canadensis leaf	0.70	Goodwin****
Cucurbita pepo apex	High	"
Lycopersicon esculentum callus	Nil	"

*Excluding papers where a membrane impedance less than 1 kΩ cm^{-2} was reported - i.e. Spitzer (1970)

**Coupling between injected cell and adjacent cell

***Coupling between two cells adjacent to injected cell

****Technique was a modification of Spanswick (1972), see Fig. 6.1.

6.5. INTERCELLULAR MOVEMENT OF INJECTED CHEMICALS

Both Spitzer (1970) and Goldsmith *et al.* (1972) injected the anionic dye Niagara sky blue into plant cells. Neither group reported movement of the dye into other cells. However, an injection technique appears essential for the determination of what compounds can pass through plasmodesmata, as only in this way can compounds be presented that cannot pass cell membranes.

A study has been made on the intercellular mobility of the coloured anionic compounds Procion Yellow (Stretton and Kravitz, 1968) and Procion Brown (Christensen, 1973). Solutions of these chemicals are deeply coloured, so that injected cells can be seen in transmitted light microscopy. They are fluorescent, or readily converted to fluorescent compounds by ultraviolet light *in vivo* and so can be detected at low concentrations by fluorescence microscopy. The dyes appear unable to pass the plasmalemma: *Elodea* leaf tissue in a solution of the dye does not take up dye intracellularly. Furthermore, *in vivo* the dyes are not absorbed by any cell component: any dye in a cell is rapidly and completely lost if the cell wall is ruptured.

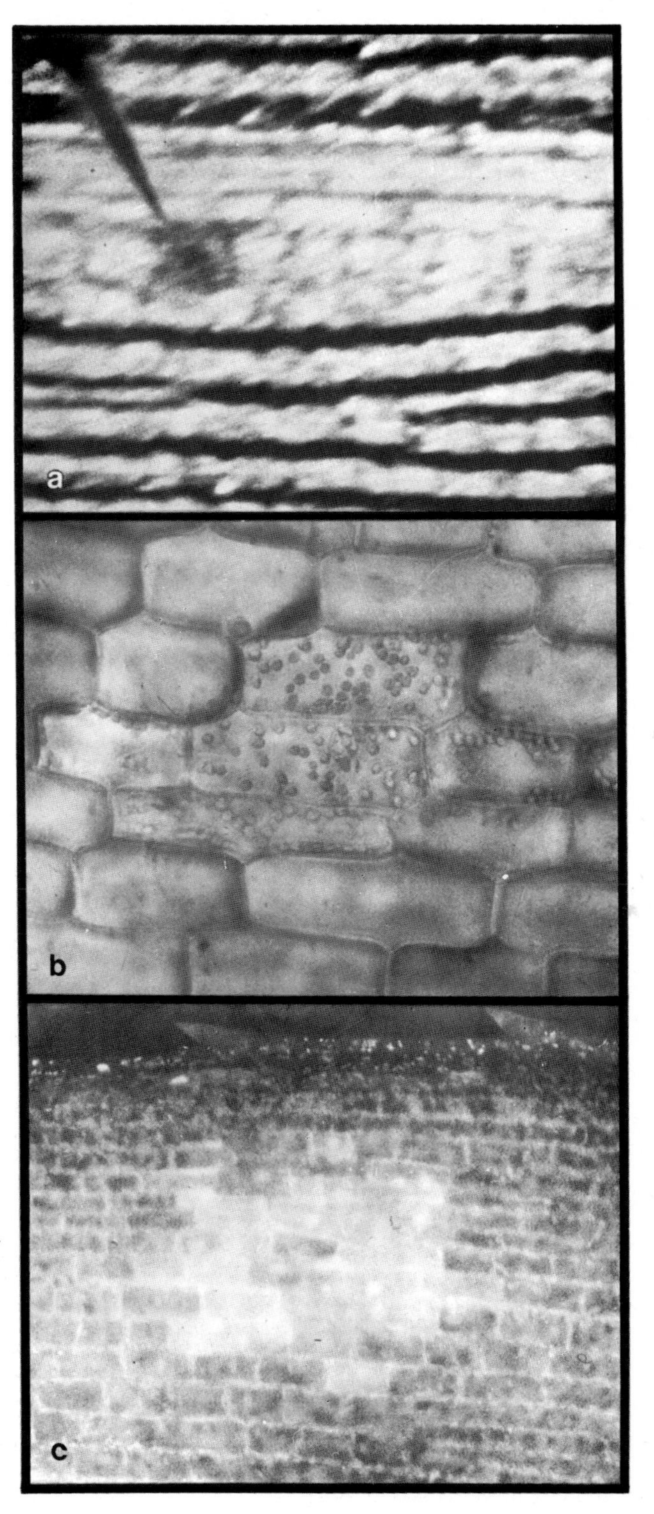

Fig. 6.2a. Iontophoretic injection of Procion Brown, 4% solution, into a leaf mesophyll cell of *Elodea*. Photograph taken during injection, about 1 minute after it had begun. Injection current: 1 mA in 30 ms pulses, one pulse per second

Fig. 6.2b. Distribution of Procion Yellow, 4% solution, after a 5 minute injection. The dye sensitised the cells containing it to ultraviolet damage. The cells were exposed to ultraviolet light for 1 minute, and then photographed in transmitted light using a 5 second exposure. In those cells containing Procion Yellow cytoplasmic streaming has stopped, and the chloroplasts are visible

Fig. 6.2c. Distribution of Procion Yellow after a 30 minute injection at 1 mA. Incident light fluorescence illumination

The dyes were injected into *Elodea* leaf mesophyll cells using standard iontophoretic techniques (Kater and Nicholson, 1973). The results are shown in Fig. 6.2. The dyes move readily from cell to cell. Since the tissue lacks a cuticle and is immersed in solution, the movement is necessarily in the symplast.

6.6. DISCUSSION

The various lines of evidence point to the conclusion that a variety of compounds *can* move from cell to cell, in a range of tissues, without coming into contact with the apoplastic compartment. Indeed, in certain situations, for instance transport through the suberised endodermis (Chapter 10), the symplast offers the *only* pathway of movement. The most critical work, that of Spanswick and Costerton (1967) indicates that there is some restriction to movement in *Nitella* plasmodesmata. This is supported by the calculations of Spanswick (1972) on *Elodea* plasmodesmata, by the work of Tyree and Tammes (1975) on *Tradescantia* stamen hairs, and that of Walker and Bostrom (1973) on *Chara*. It also gives some substance to the electron microscopy work (see Chapter 2) showing that plasmodesmata contain substructures which might reduce the effective lumen.

In essence, most of the studies in this field must be regarded as preliminary, and the central questions remain: Is the intercellular symplastic movement *via* the plasmodesmata? What types of compounds can move in the symplast and at what rates? Are there isolated symplastic territories within the plant? What are the functions of the symplast? These have yet to be convincingly answered.

ACKNOWLEDGEMENT

I am indebted to Dr. J. Silvey, Department of Neurobiology, A.N.U., for providing the expertise and equipment which made this work possible.

6.7. OPEN DISCUSSION

The simplicity and potential value of the iontophoretic dye injections attracted the first comments. ROBERTSON, noting that the procion dye had been localised by virtue of its toxicity after irradiation with ultraviolet light, asked what was known about the toxicity of the dye as applied, i.e. before irradiation. GOODWIN's reply was encouraging: the dye is acidic, but there are no sudden effects on the cells; streaming continues, and the neurobiologists find that injected nerves continue to function. He was worried, however, that it would be hard to rule out the possibility that the dye might introduce artifacts such as causing plasmodesmata to open more than *in vivo*. FINDLAY suggested that monitoring electrical coupling before and after dye injection might show whether the plasmodesmata had been affected; he also asked about the fairly large current that had been used to inject the dye - had the effects of similar currents, but in the absence of dye, been examined? Not yet, said GOODWIN, who added that the necessity for using relatively large electrodes was another hazard: however, he felt that wounding is not a problem since the dye moves through a number of cells which, unlike the injected cell itself, are presumably undamaged.

Other queries about wounding then followed. WALKER wondered whether the inability of Goldsmith *et al.* (1972) to demonstrate coupling was related to the loss of turgor which these workers had observed: had the cells pinched off their plasmodesmata? GOODWIN pointed out that Spanswick had been able to detect electrotonic coupling in the same tissue, and ANDERSON reported that he had repeated some of Goldsmith's experiments and that he too had found coupling: much depends, he said, on the method of mounting the tissue slices and the length of the waiting period before making recordings. GOODWIN, continuing, said that in his initial work he had found a high capacitance correlated with inability to detect coupling; later experiments, however, gave low capacitance and high coupling ratios; it may therefore be significant that Goldsmith too had reported high capacitances. JONES had, in his work on the electrophysiology of giant cells in roots, found that if an electrode is advanced from cell to cell across the cortex, then all the cells have the same potential, despite the fact that the cell through which the electrode has just passed is by then destroyed: in other words the plasmodesmata leading to damaged cells must seal. It was agreed that some such response to damage must occur, and that the same conclusion had been reached by Arisz and by Geiger in their experiments showing failure to wash symplastically-held solutes out of cut tissues. MARCHANT cited Fulcher and McCully's cytological observations on occlusion of plasmodesmata in wounded *Fucus*.

CARR asked how procion dye partitions between cytoplasm and vacuole, and which compartment received the injection. GOODWIN's answer was that these are yet other unknown but important aspects of the system. GUNNING described early work which should in theory, he said, long ago have demonstrated symplastic movement of injected dye: Kite (1915) was apparently the first to seek direct evidence on the site of permeability barriers in plant cells by the direct procedure of injecting dyes: various tissues were examined, including *Chara*, *Elodea*, *Tradescantia* parenchyma, and root hairs, as well as some algae like *Spirogyra* which lack plasmodesmata, but no cell-to-cell transport was reported. Neither was it in Plowe's (1931a,b) elegant experiments on cell membranes using microdissection; she also injected root hairs of *Trianea*, onion epidermal cells, and the red alga *Griffithsia*: not only could she inject without interrupting cyclosis, which in some experiments was noted to continue for many hours after the surgery, but she could routinely place the micropipette tip in either the vacuole or the cytoplasm and insert dyes which would stay in one or other compartment or partition between the two. Why was symplastic transport not detected, or commented upon? It *was* detected in other, similar, work by Worley - see 9.8. In principle the experiments were little different from those of Smith (1972), who did not inject, but observed cell-to-cell dye movement after initial uptake by the rhizoid of a fern gametophyte, or those of Tyree and Tammes (1975), who dipped the damaged end of a *Tradescantia* stamen hair into a droplet of uranin - despite the damage and the 'sealing' referred to previously, uptake occurred, and measurements of the rate of cell-to-cell transport led to the conclusion that the plasmodesmata are 99.3% occluded. GRESSEL asked whether dyes had been injected into root cortical cells and if so whether they crossed the endodermis. To this and similar queries about other tissues GOODWIN replied that his work was still preliminary, but that he had injected into stem apices, where movement was rapid, but fixation, embedding and sectioning was required for accurate localisation. See 11.4. for discussion on transport in *Vallisneria* leaves.

7

CYTOCHEMICAL EVIDENCE FOR ION TRANSPORT THROUGH PLASMODESMATA

R.F.M. VAN STEVENINCK

Department of Botany, University of Queensland, St. Lucia, Queensland, 4067, Australia

7.1. INTRODUCTION

A most appropriate review by Meeuse (1957) provides an excellent introduction to the cytochemistry of plasmodesmata covering the days of light microscopy from the 1880's until the 1950's. The main aim of such techniques was to render plasmodesmata more easily observable, and to confirm the protoplasmic nature of connections between cells. The early pronouncement by Tangl (1879) that the connections are protoplasmic was based on the great affinity of plasmodesmata for iodine. Since then a variety of staining techniques has been in use: e.g. methyl violet + orange G (Strasburger, 1923), Myer's Pyoktanin method (Jungers, 1930; Crafts, 1931; Livingston, 1935), H_2SO_4 + I_2 (Scott, Schroeder and Turrell, 1948; Scott, 1949) and silver salts (Tröndle, 1913; Pfeffer-Welheim, 1924). (See also 2.4.1. and 14.1.). The latter technique was based on the assumption that "*certain elements of plasmodesmata would absorb silver salts more strongly or reduce silver compounds more rapidly than other elements of cells and tissues*" and it was said that "*if successful, the staining would show the plasmodesmata as dark lines on a practically colourless background.*".

The advent of electron microscopy placed a very different slant on the above work. In one stroke (as evidenced by the addendum to Meeuse's review) the conclusions of early cytochemical work were placed beyond doubt, and a new era of hypotheses and experimentation needed to be established. (But note the reservations expressed by some authors - see 14.3.). This era has already provided a considerable amount of ultrastructural detail (see Chapter 2) but cytochemical evidence with respect to the function of plasmodesmata is still scarce. What is known can be grouped under three headings:

a) evidence for the presence of ions in plasmodesmata,

b) evidence for enzymes which may be involved in the ion transport process (ion specific adenosine triphosphatases),

c) evidence for particular components and molecular groupings in plasmodesmata.

TABLE 7.1.

CHLORIDE CONCENTRATIONS OF SHOOTS AND ROOTS OF 6-DAY-OLD BARLEY SEEDLINGS

	Cl^- concentration (m-equiv./kg fresh weight)	
	Shoots	Roots
Control	7.8	6.3
+ NaCl	18.0	34.0

(Seedlings were raised on half-strength Hoagland's solution in the presence or absence of 100mM NaCl.)

These differences in Cl^- concentrations were also evident in the amount of silver deposits which were encountered in mesophyll cells, and particularly in the plasmodesmata (Van Steveninck and Chenoweth, 1972; Figs. 7.1. and 7.2.). Similar differences were observed in the amount of silver deposits in plasmodesmata of root cells, but in this case quantitative differences were not easy to establish because of greater variation between the amounts of deposit in different tissues and parts of the root and because several types of plasmodesmata seemed to be present (Figs. 7.3. and 7.4.).

In mangrove mesophyll tissue plasmodesmata were very scarce between cells which contained large amounts of osmiophilic organic solute (tanniniferous cells) but they were relatively frequent between cells which were free of osmiophilic organic solutes and in conductive tissues. The plasmodesmata nearly always showed the presence of silver deposits (Figs. 7.5. and 7.6.).

Plasmodesmata in *Nitella translucens* had a distinctly different appearance from plasmodesmata in barley and mangrove tissues, being much larger in diameter (up to 200 nm), and without any contents resembling desmotubules (Fig. 7.7.). Differences in structure were also observed between plasmodesmata which occurred between nodal cells (Fig. 7.8.) and those which occurred between nodal and internodal cells

---

Figs. 7.1.-6. Plasmodesmata showing AgCl deposits. The figures indicate Cl^- concentrations based on the size of AgCl deposits. All sections unstained. The scale mark represents 1 μm. Figs. 7.1. and 2. reproduced by permission from Van Steveninck and Chenoweth (1972); Figs. 7.3. and 4. from Van Steveninck (1975); Figs. 7.5. and 6. from Van Steveninck *et al.* (in press - b) ▶

Fig. 7.1. Mesophyll cells of barley raised in the absence of NaCl
Fig. 7.2. Mesophyll cells of barley raised in the presence of 100mM NaCl.
Fig. 7.3. Root tip of barley raised in the presence of 50mM NaCl - cortical cell
Fig. 7.4. Root tip of barley raised in the presence of 50mM NaCl - stelar cell, note difference in appearance of plasmodesmata
Figs. 7.5. and 6. Mesophyll of mangrove

(Fig. 7.9.). Spanswick and Costerton (1967) have already drawn attention to these differences. A most unexpected feature, however, was the absence of silver deposits from both types of plasmodesma when the cells were fixed in OsO_4 containing Ag^+ ions (Figs. 7.10. and 7.11.). This was the case even where silver deposits occurred over the whole length of external walls of internodal cells and also where silver deposits were particularly heavy in internal walls (Fig. 7.12.). After an extensive search, some plasmodesmata were found which appeared to

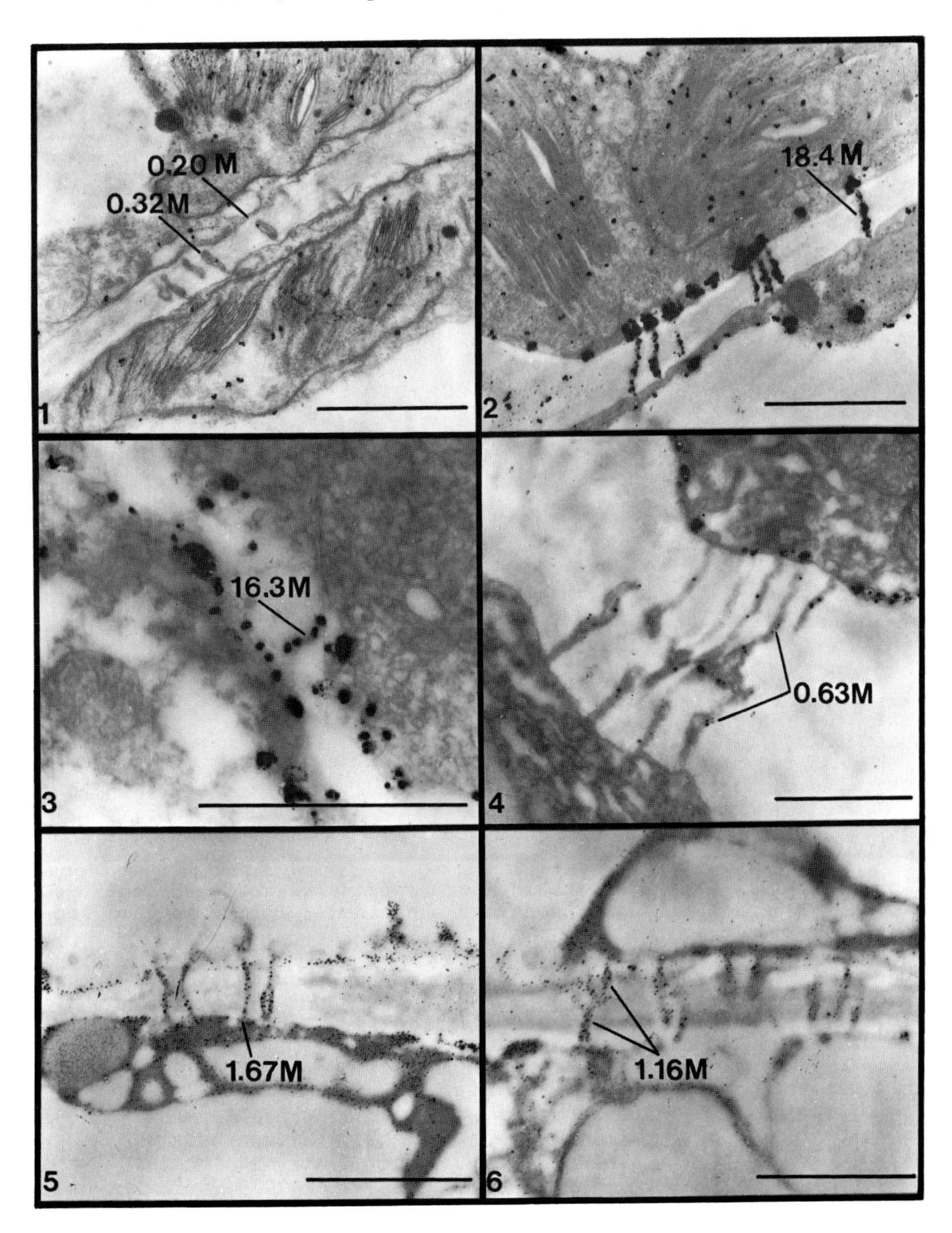

barley, mangrove and *Nitella translucens* (Table 7.2.). The ratio of chlorine to silver counts (peak-background) for deposits in barley and mangrove was significantly higher than that for the pure AgCl standards. One would have expected the reverse of this result if silver deposits in the parenchyma cells had consisted of a mixture of AgCl and other silver compounds or metallic silver. However, it was shown that the ratio could be affected by the size of the deposit, i.e. the recording of chlorine counts becomes less efficient with the increasing mass of the deposit (Van Steveninck *et al.*, in press - a). This in part explains the relatively high ratios observed for the small sized deposits in the parenchyma tissue.

7.3.1.3. Estimation of chloride concentrations Chloride concentrations were estimated on the basis of size of deposits, assuming a section thickness of 80 nm (see Van Steveninck and Chenoweth, 1972); inthe case of plasmodesmata, on the basis of the internal volume of each individual plasmodesma observed. Often the boundaries between plasmodesmata and cell walls were unclear and irregular in shape, and it then became necessary to adopt an internal diameter of the plasmodesma equal to the diameter of the largest AgCl deposit present. Results of these estimates are presented in Figs. 7.1.- 6. and 7.15. alongside the plasmodesmata and in Table 7.3.

TABLE 7.3.

ESTIMATED CHLORIDE CONCENTRATIONS BASED ON THE SIZE OF SILVER CHLORIDE DEPOSITS IN PLASMODESMATA OF BARLEY MESOPHYLL TISSUE

	Cl^- concentration (molarity)	Number of estimates
Control	0.36±0.07	6
+ NaCl	11.6±3.0	7

(Seedlings were raised on half-strength Hoagland's solution in the presence or absence of 100mM NaCl.)

It is immediately apparent from these estimates that extraordinarily high values can be reached with a possible maximum equivalent to 38M for a solid lump of AgCl. These calculations show that a considerable amount of diffusion must be involved in building up a deposit. Such problems associated with possible diffusional artifacts were considered in an earlier paper (Van Steveninck and Chenoweth, 1972). However, from considerations involving kinetics of precipitation (Nielsen, 1964; Walton, 1967) it can be expected that critical clusters providing nuclei of precipitation will form preferentially at locations where Cl^- concentrations are highest. No differences in patterns of precipitation were observed when tissues were exposed to only 3 minutes of OsO_4 + Ag^+ ion treatment instead of 60 minutes or two hours (Van Steveninck *et al.*, 1974a).

7.3.2. Ion Precipitation using Antimonate

Because of the lack of specificity of this method (Clark and Ackerman, 1971; Garfield, Henderson and Daniel, 1972; Van Steveninck *et al.*, in preparation - a) relatively few attempts have been made to use it in

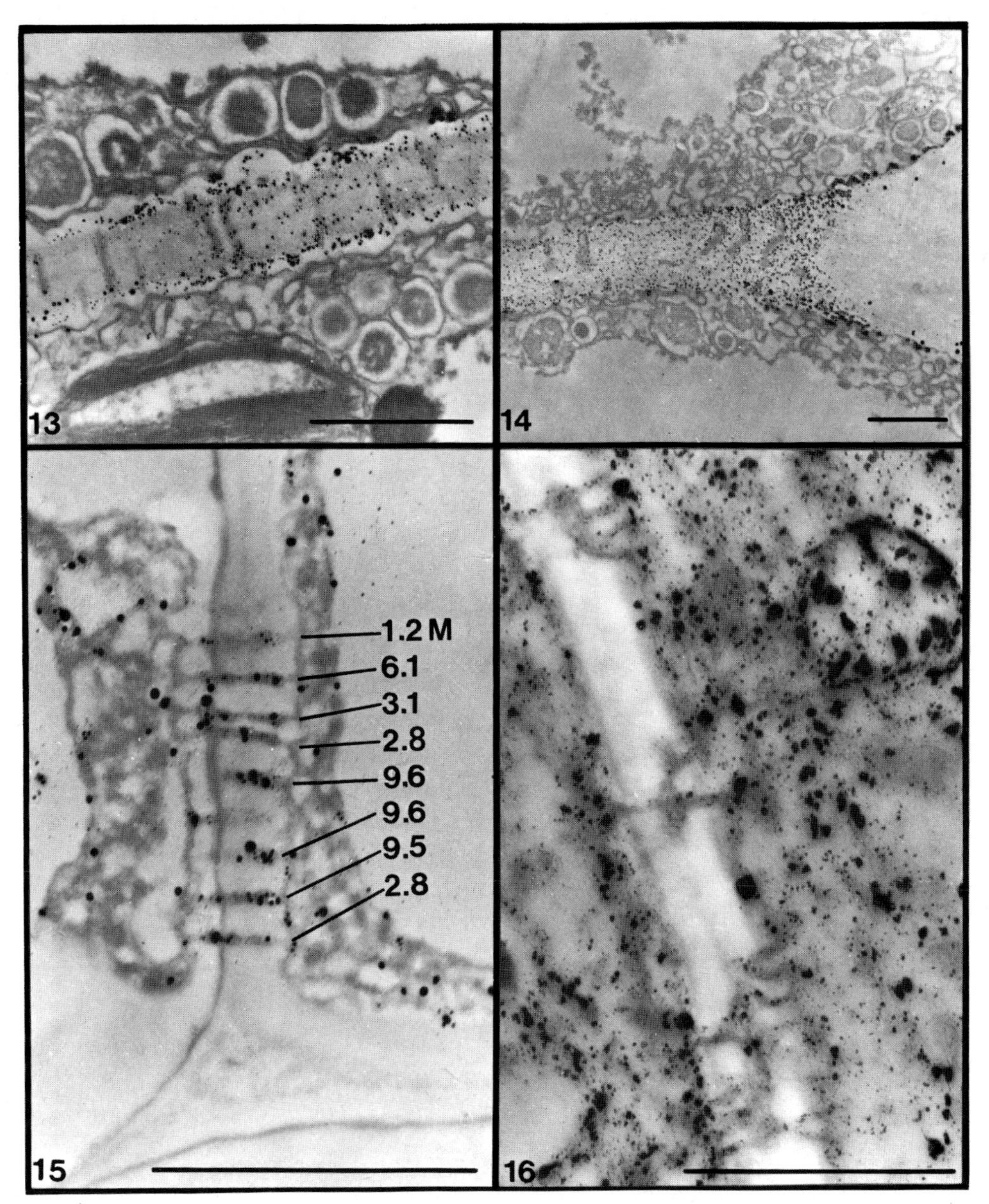

Fig. 7.13. Detail of plasmodesmata between nodal cells of *Nitella translucens* which represents a rare case in which AgCl deposits could be found inside the plasmodesmata

Fig. 7.14. Boundary showing AgCl deposits in the cell wall where plasmodesmata occur, but no AgCl in the internal phase of the plasmodesmata

Fig. 7.15. Detail of a field of plasmodesmata between a vessel and xylem parenchyma cell in barley mesophyll tissue. The figures indicate Cl^- concentrations based on the size of AgCl deposits and the dimensions of the plasmodesmata

Fig. 7.16. Precipitation products in the plasmodesmata of mesophyll tissue of *Atriplex nummularia* which was fixed in 2% OsO_4 + 2% potassium pyroantimonate

work on plant tissues. However, precipitation products appear in the plasmodesmata of salt bush (*Atriplex nummularia*) when antimonate is used with OsO_4 fixation (Fig.7.16.). These deposits may be indicative of the presence of Na^+ only when no interfering ions (especially multivalent ones) are present. However, it was not possible to show the presence of Na^+ in antimonate-induced deposits in cells of barley root tips by means of X-ray energy spectroscopy (Van Steveninck *et al.*, in preparation - a).

7.3.3. Adenosine Triphosphatase

7.3.3.1. Barley root tips Results similar to those described by Coulomb and Coulomb (1972) for *Cucurbita* root tip cells were obtained in barley root tips in the region where cells are undifferentiated and lack vacuoles. Reaction products resulting from exposure of formaldehyde-fixed tissue to the Wachstein-Meisel medium occurred frequently in plasmodesmata, were always present on the plasmalemma, and elements of the endoplasmic reticulum often showed a positive reaction in the vicinity of plasmodesmata (Fig. 7.17.). Also the amount of reaction product seemed to increase when roots of seedlings had been exposed to high concentrations of NaCl (50-100mM) prior to fixation. Other sites of activity were associated with the envelopes of plastids and mitochondria especially in the more differentiated parts of the root. These deposits however showed a greater degree of variability which might be due to differential effects associated with the fixation procedure (Moses and Rosenthal, 1968). Towards the periphery of the root tip and especially on the root cap cells a different type of deposit was observed associated with cell walls which was more opaque and crystalline than the amorphous cytoplasmic deposits (Fig. 7.18.). It first appeared that these deposits represented a non-specific phosphatase activity in the cell walls. However, the deposits shown by Robards and Kidwai (1969) on the inner side of the plasmalemma also have an opaque, crystalline appearance, while those shown by Coulomb and Coulomb (1972) and Hall (1969) have a less dense, amorphous appearance. This difference in appearance seems not to be due to the use of different fixatives since both Robards and Kidwai (1969) and Hall (1969) used glutaraldehyde as a fixative while in Coulomb's and in our work formaldehyde was used.

Figs. 7.17-22. Barley root tip or bean hypocotyl slices fixed in 3% formaldehyde + 0.5mM $CaSO_4$ in veronal acetate buffer pH 7.2 and subsequently placed in Wachstein-Meisel medium for 15 minutes at 37°C (Coulomb and Coulomb, 1972) ▶

Fig. 7.17. Barley root tip: reaction products on the plasmalemma, in plasmodesmata and the endoplasmic reticulum near the plasmodesmata
Fig. 7.18. Barley root tip: two types of deposits - dense crystalline aggregates in the cell wall and amorphous globules associated with the plasmalemma
Fig. 7.19. Bean stem slices: vascular tissue - note reaction product in the plasmodesmata of xylem parenchyma cells
Fig. 7.20. Detail of Fig. 7.19. - reaction product occurs predominantly within the plasmodesmata
Fig. 7.21. Bean stem slices: reaction product in plasmodesmata of parenchyma cells. Plasmolysis occurred frequently, but usually the plasmalemma showed a tendency to adhere to the cell wall near groups of plasmodesmata
Fig. 7.22. Bean stem slices: two types of deposits - crystalline aggregate in the cell wall and less dense amorphous globules on the plasmalemma and in the envelopes of mitochondria and chloroplasts

A more serious problem was the fact that deposits still occurred when ATP was omitted from the reaction medium, and hence the presence of deposits could not be accepted as definite proof of ATP-ase activity. Yet, deposits were absent when the exposure of the fixed tissue to the Wachstein-Meisel medium was delayed by a period of washing extending over at least 18 hours after its preliminary fixation. Enzyme activity is known to decrease with the period of washing (Kanamura, 1973), and also other endogenous organic phosphates may be removed from the tissue through enzymatic hydrolysis during this prolonged period of washing. Complications of this kind and several more which have been reported still severely limit the specificity and usefulness of the ATP-ase detection method.

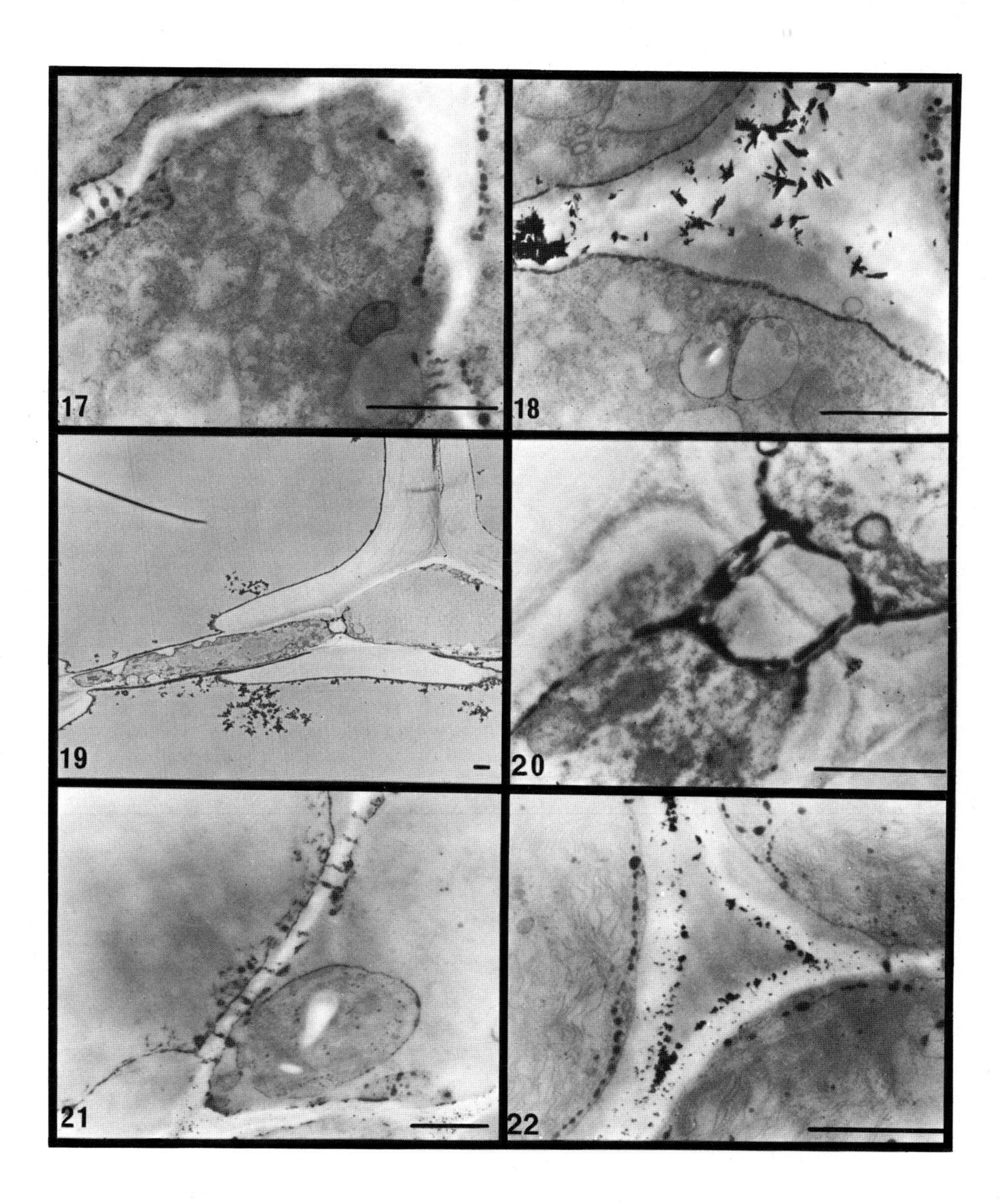

7.3.3.2. Bean stem slices Although in general more reaction product was observed in organelles (chloroplasts, plastids, nuclei, mitochondria) of cells in bean stem slices than in barley root tips, the results were essentially the same. Deposits were found in plasmodesmata, especially in xylem parenchyma (Figs. 7.19. and 7.20.), but also between other parenchyma cells (Fig. 7.21.). The formaldehyde fixation of large vacuolated parenchyma cells posed a serious problem because it was difficult to prevent plasmolysis from taking place. It was noticeable, however, that in partially plasmolysed cells the plasmalemma was adhering to the wall in places where plasmodesmata were frequent (Fig. 7.21.). Again, two types of deposit were observed (Fig. 7.22.), i.e. the granular type associated with cell walls and the amorphous type associated with cytoplasmic components.

When slices had been exposed to cycloheximide (1 μg/ml) for 6 or 24 hours no reaction products were observed, although deposits were still present after only 1-3 hours of exposure. Salt absorption, measured as ^{42}K, ^{24}Na and ^{36}Cl influx, followed a similar pattern of progressive inhibition depending on the duration of exposure to cycloheximide (Van Steveninck *et al.*, unpublished data). These results suggest that the reaction products were caused by enzymatic activity and not by the non-enzymatic hydrolysis of adenosine triphosphate (Moses and Rosenthal, 1968). However, further work is required to improve the specificity of the method.

7.4. DISCUSSION

Before the advent of electron microscopy, silver salts had been used for many years as a cytochemical stain to differentiate plasmodesmata from the rest of the cell wall (Mühldorf, 1937; Meeuse, 1957). It has now been shown by means of X-ray analytical electron microscopy that silver deposits in plasmodesmata consist of AgCl. Quantitative differences in the concentration of Cl^- in the tissue and the amounts of AgCl in plasmodesmata have been shown to exist in barley seedlings which were raised on solutions in the presence or absence of NaCl, and these results seem to indicate that, at least in barley and mangrove tissues, plasmodesmata provide an important pathway for Cl^- transport. Campbell and Thomson (1975) reached a similar conclusion for the plasmodesmata at the transfusion zone of the base of the salt glands of *Tamarix aphylla* (L.) Kerst. Although these authors could only achieve positive identification of the silver deposits by means of electron diffraction on very large aggregated deposits of the apoplast, they did observe a dramatic increase of deposits in the plasmodesmata when cut branches were placed in flasks containing 3% NaCl instead of distilled water. A startling departure from this pattern of deposition in plasmodesmata was observed between nodal and internodal cells of *Nitella translucens*. The plasmodesmata usually appeared empty while the surrounding cell walls often contained large amounts of AgCl. It suggests that, in *Nitella*, Cl^- ions can move freely through the cell wall with the plasmalemma providing a main barrier to diffusion while the plasmodesmatal contents have very little affinity for Cl^- ions, in fact less than the cell walls.

When estimates are made of the amount of Cl^- inside plasmodesmata on the basis of size of AgCl deposits encountered, it becomes immediately apparent that the values cannot represent true Cl^- concentrations, and hence it seems necessary to assume that Cl^- ions are attracted from considerable distances to the nuclei of precipitation. Accumulations of deposit at the orifice of plasmodesmata are relatively frequent (see Fig. 7.1.) and indicate that diffusional artifacts through growth of large AgCl crystals are likely to happen. It has already been pointed out (Van Steveninck and Chenoweth, 1972) that the AgCl precipitation method may meet with severe limitations because of diffusion and crystal growth under conditions when Cl^- concentrations are either very high or very low. However, it is instructive to compare the situation in plasmodesmata with that encountered by Komnick and his coworkers in the chloride cells of mayfly larvae (family *Baetidae*) (Wichard and Komnick, 1971; Komnick and Abel, 1971; Komnick and Stockem, 1973). Fixation of these cells in OsO_4 - silver lactate solution res-

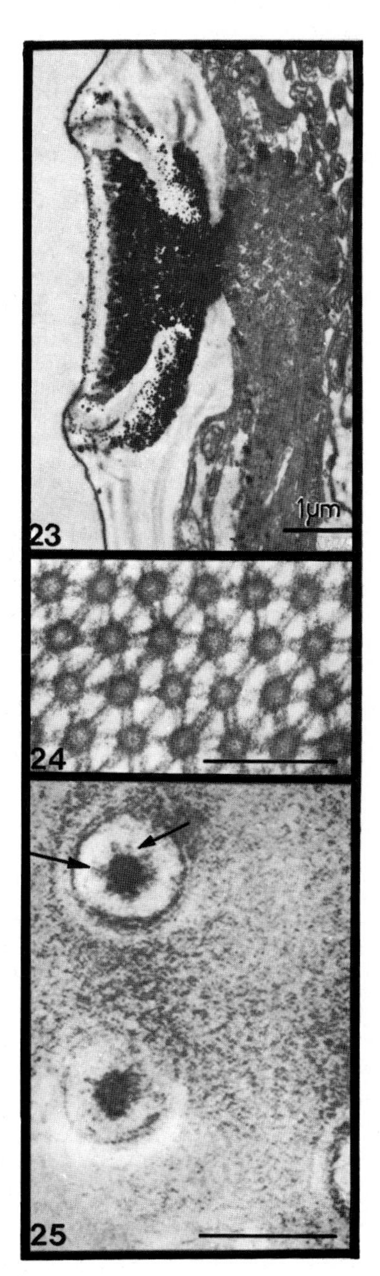

Fig. 7.23. Histochemical demonstration of Cl^- in the chloride cell complex of *Cloeon* showing very dense precipitates in the apical part of the central cell and within the porous plate (from Wichard and Komnick, 1971)

Fig. 7.24. Detail of porous plate of *Baetis* fixed in OsO_4 and stained *en bloc* in 1% aqueous phosphotungstic acid. The scale mark represents 0.1 µm (from Komnick and Stockem, 1973)

Fig. 7.25. Transverse section through plasmodesmata in the lateral wall of *Avena* coleoptile. The tubular core and spokes which radiate between core and plasmalemma lining the pore are approximately equal in dimension to one repeating unit of the porous plate of *Baetis*. The scale mark represents 0.1 µm (from Burgess, 1971)

ults in massive, dense AgCl precipitates in the central cell apex and porous plate (Fig. 7.23.). It was shown by a combination of cytochemical precipitation and autoradiography that the chloride cells absorb significant quantities of Cl^- from extremely dilute Cl^- solutions and that this property is probably due to the binding of Na^+ and Cl^- by mucopolysaccharide and glycoprotein-containing structures in the chloride cells (Komnick, Rhees and Abel, 1972). In ultrastructural detail (Komnick and Stockem, 1973) there is a similarity in dimension and appearance of elements of the porous plate of chloride cells (Fig. 7.24.) and the desmotubule with radiating spokes as seen by Burgess (1971) (Fig. 7.25.). The resemblance may be fortuitous, but it does suggest that plasmodesmata which contain structural elements in the form of a desmotubule (Robards, 1968a,b) may have a special capacity to bind relatively large amounts of ions, while absence of these structures (as in *Nitella translucens*) may indicate a simple diffusion pathway without regulatory or directional properties.

Isolation and function of the cytoplasmic contents of plasmodesmata would seem to be a first requirement before any progress can be made in characterising the chemical components. Such isolations have already been carried out successfully in the case of microtubules (Hepler and Palevitz, 1974). The fact that plasmodesmata are completely embedded in the cell wall structure may pose serious problems in developing suitable isolation techniques. On the other hand, it seems reasonable to expect that many cell wall preparations contain elements which include components of plasmodesmata. Proteinaceous components of cell wall fractions have been shown to contain S-amino acids (Mühlethaler, 1967; Thompson and Colvin, 1970), and glycoproteins have been shown to contain hydroxyproline (Lamport, 1970). The latter are of considerable interest since collagen, the only animal protein containing hydroxyproline, is known to perform a structural function. It has been implied that a plant glycoprotein (extensin) plays a role in cell wall extension, however it does not seem unreasonable to suggest other structural roles for proteins which are found to be present in cell wall fractions. Lamport (1970) has listed a wide range of groups and species of the plant kingdom in which hydroxyproline is present in the cell wall protein fraction. This list shows similarities with a list presented by Meeuse (1957) showing the occurrence of plasmodesmata in the plant kingdom, and *Nitella* presents a notable exception in having plasmodesmata and yet no hydroxyproline-containing glycoprotein in the cell wall fraction. The absence of hydroxyproline in *Nitella* is taken as an indication that *Nitella* may have a different mechanism of cell extension from that found in most higher plants, e.g. *Avena* coleoptiles. It seems equally reasonable to speculate that cytochemical differences between *Nitella* and higher plant plasmodesmata may be due to the absence or presence respectively of structural glycoprotein in the plasmodesmata.

Albersheim (1965) and Lamport (1970) list a wide range of enzymes, mostly hydrolases including ATP-ase, which have been shown to be present in plant cell wall fractions and it is possible that a significant proportion of these enzymes is situated in the plasmodesmata. Furthermore, the close similarity of the desmotubule sub-unit structure with microtubules and nuclear spindle fibres (Robards, 1968a,b) suggests that plasmodesmata may have certain enzyme activities in common with spindle fibres. Both microtubules and their associated cross-bridges (Hepler and Palevitz, 1974) and the isolated spindle apparatus of the sea urchin egg (Mazia *et al.*, 1972) have been shown to have ATP-ase activity. Evidence for ATP-ase activity in plasmodesmata has also been produced in a number of laboratories (Hall, 1969; Robards and Kidwai, 1969; Coulomb and Coulomb, 1972; Gilder and Cronshaw, 1973a,b). Although a high degree of specificity has been claimed

(Coulomb and Coulomb, 1972) we have not been able to confirm this specificity for enzymatic ATP hydrolysis. Unfortunately no information is available yet regarding possible ATP-ase activity in plasmodesmata of *Nitella translucens* but it will be interesting to establish whether differences in ATP-ase activity can be detected between the plasmodesmata of higher plants and those of *Nitella translucens*. This sort of information will help in deciding whether plasmodesmata should be classified into two types: i.e., those which provide an open cytoplasmic diffusion pathway, and those which may have a regulatory function on the passage of solutes between cells.

7.5. SUMMARY

This paper reports direct cytochemical evidence for ion transport in plasmodesmata by means of ion precipitation and X-ray analytical electron microscopy. Cytochemical evidence is also given of phosphatase activity inside plasmodesmata of barley root and bean hypocotyl tissue, but, it was not possible to confirm previous claims that the activity was specifically due to ATP-ase.

All silver deposits present in plasmodesmata of barley root, barley mesophyll and mangrove mesophyll exposed to OsO_4 and silver acetate were shown to consist of AgCl. Barley seedlings raised in the presence of 50-100mM NaCl showed more AgCl deposits in the plasmodesmata than seedlings which were raised in the absence of NaCl. Barley and mangrove tissue showed very few deposits in the cell wall compared to the massive deposits found in the plasmodesmata.

Nitella translucens provided a sharp contrast in showing considerable amounts of AgCl deposits in the cell wall while very few or no deposits could be found in the plasmodesmata. It is suggested that plasmodesmata of higher plants contain structural elements which have a great affinity for Cl^- ions while these elements seem to be absent from plasmodesmata of *Nitella translucens*. Thus a distinction may possibly be drawn between two types of plasmodesmata i.e. those which provide a simple diffusion pathway as in *Nitella* and those which contain structural units which may have a regulatory effect on transport of certain solutes through the plasmodesmata.

ACKNOWLEDGEMENTS

This work was supported by the Australian Research Grants Committee. The X-ray analytical work was carried out in collaboration with Dr. T.A. Hall and Dr. P.D. Peters in the Cavendish Laboratory, University of Cambridge, England. The work on antimonate precipitation in *Atriplex nummularia* (Fig. 7.16.) was carried out by Mrs. M.E. Van Steveninck; on ATP-ase in barley root tips (Fig. 7.17. and 7.18) in collaboration with Dr. S. Iwahori, Faculty of Agriculture, Kagoshima University, Japan; and that on ATP-ase in bean stem slices (Figs. 7.19. to 7.22.) by Mrs. D.M. Waller as part of an Honours Degree project.

7.6. OPEN DISCUSSION

FINDLAY said that he found it difficult to believe that any biological structure could contain 9.0 M chloride. (VAN STEVENINCK interjected to say that he did not believe this either!). FINDLAY continued by stating that he also knew of no forces that would cause a silver ion in solution to attract chloride ions from some distances away from it. He could only suggest that, perhaps, the silver ions were moving around and, thus, collecting chloride ions as they moved. Wouldn't this suggest, he asked, that the method was not very reliable? VAN STEVENINCK said that consideration of such comments as those made by FINDLAY had caused him to be rather pleased by the finding that silver chloride precipitation in mangrove and *Nitella* plasmodesmata differed: therefore, the precipitation seemed not to be an exclusively physical effect, such as the silver particles being swept into the plasmodesmata and becoming stuck there. FINDLAY responded by saying that if VAN STEVENINCK doubted one half of the story, then he surely didn't have good evidence for believing the other half either. He expressed his extreme scepticism about accepting the cytochemical evidence at the moment, and then asked why, if VAN STEVENINCK had used probe analysis, he had not observed 'standards' containing known amounts of chloride. VAN STEVENINCK said that it would be possible to do this sort of thing by, for example, embedding known quantities of silver chloride in gelatin; however, he also commented that in barley, for example, there was a random distribution of silver chloride crystals in the matrix of chloroplasts where there is evidence of the silver ions coming in and causing a precipitation product. FINDLAY persisted by saying that he didn't know how VAN STEVENINCK could say anything from the distribution of 'dots' of silver chloride throughout the cell. VAN STEVENINCK referred to his 1972 paper (with Chenoweth) where, he said, they had discussed the relationship of the dots to the diffusion pathway, and where it had been estimated that one dot of approximately 25 nm diameter in a cubic micrometre would represent a concentration of approximately 4 mM. Knowing such concentrations, it would be possible to deduce the mean separation of chloride ions and therefore the behaviour during crystal formation; the ions would be moving (by diffusion) in a random manner but would progressively be trapped by crystal formation. VAN STEVENINCK referred to a speed of thermal molecular motion of chloride ions of something like a thousand metres per second! (see Nielson, 1964). FINDLAY concluded by saying that he would accept that the presence of chloride in plasmodesmata had been demonstrated, but nothing further than that.

CARR asked whether model systems (such as millipore filters or the pores through mica films) had anything to offer in studying the trapping of ions and particles within small channels. VAN STEVENINCK said that something of this nature had been carried out using gelatin and was referred to by Komnick in his 1969 paper. CARR mentioned the work of Appleton (1974) who had used carboxy-methyl cellulose to study the movement of sodium and had obtained pictures from thin frozen sections of the distribution of silver chloride crystals: these seemed to be non-randomly distributed and associated with the tips of the carboxy-methyl cellulose fibres. Could this be relevant to the localisation of precipitated linear arrays by VAN STEVENINCK, he asked? VAN STEVENINCK referred to numbers of nucleation sites necessary for precipitation of silver chloride and said that there would be a solute excess in the methods that he used, and added that it was important to have experience with using the method which, he believed, had become generally recognised.

SMITH asked whether it would be possible to check the chloride concentrations by trying to work out the concentration within the vacuole. If this could be done, he argued, then it should be possible to relate absolute vacuolar concentrations to the number of silver chloride precipitates counted. VAN STEVENINCK retorted that this didn't take into account the frustrations of being an electron microscopist and the difficulties of sectioning through a whole vacuole.

GUNNING contributed further doubts by showing pictures from the work of Gullvåg *et al.* (1974) and Ophus and Gullvåg (1974) which demonstrated plasmodesmata of a moss filled with black precipitates which, it had been determined, were lead phosphate arising from the exposure of the plants to nothing more than motor vehicle exhaust fumes! He also referred to the observation that plasmodesmata in general seem to have a high affinity for heavy metals: much of the early light microscopy made use of methods such as silver impregnation to demonstrate plasmodesmata. ROBARDS noted that plasmodesmata sometimes became very heavily stained with osmium (see Robards *et al.*, 1973, Figs. 23., 24., 26.). Such results indicate the particular difficulties of inferring ion movement through plasmodesmata by precipitation of heavy elements, which could possibly be caused by reducing substances associated with the plasmodesmata.

ROBARDS referred to the recent paper of Van Iren and Van Der Spiegel (1974) which supposedly traced the pathway of potassium (using thallium as an analogue), and had demonstrated precipitation of thallium iodide within the cavity of the endoplasmic reticulum. VAN STEVENINCK thought that one of the main objections to the technique was the necessity of keeping high iodine concentrations in, for example, the fluid in the sectioning bath (knife trough) to minimise the possibility that the thallium iodide redissolved from the section. ROBARDS offered a further word of caution in saying that the use of thallium must be regarded with suspicion because it was known to be highly toxic, and physiologists had used it to demonstrate conventional absorption isotherms even though, eventually, it kills the cell. Nonetheless, the method probably had much to offer, particularly if combined with sectioning of frozen tissues, so avoiding random (Komnick) precipitation effects during fixation.

MARCHANT said that he had noted occluded material within the sinuses of *Nitella* plasmodesmata shown by VAN STEVENINCK: could this be an effect of damage during processing, he asked, and would this explain the lack of deposits within plasmodesmata? VAN STEVENINCK thought that this result might possibly be due to the fixation technique, as only osmium tetroxide had been used without pretreatment with glutaraldehyde. He added that most of the *Nitella* plasmodesmata had been found to be open, and that they had tried to minimise wounding effects by cutting firstly one internodal cell, and then the next internodal cell on the other side of the node.

WATSON questioned whether the special protein found associated with cell walls might be specific to plasmodesmata, and asked for clarification. GUNNING pointed out that a specific relationship was unlikely, particularly as hydroxyproline-rich protein had been reported in cell walls having no plasmodesmata (e.g. in unicellular algae, and cultured cells - Lamport, 1970; 1974; and in pollen tubes - Dashek and Harwood, 1971).

8

VIRUSES AND PLASMODESMATA

ADRIAN GIBBS

Department of Developmental Biology, Research School of Biological Sciences, The Australian National University, Box475, P.O., Canberra City, A.C.T. 2601, Australia

8.1. MOVEMENT OF VIRUSES IN PLANTS

It is remarkable that many viruses are able to move so freely through their host plants, despite the fact that virus particles or infective virus nucleic acids are so very much larger than the molecules, such as sugars and amino acids, for which plant translocation systems were presumably 'designed'. But they do move, and surprisingly quickly.

Early work (reviewed by Schneider, 1965) with viruses that may be sap inoculated to leaves, showed that movement in systemically infected plants is of two kinds. First there is radial spread through the leaf away from the site of infection, and then, when a leaf vein is 'encountered', there is a directional spread through the vascular tissue to all parts of the plant (Samuel, 1931). These two kinds of spread occur at different rates. For example, Samuel (1934) studied the movement of tobacco mosaic virus (TMV)[1] in tomato plants by inoculating a single terminal leaflet on each of many plants, and then at different times he cut plants into many portions, which were incubated for virus in them to multiply and then tested for virus. He found that no virus left inoculated leaflets for 3 or 4 days, but within 21 hours of first finding TMV in petioles, it was also found in roots and was in the apex within a day. These and other experiments (Uppal, 1934; Rappaport and Wildman, 1957; Dijkstra, 1962; Hatta and Matthews, 1974) have shown that viruses move through parenchymatous tissue at about 5-15 μm h^{-1} (1-4 nm s^{-1}) or one cell every 4-10 hours, whereas they move through vascular tissue at around 15-80 mm h^{-1} ($4.2x10^3$-$2.2x10^4$ nm s^{-1}). Vector-borne viruses have been recorded as moving through vascular tissue even faster, maize streak virus at 20 mm h^{-1} (Storey, 1928) and beet curly top virus at up to 1,500 mm h^{-1} (Severin, 1924; Bennett, 1934), but this difference may just reflect the greater precision that can be obtained using vectors to inoculate and test the vascular tissue directly, and indeed all viruses may move at these greater speeds in vascular tissue.

---

[1]See 8.5. for a brief portrait of each virus mentioned.

In parenchymatous tissue viruses probably move from cell to cell through plasmodesmata (Livingston, 1935; Sheffield, 1936; Martin and McKinney, 1938), and the particles of many different viruses have been found in plasmodesmata, but not in other intercellular structures; I will review some of these reports below. There is however, at least one report which suggested that the plasmodesma may not be the only path by which viruses spread between cells. Kassanis, Tinsley and Quak (1958) reported that TMV moved as quickly in cultured tobacco callus tissue as in leaf tissue, even though they could find no plasmodesmata linking the callus cells and they suggested that plasmodesmata may not be the only path by which viruses move between cells. However, later, Kassanis (1967) suggested that the methods used in the earlier work to search for plasmodesmata between the callus cells may have been inadequate.

Many lines of evidence suggest that most viruses move through vascular tissue in the phloem, rather than the xylem:-

(i) Early experiments showed that TMV will not pass from an infected part of a plant through sections of the stem killed by steam (Caldwell, 1931), even though TMV moves through the stem without infecting all the parts through which it passes (Samuel, 1934). However TMV does pass, though more slowly, through a stem that has been 'ringed' to remove phloem (Bennett, 1940); presumably living parenchyma cells provide the bridge around the 'ring'.

(ii) Virus movement is usually closely correlated with the direction and rate of movement of metabolites, and not with the transpiration rate (Bennett, 1940; Cochran, 1946).

(iii) Many viruses move readily from leaves to roots but more irregularly and slowly in the opposite direction (Price, 1938; Bennett, 1940; Fulton, 1941).

(iv) Particles of viruses have been found in sieve tube elements (Esau *et al.*, 1967; Esau and Hoefert, 1972a,b).

(v) The first parenchyma cells to be infected systemically by turnip yellow mosaic virus in Chinese cabbage are those around the phloem and not the xylem of leaf veins (Hatta and Matthews, 1974).

There is much less evidence on the role of the xylem in the movement of viruses in plants. Caldwell (1931, 1934) experimentally introduced TMV into xylem elements, and found that the leaves supplied by that xylem did not show symptoms of infection unless the xylem vessels in the leaves were deliberately damaged. Similarly Matthews (1970) reported that solutions containing radioactively labelled particles or RNA of turnip yellow mosaic virus are readily taken into a detached leaf of Chinese cabbage through the cut petiole, but the virus does not replicate in the leaf, though if manually inoculated to the surface of the leaf it does. These experiments suggest that these two viruses cannot move from xylem vessels into living cells. By contrast southern bean mosaic virus seems to be naturally translocated in the xylem, will infect the plant through the xylem, and will move through steamed sections of the stem (Schneider, 1965; Esau, 1967a). Similarly the xylem may be involved in the translocation of lettuce necrotic yellows virus for the characteristic bacilliform particles of the virus are found in the young xylem cells in leaf veins of infected plants, but not in phloem cells (Chambers and Francki, 1966), however the virus may have spread to and multiplied in the xylem before it differentiated, and this may also be the reason why the particles of dahlia mosaic have been found in developing tracheids (Kitajima *et al.*, 1969).

Viruses have a complex biochemical life cycle (reviewed by Gibbs and Skehel, 1973). An infecting virus particle entering a susceptible host cell is partly or wholly disassembled into its constituent macromolecules. These usually comprise one or more pieces of nucleic acid, one or more proteins (some of which may be enzymes), and, in some, lipids. The virus nucleic acid, or a messenger RNA transcribed from it, is translated by the host's protein synthesizing apparatus into various proteins, and the nucleic acid is replicated by a complex process involving both host and virus enzymes. Finally, the progeny nucleic acid and certain of the proteins assemble to form the progeny virus particles; those that contain lipid usually acquire their lipid-containing outer layer as they 'bud through' one of the cellular membranes of the host. Thus the virus life cycle has two contrasting stages. In one, the virus consists of passive, stable, dispersible particles, and in the other the virus is an unstable metabolically-active group of macromolecules busy sequestering the metabolites of the host. It is clear that natural spread between plants involves the particle stage of the virus life cycle, but there is some evidence that other stages may also be involved in spread within an intact plant. Siegel, Zaitlin and Sehgal (1962) have isolated a mutant of tobacco mosaic virus that produces a defective coat protein which cannot assemble correctly with the genome RNA to form particles. This mutant can be transmitted from plant to plant by extracting and inoculating its genome RNA in conditions that minimize the activity of the ribonucleases in plant sap. The spread of this mutant within plants differs from that of normal strains of TMV in that it seems to be unable to enter and spread through the vascular system; it spreads slowly through a plant, usually infects only the leaves above the inoculated leaf on the same side of the plant and one at a time, and it never spreads directly to the tip leaves. Thus it seems that for TMV at least, the uncoated genome RNA can spread from cell to cell through plasmodesmata, whereas only intact particles can spread through the phloem, but it is not yet known whether the intact particles are also able to spread through plasmodesmata even though this seems likely.

8.2. SIZES OF VIRUS PARTICLES AND GENOMES

When discussing the possible movement of viruses through the restrictive apertures of a plant such as the pores of sieve plates and plasmodesmata, it is worthwhile having some idea of their relative sizes (Fig. 8.1.).

The particles of most plant viruses, as seen, dry, in the electron microscope, are either:-

(i) isometric and between 18 nm and 80 nm in diameter, though most are 25-30 nm in diameter.

(ii) helically constructed rods 10 nm to 22 nm in diameter and up to 2.5 μm long, though mostly 300 nm to 900 nm long.

(iii) bacilliform and up to 80 nm in diameter and 400 nm long.

However, in aqueous suspension virus particles are hydrated, and may have associated with them 50-100% of their own weight of water. This will not greatly affect the overall dimensions of larger particles, but will mean that the apparent and effective dimensions of the small particles will be different. For example in dry paracrystals of TMV particles, the particle diameter as shown by X-ray diffraction is 15.2 nm, whereas in wet gels it is 21 nm (Bawden *et al.*, 1936; Bernal and

8.3.2.6. Other viruses Several individual or ungrouped viruses with isometric particles have been studied.

Some, such as belladonna mottle virus (Moline, 1973), cucumber mosaic virus (Honda and Matsui, 1974) and beet curly top virus (Esau and Hoefert, 1973) have no effect on plasmodesmata, and their particles have not been seen in plasmodesmata.

By contrast both carrot mottle virus (Murant, Roberts and Goold, 1973) and parsnip yellow fleck virus (Murant, Roberts and Hutcheson, 1975) have dramatic effects on the plasmodesmata of infected cells. The particles of carrot mottle virus are about 52 nm in diameter, and contain lipid, which they presumably acquire as they bud through the tonoplast. In infected cells, there are very long tubules attached to, and perhaps passing through, the plasmodesmata and sometimes sheathed with cell wall material (Fig. 8.2.), and these pass right through the cell, through the endoplasmic reticulum, and even indenting but not piercing the nucleus; these tubules are clearly visible in the light microscope. No particles or cores of particles can be seen in them, but frequently more particles are found budding into the tonoplast close to the free end of a tubule. Similar tubules up to 10 μm long and filled with particles occur in cells infected with parsnip yellow fleck virus.

8.3.3. Viruses with Bacilliform Particles

The largest and best-studied particles of this type are those of the rhabdoviruses which are 200-300 nm long and 60-90 nm in diameter. Lee (1967) commented that the particles of American wheat striate mosaic virus, a rhabdovirus, were too large to pass through the plasmodesmata of the host, and it seems that no-one has found particles or sub-viral particles of rhabdoviruses in plasmodesmata, nor have there been reports of unusually enlarged plasmodesmata in cells infected with viruses of this group. Hence the mode of intercellular transport of these viruses is a mystery, though Chambers and Francki (1966) did find particles of lettuce necrotic yellows virus in young xylem cells but not phloem cells, however the particles may have been produced by the cell before it differentiated.

8.4. ECTODESMATA, PLASMODESMATA AND VIRUS INFECTION

Ectodesmata are structures in the outer walls of leaf epidermal cells (Schumacher and Halbsguth, 1939; Lambertz, 1954; Franke, 1967, 1971); they are spaces that do not penetrate through the cuticle, and as they contain reducing substances they may be stained with either salts of silver or mercury or with other stains (Lambertz, 1954; Thomas and Fulton, 1968).

It has often been suggested that ectodesmata are the route by which normally inoculated virus partices enter and infect a leaf. Brants (1964; 1965) showed that various treatments, such as pressure, altered both the number of ectodesmata in leaves and the susceptibility of those leaves to virus infection. Similarly Thomas and Fulton (1968) have shown a close correlation between the numbers of ectodesmata and the virus susceptibility of leaves of different tobacco cultivars, when these were darkened, or immersed for short periods in warm water. However Merkens, de Zoeten and Gaard (1972) failed to find particles of potato virus X in the ectodesmata of inoculated leaves, but as there is at least a 10^5 fold difference in the 'ability' of the electron microscope and of a susceptible host to detect infective virus particles this result is not surprising.

Plasmodesmata may also be directly involved in the infection of plants by those aphid-borne viruses that are transmitted when migrating aphids probe plants briefly; the so-called 'non-persistent aphid-transmitted viruses'. Van Hoof (1958) and others have reported that aphids probe mostly between cells, when, presumably, the virus-containing sap they regurgitate (Garrett, 1973) comes into contact with broken plasmodesmata, however other workers (Lopez-Abella and Bradley, 1969) have reported contrary evidence.

8.5. VIRUS NOMENCLATURE

This section lists alphabetically the viruses and virus groups named in this chapter and gives some information about them to help non-virologists. More information about individual viruses can be obtained from the C.M.I./A.A.B. "*Descriptions of Plant Viruses*" published by the Commonwealth Agricultural Bureaux, Slough, England, U.K., since 1970.

Note that most plant viruses are named after the disease they cause in their commonest host, or that from which they were first isolated (e.g. tobacco mosaic virus), and many of the group names are derived by contracting the name of the type member (e.g. *toba*cco *mo*saic-like *viruses* to *tobamoviruses*), or are pseudo-classical (e.g. luteoviruses from the Latin *luteus* meaning yellow, and referring to the symptoms most commonly caused by members of the group).

In these notes some of the information for virus groups and ungrouped viruses is given in cryptograms; the meanings of the symbols used in this review are given below; a complete list is given by Gibbs (1969) and Wildy (1971).

Each cryptogram consists of four pairs of symbols (e.g. tobacco mosaic virus: R/1: 2/5: E/E: S/O) with the following meanings:

1st pair Type of nucleic acid/strandedness of nucleic acid.
Symbols for type of nucleic acid: R = RNA; D = DNA
Symbols for strandedness: 1 = single-stranded; 2 = double-stranded.

2nd pair Molecular weight of nucleic acid (in millions)/percentage of nucleic acid in infective particles.
This term gives the composition of infective particles. The genome of some viruses is divided, when different pieces of the genome occur together in one type of particle, the symbol Σ indicates the total molecular weight of the pieces in the particles (e.g. clover wound tumour virus; R/2:Σ15/20: S/S: S,1/Au), but when the pieces occur in different particles the composition of each particle type is listed separately (e.g. tobacco rattle virus: R/1: 2.3/5 + (0.6-1.3)/5: E/E: S/Ne).

3rd pair Outline of particle/outline of 'nucleocapsid' (the nucleic acid plus the protein most closely in contact with it).
Symbols for both properties:
S = essentially spherical
E = elongated with parallel sides, ends not rounded
U = elongated with parallel sides, end(s) rounded.

4th pair Kinds of host infected/kinds of vector.
Symbols for kinds of host:
I = invertebrate
S = seed plant
V = vertebrate

Symbols for kinds of vector:
Ac = mite and tick (Acarina, Arachnida)
Ap = aphid (Aphididae, Hemiptera, Insecta)
Au = leaf-, plant-, or tree-hopper (Auchenorrhyncha, Hemiptera)
Cl = beetle (Coleoptera, Insecta)
Di = fly and mosquito (Diptera, Insecta)
Ne = nematode (Nematoda)
O = spreads without a vector *via* a contaminated environment.

In all instances

* = property of the virus is not known
( ) = enclosed information is doubtful or unconfirmed
[ ] = enclosed cryptogram gives information about a virus group.

American wheat striate mosaic virus One of several viruses that cause striate mosaic in wheat and other Gramineae in North America. It is transmitted by leafhoppers and is a rhabdovirus.

Bean pod mottle virus A comovirus widespread in several legume crops in southern and eastern U.S.A. Transmitted by beetles in nature.

Beet curly top virus */*: */*: S/S: S/Au. A leafhopper transmitted virus of the western U.S.A., can infect several plant species. Isometric particles, 16 nm in diameter, are found in the nuclei of infected plants, but have not yet been characterized.

Beet western yellows virus A luteovirus of North America with a wide range of hosts. Causes damage in several crop species.

Beet yellows virus A klostervirus found in all the major sugarbeet growing areas of the world. Can cause great economic damage. Spread by aphids.

Belladonna mottle virus A tymovirus of solanaceous plants in Europe. Transmitted by beetles.

Carnation etched ring virus A caulimovirus of carnations.

Carrot mottle virus R/1: */*: S/*: S/Ap. One of the component viruses causing motley dwarf disease of carrots throughout the world. In mixed infections with carrot red leaf virus it is transmitted by aphids in the persistent manner. It has unusual lipid-containing particles, 53 nm in diameter, that bud through the tonoplast into the vacuole, and most closely resemble those of the mosquito-borne togaviruses that cause fevers and encephalitides of vertebrates throughout the world (e.g. yellow fever virus).

Cauliflower mosaic virus A caulimovirus of crucifers throughout the world.

Caulimoviruses [D/2: 5/15: S/S: S/Ap] A widespread group of about six viruses, which cause mosaic symptoms and infect a narrow range of hosts; named after cauliflower mosaic virus. Transmitted by sap and by aphids in the non-persistent manner. Its type member, cauliflower mosaic virus, was the first plant virus shown to have a DNA genome, which is circular and double-stranded. Caulimovirus particles are 50 nm in diameter, and resemble those of the papovaviruses of mammals in composition and appearance.

Cherry leaf roll virus A virus with a wide host range found in several perennial woody hosts in Europe and North America. Transmitted by nematodes, possibly a nepovirus.

Clover wound tumour virus This virus has been isolated only once and from a leafhopper. Infects certain leafhoppers, in which it causes no symptoms, and also a wide range of plants, in which it causes tumours of the vascular bundles. The tumours consist of pseudo phloem cells. Temporarily grouped with the reoviruses of animals, because of the similarities of their unusual particles.

Comoviruses [R/1: 2.3/34 + 1.5/23: S/S: S/Cl]. A cosmopolitan group of about ten plant viruses with restricted host ranges; named from cowpea mosaic virus. Transmitted by beetles, by sap inoculation and occasionally by seed. Their angular isometric particles are of three kinds; one, just the protein shell, the others containing the two pieces of the divided genome. Comoviruses share many properties with nepoviruses, and may be very distantly related to them.

Cowpea mosaic virus A common and economically important comovirus of cowpeas throughout the tropics.

Cucumber mosaic virus R/1: 1.3/19 + 1.1/19 + 0.8/19: S/S: S/Ap. A cosmopolitan virus with a particularly wide host range. Transmitted by aphids in the non-persistent manner, by seed, and, experimentally, by sap inoculation. Its genome is divided into three parts which are packaged in different particles.

Dahlia mosaic virus A world-wide caulimovirus.

Klosterviruses R/1: (2.3-4.3)/5: E/E: S/Ap. A widespread group of five to seven plant viruses with very flexuous particles 0.6-2.0 μm long. Most cause symptoms of yellowing with necrotic flecks. Some are transmitted in a semi-persistent manner by aphids, but only with difficulty by sap inoculation. Named from the particle shape; Greek, *kloster*, a thread.

Lettuce necrotic yellows virus An Australian aphid-transmitted rhabdovirus.

Luteoviruses R/1: 2/*: S/S: S/Ap. A world wide group of more than 20 viruses that cause economically damaging yellowing diseases to a wide-range of crops. Includes potato leaf roll and barley yellow dwarf viruses. Transmitted by aphids in the persistent manner, and confined to the phloem tissue of plants. Not transmitted by sap inoculation. Named from the symptoms caused; Latin *luteus*, yellow.

Maize rough dwarf virus A virus with many similarities to the reoviruses of animals. Found in maize throughout Europe, transmitted by leafhoppers.

Maize streak virus (R)/1: */*: S/S: S/Au. A virus which causes economic damage in several graminaceous crops in Africa and South-east Asia, particularly in maize. Transmitted only by leafhoppers. Its particles are about 20 nm in diameter, usually in closely appressed pairs measuring 30x20 nm.

Nepoviruses [R/1: 2.4/42 + 1.4/29: S/S: S/Ne] A widespread group of soil-borne nematode-transmitted viruses with very wide host-ranges, causing damage in temperate fruit crops. World-wide in distribution. The genome of the viruses is divided into two parts. Named from their properties; *ne*matode-borne *po*lyhedra particle *viruses*, *nepoviruses*.

Parsnip yellow fleck virus R/1: 3.7/*: S/S: S/Ap. An unusual virus of umbellifers in western Europe. Has isometric particles 30 nm in diameter. Transmitted by aphids in the semi-persistent manner, but only in the presence of a helper virus.

Potato virus X A potexvirus widespread in potatoes throughout the world.

Potato virus Y A potyvirus common in potatoes and other solanaceous plants throughout the world. Potato virus C is a variant of potato virus Y, that has lost the ability to be transmitted by aphids except with the aid of a helper virus such as potato virus Y.

Potexviruses [R/1: 2.3/6: E/E: S/O] A common and widespread group of viruses named after the type member. Very infectious, spreads when plants touch.

Potyviruses [R/1: 3.75: E/E: S/Ap] The largest known group of plant viruses with over 100 members. Cause mosaics and damage in a wide range of plants, though individual members may have restricted host ranges. Transmitted by aphids in non-persistent manner, some by seed, and all by sap inoculation. Named after the type member.

Reoviruses [R/2: Σ15/15: S/S: V,I,S/O, Au,Di] A group of viruses of plants, and animals. They produce particles about 50-80 nm in diameter which contains a double-stranded RNA genome that is divided into about 12 pieces, each of which codes for a single protein. Plant reoviruses are widespread, mostly in Gramineae. They are transmitted by, and infect, leafhoppers, and usually cause tumours in infected plants.

Rhabdoviruses [R/1: 3.5/2: U/E: V,I,S/O,Ac,Ap,Au,Di] A world wide group of viruses of plants and animals. They are named after their particles (Greek, *rhabdos*, a rod) which are bacilliform. The plant rhabdoviruses cause yellowing diseases and are transmitted by aphids or by leafhoppers and multiply in their vectors.

Rice dwarf virus A rice virus of the reovirus group found in Japan and Korea. It was reported by Takata in 1895 that rice dwarf disease was leafhopper-borne; the first report of a vector-borne virus, and later it was the first to be shown to be transmitted by a vector to its progeny.

Southern bean mosaic virus R/1: 1.4/21: S/S: S/Cl. A beetle-borne virus of beans from the southern U.S.A.

Strawberry latent ring spot virus A nepovirus found in the U.K.

Tobacco etch virus A potyvirus common in North and South America. Confined to Solanaceae in nature, but has a wider experimental host range.

Tobacco mosaic virus One of the best known and most fully studied viruses. In 1898 the pathogens causing mosaic of tobacco and foot and mouth disease of cattle were shown to pass through bacteria-proof filters, thus establishing a new group of pathogens which became known as viruses.

The third conclusion means in effect that the process determining the rate of symplastic transport will be diffusion through the plasmodesmata. This conclusion was based on incorrect reasoning, and is itself incorrect. A correct argument, given here, leads to the opposite conclusion, at least for charophyte plants, a possibility acknowledged by Tyree, Fischer and Dainty (1974) and referred to by Walker and Bostrom (1973).

9.3.1. Rate-of-Arrival

The junction of two *Chara* internodes (1 and 2) is shown diagrammatically in Fig. 9.1., which also represents the experimental arrangement. We consider first how a solute (e.g. ^{36}Cl) reaches the entrances of the plasmodesmata. The possible routes are: (i) by bulk movement of cytoplasm, (ii) by diffusion in cytoplasm and (iii) by transport across the tonoplast. These processes have the following consequences.

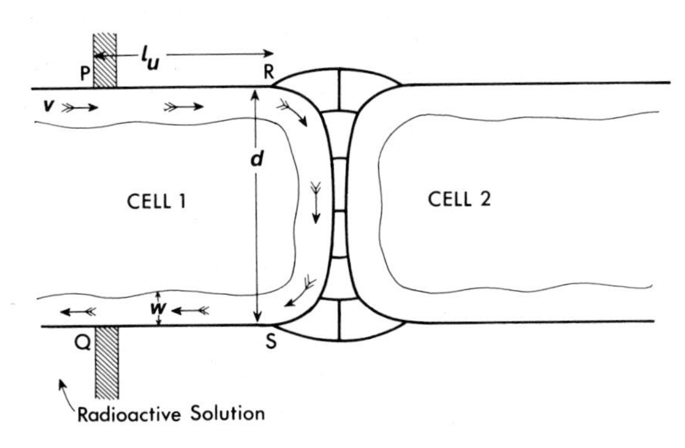

Fig. 9.1. Cross-sectional, diagrammatic view of a *Chara* node, illustrating the physical arrangement used in experiments

9.3.1.1. Lateral diffusion in the cytoplasm A ribbon of cytoplasm will, if exposed to a new environment on one side, come to 0.9 of equilibration in a time $t_w \simeq w^2/D$. Taking w to be 10 μm and D to be 2 x 10^{-9} m^2 s^{-1}, $t_w \simeq$ 50 ms. This can be compared with the time t_n for the transit of the cytoplasm across the node (from R to S in Fig. 9.1.), which if d is 1 mm and v is 100 μm s^{-1}, is 10 s. Thus there is plenty of time, during the passage of an element of cytoplasm across the node, for it to equilibrate with that in the next cell if the wall is permeable enough. In other words the cytoplasm will be well-mixed laterally by diffusion (but see Chapter 5).

9.3.1.2. Longitudinal diffusion and bulk flow The distance between the labelled solution that raises the specific activity of the cytoplasm and the node (PR in Fig. 9.1.) is 5 mm in the experiments to be quoted, and for this distance the diffusion equilibration time is $t_l \simeq l_u^2/D \simeq$ 12 ks, while the flowing cytoplasm traverses this distance in 50 s.

If we consider the transport of a solute from Q to S in Fig. 9.1., by diffusion and by bulk flow, the steady-state equation is:

$$\underline{\phi} = -v \frac{[c_S - c_Q exp(-v l_u/D)]}{[1 - exp(-v l_u/D)]} \qquad ....(1)$$

if $\underline{\phi}$ is positive from Q to S and v is positive from S to Q. Substituting the values already given for v, l_u and D, we get $exp(-vl_u/D)$ as 2.7×10^{-109}, so that essentially, whatever value c_Q has,

$$\underline{\phi} = -vc_S$$

as would be predicted from bulk flow alone. For transport from P to R the same conclusion holds, eq. (1) being appropriate with the sign of v changed. Thus there is no route for solute to reach S from Q except by streaming the long way round the cell.

9.3.1.3. The vacuole as source The contents of the vacuole are also brought up to the node by bulk flow, so that in principle a significant quantity of solute could cross the tonoplast into the cytoplasm compared with the quantity leaving it as it passes the node. But if we take the tonoplast flux to be of the order of 0.3 μmole m^{-2} s^{-1} for *Chara* (Hope and Walker, 1975) and the node flux to be 50 μmole m^{-2} s^{-1} (Bostrom and Walker, 1975) the tonoplast flux can hardly be significant. This is true *a fortiori* when we are considering experiments with recently-applied labelled solutes, which will have lower specific activities in the vacuole than in the cytoplasm. We can reach the same conclusion by comparing the maximum vacuolar contribution to the rate-of-arrival at the node, with the minimum cytoplasmic contribution:

$$R^* = s_c c(\pi d/2)wv + s_v \underline{\phi}_{vc}(\pi d^2/4) \qquad \text{....(2)}$$

Taking c for chloride to be at least 3 mole m^{-3} (Hope and Walker, 1975) and the other values as before, the ratio of the second term in equation (2) to the first is less than 0.08 (s_v/s_c). So in the following the vacuole is neglected as a direct source of solute for the node, as is longitudinal diffusion in the streaming cytoplasm. The rate at which labelled solute arrives at the node is taken to be given by the first term only in equation (2).

We have come then to a picture of a ribbon of streaming cytoplasm which picks up label as it streams from Q to P, exchanging it with the vacuole and plastids, and which loses it in passing the node. The effective isolation of the cytoplasm as it passes the node means that a sufficiently permeable node wall could completely equilibrate the cytoplasm here with that of the neighbouring cells - a conclusion contrary to that of Tyree (1970), who stated that the cytoplasm would be a well-mixed compartment if the streaming velocity were greater than the mean drift velocity of solute particles in the plasmodesmata. This condition is satisfied in *Chara*, but Tyree's conclusion is not correct because his argument neglected the fact that the same isolated ribbon of cytoplasm must sweep across the entrances of many plasmodesmata in succession.

9.3.2. The Counter-Current Node

The greatly simplified model of the node which will form the basis of this discussion is shown in Fig. 9.2a. It represents the internodal cells as rectangular in section, and it does not include the node cells which form a layer between the internodes. This latter is the more risky simplification, whose effect will be evaluated later.

We will consider a preparation of two cells, with no net flow of solute between them, but with an exchange of tracer. The tracer will for simplicity be regarded as a solute for which all fluxes are proportional to concentration. The node wall RSTU will be regarded as a membrane of permeability P given by $P = \underline{\alpha}D/l_p$. If we take a horizontal slice through the node wall between $x = \underline{x}$ and $x + dx$, it contains two strips of cytoplasm (shown end-on in Fig. 9.2b.) for each of which we can write a conservation equation:

$$v_1 b w_1 c_1(x) - v_1 b w_1 c_1(x + dx) - dJ_N = 0 \qquad \text{....(3)}$$

$$v_2 b w_2 c_2(x) - v_2 b w_2 c_2(x + dx) - dJ_N = 0 \qquad \text{....(4)}$$

where dJ_N is given by:

$$dJ_N = Pb\ dx\ [c_1(x) - c_2(x)] \qquad \text{....(5)}$$

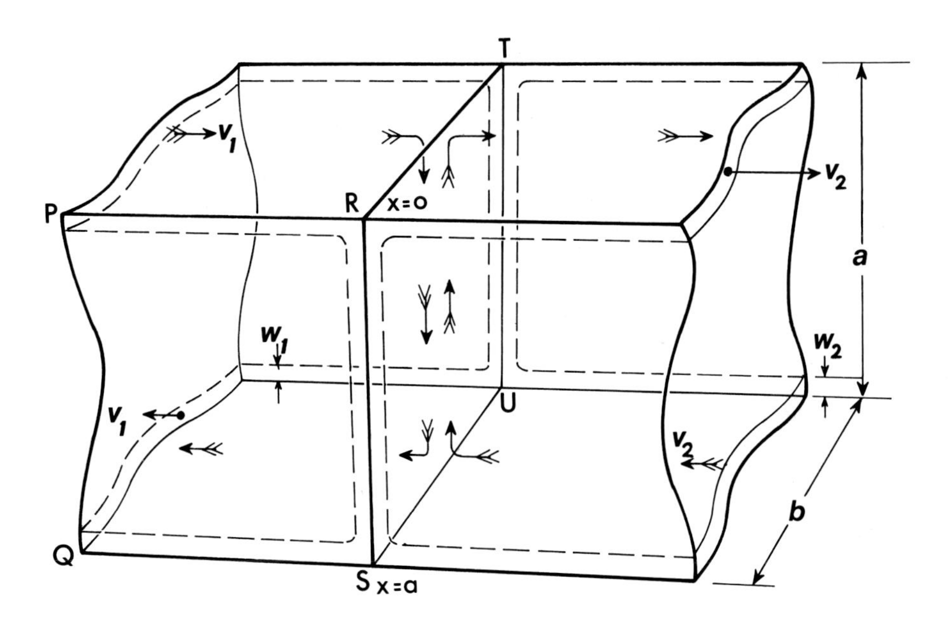

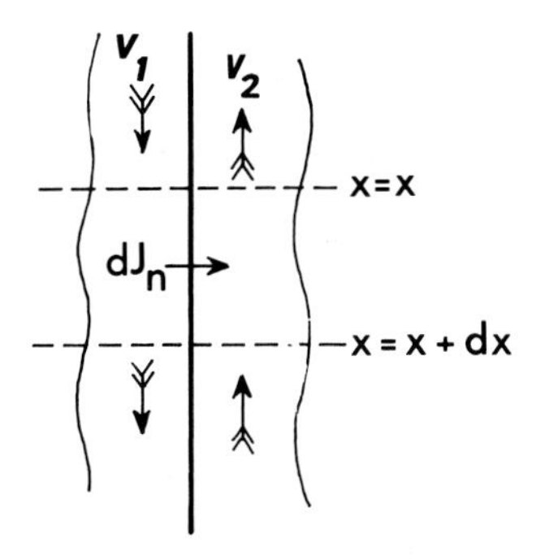

Fig. 9.2a. The model of the *Chara* node used in setting up equations. A stream of cytoplasm in the left-hand 'cell' flows from P to R, down the node wall to S, and thence to Q and around the entire cell. A similar stream in the right-hand cell moves up the wall

Fig. 9.2b. Section through the node wall and the two moving streams of cytoplasm

These equations can be integrated from $x = 0$ to $x = a$. With the simple boundary conditions $c_1(0) = c_o$ and $c_2(a) = 0$, the solutions for c_1 and c_2 read:

$$c_1(x) = c_o \frac{V_1 - V_2 exp[-Q(a-x)/a]}{V_1 - V_2 exp(-Q)} \qquad \text{....(6)}$$

$$c_2(x) = c_o \frac{V_1\{1-exp[-Q(a-x)/a]\}}{V_1 - V_2 exp(-Q)} \qquad \text{....(7)}$$

where we have replaced v_1w_1 by V_1, v_2w_2 by V_2, and $Pa(V_1 - V_2)/V_1V_2$ by Q. The transfer efficiency of the system, T_1, is given by

$$T_1 = [c_o - c_1(a)]/c_o \qquad \text{....(8)}$$

$$= \frac{V_2[1 - exp(-Q)]}{V_1 - V_2 exp(-Q)} \qquad \text{....(9)}$$

and the rate of transfer of solute from cell 1 to cell 2 is:-

$$J_N = c_o T_1 V_1 b \qquad \text{....(10)}$$

There is also a reverse transfer efficiency, T_2, in general different from T_1:

$$T_2 = \frac{V_1[1 - exp(-Q)]}{V_1 - V_2 exp(-Q)} \qquad \text{....(11)}$$

When the system is symmetrical $V_1 = V_2$, and the limiting values are:

$$c_1(x) = c_o \frac{P(a-x) + V}{Pa + V} \qquad \text{....(12)}$$

$$c_2(x) = c_o \frac{P(a-x)}{Pa + V} \qquad \text{....(13)}$$

$$T = \frac{Pa}{Pa + V} \qquad \text{....(14)}$$

$$J_N = c_o TVb \qquad \text{....(15)}$$

$$= c_o b \frac{PaV}{Pa + V} \qquad \text{....(16)}$$

The equations for c_1 and c_2 in the symmetrical case, (12 and 13), predict that each will have a linear gradient in the x direction; these gradients are equal and are given by $c_oPV/(Pa + V)$. Under the same condition of symmetry, equation (15) for J_N reduces to two limiting forms, in situations for which (i) $V >> Pa$ or (ii) $Pa >> V$:

(i) $J_N \simeq c_o bPa \simeq c_o PA$ ....(17)

(ii) $J_N \simeq c_o bV \simeq c_o bwv$ ....(18)

These cases correspond to J_N being limited (i) by diffusion in the plasmodesmata, or (ii) by the rate of arrival of solute, at the intercellular wall, in the stream of moving cytoplasm. Which of these cases obtains, for a given plant system, will be determined by the relative magnitudes of Pa and V; that is, by the relative magnitudes of $\underline{\alpha}aD/l_p$ and vw. The transfer efficiencies in these limited cases are given by

(i) $T \simeq Pa/V \simeq \underline{\alpha}aD/l_p vw$ ....(19)

(ii) $T \simeq 1.0$ ....(20)

Equation (20) shows the well-known property of counter-current systems, that all the solute in one stream may be transferred to the other.

In the unsymmetrical case there are also two limiting situations, corresponding to diffusion- or to streaming-limited transfer. Depending on the nature of the asymmetry in the streams, T may have limiting values less than 1.0 when the receiving stream is rate-limiting. If $V_2 < V_1$, and $V_1 < Pa$:

(ii) $T \simeq V_2/V_1$ ....(21)

9.3.3. The Co-Current Node

An exactly similar analysis, for a transfer system in which the streams run in the same direction down the transfer wall, shows that again there may occur diffusion-limited or streaming-limited transfer. In the latter situation T will be given by:

(ii) $T \simeq V_2/(V_2 + V_1)$ ....(22)

which for symmetry reduces to:

(ii) $T \simeq 0.5$ ....(23)

In the case of diffusion-limited transfer, co- and counter-current systems will give similar fluxes and transfer efficiencies.

9.4. THE REAL *CHARA* NODE

The internodal cells of charophyte plants do exhibit counter-current streaming wherever they meet at a node, whether in the main axis or in a lateral axis ('leaf'). There are several important reasons why the simple theory outlined here might not be applicable to the transfer across these nodes.

The actual node is hemispherical rather than flat and rectangular, so that the length of the path of streaming across the node (a) is not the same for different lateral elements of the cytoplasm. In case (i) we could calculate the rate of transfer (J_N) from PAc_o (equation 17) where A is now the area of the hemispherical surface; in case (ii) a does not enter the expression for J_N unless the value Pa at the edges falls far enough to produce diffusion-limited transfer. Also, the edges of the moving ribbons of cytoplasm are not isolated, as shown in Fig. 9.2a., rather they butt together at the 'white line'. Although Fritsch (1935) quotes a report that an inwardly projecting flange of cell-wall separates the two streams, no recent observer has found any sign of such a thing. So there is a direct path for solute to move from P to Q (in Fig. 9.2a.) without passing over the wall RS. This again affects the edge of the ribbon most, and would tend to lower T in the streaming-limited situation.

More seriously, the node itself interposes a number of small cells between the two internodes (Fritsch, 1935; Spanswick and Costerton, 1967; Fischer, Tyree and Dainty, 1974). Observation shows that in each of these cells the cytoplasm rotates around a single axis, but that the axes are very variously oriented in space (Fig. 9.3.). Between the counter-current flows then is not simply a wall perforated with plasmodesmata, but a layer of small cells of various flow directions (given in Fig. 9.3. for one particular isolated node) and speeds. As before, this will be significant only in the case of streaming-limited transfer. The quantitative effect in this case is not easy to analyse, but it seems possible that the limiting value for T might still be 0.5 or more.

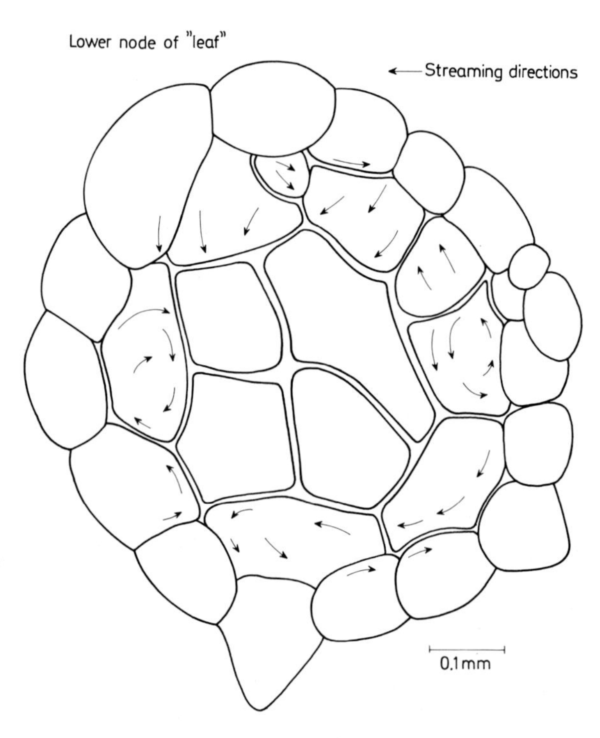

Fig. 9.3. End-view of an actual node of *Chara*, showing the directions of streaming observed in that node. When seen to stream, the central cells do so around axes parallel to the length of the internodes (perpendicular to the page)

The theory stated here will apply generally to all solutes (and to the solvent as well), since it has been quite general except in the argument about the contribution of the tonoplast flux to transfer at the node. Although specific values of fluxes were quoted here, the same argument probably applies to many different solutes.

The remainder of this paper describes the attempt to determine the rate of transfer of chloride across the node, and the attempt to determine, for chloride and for DMO[1], whether transfer is or is not streaming-limited, and what the value of T is.

It is taken for granted here that solutes reaching the plasmodesmata do so from the ground-plasm, and not directly from the endoplasmic reticulum. No association between endoplasmic reticulum and plasmodesmata has been seen in mature nodes of charophyte plants (Pickett-Heaps, 1967; Fischer, Tyree and Dainty, 1974; Bostrom and Walker, 1975; and Chapter 3).

9.5. EXPERIMENTAL EVIDENCE

The method by which transfer experiments are made is described in detail elsewhere (Bostrom and Walker, 1975). An isolated pair of joined internodes, from the main axis of a *Chara* shoot, is set in a chamber which bathes both cells in solution. One cell is bathed in a solution of a radioactive solute which does not make contact with the node or with the 5 mm of the cell surface nearest the node. At the end of the uptake period the distribution of tracer is determined: one determines the radioactivity of the sap and of the cytoplasm of the first cell, and the radioactivity transferred out of this cell to the node, the second internodal cell and to the unlabelled bathing medium surrounding it.

In a study of the effect of streaming speed on the rate of transfer of chloride, Bostrom has measured the fraction of the total chloride uptake transferred out of the first cell (Walker and Bostrom, in preparation). The speed of streaming was set by a pretreatment of the first cell with cytochalasin B for several hours at one of a number of low concentrations. This affects the streaming speed of the first cell more than that of the second cell or the node cells, though clearly at the higher speeds cytochalasin B will be transferred to all these cells too. Uptake of ^{36}Cl was allowed during the last 30 minutes. Streaming speeds of 6-110 $\mu m\ s^{-1}$ were obtained, although it was not always possible to hold the speed constant during an experiment. In 32 determinations, no correlation was found between total chloride uptake and streaming speed, but there was a significant correlation between the fraction transferred and the streaming speed. There was no sign that the fraction transferred might reach a limiting value at the higher speeds. This would have been expected if the rate of transfer were streaming-limited at low speeds but diffusion-limited at higher speeds. Thus it was concluded that the streaming speed needed to give diffusion-limited transfer ($v > Pa/w$) lies above 100 $\mu m\ s^{-1}$, the normal maximum speed. No estimate of the value of T has been attempted from these results, since this would require the measurement of the concentration of chloride in the flowing cytoplasm, for which there is no method free of doubt (see Walker, 1974; Hope and Walker, 1975).

If the transfer of Cl^- is streaming-limited, the same should be true of most small solutes. The value of D might depend on the electric charge if the plasmodesmata contain a high concentration of fixed charges, so that what is true of Cl^- might possibly not be true of small cations. In an attempt to measure T, work has been started on the rate of transfer of the weak acid DMO. This substance penetrates *Chara* cells readily, shows an efflux half-time of about 20 minutes, and

[1]DMO: 5-,5-dimethyl-oxazolidine-2-,4-dione.

distributes itself between cytoplasm and vacuole in proportions that at equilibrium reflect their local pH (Walker and Smith, 1975). It reaches concentrations in the cytoplasm some 30 times greater than those in the vacuole, so that the measurement of its concentration in the cytoplasm can be made without gross errors.

One cell of the pair was exposed to 30 μM ^{14}C-DMO for periods of 10-250 minutes and at the end of this period the preparation was cut up for counting. DMO penetrates so rapidly that much of the transferred tracer is found in the 4 ml of solution bathing the second cell: this solution acts as a sink for the transferred tracer. The time-course of the fraction transferred is shown in Fig. 9.4., each point representing one cell pair. Although these results are preliminary, it is possible to derive some conclusions from them.

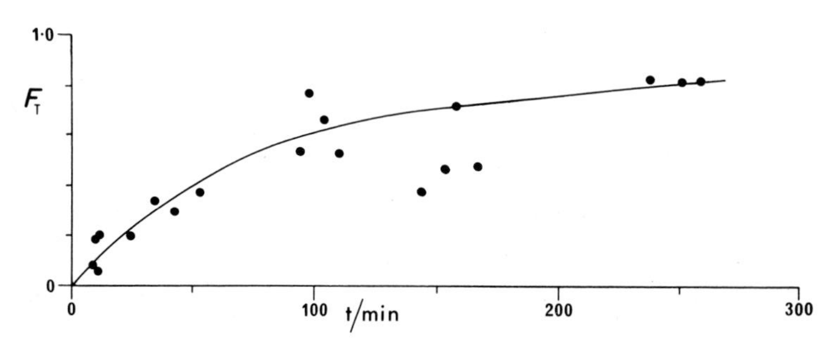

Fig. 9.4. Experiments with DMO: time-course of the fraction transferred to the second cell and its bathing solution (F_T). Each point represents one cell pair. The line is a calculation based on streaming-limited transfer at the node, with T = 0.6

If the vacuole of the first cell equilibrates rapidly with the cytoplasm, as is suggested by the efflux half-time of 20 minutes accounting for most of the radioactivity in the cell, and if the unlabelled solution and the second cell constitute a large sink, then there will be a quasi-steady state of transfer of DMO during which the content of the first cell is constant and the quantity transferred rises linearly with time. During this quasi-steady state, the fraction transferred is given by:

$$F_T \simeq [\underline{\phi}_{oc}A_c t - (A_cY_c + A_cY_v)]/\underline{\phi}_{oc}A_c t \qquad \text{....(24)}$$

where $\underline{\phi}_{oc}A_c t$ is, approximately, the total uptake, and A_cY_c and A_cY_v are the (approximately constant) quantities in the cytoplasm and the vacuole of the first cell. But in the quasi-steady state:

$$\underline{\phi}_{oc}A_c \simeq \tfrac{1}{2}T\pi dvwc \qquad \text{....(25)}$$

From (24) and (25):

$$(1-F_T)^{-1} = T\left[\frac{vwc}{2l(Y_c + Y_v)}\right]t \qquad \text{....(26)}$$

where w, now that we are considering a real *Chara* rather than, as previously, a theoretical model, is the width of the flowing *endoplasm*, so that wc does not equal Y_c, the quantity per unit area in the *whole* cytoplasm.

Equation (26) suggests that once the quasi-steady state is established, the plot of $(1-F_T)^{-1}$ against t should have a slope proportional to T, while the other factors are relatively constant, and can be determined. Only w is not determined directly; Bostrom (Bostrom and Walker, 1975) has found the total thickness of the cytoplasm in *Chara* to have a mean of 14.8 μm, so w is taken to be 7.4 μm while c is calculated from Y_c using the total thickness of 14.8 μm.

In the experiments of Fig. 9.4., the mean values for those points at times longer than 60 minutes are:

$l = 29 \pm 2$ mm

$d = 1.0 \pm 0.1$ mm

$(Y_c + Y_v)/Y_c = 3.1 \pm 0.6$

$v = 90 \pm 2$ μm s^{-1}

Slope of $(1-F_T)^{-1}$ against $t = (1.6 \pm 0.3) \times 10^{-2}$ min^{-1}, which yield a value for T of 1.1, with an accuracy of perhaps ± 30%.

An iterative simulation programme for a desk calculator has been set up[1], and used to check the approximate validity of this analysis. For a number of runs it was found that equation (26) held with an accuracy of about 10%. The best fit to the data of Fig. 9.4. is shown as a continuous line on that figure: it was calculated with $T = 0.63$.

Although these results are preliminary, they are a useful indicator that T lies between 0.5 and 1.0, and offer the hope of a reasonably accurate measure of T in the near future. The high value found requires that DMO^- transfer, like Cl^- transfer, be streaming-limited.

Bostrom and Walker (1975) have attempted to measure the rate of transfer (J_N) of chloride, in spite of the difficulty that the specific activity of chloride in the flowing cytoplasm is not accessible to measurement. They measured the time course of transferred ^{36}Cl, and for each cell pair determined the specific activity of the whole cytoplasmic sleeve at the end of the uptake time. The sleeve was prepared by the passage of an air-bubble through the cell (MacRobbie, 1964), and check experiments suggested that it contained less than 0.05 of the cell vacuole. They calculated an apparent rate of transfer of chloride, Φ_N, from this measured mean specific activity of the sleeve, and from the quantity of transferred radioactivity. The time course of Φ_N is shown in Fig. 9.5. It was expected that Φ_N would fall with increasing uptake time, reaching at long times equality with J_N: for the specific activity of the sleeve should be initially much lower than that of the flowing cytoplasm, and should approach it at long times as the plastids and the cytoplasm equilibrate. In the experiments of Series I, up to the maximum time of 360 minutes, there was little sign of Φ_N reaching a lower limit - the value at 360 minutes was 40 ± 20

[1]It is a straightforward matter to set up programmes for the iterative calculation of the transport of solutes in a pair of streaming *Chara* internodes, provided that the node can be treated as a wall with a transmission coefficient T. Programmes used by the author provide for the influx of solute to a moving cytoplasmic layer, its exchange with a stationary plastid layer and with a well-mixed vacuole, and its transfer across a node to a second cell. Copies of such a programme in the Fortran Language are available, as is a reduced version for the HP9810A calculator.

pmole s^{-1}. However in the second Series (II) extending the time to 720 minutes, $\underline{\Phi}_N$ seemed to have decreased little from its value at 360 minutes (75 ± 20 pmole s^{-1}) to that at 720 minutes (60 ± 10 pmole s^{-1}). These figures are suggested as upper limits on J_N, being too high if at 720 minutes the sleeve contains unequilibrated plastids or a significant amount of vacuole. It was suggested by the authors that the values of $\underline{\Phi}_N$ (40-60 pmole s^{-1}) might be overestimates of J_N by up to 10 times.

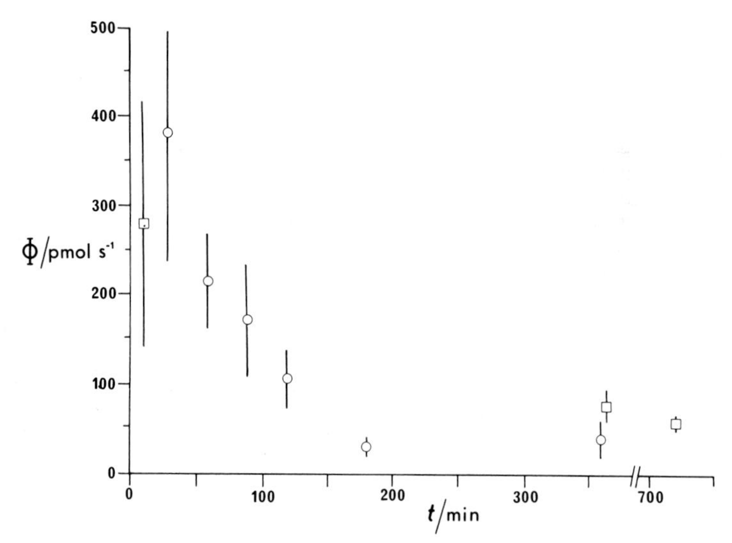

Fig. 9.5. Experiments with Cl^-: time-course of the apparent flux ($\underline{\Phi}_N$) from the cytoplasm of one cell to the node and next internode. Points are means for the number of cells shown, with standard errors of the mean. Round symbols, Series I; square symbols, Series II experiments

It is of interest that in the Cl^- experiments the value of F_T was essentially constant with time at 0.29 (Series I) and 0.45 (Series II), being lower only at the shortest time (10 minutes, Series II, 0.27). This reflects a fundamental difference between Cl^- and DMO^-. For Cl^- the sinks are the second cell, the vacuole of the first cell, and in effect the plastids of the first cell: for DMO^- the sink is the second cell and its bathing solution.

The dimensions of the plasmodesmata in the *Chara* cells studied are given in Table 9.1. (Bostrom and Walker, 1975).

9.6. DISCUSSION AND CONCLUSIONS

It has been assumed here that the movement of solutes in the plasmodesmata proceeds by diffusion in a medium that is basically aqueous. If we consider the intercellular wall at the end of the *Chara* internode, it should have a permeability $P = 3.7 \times 10^{-5}$ m s^{-1} for a solute whose diffusion coefficient was equal to that of KCl in water (2×10^{-9} m^2 s^{-1}). This would give a diffusion-limited rate of transfer of chloride of 400 pmole s^{-1} if the concentration in the cytoplasm is 10 mole m^{-3}, or 2.4 nmole s^{-1} if it is 60 mole m^{-3}. The rate of streaming-

TABLE 9.1.

DIMENSIONS OF PLASMODESMATA IN *CHARA* CELLS

Wall	Wall Thickness $l_p \times 10^{-6}$	Plasmodesma diameter $d_p \times 10^{-9}$	Fraction of wall area occupied $\underline{\alpha}$
Internode/central cell	2.3±0.1	110±4	0.043±0.007
Internode/peripheral cell	3.4±0.2	93±2	0.041±0.005
Node cell/node cell	2.0±0.2	97±3	0.027±0.003

limited transfer, according to the work quoted here, is 60 pmole s^{-1} or less. Thus there is no need to postulate a bulk flow mechanism in the plasmodesmata.

Measurements of the electrical resistance of the charophyte node give values of the 'hindrance factor' of about 350 (Spanswick and Costerton, 1967; Bostrom and Walker, 1975). It is not clear how this is to be proportioned among the various conducting species in the cytoplasm, nor how much charge-selectivity the plasmodesmata might show.

Although the measurements of the rate of chloride transfer do not at all closely define the value of the 'hindrance factor', D_w/D_p, the other experiments discussed here may help to define its value. The diffusion-limited transfer rate would be, if H is written for D_w/D_p,

$$J_N = \underline{\alpha} D_p A c / l_p$$

$$= \underline{\alpha} D_w A c / l_p H$$

$= (40c/H) \times 10^{-12}$ mole s^{-1}, if A is 1.4×10^{-6} m^2 and if $\underline{\alpha}$, corrected for the thickness of the plasma membrane, is 0.032.

The DMO experiments suggest that for chloride as well as DMO, $T = 0.6$, so the streaming-limited transfer rate should be:

$$J_N = T.\tfrac{1}{2}.\pi.dvwc$$

$$= (0.6\ c) \times 10^{-12} \text{ mole s}^{-1}$$

The latter value for J_N must be smaller, if the actual transfer rate is streaming-limited, so that

$40\ c/H > 0.64\ c$, whence

$$H < 63$$

If the electrical value of H is in fact 350 or thereabouts, we have a discrepancy of about a factor of 6.

Diffusive transfer would not be expected to show any inherent polarity; this is in line with the finding of Bostrom and Walker (1975) that the transfer of chloride did not depend on whether the apical or the basal cell of the pair were labelled.

The present discussion leads one to the conclusion that the *Chara* plant is indeed a syncytium, as Spanswick and Costerton (1967) have already said. More than half the solutes, and more than half the actual water molecules, in the stream of cytoplasm that approaches a node, enter it. In the real *Chara* plant these solutes will be distributed to the laterals ('leaves') as well as to the main axis, in proportions that remain to be determined.

An implication of the present work for studies of higher plants is that it clearly can no longer be assumed that each cell interior will be well mixed by cyclosis, with the major barrier to intercellular movement being the intercellular wall and its array of plasmodesmata. As a rough check on the situation, an estimate should be made of vw for comparison with $a\underline{\alpha}\underline{D}/l_p$; although it remains uncertain what value to attribute to D.

9.7. OPEN DISCUSSION

The discussion opened with challenges from GOODCHILD, who asked what direct evidence there was for plasmodesmatal transport of Cl^-, and GUNNING, who pointed out that every cell in a *Chara* plant was bathed in pond water that contains Cl^-, so why should the plant bother to translocate it? The same answer demolished both points: WALKER said that measurements of trans-plasmalemma fluxes of Cl^- had often been made for *Chara*, and that such fluxes could not cope with the rates of consumption observed in young, growing cells - hence *translocation* is a necessity; further, the rates of translocation calculated from growth rates or observed in experiments such as he had described would require trans-membrane fluxes far in excess of anything that had been measured - hence translocation *through a low-resistance pathway* is a necessity. He expressed his data as transfer rates from cell to cell and not on an area basis (it would be about 4×10^{-5} mole m^{-2} s^{-1}) because it is invalid to do so when the flux is not area-limited: in *Chara* it depends upon streaming velocity and cytoplasm thickness - to make the node bigger would have little effect in that the plasmodesmata already handle everything that cyclosis delivers to them. Apart from the argument based on fluxes, there is additional evidence that *Chara* translocates *in vivo*: starch is stored in the rhizoids and gets utilised, and there are growth correlations which imply movement of 'influences' from main growing points to lateral apices.

GIBBS suggested that if translocation is really important to *Chara* then an optimum balance between rate of streaming, the geometry of the node junctions, and the plasmodesmatal frequency and dimensions might have been worked out in the course of evolution. WALKER had not investigated this, but no doubt models could be set up to do so. FINDLAY thought that *Chara* would be far too silly to rationalise its construction in this way, but brought up the possibility that the plasmalemma near the node might have an unusually high permeability. ROBERTSON asserted that no cell would be silly enough to have uniform function over large expanses of membrane and cited examples of regional specialisation to reinforce this point, to which WALKER riposted that *Chara* may not have any brains but it has been around for 4×10^{8} years and so cannot be all that badly designed. He reiterated that, if translocation is to be accounted for by a regionally high trans-membrane permeability at the node, then fluxes higher by several orders of magnitude than any that had been found would be needed; all in all he had a teleological faith that *Chara* would not have put holes in its node walls if they were not useful for something.

MARCHANT brought the enquiry into the intelligence or otherwise of *Chara* to an end by asking how it was known that cytochalasin, when used to slow cytoplasmic streaming, was not also affecting the rate of transfer as a *direct* effect. WALKER said it was not possible to make this distinction, but there was certainly no evidence that the plasmodesmata harbour some sort of 'streaming machine' that might make them a selective target for cytochalasin - in fact, under normal circumstances the plasmodesmata *hinder* transport by a factor of about 60 as compared with what would be expected from their dimensions. This hindrance to Cl^- diffusion was itself a problem in that it is so much less than the hindrance that Spanswick and Costerton (1967) had observed for the passage of electric current through the equivalent walls in *Nitella*: obviously the assignment of realistic diffusion coefficients to solutes in plasmodesmata is a major problem, exacerbated in *Chara* because it is not plasmodesmatal diffusion that is limiting, but the velocity of cytoplasmic streaming.

This led on to questions about the role of streaming in other situations. GRESSEL pointed out that many cells are not much larger than the thickness of the cytoplasm in *Chara*: in such cases would not the distances involved be small enough for diffusion to be highly effective and streaming correspondingly less important? WALKER replied that this would be true for intracellular transport, but transcellular transport in the symplast must involve appreciable distances, and if there is no streaming the whole tissue will behave somewhat as a block of water, with time needed for diffusion proportional to the square of the distance. SMITH referred to Tyree (1970), who estimated that streaming would become important once the pathway exceeded about 40 μm. GRESSEL asked if cytochalasin had been used on tissues consisting of small cells to see if streaming really is important. WALKER cited Goldsmith and Ray (1973) who had found no effect of cytochalasin on the rate of auxin transport, but there was the difficulty of knowing how far the cytochalasin itself penetrates into a tissue - if it stops streaming in the outermost cells it may inhibit its own further progress into the tissue, where normal transport might thus continue. GUNNING described the work of Worley (1968) on transport of fluorescein absorbed or injected into cortical cells (small) and fibre cells (very elongated) in stem internodes: dinitrophenol reversibly stops streaming in both cell types, but plasmolysis shows that semi-permeability is retained; the rate of cell-to-cell transport of the dye is hardly affected by dinitrophenol in the tissue composed of relatively small cells, but is greatly retarded in the case of the long fibres, in fact, in the presence of the inhibitor the diffusion fronts advance uniformly in cortical tissue and adjacent fibres, where the cells are some twenty times longer. Worley claims to have seen similar plasmodesmata in the two tissue types by electron microscopy, and his conclusion that the rapid transport normally seen in non-inhibited fibres depends upon streaming, while the slower transport seen in the cortical cells does not, seems very reasonable. Earlier reports that transport of caffeine and lithium salts along *Vallisneria* leaves is independent of streaming (Kok, 1933, see 11.2.1.) are suspect in view of the likely participation of phloem transport. WALKER and CARR commented that it is clear that if rapid transport through the symplast is required, then elongated cells containing streaming cytoplasm, where there are small concentration drops at each cell junction, and no drop along the successive intracellular pathways, represent a highly efficient mode of construction - so once again *Chara* turns out to be not so silly after all.

10

THE ROLE OF PLASMODESMATA IN THE TRANSPORT OF WATER AND NUTRIENTS ACROSS ROOTS

A.W. ROBARDS[1] AND D.T. CLARKSON[2]

[1]Department of Developmental Biology, Research School of Biological Sciences, The Australian National University, Box 475, P.O., Canberra City, A.C.T. 2601, Australia[3]

[2]A.R.C. Letcombe Laboratory, Letcombe Regis, Wantage, England

10.1. INTRODUCTION

It appears probable that *all* cells of the meristematic region of plant roots are joined together by plasmodesmata. Clowes and Juniper (1968) have estimated that even the smallest meristematic cells would have between 10^3 and 10^5 connections. The patterns of loss and 'dilution' (through wall expansion) lead to the distribution of plasmodesmata found within the mature parts of the root system. Connections remain between living cells, but not through the walls of mature xylem vessel elements.

In studying movement of water and solutes from cell to cell, it is often impossible to distinguish between the contribution of two possible routes: (i) outwardly directed transport across the plasmalemma of one cell, through the intervening wall, and across the plasmalemma into an adjacent cell; or, (ii) movement from cell to cell *via* cytoplasmic connections - the plasmodesmata. Thus, work with higher plants has frequently been done with cells or tissues where there have been prior grounds for suspecting that intercellular movement of water and solutes *must* be confined to the symplastic pathway, and hence go *via* plasmodesmata. A case in point is the root endodermis of cereal plants which deposits an apparently impermeable layer of suberin over the whole internal face of the cell walls. Cortical cells of roots are well endowed with plasmodesmata, but it is not easily possible to attribute directly any particular flow to them because communication between them can be effected by the first of the two contributions (i) des-

[3]Written while A.W.R. was on Sabbatical leave from the Department of Biology, University of York, Heslington, York, YO1 5DD, England (address for correspondence).

cribed above (but see 10.5.1.2. and 10.5.1.3.). It is perhaps reasonable to suppose, however, that plasmodesmata which can be demonstrated to function in some situations, function similarly in others. Although it is hoped that much of what will be stated in this Chapter will be of general application, the detailed discussion necessarily draws on observations and results from relatively few species.

10.2. THE PLASMODESMATA OF ROOTS

10.2.1. Structure

The structure of young root plasmodesmata is generally of the 'simple' type (Fig. 2.4.; Helder and Boerma, 1969; Robards *et al.*, 1973), although a median nodule may develop later; in phloem cells typical branched plasmodesmata can be found. If phloem plasmodesmata are ignored (as they are not involved in the present discussion), then the plasmodesmata through the walls of all cells from the cortex to the xylem parenchyma appear to have a similar structure. A desmotubule is normally present, and is commonly seen to be associated with endoplasmic reticulum. It seems increasingly probable that meristematic and young plasmodesmata are narrower than the more mature structures: diameters of 25-45 nm are common for the young connections; 50-60 nm is more frequently quoted for the older strands (this difference seems largely to be a function of the size of the gap between desmotubule and plasmalemma) (Tables 2.1. and 2.2.). As will be seen later, and as has already been emphasized in 1.2.1., it is of critical importance that the true dimensions are established, as they have such a profound effect on calculations of the capacity of the plasmodesmata for transport. An interesting, and unresolved, feature is the tightness of the fit of the desmotubule within the plasmalemma-lined tube. In many species there appears to be a tight neck around the desmotubule at the ends of a plasmodesma (e.g. barley root or willow xylem ray cell plasmodesmata - Fig. 2.4.); in other plants (e.g. *Azolla* roots or *Abutilon* nectary hairs - Figs. 2.4. and 1.1.) the desmotubule is a much looser fit throughout. Once again, the different opportunities for symplastic transport offered by these two situations either through the desmotubule alone, or through both desmotubule *and* cytoplasmic annulus (Fig. 2.10.) are intriguing, but it seems certain that real and consistent differences in plasmodesmatal structure do exist: whether these are permanent or transitory is another interesting problem remaining to be clarified.

10.2.2. Distribution

If a radial file of cells is considered, running from root epidermis, through the cortex, across the endodermis and pericycle to the xylem parenchyma, then all of these cells are interconnected by plasmodesmata. There is a difficulty in calculating frequencies for cortical cells because such a large fraction of the wall surface abuts intercellular spaces, but cited data are of the order of 1-2 plasmodesmata μm^{-2}, although Helder and Boerma (1969) referred to their occurrence as 'rare'. It would be useful to obtain accurate frequencies for the cortex, taking into account the true cell to cell wall area.

Frequencies of plasmodesmata in endodermal cell walls have been comparatively well studied in relation to those of the root cortex. All endodermal walls have plasmodesmata, although whether there are generally significant differences between tangential and radial surfaces remains to be seen. Some authors (e.g. Helder and Boerma, 1969; Kurkova, Vakhmistrov and Solovyev, 1974) have stated the connections across radial longitudinal walls to be either very rare, or absent. We believe this to be an erroneous conclusion arising from sampling problems, as the frequencies of plasmodesmata across the wall are comparable with those in the other endodermal walls (Stamboltsyan, 1972 - tomato and wheat; Robards *et al.*, 1973 - barley; Klasova, 1975 - maize; Haas and Carothers, 1975 - maize; Pate, Gunning and Briarty, 1969 - *Pisum* root nodule endodermis; Table 2.1. - *Cucurbita*). There has been the further argument as to whether plasmodesmata traverse the Casparian band of the endodermal cell. This partly arises from the suggestion of Scott (1956), that the plasmalemma remains firmly attached to plasmodesmata in the radial wall during plasmolysis. This phenomenon is now known to be brought about by the firm bonding of the plasmalemma to the Casparian band itself (Bonnett, 1968; Robards *et al.*, 1973). There appears no *a priori* reason why plasmodesmata should *not* remain within the region of the wall destined to become the Casparian band, and recent micrographs do show connections within this zone (Robards *et al.*, 1973; Klasova, 1975; Haas and Carothers, 1975). The endodermal plasmodesmata are confined to pit fields which, in the case of those species that form tertiary endodermal walls, become the sites of pits. Endodermal cells are, therefore, in symplastic continuity with all surrounding cells, with a single exception. This exception occurs in plants where a protoxylem pole cell (in the pericycle) immediately abuts upon an endodermal cell (Fig. 10.1.). The protoxylem pole cells differentiate extremely rapidly (at least in barley - Robards and Jackson, 1975; Robards, in press) and, by 10 mm from the root tip are empty, functional, xylem vessel elements. The plasmodesmata appear to be lost at an early stage of differentiation in both protoxylem and metaxylem vessel elements, but whether this feature is the partial cause of, or effect from, vessel element differentiation is not yet known. Apart from the protoxylem cells, the other cells of the pericycle are joined by plasmodesmata to each other and to cells both external (endodermis) and internal (stelar parenchyma).

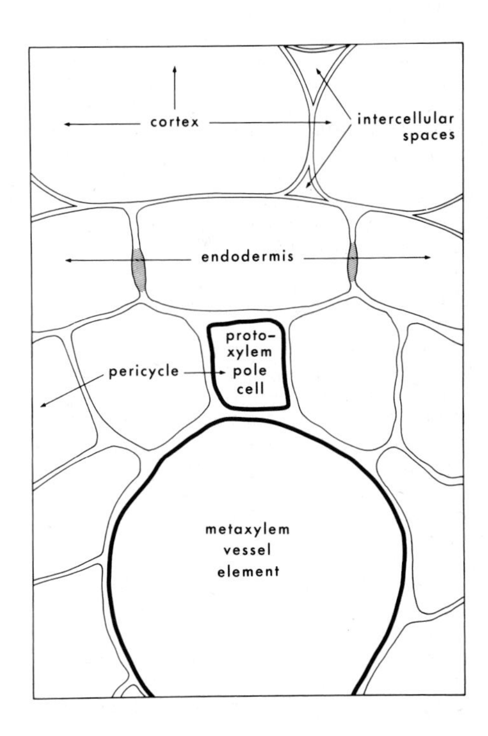

Fig. 10.1. The *critical zone* for the radial movement of water and ions from cortex to xylem in a barley root. The apoplastic barrier of the Casparian band (shaded) means that all solutes must enter the cytoplasm of the endodermis. This may be achieved either by symplastic transport from cortical cells, or by movement from outer endodermal/cortical apoplast to endodermal symplast. This latter route is blocked once the suberin lamella has been deposited between the endodermal wall and plasmalemma, and all radial movement through the endodermis is then presumed to be symplastic

An interesting, but as yet statistically unsubstantiated, feature of barley pericycle cells is the rather regular appearance of groups of plasmodesmata through the radial longitudinal walls closer to the internal than to the outer tangential wall. However, *no* firm evidence has been forthcoming to support the view that frequencies of plasmodesmata in the anticlinal walls of pericycle cells might reflect a tangential 'funnelling' of water and ions towards the protoxylem pole cells.

The xylem parenchyma cells of the root are also well endowed with plasmodesmata, linking these cells to each other and to pericycle cells (Läuchli *et al.*, 1974c). In connection with this last paper, it is relevant to comment here that the barley roots used by Läuchli *et al.*, seem to differentiate more rapidly than those that we have used. This means that it is not possible to make direct comparisons, between structure and function, at any given point. For example, in much of our work the metaxylem vessel elements have not fully differentiated by 10 mm from the tip, whereas Läuchli *et al.*, refer to such differentiated cells only 3-4 mm from the apex. These are obviously real variations, and they stress the need to relate physiological experiments to the structure of exactly the same root system under study.

The conclusion to be drawn from the distribution of plasmodesmata is that a continuous symplastic pathway is provided between epidermis and cortex, all the way through to the xylem parenchyma cells adjacent to the vessel elements. Essentially similar conclusions were reached by Stamboltsyan (1972). So far, no dramatic variations in frequency have been reported that would indicate the possibility of preferential flow in specific directions.

10.3. THE SYMPLASTIC COMPARTMENT

As described above, continuity of endoplasmic reticulum through plasmodesmata offers at least two theoretically possible channels of direct intercellular communication: either through the cytoplasm in general, or through the closed compartment bounded by the membrane of the endoplasmic reticulum connected *via* desmotubules. Whether either or both of these compartments are part of the symplast as envisaged by transport physiologists remains to be seen, but there is a growing awareness that ions in the cytoplasm may be located in several compartments (MacRobbie, 1971b; Pallaghy, Lüttge and von Willert, 1970). The view that the symplast is a compartment within the cytoplasm rather than the whole of the cytoplasm does not, therefore, invoke any new concepts (see also 2.5.). It has been suggested that small vacuoles or pinocytotic vesicles formed at the plasmalemma may transport ions directly, either to the vacuole, or to the endoplasmic reticulum, without mixing their contents directly with the bulk cytoplasm (MacRobbie, 1969; Baker and Hall, 1973). The endoplasmic reticulum in such a model is envisaged as a general distribution network through which ions can move rapidly about the cell; if, indeed, there is an intimate association of the endoplasmic reticulum and the plasmodesmata, this idea can be extended readily to the whole of the symplast.

Pallaghy *et al.* (1970) concluded, from a study of elution rates, that parallel pathways exist for fluxes of ions across the cytoplasm between the vacuole and the external medium; these authors speculated that transport of part of the ion flux occurred in vesicles which were derived from endoplasmic reticulum.

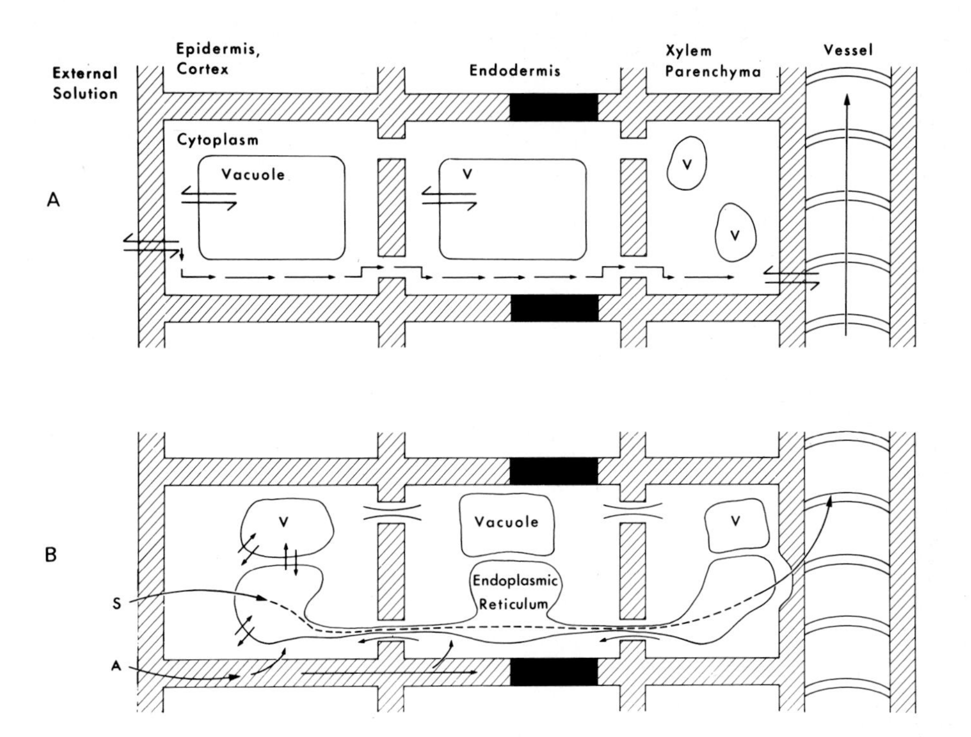

Fig. 10.2A. A conventional, physiological, model of a plant root showing transport between cells *via* open cytoplasmic channels. (Redrawn from Lüttge, 1974). The Casparian band in the anticlinal endodermal walls is shown in black

Fig. 10.2B. Some additional opportunities for symplastic (S) or apoplastic (A) transport are suggested in a modification of the above model, showing a hypothetical concept of a symplastic compartment (the endoplasmic reticulum) which may be continuous from cell to cell *via* plasmodesmata. A similar model has recently been put forward by Bräutigam and Müller (1975b) in relation to transport through *Vallisneria* leaves

Plant physiologists have usually shown plasmodesmata as completely open channels of communication between cells (Fig. 10.2A.). The firm knowledge that most root plasmodesmata are traversed by a desmotubule which may be connected to endoplasmic reticulum encourages the proposal of an alternative hypothesis for symplastic compartmentation as shown in Fig. 10.2B. A similar model has been proposed by Bräutigam and Müller (1975b) to support their demonstration that α-aminoisobutyric acid moves symplastically through *Vallisneria* leaves at rates greater than would be expected from diffusion alone.

10.4. THE ROOT ENDODERMIS

The roots of higher plants have a cylindrical layer of cells - the endodermis - which is located between the cortex and the stele (Fig. 10.1.). The walls of endodermal cells undergo changes that are generally considered to reduce their permeability to water and solutes:

firstly (State I), a band of suberin and lignin (the Casparian band) develops in all radial (anticlinal) walls; later, (State II) a suberin lamella may be deposited over the whole internal wall surface of the endodermal cells; a thick cellulosic wall is produced internal to the suberin lamella in some species (State III) - this wall layer does not appear to be of great direct importance to physiological function, as pits through it coincide with the pit fields of State I and State II. The Casparian band appears to block movement towards the stele through the free space of the cell walls (apoplast), therefore any solutes that are to move further inward must traverse the endodermal plasmalemma, and so enter the symplast.

The suberin lamella is thought to restrict severely the movement of water and ions from the apoplast by placing a barrier of low permeability between the free space and the plasmalemma of the endodermal cells (see Robards and Robb, 1972; 1974, for evidence of this from the use of electron-opaque apoplastic tracers; also see Clarkson and Robards, 1975, for a fuller discussion of this point). The actual permeability of the suberin lamella remains a matter of conjecture, but recent work to isolate the endodermis (Robards, Payne and Gunning, in press) should lead to a better understanding of both chemistry and permeability of root endodermal suberin. There is, however, much evidence to show that water and some ions move across endodermal cells with fully elaborated suberin lamellae and it has therefore been suggested that the observed transport across the endodermis is *via* the plasmodesmata (Clarkson and Robards, 1975).

In addition to the normal constriction of the desmotubule by the plasmalemma in the neck region of root plasmodesmata, the formation of the suberin lamella appears to create a further constriction on each traversing plasmodesma (Karas and McCully, 1973; Robards *et al.*, 1973, Fig. 27; Haas and Carothers, 1975); a similar feature occurs where plasmodesmata are found through suberized layers of leaf bundle sheath cells (O'Brien and Carr, 1970; Kuo, O'Brien and Canny, 1974).

It has been noted that the suberin lamella may not form over the Casparian band, or its development is delayed in this region (e.g. Klasova, 1975; Haas and Carothers, 1975). However, providing that the seal between the plasmalemma and Casparian band is a good one, no functional difference would be expected between an endodermal cell where the suberin lamella continued across the Casparian band, and one where the lamella was tightly abutting the edges of the band. A different situation would prevail in those plants where an incomplete suberin lamella is deposited (Priestley and Radcliffe, 1924).

Haas and Carothers (1975) have published a detailed appraisal of endodermal development in *Zea mays* roots: there is a close correlation between the observations of these workers and those of Robards *et al.* (1973) on *Hordeum vulgare*.

10.5. THE MOVEMENT OF WATER AND IONS ACROSS THE ROOT

10.5.1. Some Published Assessments

10.5.1.1. Helder and Boerma Helder and Boerma (1969) considered the possibility that plasmodesmata may be significant in intercellular transport across the inner tangential endodermal wall (ITEW) of barley roots. They calculated that such a single wall is traversed by about 20,000 'simple' plasmodesmata, with tight necks, an average length of 0.25 μm, and a combined desmotubular area of about 5.0 μm^2 (this is, in fact, the total area occupied by all the desmotubules, *including their walls).* Previous studies (Helder, 1964) had revealed that approximately 3×10^{-4} μeq chloride ions passed across this wall per day. Making the assumption that the transport involves simple diffusion through the desmotubules, and that the diffusion constant is the same as for normal solutions, they state that *"the concentration gradient from one end of the tubule to the other can be calculated. We obtained a value of about 0.1 μeq per litre".* This concentration gradient (or, presumably, 'difference') across the desmotubule is very small indeed. However, reduction of free diffusion in the desmotubules, say Helder and Boerma, would necessitate an increased concentration gradient (difference) and, after citing evidence from Spanswick and Costerton (1967) relating to the possible restrictions upon ion flow through plasmodesmata, they conclude that *"one may even feel that diffusion is almost impossible and that permeability may be as low as that for the cell membrane. If this is correct, the tubules would constitute a serious barrier for intercellular transport and substances taken up into the cisternae of the endoplasmatic reticulum would become sequestered. (O'Brien and Thimann 1967a). As a consequence the plasmalemma would remain as the only means of intercellular transport".* The plasmalemma is considered to function both by absorption across the membrane as well as by translocation along its inner surface; the authors thus minimize the role of the plasmodesmata, and merely refer to the potential significance of their valve-like structure.

Despite the pessimism of Helder and Boerma, the majority of the most recent papers come to conclusions supporting the idea of symplastic transport, whether or not this takes place through open plasmodesmatal pores. Jarvis and House (1970), for instance, plasmolyzed maize root cells with mannitol and suggested that the reduced rate of ion transport could be accounted for by the severance of symplastic connections. In a similar, indirect, manner Pitman *et al.* (1970) suggest that their results, from electrical potential difference experiments on barley roots, could be explained by assuming that leakage of ions from freshly cut root surfaces could be through plasmodesmata exposed to the external solution. Subsequently, a reinstatement of a higher potential difference could arise from sealing of the exposed plasmodesmata. These authors clearly assume a functional role in ion transport for plasmodesmata in forwarding this explanation.

10.5.1.2. Tyree No assessment of the importance of plasmodesmata from cell to cell across the root would be complete without reference to the paper of Tyree (1970), whose main, relevant, conclusions were:-

(a) plasmodesmata constitute the pathways of least resistance for the diffusion of small electrolytes and non-electrolytes;

(b) diffusion will be the predominant mechanism of transport for small non-electrolytes, and also for small electrolytes, provided that there are no substantial electrostatic forces within the pore;

(c) solute distribution in the bulk cytoplasm ought to be a combination of self-diffusion and cyclosis;

(d) there would be almost perfect mixing of the bulk cytoplasm (but see Chapter 5); and

(e) there can be significant coupling of water transport to the symplastic transport of solutes.

Tyree, as opposed to Helder and Boerma, clearly considers that plasmodesmata are functional. However, his data were drawn from previously published works of other authors concerning widely different plants, methods and interpretations; they thus represent only very crude 'average' estimates of the essential parameters (see also Table 2.1.). For example, his calculations of concentration differences across onion root plasmodesmata (p. 204), are based upon a pore radius of 44 nm, which is probably twice as great as the true situation. Again, despite his recognition of the presence of a desmotubule in many plasmodesmata of higher plants, Tyree's assumption of plasmodesmatal structure is generally of a pore of radius 30 nm or 60 nm traversed by a central rod (*not* a desmotubule) of radius a mere 3 nm (p. 192). Hence the lumen and wall of the desmotubule appear to be ignored. The consequences of such assumptions can be important. For example, Tyree uses the 44 nm pore radius which leads to a relative pore area of 0.009, calculated from the shadowed onion cell wall preparations of Scott *et al.* (1956) and the data of Strugger (1957b) (Tyree, p.204). However, assuming the same plasmodesmatal frequency as Tyree, of 1.5 μm^{-2}, and more reasonable estimates of plasmalemma-bound, or desmotubule-bound, pores of 20 or 10 nm diameter respectively, the relative area would, in fact, be only 0.0005 or 0.0001.[1] Tyree goes on to calculate the change in solute concentration that would be needed to maintain experimentally observed salt (K^+) fluxes between root cells, and concludes that a difference of about 0.1 mN (= 0.1 mM or 0.1 mole m^{-3}) per cell would be needed; and so, perhaps, a change of approximately 1.0 mole m^{-3} over a whole file of about ten cells from cortex to stele. Using the lower relative pore areas of 0.0001 or 0.0005, while keeping other parameters the same as used by Tyree, the concentration difference (ΔCs) required between two cells would be 7.0 or 2.0 mole m^{-3} for the smaller or larger pore radii respectively, as opposed to Tyree's 0.1 mole m^{-3}. A file of ten cells would therefore need a difference of about 70 or 20 mole m^{-3}, a level that could still reasonably be expected to be maintained within the root, although considerably higher than Tyree's estimate.

Another feature highlighted by Tyree is the difficulty of estimating the viscosity of the pore fluid which, he assumes (ignoring any desmotubule), is "*much like the bulk cellular cytoplasm*", and to which he assigns the rather high values of 0.5 or 2.0 poise. These correspond to: (i) the assumed minimum viscosity of the pore contents; and (ii) a higher value, having regard to the possibility for partially gelled cytoplasm. Clarkson, Robards and Sanderson (1971) used the lower of Tyree's figures, as well as a cited viscosity of vacuolar sap (2×10^{-2} poise) in their calculations. The problems of determining intracellular viscosity have been referred to by Cooke and Kuntz (1974), who quote values of 14×10^{-2} and 6×10^{-2} poise for yeast and *Euglena* respectively. If the desmotubule contains part, or all, of any transcellular fluxes, then the viscosity of this compartment, which may well be yet lower than the previous estimates for cytoplasm, will need to be determined.

---

[1] In the examples given here, and those that follow, two estimates of pore diameter have been used: 10 and 20 nm. The former is a reasonable assumption for the lumen of the desmotubule (but ignoring any central rod), while the latter is an arbitrarily chosen figure for a narrow, unimpeded, plasmalemma-bound pore. To the extent that the exact nature and dimensions of the conducting pores remain unknown, neither estimate can have any direct bearing on our present understanding of physiological function, but they do allow calculations to be made to evaluate the possibilities for transport through pores of a size that may well exist in nature.

Despite such considerations, Tyree's conclusions were both valuable and pertinent. It seems that the effects of his over-estimate of pore size may have been diminished by the choice of high values for viscosity. His calculations demonstrate that plasmodesmata could indeed function in symplastic transport, but at least two interesting facets need to be investigated further: the endoplasmic reticulum is commonly seen and referred to within plasmodesmata, so posing the question of whether the cavity of the endoplasmic reticulum constitutes a symplastic pathway (an aspect totally ignored by Tyree, although raised previously by O'Brien and Thimann (1967a)); and, in some cases, the bulk flow of water or solute flux between cells is so great that an explanation in terms of diffusion alone must be questioned. Perhaps these two observations can be combined. If there is bulk flow between cells, then it is difficult to envisage this occuring through the highly vacuolate cells of the root without disruptive effects. If, however, such symplastic transport were to be confined to an internal compartment, then the process might more easily be understood (Fig. 10.2B.). The narrower pore size would imply considerably higher rates of flow but, as will be seen (10.5.2.; 10.5.3.), not unreasonable ones. It is not possible to state unequivocally that bulk flow is required to explain experimentally determined rates of transport across roots: diffusion is a highly efficient process, particularly if the narrow pores through the walls are not too long (1.2.1.).

10.5.1.3. Ginsburg and Ginzburg H. Ginsburg and B.Z. Ginzburg have published a series of papers concerning water and solute movement, and electropotentials, in roots. They have mainly used *Zea mays*, and have commonly employed the technique of removing the stele from the cortex to produce cylindrical 'cortical sleeves' (the fracture occurs through the anticlinal walls of the endodermal cells). Using such 'sleeves', suitable connections can be made *via* polyethylene tubes so that the inside and outside of the cortical tube can be exposed to different solutions (e.g. of different osmotic pressures). In relation to water, Ginsburg and Ginzburg (1970a) conclude "*....it is probable that water passing radially across the root moves by some route additional to the cell walls. It is possible that an important factor accelerating water movement is a convective flow, e.g. of the cytoplasm, coupled to some chemical reactions*". Similarly (1970b), they consider that radial movement of ions must be through, rather than between, cortical cells; and that anions are pumped, whereas cations move passively (although they admit the possibility that a cation pump could exist, for example, within the intact endodermis which is lost from their preparations). The active transport of water is reiterated by Ginsburg (1971) and Ginsburg and Ginzburg (1971) who envisage "*water reaching the endodermis or xylem tubes to be discharged by a shrinking of the gel part of the cytoplasm in the neighbourhood of the plasmodesmata*". Such a scheme would not be in accord with the ideas of other workers (see Anderson, 1975c) who believe that water flow normally follows solute movement; neither would it be easy to understand how shrinkage of cytoplasm could bring about a directional flow of water through the plasmodesmata of a radial file of cells.

The importance of relating such results to known structures (and also the difficulties of interpreting data from 'wounded' tissues) is apparent, and is equally relevant in the work of Ginsburg (1972), and Ginsburg and Ginzburg (1974) on root electropotentials. In the first paper a model is proposed for the analysis of steady-state electropotentials in plant roots. Radial symplastic continuity is assumed, although some curious estimations of relative pore (plasmodesmatal) areas through epidermal and endodermal walls arise from the comment by Helder and Boerma (1969) that plasmodesmatal frequencies in cortical cells are much lower than those in endodermal cells - a statement that is probably inaccurate in relation to true *intercellular* wall area (see 10.2.2.). However, in Ginsburg and Ginzburg (1974) it is concluded, among other things, that the potential difference profile across the root indicates that all cytoplasmic phases are equipotential, which suggests a continuity within and between cortical and stelar symplasms. This last statement is an important one in relation to proposed ideas of symplastic continuity and depends heavily on a knowledge of where the monitoring electrode is lodged within a cell. (Bowling (1972) and Jones, Novacky and Dropkin (in press) have also been unable to find any significant gradient in potential difference between the epidermis and stele of the roots of *Helianthus annuus* or *Impatiens balsamina* respectively). Mature cortical and endodermal cells are highly vacuolate; pericycle and xylem parenchyma cells slightly less so. One could imagine an equipotential situation *either* if the symplastic pathway is completely open (Fig. 10.2A.) with the electrode in the 'cytoplasm'; *or* if the plasmodesmata are sealed (Fig. 2.10C.), but with the electrode in the symplastic compartment (endoplasmic reticulum?). This emphasises the need for electrophysiological work to be carried out with closely correlated, and detailed, structural studies.

The conclusions to be drawn from the papers of Ginsburg and Ginzburg are ones of general support for the concept of symplastic continuity across roots and, particularly, across the cortex.

TABLE 10.1.

BARLEY ROOTS - DIMENSIONS, MEASUREMENTS AND ASSUMPTIONS USED IN CALCULATIONS FOR TABLES 10.2. - 10.5. AND 10.7.

Inner diameter of endodermal cylinder of barley seminal root axis	200 μm
Pit diameter	1.6 μm
Endodermal cell length	206 μm
Endodermal cell breadth (tangential)	21 μm
Endodermal cell breadth (radial)	8 μm
Pit frequency per square micrometre	0.0194 μm
Plasmodesmata per pit	54 μm
Length of plasmodesmata	0.5 μm
Diameter of plasmodesmatal pore*	
a) enclosed by desmotubule (excluding central rod)	10.0 nm
b) enclosed by plasmalemma (at 'neck')	20.0 nm
Uptake of water by a 3.5 mm long segment of barley seminal root axis, 440 mm from root tip (all endodermal cells at State III)**	5.56×10^{-14} m^3 s^{-1}
Estimate of plasmodesmatal pore viscosity***	
a) upper limit	50×10^{-2} poise
b) lower limit	2×10^{-2} poise

* See Footnote 1 in 10.5.1.2. for a comment on the choice of these dimensions.
** See Clarkson, Robards and Sanderson (1971), Table 2.
*** See 10.5.1.2. for a discussion of pore viscosity.

10.5.1.4. Clarkson, Robards and Sanderson Another contemporary attempt to relate plasmodesmatal structure and frequency to observed rates of intercellular movement was made by Clarkson, Robards and Sanderson (1971), using barley roots, in which transport of water and phosphate ions from the external solution into the vascular stele across a thick-walled endodermis could be demonstrated. The view was taken by Clarkson *et al.* that the States II and III endodermal walls are probably impermeable, or only slightly permeable to water and solutes, and that, as the plasmodesmata are the only apparent channels of communication, the observed flows of water and phosphate ions must have passed through them.

10.5.2. Water

Some data from Clarkson *et al.* (1971) have been revised, and are presented here (Tables 10.1.-6.). The slightly higher plasmodesmatal frequency across the ITEW (1.05 plasmodesmata μm^{-2}, compared with the earlier 0.67) arises from an improvement in the estimation of pit frequency obtained by using isolated endodermal cells (Robards, Payne and Gunning, in press). Tables 10.1. and 10.2. simply state the data and their simple derivations in relation to water movement across the endodermis. By using the Poiseuille formula (which, while open to criticism, is the best method at present available - 1.2.1.) the pressure needed to sustain the observed rates of flow can be calculated for limiting cases

TABLE 10.2.

BARLEY ROOTS - DERIVATIONS OF DATA FROM TABLE 10.1.

Surface area of inner tangential endodermal wall (ITEW) of a 3.5 mm long segment of barley root	$2.2x10^{-6}$ m^2
Number of pits in a 3.5 mm segment (ITEW)	$4.28x10^4$
Floor area of 1 pit	$2.01x10^{-12}$ m^2
Number of plasmodesmata in a 3.5 mm segment (ITEW)	$2.31x10^6$
Surface area of plasmodesmatal pores	
a) enclosed by desmotubule	$7.85x10^{-17}$ m^2
b) enclosed by plasmalemma	$3.14x10^{-16}$ m^2
Total surface area of pores in a 3.5 mm segment (ITEW)	
a) enclosed by desmotubule	$1.81x10^{-10}$ m^2
b) enclosed by plasmalemma	$7.25x10^{-10}$ m^2

$5.56x10^{-14}$ m^3 s^{-1} water cross the inner tangential wall of State III endodermal cells. If this passage is restricted to the plasmodesmata, then the total flow of water through them is	
a) (desmotubule)	$3.07x10^{-4}$ m^3 m^{-2} s^{-1}
b) (plasmalemma)	$7.67x10^{-5}$ m^3 m^{-2} s^{-1}
or the flow through each single plasmodesma is	$2.41x10^{-20}$ m^3 s^{-1}

of pore radius and viscosity. Only in the case where minimum pore size is combined with maximum viscosity does the pressure appear untenably high (Table 10.3.). (Difference in salt concentration to account for the pressure drop may also be calculated as a crude approximation, and varies between about 3.0 millimolar and 1.0 molar (3-1,000 mole m^{-3}) for cases I to IV). The hydraulic conductivity of *individual* plasmodesmata is much greater than that of an equivalent area of membrane, but the conductivity of the overall plasmodesmatal pathway through the ITEW is in the same range as for the plasmalemma of plant cells (Table 10.4.). Thus the ITEW plasmodesmata are too few and too small to do more than merely compensate for the presence of the impervious cell wall.

TABLE 10.3.

PRESSURE DIFFERENCES NEEDED TO SUPPORT WATER FLOW THROUGH THE ENDODERMAL PLASMODESMATA OF BARLEY

The pressure difference to support the rates of flow (Table 10.2.) may be calculated (Poiseuille), using two estimates of viscosity, and the two different pore diameters.

Case	r (nm)	η (poise)	ΔP (bar)
I	10	$2x10^{-2}$	$6.06x10^{-2}$
II	10	$5x10^{-1}$	$15.10x10^{-1}$
III	5	$2x10^{-2}$	$9.69x10^{-1}$
IV	5	$5x10^{-1}$	$2.42x10^{1}$

TABLE 10.4.

HYDRAULIC CONDUCTIVITIES OF THE PLASMODESMATAL PATHWAY THROUGH THE INNER TANGENTIAL ENDODERMAL WALL OF BARLEY ROOTS

The calculated pressure differences (Table 10.3.) may be used to determine the hydraulic conductivity of the plasmodesmatal pathway.

Case	r (nm)	ΔP (bar)	Hydraulic conductivity (Lp) - (m s^{-1} bar^{-1})
I	10	$6.06x10^{-2}$	$4.18x10^{-7}$
II	10	$15.10x10^{-1}$	$1.68x10^{-8}$
III	5	$9.69x10^{-1}$	$2.61x10^{-8}$
IV	5	$2.42x10^{1}$	$1.05x10^{-9}$

Whereas the data of Clarkson *et al.* (1971) show hydraulic conductivities on the basis of individual plasmodesmata, the values here take frequency into account and it emerges that the hydraulic conductivity of the plasmodesmatal pathway is in the same range as for the plasmalemma of plant cells (e.g. $1\text{-}3x10^{-7}$ m s^{-1} bar^{-1} for *Nitella* (Dainty and Ginzburg, 1964); $2x10^{-11}$ m s^{-1} bar^{-1} for *Zea mays* root cortex - section 5.3.).

TABLE 10.5.

WATER VOLUME CHANGE RATE AND FLOW VELOCITY THROUGH ENDODERMAL PLASMODESMATA OF BARLEY ROOTS

The flow of water through a single plasmodesma is approximately $2.41x10^{-20}$ m^3 s^{-1} (Table 10.2). The volume change rate and flow velocity, as ΔP, for each plasmodesma, is greatly influenced by pore size.

Case	Assumed pore diameter	Surface area of pore (m^2)	Volume of pore (m^3)	Volume changes (s^{-1})	Flow velocity ($m\ s^{-1}$)
I	10 nm	$7.9x10^{-17}$	$3.9x10^{-23}$	618	$3.05x10^{-4}$
II	20 nm	$3.1x10^{-16}$	$1.6x10^{-22}$	151	$7.7x10^{-5}$
III	40 nm*	$1.3x10^{-15}$	$6.3x10^{-22}$	38	$1.9x10^{-5}$

*(A hypothetical case, given for the sake of example, assuming shrinkage during processing).

TABLE 10.6.

CALCULATED RATES OF WATER UPTAKE BY 4 WEEK-OLD BARLEY PLANTS FROM OBSERVATIONS ON SEGMENTS AND FROM WHOLE PLANTS OF COMPARABLE SIZE*

		Length** (m)	Weight** (g)	Rate of water uptake (m^3x10^{-10} g^{-1} root s^{-1})	(m^3x10^{-10} s^{-1})
Water uptake from root segments	Axes				
	Apical 60 mm	1.54	0.38	6.7	2.5
	Remainder	6.48	1.63	1.4	2.3
	Laterals				
	Apical 4 mm	6.40	0.22	11.1	2.4
	Remainder	39.80	1.52	5.0	7.6
	Total root system	54.22	3.75	3.9***	14.8
Water uptake measured from whole plants				2.2	8.3

* Table adapted from Graham, Clarkson and Sanderson, 1974.
** Data from Hackett, 1968.
*** This total is obtained by adding the average contribution to total water uptake from each part of the root system. The total weight of the root system is 3.75 g, therefore (for example) the average rate per gram of apical 60 mm of main root axis is:-

$$\frac{0.38}{3.75} \times 6.7 = 0.68x10^{-10}\ m^3\ g^{-1}\ s^{-1}$$

Recently Newman (1975) has recalculated the resistances of the apoplast (cell wall) and symplast pathways for water movement across roots and has suggested that in all parts of the root the latter offers less resistance. This now implies that there is mass flow of water through the symplast and through plasmodesmata. The flow of water through plasmodesmata leads to some startling figures for flow velocity and volume change rate (Table 10.5.), although not greater than those determined experimentally for other locations (e.g. across plasmodesmata of *Abutilon* nectary hairs - Chapter 11). Although they appear extremely high, there is no reason to suppose that such flows could not be accommodated by the plasmodesmata. However, the data in Tables 10.3.-10.5. accentuate the 'fourth power' (r^4) relationship of pore radius to ΔP, Lp, volume changes, and flow velocity: a further reminder of the importance of obtaining accurate dimensional data.

Graham, Clarkson and Sanderson (1974) determined the rates of water uptake by four week old barley plants using two different methods: (i) by extrapolating data from water uptake by short root segments (micropotometry); (ii) by determining water loss from whole plants (Table 10.6.). The total water uptake by a root system, expressed per gram freshweight, shows good correspondence between the two methods. The data show that older (more suberized) parts of lateral roots are more permeable to water than suberized parts of main root axes; it is also clear that the basal (suberized) zones of both main and lateral root axes contribute substantially to total water uptake. The concept of an apical absorbing zone for water is therefore misleading, and the presumption that water moves through the plasmodesmata in the older zones is reinforced.

10.5.3. Phosphate and Potassium

When turning to the situation relating to the translocation of ions across the suberized endodermis, a less definite case for plasmodesmatal involvement is seen. Both phosphate translocation (in barley - Table 10.7.) and potassium translocation (in marrow - Table 10.8.)

TABLE 10.7.

PHOSPHATE TRANSLOCATION BY BARLEY ROOTS

Phosphate - translocation from 3.5 mm barley root segment, 440 mm from root tip, all cells at State III	$8.3\text{x}10^{-15}$ mole seg^{-1} s^{-1}*
Flux across inner tangential endodermal wall	$3.8\text{x}10^{-9}$ mole m^{-2} s^{-1}**
Flux through plasmodesmata is therefore:-	
a) enclosed by desmotubule	$4.6\text{x}10^{-5}$ mole m^{-2} s^{-1}
b) enclosed by plasmalemma	$1.1\text{x}10^{-5}$ mole m^{-2} s^{-1}
The flux through a single plasmodesma would be	$3.58\text{x}10^{-21}$ mole s^{-1}

* Taken from Clarkson, Robards and Sanderson (1971), Table 2.

** Such values may be compared with those obtained for fluxes across cell membranes: e.g. $1.2\text{x}10^{-8}$ mole m^{-2} s^{-1} for the plasmalemma of *Nitella* (Smith, 1966); (see also Table 10.8., Footnotes).

across the ITEW occur at rates that are consistent with the reported fluxes of these ions across cell membranes. However, it must still be assumed that these fluxes move *via* the plasmodesmata because the suberin lamella blocks access to the endodermal plasmalemma from the cortical (outer endodermal) apoplast: if the suberin lamella is as impermeable as usually supposed, then the only pathways between cortex and endodermal symplast are the plasmodesmata.

TABLE 10.8.

POTASSIUM TRANSLOCATION BY MARROW ROOTS

Pore diameters (as for barley - Table 10.1.).	10 or 20 nm
Plasmodesmatal frequency (μm^{-2})	6.2*
Assumption of potassium translocation from 3.5 mm segment in 6 h	7.0 mole m^{-3}**
Volume of treated segment	2.4×10^{-9} m^3
Potassium crossing endodermis in 6 h	1.68×10^{-8} mole
Radius of stele	0.26 mm
Surface area of inner tangential endodermal wall	5.72×10^{-6} m^2
Potassium flux across inner tangential wall of endodermis of segment	7.8×10^{-13} mole s^{-1}
which is	140×10^{-9} mole m^{-2} s^{-1}***
Number of plasmodesmata in a 3.5 mm segment (ITEW)	3.6×10^{7}
Total surface area of pores in a 3.5 mm segment (ITEW)	
a) enclosed by desmotubule	2.78×10^{-9} m^2
b) enclosed by plasmalemma	1.11×10^{-8} m^2
Therefore flux across plasmodesmata is	
a) enclosed by desmotubule	2.8×10^{-4} mole m^{-2} s^{-1}
b) enclosed by plasmalemma	7.0×10^{-5} mole m^{-2} s^{-1}
The flux through a single plasmodesma would be	2.19×10^{-20} mole s^{-1}

*Data obtained by Miss Sally Wingrave at the University of York.

**From Harrison-Murray and Clarkson, 1973, Table 2. The 'translocated' values for potassium have had 30% of the amount found in the treated segment added to them, this being presumed to have crossed the endodermis. The value of 7.0 mole m^{-3} treated segment is therefore a very approximate one.

***These values may be compared with those obtained for potassium fluxes across cell membranes by other authors: e.g. $0.96\text{-}7.2 \times 10^{-9}$ mole m^{-2} s^{-1} in roots of *Zea mays* (Ginsburg and Ginzburg, 1970a); 89×10^{-9} mole m^{-2} s^{-1} in *Valonia* (Gutknecht and Dainty, 1968); 150×10^{-9} mole m^{-2} s^{-1} in *Griffithsia* (cited by Gutknecht and Dainty, 1968); $10\text{-}100 \times 10^{-9}$ mole m^{-2} s^{-1} as an upper estimate of the capacity of most plant cell membranes (MacRobbie, 1971a). The potassium fluxes across the endodermis of marrow appear to be approximately 40 times greater than phosphate fluxes across the endodermis of barley (Table 10.7.).

10.5.4. Relevance of Cytochemical Studies to the Function of Root Plasmodesmata

Cytochemical aspects of plasmodesmata are considered fully elsewhere (Chapter 7), but it is relevant to make certain comments here. Phosphatase activity can be demonstrated to be associated with plasmodesmata, so suggesting dynamic processes as yet not understood. There have been many suggestions that chloride ions pass through plasmodesmata (Van Steveninck and Chenoweth, 1972; Ziegler and Lüttge, 1966; 1967) and, more recently, the demonstration of silver chloride deposits within plasmodesmata and the endoplasmic reticulum in barley roots after precipitation with silver nitrate (Läuchli, Kramer and Stelzer, 1974; Stelzer, Läuchli and Kramer, 1975). Van Iren and Van der Spiegel (1975) have loaded young roots with thallium sulphate, and followed this with ammonium iodide. The resulting precipitate of thallium iodide can be seen in the electron microscope, and is, in part, located within the endoplasmic reticulum. Such results as these, while open to the criticism that high concentrations of toxic salts were used, do lend support to the idea that, at least in some circumstances, ions can move through the plasmodesmata, and within the endoplasmic reticulum.

10.5.5. Bidirectional Movement Through Plasmodesmata

Although water and ion movement in roots is considered largely as centripetal processes, the continuing respiration of the cortex in basal zones implies the necessity for some movement of respiratory substrates in the opposite direction. Dick and ap Rees (1975) have provided further evidence for believing that cortical and apical cells of pea roots receive their carbohydrate as sucrose *via* the symplasm from the stele. This points to the requirement for bidirectional movement through plasmodesmata; it remains to be seen whether this is possible through single plasmodesmata or whether there are plasmodesmata oppositely 'polarized' with respect to the transport which takes place through them. In connection with the second possibility, it has been suggested that, in the case of the mestome sheath (Kuo, O'Brien and Canny, 1974), plasmodesmata opposite the xylem and phloem operate in opposite directions; such a situation could equally well be the case in the endodermis, although we have no evidence of this.

In the particularly interesting situation of the fluxes across the endodermis to and from bacterial nodules on leguminous roots, the problem is accentuated by having relatively high fluxes operating in both directions (Table 10.9.). Although the walls are not suberized, the rapid fluxes imply movement through plasmodesmata (Pate, Gunning and Briarty, 1969).

10.5.6. Passage Cells

Earlier authors have placed some stress on the presence, and possible function, of passage cells in the endodermis (e.g. Lundegårdh, 1950). The situation in barley roots is that the Casparian band can be detected in the radial walls of *all* endodermal cells from 5.0-7.0 mm behind the tip; various stages of endodermal wall thickening can be found from about 120 mm to 300 or 400 mm back. In this intermediate region it is very common to find endodermal cells with relatively well thickened walls adjacent to cells with thin walls and Casparian bands. Such cells have been interpreted as passage cells. Our results make it quite clear that one must carry out a complete developmental study within the root before attempting to relate structure to function, because *all* endodermal cells in barley eventually develop to State III (i.e. they have a suberin lamella and thick cellulosic wall). (Similar conclusions were reached by Haas and Carothers,

1975). In any case, as shown in Clarkson *et al.* (1971) the presence of a few passage cells would only act as an acceptable alternative for intercellular movement of water and ions if extremely high flux rates could be contemplated across their membranes, and there is little likelihood that this occurs.

TABLE 10.9.

FLUXES ACROSS THE ENDODERMIS TO AND FROM LEGUMINOUS ROOT NODULE BUNDLES

Diameter of stele*	100 μm
Total strand length per plant (average over 9 days)*	1194 mm
Surface area of inner tangential endodermal wall per plant*	3.75×10^{-4} m^2
Average export per plant over same 9 day growing period*	109.6×10^{-6} mole day^{-1}
Averaging export over 24 h, flux across ITEW is	3.38×10^{-6} mole m^{-2} s^{-1}**
As the flux varies during the day, and analysis of bleeding sap underestimates total export of fixed N *in vivo*, *** rates exceeding 200 μg N h^{-1} plant^{-1} can be reached. By using a factor of 3.87×10^{-2} to convert g to mole (Gunning *et al.*, 1974, Table 7) this becomes	5.73×10^{-6} mole m^{-2} s^{-1} *export*
However, the nodule also *imports* 0.77 mole sucrose (assayed as C) for every mole of N fixed, **** which is	4.42×10^{-6} mole m^{-2} s^{-1} *import*

*Data from Gunning, Pate, Minchin and Marks, 1974. The exported material is a mixture, mainly of asparagine and glutamine.
**An equivalent figure, for movement across the mestome sheath of wheat, is 2.3×10^{-6} mole (sucrose) m^{-2} s^{-1} (Kuo, *et al.*, 1974).
***Minchin and Pate, 1974.
****Minchin and Pate, 1973 (4.1 mg C is used for every mg N exported. This is equivalent to 9.7 mg sucrose. 0.77 mole sucrose are imported for each mole N exported).

We are grateful to Prof. Brian Gunning for collating the data for this Table.

10.6. MOVEMENT OF WATER AND IONS INTO THE XYLEM

Anderson, Aikman and Meiri (1970) have made the point (reiterated by Anderson, 1975c), that the magnitude of the potassium flux across the membranes of the cells loading the xylem vessel elements in maize is much higher than normally found across plant cell membranes: 4.36×10^{-7} mole m^{-2} s^{-1} were calculated to traverse the walls surrounding the stele[1]; or 3.32×10^{-6} mole m^{-2} s^{-1} across the walls of the xylem vessel elements. Symplastic transport across the endodermis could account for the high flux in the former case, although loading of the xylem vessels across a membrane appears inevitable. (In the young root, the apoplast of the stele is probably fairly freely permeable: once past the barrier of the endodermis, ions could be moved out of *any* cells [endodermal, pericycle, xylem parenchyma] into this apoplastic compartment, so reducing the total efflux per unit area required across cell membranes. *However*, the walls of xylem vessel elements and xylem parenchyma soon become lignified, and movement into vessel elements must then presumably be direct from surrounding parenchyma cells). Anderson (1975c) compares the K^+ fluxes (above) with those reported from other situations and concludes that they are *"many times larger than 'typical' values for other root cells"*. (He also comments that the fluxes must cross living xylem vessel membranes. The effective membrane would, in fact, be that of the xylem parenchyma, not the conducting vessel elements which no longer possess a membrane. This, however, does not affect the validity of his argument *per se*). MacRobbie (1971a) regards ion fluxes of between 1 and 10×10^{-8} mole $m^{-2}s^{-1}$ as about the upper general order of magnitude that plant cells should be expected to maintain. The fluxes calculated by Anderson *et al.* are somewhat (4-330 x) greater than this; indeed, the situation in maize could well be worse than supposed by Anderson, in that only part of the total xylem surface area would be available for loading in the older (lignified) parts of the stele. The same problem does not appear to apply to barley or marrow for the phosphate or potassium fluxes measured there (Tables 10.7 and 10.8.). Taking the example of phosphate translocation from the cortex to the xylem in barley (Table 10.7.), it will be seen that the flux across the endodermis is only 3.8×10^{-9} mole $m^{-2}s^{-1}$. The surface area of the xylem of a seminal root may be calculated approximately from: 1 central metaxylem vessel element - 50 µm diameter; + approximately 8 peripheral metaxylem elements - 15 µm diameter; and + 8 protoxylem pole cells which each have a transverse wall length of 18 µm adjacent to potentially loading cells(i.e. excluding the interface with the metaxylem). This gives a total area of 2.37 mm^2 in a 3.5 mm long segment - very similar to the area of the inner tangential endodermal wall (2.2 mm^2). Even assuming that only 25% of this surface was available for fluxes between vessel elements and parenchyma cells, then the *available* area would be 0.59 mm^2, and the phosphate flux across the loading membranes is 1.4×10^{-8} mole m^{-2} s^{-1} - an acceptable figure for plant cell membranes. Similar results may be derived for potassium fluxes across the endodermis and stelar parenchyma membranes of marrow (Table 10.8.).

The results of Anderson *et. al.* (1970) were from excised root segments; the structure of the cells in the uptake zone was not determined. The general opinion at the present time is that water and ions are moved into empty xylem vessel elements from adjacent parenchyma cells. This occurs in young or old parts of the root alike and it seems inevitable that a membrane (of the parenchyma cells) must be crossed. No plausible alternative theory has been put forward. It seems inevitable that high flux rates will often be involved, and it may be necessary to revise our estimate of maximum transmembrane ion fluxes unless an explanation can be provided in other terms.

---

[1] This figure was given incorrectly by Anderson *et al.* (1970), but is revised in Anderson (1975c).

10.7. CONCLUSIONS

There is no reason to suppose that, at the sites where it seems probable that they must function, plasmodesmata cannot carry the water flow and solute fluxes that can be attributed to them. Using all but the most pessimistic assumptions, it is possible to regard plasmodesmata as the most probable routes for intercellular transport. Whether bulk flow can take place through the plasmalemma-lined tube, or through the desmotubule itself (with the implication that the endoplasmic reticulum is a separate symplastic compartment), or whether diffusion can account for all cases, remains to be seen. More experiments are required where the uptake and translocation of water and solutes are precisely related to the structure of the cells in the uptake zone.

ACKNOWLEDGEMENTS

We would like to thank John Sanderson and Margaret Jackson for their excellent collaboration in the experiments and observations documented here, Dr. R. Scott Russell for his continuing encouragement of this work, and the Agricultural Research Council for financial assistance to A.W.R.

10.8. OPEN DISCUSSION

GRESSEL commenced the discussion by commenting that the model of the endoplasmic reticulum as the symplast certainly disregards any effects of cytoplasmic streaming. Not necessarily, said ROBARDS, because, although electron micrographs show a static picture, they only represent an infinitely small moment in time and give no impression of dynamic cellular activities. In any case, he added, in many root cells there would appear to be problems of cytoplasmic streaming anyway because, if the cytoplasmic reticulum does traverse the plasmodesmata, then the ends are exposed to any streaming movement. GUNNING asked whether anyone had seen cytoplasmic streaming within the root cortex. ANDERSON did not think so, and ROBARDS added that he had been worried about the tacit assumption that all cells stream (see also 9.7.).

WALKER said that he had been rather confused over the concept that plasmodesmata could work in two directions at once: if water does not flow through them, then they will operate bidirectionally by diffusion in any case. Was ROBARDS thinking of two-way transport by bulk flow, he asked? There was some discussion about the various opportunities and routes for bulk flow as compared with diffusion (see also 11.4.). ROBARDS suggested that bulk flow could operate through the desmotubule, and diffusion, where appropriate, through the cytoplasmic annulus. SMITH asked if it was not more likely that the situation would be the other way round, because to obtain bulk flow through the desmotubular pathway another membrane would have to be traversed. ROBARDS replied that, if the symplastic pathway operates across a number of cells, then the addition of two extra membranes, and therefore two extra resistances, need not make a large extra contribution to the whole energy requirement. The extra resistances are not met in every cell, but only at the beginning and the end of each pathway. GRESSEL commented that the endoplasmic reticulum might provide a very large membrane loading area for the symplast within a single cell, and GUNNING said that the areas would be in a range from somewhat less than 1 μm^2 μm^{-3} of cytoplasm to about 10 μm^2 μm^3 of cytoplasm (Gunning and Steer,

1975). He also added, that if bulk flow occurs to any extent through the gross cytoplasm, then this could have disruptive effects; it would be sensible to confine bulk flow to an isolated compartment.

ANDERSON asked whether the concept of movement through the endoplasmic reticulum did not mean movement through very long lengths of narrow pipes: wouldn't this require a very large pressure drop? ROBARDS replied that this may be so, although it would depend on the effective diameter of the membrane-bound cisternae, as well as the location within the cortex where the symplastic compartment is loaded. In any case, said ROBARDS, whatever the path, it is likely to be extremely tortuous as the mature cortical and endodermal cells are highly vacuolate. GUNNING asked to what extent the solute fluxes and water fluxes are coupled together; would they have to move in the same compartment? ANDERSON replied that he did not think so.

CARR recalled the two directions of flow necessary across the root endodermis: sugars outwards from the phloem; and water and ions inwards into the xylem. The two directions may be anatomically separated, he said. CARR thought that it would be easier to envisage two directions being separated anatomically rather than as depicted in the model of bidirectional movement through a single plasmodesma. ROBARDS said that this may well be the case for the root endodermis and the mestome sheath, but could not, presumably, operate in such situations as the *Abutilon* nectary hairs. Therefore, if bidirectional plasmodesmata appeared necessary for one situation, it was less difficult to think that they might exist in others. WALKER asked for confirmation that there was not a place in the root where it could be said that there was bulk flow in one direction and diffusive movement in the other. ROBARDS replied that he did not think there was firm evidence for this.

How permeable is the suberin to water, asked CARR? ROBARDS agreed that the impermeability of the suberin lamella is a major feature of the argument that plasmodesmata were of paramount importance in accommodating the flow across the root endodermis. He knew of no evidence directly relating to the permeability of the suberin lamella in this situation, although he recalled that the blocking of apoplastic/symplastic exchange by the suberin lamella had been implicitly questioned in the paper of Vakhmistrov (1971) as, indeed, had the impermeability of the Casparian band (Huisinga and Knijff, 1974). ANDERSON, also, thought that the suberin is very impermeable and, in any case, the resistance of the two membranes as well as the cell wall would provide a formidable barrier if solutes are to move from cell to cell in this way. ANDERSON then asked what happens to the epidermis of barley roots at the time that the endodermis is maturing? ROBARDS replied that the epidermis does not become suberized in barley, although the situation is different in (for example) marrow and maize.

QUAIL asked whether there is any evidence that the reduction in calcium uptake in relation to ageing of the root could not be related to a reduction in the efficiency of a specific active uptake system in the membranes of the cells, rather than the structural barrier outlined in the paper. GOODWIN expressed interest in the same question and commented that calcium does at least get into the State I cell. BOSTROM added that calcium moves very slowly in *Chara* and *Vallisneria* and, therefore, appears not to be transported well symplastically. WALKER added that the concentration of calcium in the ground plasm is extremely low, and that mitochondria accumulate calcium at the expense of the ground plasm, for instance Williamson (1975) in preparing perfusion fluids for his *Chara* cells, found that cyclosis is inhibited if the level of free Ca^{++} is $>10^{-7}$ M. CARR found this rather interesting because it had been suggested that calcium in animal gap junctions regulates their permeability: in fact, there are models based upon an interaction between calcium and cyclic AMP.

GUNNING asked whether the effects of cycloheximide and other protein synthesis inhibitors in blocking ion transport to the root xylem could lie in effects upon plasmodesmata. ROBARDS said that Lüttge *et al.* (1974) interpreted the blocking as occurring in the xylem parenchyma and that plasmodesmata need not be involved; also that Benbadis, Levy and Deysson (1974) had demonstrated survival of plasmodesmata in onion meristematic cells following 6 h in 10 $\mu g\ ml^{-1}$ cycloheximide, though the published pictures were at too low a magnification to see details of ultrastructure.

11

THE ROLE OF PLASMODESMATA IN SHORT DISTANCE TRANSPORT TO AND FROM THE PHLOEM

B.E.S. GUNNING

Department of Developmental Biology, Research School of Biological Sciences, The Australian National University, Box 475, P.O., Canberra City, A.C.T. 2601, Australia

11.1. INTRODUCTION

The role of plasmodesmata in transport to and from the phloem is not easily discussed, for relevant evidence is both meagre and circumstantial. They do occur, or course, in source and sink tissues, and it is indisputable that a rather specialised type of plasmodesma is found on the side walls of sieve elements and sieve cells. However, preoccupation with the problems of long-distance transport along files of sieve cells and sieve tubes has overshadowed work on the short distance processes of loading and unloading the phloem - this despite the emphasis placed upon symplastic continuity between sieve elements and their donor and receptor tissues by Münch (1930): very little indeed is known about the distribution, frequency, and function of plasmodesmata along these short-distance pathways.

This review will start with the leaf tissues from which phloem loading occurs, moving from the mesophyll towards the veins of the leaf, and will then consider briefly certain selected aspects of unloading processes.

11.2. SYMPLASTIC PATHWAYS BETWEEN DONOR TISSUES AND SIEVE ELEMENTS

11.2.1. The Leaf Mesophyll

Haberlandt (1914) in his monumental work on relationships between structure and function in plants, enunciated two principles which he considered to dominate the anatomy of the leaf mesophyll. As translated by Drummond these are the *Principle of Maximum Exposure of Surface* and, more relevant, the *Principle of Expeditious Translocation*, which refers to the arrangement of mesophyll cells in chains or plates, fanning out from the veins of the leaf in patterns suggestive of a sequential cell-to-cell delivery system (Fig. 11.1.). Plasmodesmata were of more interest to Haberlandt as channels for the transmission of stimuli, and he did not emphasise their possible role in translocation of assimilates. Nevertheless his Principle is readily interpreted as evidence favouring the idea that the sieve tubes of the main, second, and finer orders of leaf veins constitute a symplastic system that is in turn connected beyond the veins to yet finer tributaries, bringing the entire mesophyll into the catchment area.

Münch (1930) clearly accepted this view, and even used Poiseuille's Law to estimate the magnitude of the mass flow that would, on his theories, pass through the plasmodesmata that connect successive palisade mesophyll cells along the tributaries. He used the only quantitative data he could find, namely Kuhla's count (Kuhla, 1900) of 7 plasmodesmata per palisade cell-to-cell junction in *Viscum*, the total area of which was 100 μm^2. Knowing various parameters of the export of assimilates, and raising Kuhla's estimate from 7 to 10 on the grounds that *Viscum*, being a parasite, might have a poorly developed translocation system in its leaves, he suggested that a mass flow carrying about 4×10^{-4} pg s^{-1} of assimilate at a velocity of 5-8 nm s^{-1} through the plasmodesmata would be required. Münch obviously considered his result reasonable, and one wonders if he would have been discouraged had he known that the true radius of a plasmodesma is likely to be 20 nm, rather than the 250 nm which Kuhla had cited (Münch considered that plasmodesmata become constricted by cell wall swelling during specimen processing, so that the observed dimensions underestimate the *in vivo* state). On the basis of the Poiseuille formula the discrepancy makes Münch's result an underestimate by a factor of more than 150, and implies flow velocities of about 1 $\mu m\ s^{-1}$, equivalent to some 1-2 volume changes s^{-1} in the plasmodesmata. We are not yet much better able to quantify plasmodesmatal frequencies in mesophyll, but pit fields in that tissue often contain many more than ten (e.g. Gunning and Steer, 1975, Plate 14c,d). Also Brinckmann and Lüttge (1974) find that plasmodesmata occupy 0.38% of the cell wall of mesophyll cell junctions in *Oenothera*, and and assuming a radius of 20 nm, this implies a frequency of about 3 μm^{-2} where the mesophyll cells are in contact, thereby reducing the estimates of the required fluxes (above) by a factor of thirty.

Other points made by Haberlandt concerning the arrangement of mesophyll cells and translocation between them have been followed up in some detail. Painstaking measurements of surface areas exposed to air spaces in the leaf have been made by Turrell (1936), and of tissue layer dimensions by Wylie (1939). Wylie suggested that a correlation exists between interveinal spacings and the relative proportions of palisade and spongy mesophyll. With some exceptions palisade cells (although superficially compactly organised) have little lateral contact with one another and hence are unfavourably oriented for conduction in the plane of the leaf. Thus if the palisade : spongy mesophyll ratio is high, the interveinal spacing tends to be smaller. Where the more laterally-oriented spongy mesophyll predominates, the vein spacing is greater.

There are exceptions to the above correlation, some of which relate to another important concept due to Haberlandt - the *collecting cell*. Haberlandt described several types of collecting cell, amongst them the bundle sheath cells of what are now known to be C_4 plants, where the term collecting cell is very apt (Chapter 12). Other examples consist of lobed cells in the interveinal regions, each lobe connecting to a palisade cell (or file of palisade cells): the flow of assimilate from the ultimate tributaries was pictured as converging in these cells, prior to onward movement towards the veins (Fig. 11.1.).

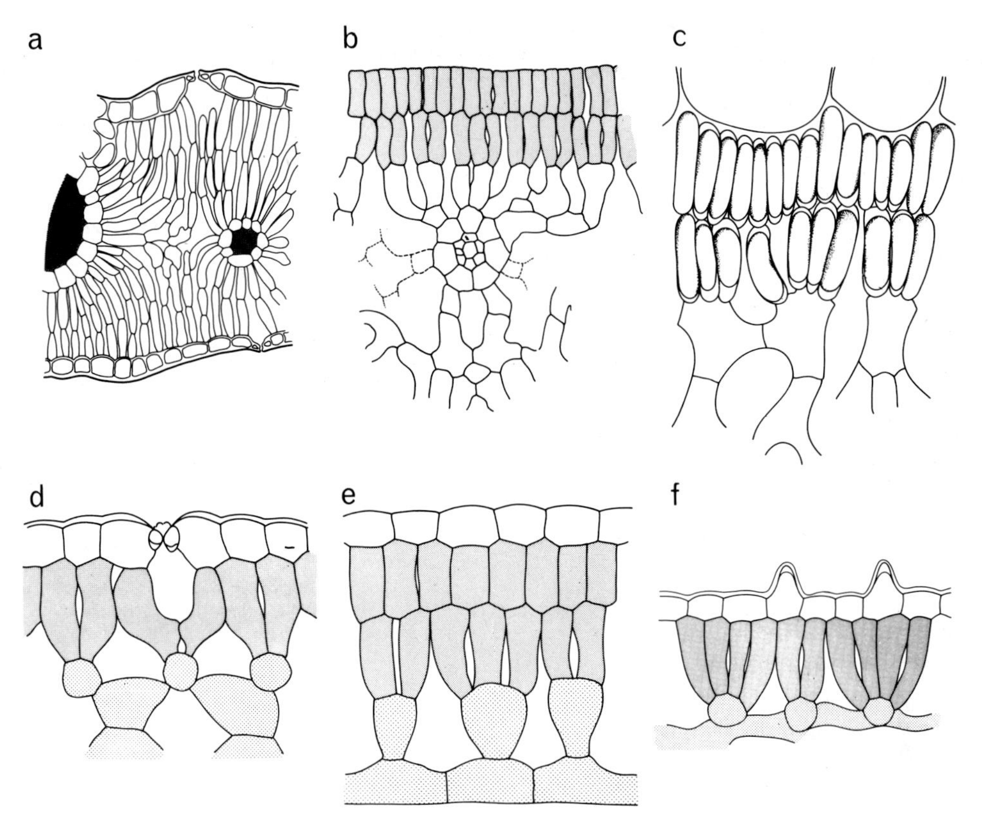

Fig. 11.1. Diagrams of leaf anatomy illustrating Haberlandt's Principle of Expeditious Translocation. (a) *Scabiosa ucrainica*, with files of mesophyll cells fanning out from the veins; (b) *Ficus elastica*, with cells converging on the veins from the palisade mesophyll layers; (c) *Ficus elastica* following 1 hour in 0.6 M sucrose, showing that the palisade mesophyll cells plasmolyse at a lower osmolarity than do the 'collecting cells' they drain into; (d) *Ornithogalum umbellatum*, (e) *Eleagnus angustifolia*, and (f) *Pulmonaria officinalis*, all showing 'collecting cells' attached to two or more palisade mesophyll cells

(a), (b) re-drawn from Haberlandt (1914); (c) re-drawn from Roeckl (1949); (d), (e), (f) re-drawn from Meyer (1962)

Glycine max (soybean) is one example of a plant which has larger interveinal spacing than would be expected for its palisade : spongy mesophyll ratio (Fisher, 1967). The discrepancy can be rationalised, in that soybean leaves possess a flattened network of horizontally extended pale green cells that is strategically located in contact with the lowermost palisade layer on its adaxial face, in contact with the three-dimensional reticulum of spongy mesophyll cells on its abaxial face, and linking with the border parenchyma of the surrounding veins at the level of the phloem. Fisher (1967) supports the suggestion of Thaine (1965) that this cell-layer might function in horizontal translocation in the interveinal islets. Weston and Cass (1973) describe the development of the layer and cite records of similar structural adaptations in other genera. Yet other examples are described in Meyer (1962).

Anatomical patterns are suggestive, but what other evidence is there for symplastic transport in the mesophyll? The plasmodesmata in *Oenothera* leaves that have been counted by Brinckmann and Lüttge (1974) are considered to provide a pathway that accounts for the observed electrical coupling within the mesophyll tissue, not only between green cells, but also between green and non-green cells in variegated material. Plasmolysis by a mannitol solution chosen so that the mesophyll but not the companion cells or sieve elements would be affected (Geiger, Giaquinto, Sovonick and Fellows, 1973) disrupts translocation of assimilate to the veins, a phenomenon that is interpretable in terms of the partial rupturing of plasmodesmata (Geiger, Sovonick, Shock and Fellows, 1974). Bräutigam and Müller (1975b) also demonstrate inhibition of translocation in leaf tissue by plasmolysis, and Smith (1972) gives direct evidence for plasmolytic inhibition, though not in mesophyll. Although there does appear to be symplastic continuity in the mesophyll, this, perplexingly, does not prejudice the setting up of turgor gradients. Roeckl (1949) has found, for instance, that the osmotic value of palisade mesophyll cells is lower than that of the collecting cells they are supposed to drain into (Fig. 11.1.), and that the sieve tube sap itself has a yet higher osmotic value (see later). Münch (1930) envisaged that plasmodesmata would have to offer a resistance to flow, so that turgor gradients could be maintained, but he postulated a gradient in the opposite direction to that found by Roeckl.

For the most comprehensive studies of symplastic transport in leaf parenchyma we must turn to the somewhat special case of the submerged leaves of a water plant.

Kok (1933) exploited the anatomical advantages offered by the long strap-shaped submerged leaves of *Vallisneria* to study the relationships between cytoplasmic streaming and translocation. She was unable to find any correlation between the rate of movement of lithium salts and caffeine and the occurrence or otherwise of streaming. In retrospect, however, her main contribution may have been to stimulate in the Botanical Laboratory at Groningen a programme of investigations on transport in the mesophyll of *Vallisneria* leaves, investigations which are still under way there and elsewhere.

The work on symplastic transport in *Vallisneria* has been reviewed on several occasions (Arisz, 1952; Helder, 1967; Arisz, 1969; Spanswick, 1975), and only a few relevant points will be raised here. Arisz and his co-workers showed that many substances are mobile over long distances within a non-leaky compartment of the *Vallisneria* leaf tissue. Mobile substances include organic acids, amino acids, amides, indoleacetic acid, bicarbonate, and a range of mineral anions and cations. Calcium is non-mobile, though it is not known whether this is because it is sequestered into intracellular compartments, or whether there are specific blocks to calcium transport. It was realised in the earliest post-war papers that, to enter this trans-cellular transport compartment, the materials have to be absorbed from the external solution, and that once inside, an alternative to transport is that they can be diverted into the vacuole, where they are not immediately available for symplastic transport. Comparable processes are seen in roots.

Arisz, unlike Münch, for long refrained from assuming that a semipermeable plasma membrane bounds the symplast. Prior to the early 1960's he had claimed only that cytoplasmic continuity from cell to cell (together with some provision, which might be a membrane *or* a means of binding solutes to mobile carriers to prevent their escape into the apoplast) was necessary to account for his results. Failure to detect leakage into the apoplast or bathing medium from intact or cut portions of leaf containing mobile substances demonstrated that there is some sort of seal around the symplast. Also, when cells are damaged, it is implied that the intercellular connections leading to them from the rest of the tissue become closed off. Geiger, Sovonick, Shock and Fellows (1974) have also concluded that mesophyll cell plasmodesmata seal if the cells are damaged, as did Pitman *et al.* (1970) for root cells.

It was in 1937 that Arisz and Oudman obtained the first hint that a pathway for intercellular transport exists outside the vascular bundles (see Arisz and Schreuder, 1956a); by 1944 the compartmentation of the system had been uncovered and soon the results were being discussed in terms of the symplast and plasmodesmatal connections (Arisz, 1948). A host of supplementary experiments has been carried out subsequently both by Arisz and his colleagues, and, more recently, by Bräutigam and Müller (1975a,b,c), confirming again and again that symplastic transport does occur, and defining more and more attributes of the system. In only one of the many papers (Arisz and Schreuder, 1956b) is it reported that any attempt to actually visualise the plasmodesmata was made. When subjected to a technique devised to detect ectodesmata, "*very fine plasmodesmata*" were seen by light microscopy "*in some suitable preparations*" of unspecified tissues. Even in 1966 Arisz and Wiersema (1966a) could be no more specific about the presence of plasmodesmata than to write "*their presence in* Vallisneria *tissue is likely, but they are not actually mentioned in the literature*". Bräutigam and Müller (1975b) nevertheless envisage not merely plasmodesmatal transport, but transport through desmotubules, together with the necessary equilibria between free space and cytoplasm, and between cytoplasm and the endoplasmic reticulum within it. Considering the key roles *presumed* structures have in the interpretation of the decades of physiological work on *Vallisneria* leaves, it is regrettable that so little effort should have been made to examine their *actual* structure.

The key experiments showing that a symplastic pathway exists in the interveinal region were made by excising lengths of vein, leaving parenchyma tissue as bridges connecting donor to receptor expanses of leaf. Transport of chloride and asparagine across such bridges was observed (Arisz and Schreuder, 1956a) and it was later demonstrated that sodium azide in concentrations which inhibit secretion into the vacuoles does not block symplastic transport when applied to the parenchyma bridges or to transporting zones of intact leaves (Arisz, 1958). This, of course, is as expected for a passive transport process. Experiments on parenchyma bridges versus vascular bundle bridges highlight a problem that applies to most of the work on *intact Vallisneria* leaves, namely that there are clear indications of long distance transport in the phloem in addition to transport through the parenchyma. Vein loading and vein unloading are in turn involved (Arisz, 1960), at least the former of these processes seeming to be specifically affected by inhibitor treatments (Arisz, 1958) thus further complicating analysis of a situation already rendered complex by equilibrium shifts between uptake, symplastic transport, secretion into vacuoles, and withdrawal from vacuoles.

Attention is drawn to the existence of parallel pathways in the phloem and in the interveinal regions (both symplastic), as the fact unfortunately blurs what would otherwise be perhaps the only clear cut demonstration of bidirectional flow in the symplast - experiments showing that asparagine and chloride can be transported simultaneously in opposite directions in intact leaves (Arisz, 1960). It would be most valuable to repeat these experiments using parenchymatous bridges. Also, since phloem transport occurs it is not clear how to interpret the estimated velocity of transport along the leaves. The observed 5-12 $\mu m\ s^{-1}$ is presumably a function of detector sensitivity and an amalgam of transport through plasmodesmata, across intracellular compartments, and along sieve tubes. Certainly it is too rapid, and occurs over too long distances, to be accounted for by simple diffusion and it is noteworthy that where phloem transport is held to be eliminated, as in the parenchyma bridge experiments, the amount of transport is markedly reduced if the length of the bridges exceeds 4 mm - this perhaps being a measure of the relative inefficiency of diffusive symplastic transport where the number of cells to be traversed is large or the distances involved exceed the millimetre range.

Finally, although the very important work of Arisz and his associates has been included here in a section concerned with symplastic transport in the mesophyll, it is clear that even after simplification of the system to the form of non-vascular tissue bridges, both epidermal and centrally located cells remain. The extent of the symplastic connections in *Vallisneria* is uncertain, but enough is known about plasmodesmatal distributions in the leaves of the other water plants *Elodea* and *Potamogeton* to indicate that the epidermal cells are symplastically connected, and could be a transport pathway (Spanswick, 1972; Helder, 1975).

11.2.2. The Vein Boundary

11.2.2.1. Bundle sheath and border parenchyma In some angiosperm leaves the cells lying adjacent to the xylem and phloem of the small veins are little if at all different from mesophyll cells (e.g. the cross veins of wheat leaves (Kuo, O'Brien and Zee, 1972), and the minor veins of *Cucurbita* leaves (Turgeon, Webb and Evert, 1975)). In most, however, the cells peripheral to the veins are specialised to some degree, being anatomically, and presumably functionally, intermediaries between the mesophyll and the vascular tissue.

The minor veins of dicotyledonous leaves are usually sheathed by elongated cells with few, small, chloroplasts. Sometimes this border parenchyma extends from the minor veins back to larger categories of vein, and according to a survey made by Armacost (1944), it covers 99% of the total vein length. The rapid cytoplasmic streaming observable by phase-contrast microscopy of border parenchyma in hand sections of (for example) *Nicotiana* leaves, betrays the high metabolic activity of these cells. The special metabolic functions of the bundle sheath cells of monocotyledonous and dicotyledonous C_4 plants are dealt with in Chapter 12. The border parenchyma, or bundle sheath, is usually continuous, even enclosing the blind endings of veins (Armacost, 1944; Esau, 1965), but occasional intercellular spaces are seen to pass from the mesophyll through fissures in the sheath to the surface of the contained phloem parenchyma (Gunning, Pate and Briarty, 1968; Turgeon, Webb and Evert, 1975). Two adaptations exist, however, which may provide an apoplastic barrier between mesophyll and vein: the suberised lamella of the mestome sheath of foliar veins in many members of the Gramineae, Cyperaceae, and Juncaceae (O'Brien and Carr, 1970; Kuo and O'Brien, 1974b), and the Casparian bands seen in the radial walls of the bundle sheath of certain members of the Primulaceae and Plantaginaceae (Trapp, 1933; Marks, 1973).

11.2.2.2. The outer tangential wall The presence of plasmodesmata in the outer tangential wall of bundle sheath cells is poorly documented, except in the case of C_4 plants (Olesen, 1975; and Chapter 12.). In the C_3 plant *Nicotiana*, however, their presence and function *in vivo* is inferred from work in which veins were isolated by enzymatic digestion. The conditions could be manipulated so as to hydrolyse cell walls within the mesophyll and down to the outer tangential wall of the sheath, leaving veins with attached border parenchyma (Cataldo and

Berlyn, 1974). Such preparations were less able to accumulate sucrose from solutions than were veins in intact leaf discs, implying that *in vivo* at least some of the assimilate may enter the vein by another route, namely plasmodesmata which, it is presumed, connected the mesophyll to the border parenchyma prior to digestion (Cataldo, 1974).

11.2.2.3. The inner tangential wall Esau (1967b) has drawn attention to the tendency for the chloroplasts of border parenchyma cells to lie on the outer face of the cell, and has speculated that this arrangement leaves the wall next to the vascular tissues free for the exchange of materials between the sheath and the conducting tissues. Plasmodesmata do occur here and have been studied very much more than those connecting the sheath to the mesophyll. The available information is, however, rather confusing in that a high degree of taxonomic and anatomical variation in pit and plasmodesmatal distribution is becoming apparent.

In wheat, most pits occur where the mestome sheath and the vein phloem are contiguous, and pits are sparse adjacent to the xylem region and at the ab- and ad-axial poles of the veins (where there is little or no mesophyll outside the sheath) (Kuo, O'Brien and Canny, 1974). The numbers of these plasmodesmata will be considered later, but it is relevant to point out here that they are of the type where the plasma membrane is constricted at the cytoplasmic extremities, and that in addition there is another constriction at the level of the suberin lamella (O'Brien and Carr, 1970). There is also an intriguing spatial association between sheath cell mitochondria and the pits.

The minor veins of *Cucurbita pepo* are especially interesting in that they are bicollateral along their entire length. The minor veins contain sieve elements, companion cells, large abaxial parenchymatous cells described as 'intermediary' and which probably are companion cells, and a parenchyma cell between the xylem and the adaxial phloem. All of these living cells abut the border parenchyma (which is nearly indistinguishable from mesophyll), and there are very marked differences in the frequency of plasmodesmata between the various classes of 'sheath' -vein cell junction. The intermediary cells possess very high frequencies, with individual clusters containing over 300. The adaxial companion cells have far fewer. The central parenchyma cell has only occasional ones (Turgeon, Webb and Evert, 1975).

The situation in those dicotyledons where the vein phloem parenchyma is modified by the development of cell wall ingrowths to become transfer cells (see subsequent section) may be different again. At first it was reported that plasmodesmata had not been found between the modified companion cell ('A-type' transfer cell) and the sheath, and the lack of symplastic continuity was discussed in relation to its implication, which is that the mechanism of vein loading involves retrieval from the apoplast (Gunning, Pate and Briarty, 1968). It was later shown for *Impatiens* (Gunning and Pate, 1969) and *Vicia* (Gunning, Pate, Minchin and Marks, 1974) that these cells *can* have plasmodesmata leading out to the sheath, though there are very few of them. Further, a more recently recognised type of vein transfer cell, the 'B-type', corresponding to a type of phloem parenchyma that is probably not a companion cell, can be relatively well endowed, as can its unmodified counterpart in a species that has only the 'A-type' of transfer cell (Table 11.1.). Considering plasmodesmata possessed by *all types* of vein parenchyma together, in *Vicia* (with only the A-type transfer cell) about 6% of the plasmodesmata were found to lead out to the sheath, while in *Tussilago* (with A- *and* B-types) the corresponding figure was 25%: frequencies per unit area were not determined (Gunning *et al.*, 1974).

There is an extraordinary dearth of information on plasmodesmata in the inner tangential wall of bundle sheath or mestome sheath cells of C_4 plants. They are visible in electron micrographs of corn leaf veins (Singh and Srivastava, 1972) but by and large most investigations have concentrated on the outer tangential wall and have ignored the problems of phloem loading. The translocatory fate of assimilates produced by Kranz-type cells that are remote from vascular tissues (Brown, 1975) is even more obscure.

Attempts to provide functional interpretations for the above plasmodesmatal distribution patterns are postponed until later. The range in observed structure would seem to indicate that a diversity of vein loading pathways will ultimately be disclosed.

11.2.3. The Phloem

11.2.3.1. Cell types At least two distinct types of phloem parenchyma cell occur in minor veins (Esau, 1973). One is probably a companion cell, in functional if not in ontogenetic terms. It has dense cytoplasm with abundant endoplasmic reticulum and well-developed energy-producing systems in the form of mitochondria and chloroplasts. It invariably abuts a sieve element. The other is more vacuolate, shows a much less dense fixation image with fewer organelles when prepared for light or electron microscopy, and is not always contiguous with sieve elements. The former, which from now on will be referred to as a companion cell, can become modified as an A-type transfer cell; the latter, which will be designated simply as phloem parenchyma, as a B-type transfer cell (Pate and Gunning, 1969). The two cell types are recognisable in vein phloem of plants as primitive as the horsetail *Equisetum* (Gunning, Pate and Green, 1970). The two differ in physiology as well as structure, with the companion cells (and sieve elements) having a high osmotic pressure, while in the phloem parenchyma the osmotic pressure is very much less (Shih and Currier, 1969), lower even than in the mesophyll (Geiger, Giaquinta, Sovonick and Fellows, 1973). The osmotic difference no doubt accounts in part for the very different response to cytological fixatives, and for the susceptibility of the phloem parenchyma to ice crystal damage during freeze-substitution in media which protect the companion cells (Geiger, Giaquinta, Sovonick and Fellows, 1973).

11.2.3.2. Symplastic connections, general Symplastic continuity between adjacent companion cells, between phloem parenchyma and companion cell, and between companion cell and sieve element, has been recognised in cotton (Shih and Currier, 1969), *Tetragonia* (Esau and Hoefert, 1971b), *Beta* (Esau, 1967b, 1972; Geiger, Giaquinta, Sovonick and Fellows, 1973), *Mimosa* (Esau, 1973), *Vicia* and *Tussilago* (Gunning, Pate, Minchin and Marks, 1974), and *Cucurbita* (Turgeon, Webb and Evert, 1975). The two types of phloem parenchyma cell cannot be distinguished in the lower vascular plant *Lycopodium lucidulum*, but the phloem parenchyma cells are all interconnected, with the frequency of plasmodesmata increasing approaching the sieve cells (Warmbrodt and Evert, 1974b). Sieve elements very seldom come into direct contact with the surrounding photosynthetic tissue, but in one case where this does happen, no plasmodesmata were found (Kuo, O'Brien and Zee, 1972).

Whilst there is variation in the extent of outwardly directed plasmodesmatal connections to the border parenchyma (see previous section), there are certain consistencies in the distribution pattern of intraveinal plasmodesmata (Table 11.1.). Thus the phloem parenchyma (as designated here) very rarely connects with the sieve elements, indeed where the cell type is modified as a B-type transfer cell, it is on the sieve element-parenchyma cell junction that the labyrinth of wall

TABLE 11.1.

RELATIVE FREQUENCIES OF PLASMODESMATA BETWEEN CELL TYPES IN LEAF VEINS

(a) Sugar beet (from Geiger, Giaquinta, Sovonick and Fellows, 1973)

	Companion cell	Phloem parenchyma	Bundle sheath	Sieve element
Companion cell	rare	abundant	*	abundant
Phloem parenchyma		common	common	rare

*cell combination rare or absent

(b) *Vicia faba* (from Gunning *et al.*, 1974) (percentages)

	A-type transfer cell	Phloem paren-chyma	Xylem paren-chyma	Bundle sheath	Sieve element
A-type transfer cell	1.9	25.5	1.9	1.9	39.2
Phloem parenchyma		17.6	4.0	4.0	0
Xylem parenchyma			4.0	0	0

(c) *Tussilago farfara* (from Gunning *et al.*, 1974) (percentages)

	A-type trans-fer cell	Phloem paren-chyma (unspeci-alised)	B-type transfer cell	Xylem paren-chyma	Bundle sheath	Sieve element
A-type transfer cell	1.6	1.9	8.2	9.0	7.1	40.3
Phloem parenchyma (unspecia-lised)		0.5	3.0	0	1.5	0
B-type trans-fer cell			1.8	0	11.5	1.5
Xylem parenchyma				7.7	4.5	0

ingrowths develops (Pate and Gunning, 1969). The great majority of the intraveinal plasmodesmata interconnect the companion cells and sieve elements - 40% of the total in *Vicia* and *Tussilago* (Gunning *et al.*, 1974). The next most abundant type lies between the companion cells and the phloem parenchyma, despite the large osmotic difference between these two cell types. Thus, even where direct continuity between companion cell and border parenchyma is slight (see above), there is an indirect route *via* the phloem parenchyma. Often the plasmodesmata that lie within and between the two cell types are complex branched structures (e.g. Gunning, Pate and Briarty, 1968; Evert, Bornman, Butler and Gilliland, 1973).

11.2.3.3. Plasmodesmata between companion cell and sieve element
The sieve element-companion cell plasmodesmata have received much attention since their first description in 1880 by Wilhelm (see historical survey in Esau, 1969, p. 128). Their structure seems to be fairly consistent throughout the phloem tissues that have been examined in detail, and the following information has been taken from a wider literature than that concerned solely with leaf veins.

Aspects of their development have been considered in Chapter 4. To begin with they are simple plasmodesmata, but in angiosperms they later branch on the companion cell side of a median cavity (the word

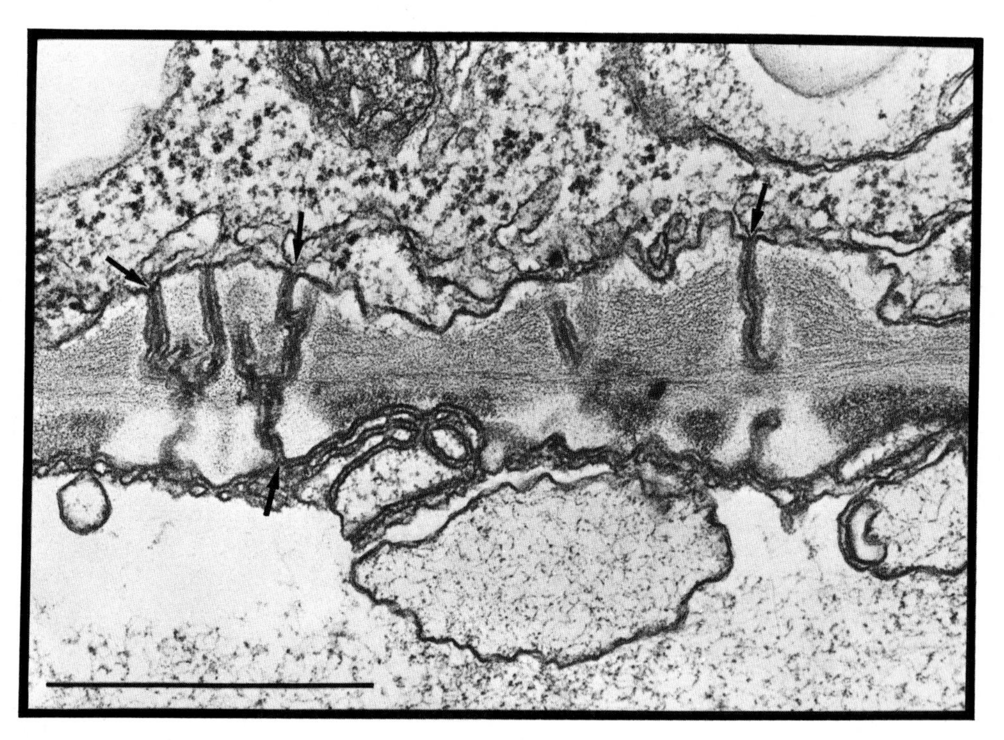

Fig. 11.2. Electron micrograph showing portions of branched plasmodesmata leading from companion cell (above) to sieve element (below) in internode phloem of *Cucurbita maxima*. Arrows point to probable continuity of the desmotubule and the endoplasmic reticulum systems of the two cells. Electron lucid sleeves of callose are seen round the plasmodesmata on the sieve element side of the wall. Reproduced by permission from Esau and Cronshaw (1968). Scale marker 1 μm

'cavity' has been recommended because in this case the term median 'nodule' does not convey the complexity of the structure (Esau, 1973)), while callose comes to line the lumen, which enlarges somewhat, on the sieve element side. It is often said that they resemble sieve plate pores on the sieve element side, connected to conventional plasmodesmata on the companion cell side (see review by Esau, 1969). Moving down the plant kingdom, the same general structure is seen in gymnosperms, where, however, the 'albuminous cell' replaces the companion cell (Wooding, 1968, 1974; Behnke and Paliwal, 1973; Evert, Bornman, Butler and Gilliland, 1973; Sauter, 1974; Carde, 1974). However, in examples of homosporous (Liberman-Maxe, 1971; Evert and Eichhorn, 1974) and heterosporous (Kruatrachue and Evert, 1974) ferns, and in *Lycopodium* (Warmbrodt and Evert, 1974a), and *Selaginella* (Burr and Evert, 1973), the branching system no longer develops, though the callose sleeve may remain on the sieve cell side of the wall. In a sense, therefore, the development of branches from single pores in angiosperms is an ultrastructural example of ontogeny recapitulating phylogeny, and indeed there is also an ultrastructural example of the notion that juvenile structures may show 'primitive' features, for the plasmodesmata in the side walls of root protophloem elements apparently remain single (Esau and Gill, 1972) or nearly so (Esau and Gill, 1973).

It has already been mentioned that some 40% of the plasmodesmata found in small veins in leaves can be of the type connecting companion cell to sieve element. Their frequency in this wall was in fact noted very early on by Kuhla (1900) who counted 50 in one such cell junction in *Viscum*. Alexandrov and Abessadze (1927) picture a number of other striking examples in a paper devoted to the side wall of the sieve tube. Gunning *et al.* (1974) estimate for *Tussilago* minor veins that each 100 μm^2 of junction carries 595 branches on the companion cell side, joining with 214 on the sieve element side.

Whereas the plasmodesmatal branching on the companion cell side of the wall is a feature of gymnosperms and angiosperms (see above), the high frequency seems to have evolved much earlier, being already evident in *Lycopodium* (Warmbrodt and Evert, 1974a), yet absent from those mosses that have been studied, where the leptoids (equivalent to sieve elements) are only occasionally (if at all - Stevensen, 1974) connected to the neighbouring parenchyma (Hébant 1967, 1969, 1970a, 1970b, 1974).

Certain structural details of the sieve element-companion cell plasmodesmata vary. Sometimes they occupy a swelling in the wall, which is amply large enough to be seen in the light microscope. The amount of callose varies. When a thick nacreous wall is formed in the sieve element a depression leading to the plasmodesmatal orifice is left (Behnke, 1971). They also survive lignification of the sieve element wall (Kuo and O'Brien, 1974a).

The degree of branching, and the extent of the median cavity and the tortuosity of its membranous contents also vary. More constant features are the presence of callose on the sieve element side, and the existence of desmotubules which are observable in all groups of vascular plants (see above references).

Shih and Currier (1969) and Esau and Cronshaw (1968) have published outstanding electron micrographs showing continuity between companion cell endoplasmic reticulum and the desmotubule, and also strongly suggesting continuity between the desmotubule and the special form of endoplasmic reticulum possessed by sieve elements (Fig.11.2.). Although they may not have demonstrated continuity, many other authors have commented upon the spatial association between the sieve element endoplasmic reticulum and the sites of the plasmodesmatal orifices (e.g. Wooding, 1968; Evert, Eschrich and Eichhorn, 1971; Esau, 1972; Behnke and Paliwal, 1973; Ledbetter and Porter, 1970; Kruatrachue and Evert, 1974), and some have speculated from their micrographs that the companion cell and sieve element endoplasmic reticulum systems, interconnected as they are by the desmotubule, could be a pathway for entry of assimilate into the sieve elements (Wooding, 1968; Shih and Currier, 1969). The desmotubule may not always be as narrow a tube as is inferred from micrographs of less specialised plasmodesmata, thus Wooding (1968) illustrates by both longitudinal and transverse sections a most remarkable swelling of the desmotubule in regions where the plasmodesmatal canals themselves are distended, approaching and at the median cavity; the same applies to albuminous cells (Carde, 1974).

It is necessary to add a warning that the desmotubule is not always seen. Fellows and Geiger (1974) show one example of what is probably a plasmodesmatal canal on the sieve element side of the wall in a *Beta* leaf vein prepared for electron microscopy by freeze-substitution: the lumen is unusually wide in relation to the thickness of the plasma membrane[1], and there is no sign of a desmotubule. Freeze substituted phloem in *Vicia* and *Abutilon* nectaries, however, has plasmodesmata that are much more like their chemically-fixed counterparts, and desmotubules are preserved (A.J. Browning, personal communication). Other convincing electron micrographs showing no sign of a desmotubule are presented by Esau, Cronshaw and Hoefert (1967), for plasmodesmata through which beet yellows virus is being transmitted. It may be that the effective lumen for virus transport is enlarged by rupturing the desmotubule (which most certainly *is* present in uninfected material (Esau, 1972)).

The transport of assimilates to and from the sieve element is the most obvious possible role for the desmotubule and/or its surrounding annulus (the latter being especially well imaged in Liberman-Maxe, 1971, Fig. 9b). Gilder and Cronshaw (1973a,b) show that some adenosine triphosphatase activity may be found in the plasmodesmata, but a specific role cannot be assigned to the enzyme, which occurs in many other locations in the sieve elements. There is, however, at least one other possible function that should not be overlooked. Most sieve elements are anucleate, and are considered to be maintained by their companion cells (or by other neighbouring cells where companion cells are absent, as in much of the protophloem (Esau, 1969) and in lower plants). It is not entirely clear what is implied by 'maintenance', but one aspect could well be the passage of informational molecules or of plasma membrane and/or endoplasmic reticulum by either membrane flow through the plasmodesmata, or by translational diffusion of membrane molecules. A similar idea applies to 'maintenance' of anucleate rhizoids in the chytrid described by Powell (1974), as well as in the experiments of Townsend described in 13.3.1. Certainly most sieve elements and their companion cells are said to die in synchrony (see Esau, 1969, 1973). One exception to this generalisation has been observed in *Smilax* (Ervin and Evert, 1967). There, when some of the larger sieve elements die and lose turgor, their associated companion cells can remain alive and balloon through the pit areas into the former sieve element lumen, forming tylosis-like structures.

---

[1] According to the scale marker and the caption, the diameter is nearly 200 nm. According to the text the diameter is 20 nm, but this (Fellows, personal communication) is an error.

11.2.4. Loading Fluxes and Pathways

11.2.4.1. Fluxes Calculations of assimilate fluxes from cell to cell have been made for plasmodesmata in the mesophyll tissue (Münch, 1930), for plasmodesmata in the inner tangential wall of the mestome sheath in wheat (Kuo, O'Brien and Canny, 1974), for the postulated uptake process at the companion cell plasma membrane (Gunning *et al.*, 1974; Sovonick, Geiger, and Fellows, 1974) and for the sieve element-companion cell plasmodesmata (Gunning *et al.*, 1974). The two calculations for the trans-companion cell plasma membrane flux yield remarkably similar values that are regarded as reasonable, and will not be considered here. The studies of plasmodesmatal fluxes will be examined separately and then a synthesis will be attempted.

The basis of Münch's calculations has already been given (11.2.1.). He estimated that a cell-to-cell flux of 1.17×10^{-18} mole s^{-1} in the mesophyll would pass through a total of 10 plasmodesmata (erroneously large ones) in a wall of surface area 100 μm^2. If instead we employ Brinckmann and Lüttge's (1974) value of 0.38% for the percentage cover of plasmodesmata in *Oenothera* mesophyll cell junctions and derive from this a frequency of 300 per 100 μm^2 (see 11.2.1.), the flux would be 3.8×10^{-21} mole s^{-1} per plasmodesma. Assuming that a 5% sucrose solution is being transported and a plasmodesmatal length of 0.5 μm and radius of 20 nm, this flux would be equivalent to a mass flow through the plasmodesma of velocity 2 μm s^{-1} and 4 volume changes s^{-1}. These values are increased 16-fold if the flux is imagined to be constrained to a desmotubule of radius 5 nm. Alternatively, the flux could be maintained by a concentration drop along the plasmodesmata of a mere 3 mM m^{-3} (3 μmolar) - a value so low that the need to postulate mass flow at this part of the pathway seems unnecessary.

The next calculations are based on painstaking counts of pits and plasmodesmata in the inner tangential wall of the mestome sheath of wheat. Kuo *et al.* (1974) point out that the assimilate that moves from mesophyll to phloem in leaves with a mestome sheath that is sealed by a suberised lamella probably *has* to traverse the plasmodesmata in the tangential walls of the sheath cells. They assume that the suberised lamella is indeed a barrier to movement in the apoplast and calculate that the flux of sugar through the mestome sheath could be up to 5.8×10^{-13} mole s^{-1} mm^{-2} of leaf lamina. In terms of surface area of mestome sheath this flux is equivalent to 2.3×10^{-12} mole s^{-1} mm^{-2} of inner tangential wall, or, knowing that in 1 mm^2 of leaf lamina the inner tangential wall adjacent to the phloem carries 2×10^6 plasmodesmata, the flux per plasmodesma is 2.9×10^{-12} mole s^{-1}. The lumen diameter is 50 nm, and the total proportion of lumen summed over the whole wall is therefore 1.5%.

If the flux through the plasmodesmata is imagined first of all to be a volume flow of 5% sugar solution, the flux is equivalent to nearly 2 volume changes per second for the cytoplasmic annulus (flow velocity 1 μm s^{-1}, taking the length of each plasmodesma as 0.5 μm). On the other hand if the flux is considered as a diffusive process, the concentration drop from the sheath end of the plasmodesmata to the vein phloem end that is necessary to drive the required flux would be 0.146 mole m^{-3} (0.146 mmolar) this being a 0.1% concentration change from a starting concentration of 5%. Since Roeckl (1949) and others have been able to detect osmotic differences between leaf cells by relatively insensitive methods, Kuo *et al.* argue that such minute concentration differences could easily exist or even be exceeded. Indeed they estimate that 50 chloroplasts photosynthesising for as little as 2 minutes could generate enough sugar to create a gradient of the required magnitude.

Kuo *et al.* have made the above calculations on the basis of an open lumen of 50 nm diameter. It has however, been found that the cytoplasmic annuli of the plasmodesmata in question have constrictions at both orifices as well as at the level of the suberised lamella (O'Brien and Carr, 1970). It is therefore a very real possibility that the fluxes may have to pass through the desmotubule. A desmotubule lumen of radius 5 nm would have a cross sectional area of one twentyfifth that of the 50 nm diameter lumen, so for desmotubular transport all of the above results would have to be multiplied by 25. Assuming mass flow there would have to be 50 volume changes s^{-1}, at a flow velocity of 25 μm s^{-1}, and assuming diffusion, the necessary concentration drop would have to be 3.65 mole m^{-3} (3.65 mmolar), that is a 2.5% concentration difference at an average of 5% sugar.

The final data apply to the companion cell-sieve element plasmodesmata of *Tussilago* (Gunning *et al.*, 1974). It was estimated that the area of this wall junction in 1 mm^2 of leaf is 0.1 mm^2, and that it carries 5.95×10^5 plasmodesmatal branches leading into 2.14×10^5 channels on the sieve element side of the wall. If these plasmodesmata are the sole pathway for loading the sieve elements, and if they carry a solution that is as concentrated as sieve tube sap - say 15% sucrose - then the loading at a maximum photosynthetic rate would be equivalent to about 7 volume changes per second (flow velocity 2.5 μm s^{-1}) in the channels at the sieve element side of the wall, provided that the flow is through a cytoplasmic annulus with outer and inner radii of 31 and 10 nm. The pressure drop needed for such a flow would be less than one thousandth of a bar, even when the resistances to Poiseuille flow are summed for the whole branching system of the plasmodesmata. If, on the other hand, the flow is imagined to be constrained to a desmotubule of radius 5 nm, the required rate of volume changes would be nearly 250 s^{-1} (flow velocity 90 μm s^{-1}) and the required driving pressure 0.26 bar. If the flux is diffusive rather than convective, the concentration drop needed would be about 1.5 mole m^{-3} (1.5 mmolar) (or 0.3% of the 15% sucrose solution) for the cytoplasmic annulus, or 0.04 mole m^{-3} (40 mmolar) (or 9% of the 15% sucrose solution) for the desmotubule.

The above sets of data have been taken, with minor modification, from the original papers. By constructing an imaginary chimaera which has vein and mesophyll anatomy as in *Beta* (Geiger and Cataldo, 1969), mesophyll plasmodesmata as in *Oenothera* (Brinckmann and Lüttge, 1974), a bundle sheath as in *Triticum* (Kuo *et al.*, 1974), and companion cells and sieve elements as in *Tussilago* (Gunning *et al.*, 1974) it is possible to bring the separate figures to a common basis and so to learn more about quantitative aspects of plasmodesmatal transport along the overall pathway. In computing their flux values, Gunning *et al.* assumed an export rate of 6.4×10^{-13} mole sucrose s^{-1} mm^{-2} of leaf. Kuo *et al.* used 5.8×10^{-13} mole s^{-1} mm^{-2}, and Sovonick, Geiger and Fellows (1974) used 2.9×10^{-13} mole s^{-1} mm^{-2}. Let us begin by normalising to a basis of 4.83×10^{-13} mole s^{-1} mm^{-2} (i.e. 1 microgram sucrose per minute per square centimetre of leaf).

Mesophyll: the anatomical studies of Geiger and Cataldo (1969) show that in *Beta* a 33 μm length of minor vein services 29 mesophyll cells, and that the total length of the veins is about 7 mm mm^{-2} of leaf. The cellular dispositions will be such that the photosynthate moving from the most distal cell will not have to pass through 29 cell-to-cell junctions. The mesophyll symplast is in the form of a tributary system. The flux at the distal extremities will be less than that in the larger branches where several inflowing streams have converged. If the Principle of Expeditious Translocation (11.2.1.) applies to the construction of *Beta* mesophyll, there might be as few as, say, 10 junctions for every 29 cells. Let us further assume, with Münch (1930), that each junction averages 100 μm^2 in area. Then the total area of mesophyll cell junctions per mm^2 of leaf is 0.23 mm^2. If each of the plasmodesmata that Brinckmann and Lüttge (1974) found to occupy 0.38% of the cell junctions has a radius of 20 nm, the total junctions will carry 6.9×10^5 plasmodesmata mm^{-2} of leaf. Hence the *average* flux of photosynthate per plasmodesma is 4.83×10^{-13} divided by 6.9×10^5, or 7×10^{-19} mole sucrose s^{-1} plasmodesma^{-1}.

We can estimate the magnitude of the likely *minimum* flux by calculating the photosynthetic output per cell, and assuming that it will equal the flux at the distal part of the mesophyll tributaries. The value works out at 2.6×10^{-19} mole sucrose s^{-1} plasmodesma^{-1}.

Bundle Sheath: The data of Kuo *et al.* (1974) are easily modified to the normalised export rate. At the junction of the inner tangential wall and the vein phloem, the total wall area is 0.26 mm^{-2} mm^{-2} leaf with 2×10^6 plasmodesmata, each carrying 2.42×10^{-19} mole sucrose s^{-1} plasmodesma^{-1}. Thus there is less wall junction than in the mesophyll, butcompensation in the form of increased frequency restores the flux per plasmodesma to slightly less than that in the mesophyll.

Companion cell-sieve element: Here the wall area is still less, at 0.1 mm^2 mm^{-2} of leaf. The branching of the plasmodesmata makes comparisons difficult, but it was calculated (Gunning *et al.*, 1974) that it is the channels on the companion cell side which are likely to be rate limiting, for although there are more of them, they are longer and narrower than the short wide channel on the sieve element side. The rate-limiting branches were found to number 5.95×10^5 mm^{-2} of leaf, hence the average flux is 8.16×10^{-19} mole sucrose s^{-1} plasmodesma^{-1}. Once again it would seem that the decrease in wall area and the increase in plasmodesmatal frequency have combined to maintain the flux per plasmodesma close to the value found in the mesophyll and bundle sheath. Other workers have, however, noted a considerably larger number of branches on the companion cell side than did Gunning *et al.*, whose method of counting may have led to an underestimation. It is also possible that the companion cell might, because of its special metabolic properties (see below) be able to generate greater driving forces than in the mesophyll or bundle sheath.

The main message to emerge from these calculations is that there is a remarkable uniformity in the flux per plasmodesma, considering the diversity of tissues and species involved in the comparisons. Not enough basic information is available as yet to rule out coincidence, but at least our attention has been drawn to the possibility that the plant can regulate its plasmodesmatal frequencies per unit area of wall in response to demands for different cell-to-cell transport capacities - just as transfer cells are thought to be able to increase their plasma membrane area in response to the presence of solutes to be transported (Pate and Gunning, 1972; Gunning *et al.*, 1974). It would be fascinating to test this idea further by measuring frequencies of plasmodesmata along the tributaries of the mesophyll symplast to see if they are less densely packed moving away from the veins towards the centres of the vein islets where the catchment of assimilate begins and the cell-to-cell fluxes are less. Another test might be to compare plants at different stages of sink to source conversion during leaf growth (Fellows and Geiger, 1974) or plants grown under conditions that yield markedly different export rates (Thorne and Koller, 1974; Servaites and Geiger, 1974). It is interesting that Habeshaw (1969) was able to saturate the translocation system of sugar beet by raising the light intensity, but it is not known what component of the system became limiting. It is, of course, a reasonable generalisation that many plasmodesmata are found where symplastic transport is intensive (see 2.3.1.), and the question of whether additional plasmodesmata can be formed secondarily, or whether the plant can predict at the time of cell plate formation where particular frequencies will be required (see 4.3.) is raised once again.

11.2.4.2. Pathways It would be out of place to consider here the large amount of physiological evidence that bears upon the subject of phloem loading in leaves. A comprehensive account is given by Geiger (1976). Two extreme views of the process may be taken. One is that assimilates move symplastically from mesophyll to sieve element, perhaps in a protected compartment such as the endoplasmic reticulum. The other is that there is a symplastic discontinuity along the pathway, and that the assimilates temporarily enter the apoplast and then are selectively and actively concentrated into the symplast again prior to export in the sieve tubes.

We have seen that the structural evidence does not rule out the former of these hypotheses, for there are plasmodesmata all along the route, though in certain plants those junctions with few plasmodesmata, such as the outer face of the A-type transfer cell, might have to be circumvented. On the other hand if the symplastic route is indeed open and is followed, it would be anticipated that the assimilates would be diffusing down a concentration gradient from their source, and there is weighty evidence that this smooth gradient does not exist over the whole pathway, either in the plants studied plasmometrically by Roeckl (1949) (see 11.2.1), or, more especially, in the sugar beet leaves studied so intensively by Geiger and his associates. Roeckl found that the palisade mesophyll cells in *Robinia* had an average osmotic pressure of nearly 18 bar, while the sieve tube sap was equivalent to more than 30 bar. Geiger, Giaquinta, Sovonick and Fellows (1973) recorded an even more striking adverse gradient for diffusion from mesophyll to sieve element, and as already noted (11.2.3.1) added the finding that the companion cells and sieve elements had almost equally high osmotic pressures. These two cell types also accumulate comparatively high levels of radioactivity during steady state photosynthesis and translocation in labelled carbon dioxide (Geiger and Cataldo, 1969). This and other evidence favours the second view of phloem loading, suggesting that assimilates leave the symplast, and are retrieved in a highly concentrated form from the apoplast at the sieve element-companion cell surface - a conclusion foreshadowed more than 40 years ago by Phillis and Mason (1933). Measurements of the relevant surface areas indicate that for the sieve element itself to do the loading would require an unreasonably high flux per unit area of plasma membrane (MacRobbie, 1971a), but that adding the comparatively large surface area of the companion cell membrane (especially if it is a transfer cell) makes the trans-membrane flux much more feasible (Gunning *et al.*, 1974; Sovonick, Geiger and Fellows, 1974).

The final loading process will be examined again later, meanwhile let us return to the earlier steps. Where might the assimilates leave the symplast? The low osmotic pressure of the phloem parenchyma cells in the veins makes them attractive candidates for the end of the symplastic road: they would certainly be strategically located near to the companion cell plasma membrane and plasmodesmata connect them to the bundle sheath (Table 11.1.). There are at least two pieces of evidence that the symplastic pathway through the mesophyll is used (apart from the observations of suggestive patterns of cell and tissue disposition already considered): one is that plasmolysis of the mesophyll reduces translocation (Geiger, Sovonick, Shock and Fellows, 1974): the other that isolated veins absorb sucrose less well than do veins that are symplastically connected to the mesophyll (Cataldo 1974). Added to these points there is the very telling fact that leaves which possess presumed permeability barriers (Casparian bands, suberised lamellae, see 11.2.2.1.) in the apoplast surrounding the veins seem well able to function: it would make very little sense for such plants to release their photosynthate on the wrong side of their barriers, i.e. distal to the phloem. On the other hand, wash-out experiments indicate that leakage into the apoplast *can* occur in the mesophyll (Kursanov and Brovchenko, 1970).

No definite conclusion regarding the role of the symplast in mesophyll-vein transport can be reached at present. It is probably used to differing extents in different plants, and probably not to the exclusion of apoplastic movement. Perhaps plants with peri-veinal apoplastic barriers release assimilate intraveinally, whilst others, such as those studied by Geiger, and Kursanov and Brovchenko, can release over a wider area. In the former type a major function of the barriers may be, as Kuo, O'Brien and Canny (1974) suggest, to prevent the transpiration stream from washing assimilates that have been released into the apoplast away from the vein phloem, instead confining them to a compact region from which they can efficiently be retrieved. Schoolar and Edelman (1971) describe promotion of leakage from the veins of a C_4 plant by iodoacetate, and since it is sucrose that leaks out, the simplest interpretation is that the products of re-fixation in the bundle sheath cells pass on into the vein and *then* into the apoplast. The likely site of release cannot be pinpointed, as very little is known about plasmodesmatal frequencies in cells lying within the mestome sheath of any plant, whether C_4 or C_3.

There is evidence that cells of the vein rib, or bundle sheath extension, described by Alexandrov (1925), Armacost (1944) and Wylie (1952) can aid in passing the inflowing transpiration stream towards the inner face of the epidermis, where spread in the plane of the leaf takes place, followed, no doubt, by smaller scale fluxes which replenish evaporative losses wherever they have occurred (Wylie, 1943; Crowdy and Tanton, 1970; Tanton and Crowdy, 1972b; Sheriff and Meidner, 1974). Such a flow pattern would make the retrieval of apoplastic assimilates much more feasible than if a bulk flow of water was passing through the walls from vein to mesophyll (Gunning *et al.*, 1974). Sovonick, Geiger, and Fellows (1974) estimate that the sucrose concentration in the apoplast in the vicinity of the vein phloem would have to be up to 20 mM in order to support observed rates of translocation.

Once the assimilates have been retrieved, the final symplastic step along their road to the phloem is from companion cell to sieve element. We do not know what degree of compartmentation is involved here, what carbohydrate interconversions might participate, and whether sugar derived from stores (vacuolar sugar or insoluble polysaccharide) follows the same path as new photosynthate (see Humphreys and Garrard, 1971; Wardlaw, 1974; and Geiger, 1976). The relationships between companion cell, endoplasmic reticulum, desmotubule and sieve element endoplasmic reticulum have already been stressed. That this membrane-bound system may be more than just a *potential* pathway for sugar transport is suggested by the observation of swollen cisternae in the companion cells near the plasmodesmata, as if their contents had had a low water potential when fixation commenced (Shih and Currier, 1969).

Unfortunately, the picture cannot be left in the above relatively simple state. Numerous problems remain. One of the most puzzling is posed by the existence of plasmodesmata in walls of companion cells other than the wall contiguous with the sieve element. Can the companion cell somehow direct the loading flux through the sieve element plasmodesmata and avoid losses to other cells such as the phloem parenchyma? What of the 'intermediary' companion cells in the abaxial phloem of *Cucurbita pepo* veins? They have large numbers of plasmodesmata leading out to the mesophyll (Turgeon, Webb and Evert, 1975) and it is hard to see how they could function in accumulation from the apoplast and loading the sieve elements unless there are ways and means of selectively sealing plasmodesmata or fitting them with one-way valves. All leaves are translocatory sinks before they become sources of assimilate, and it is not impossible that some of the plasmodesmata that can be seen in the veins of mature leaves are no-longer-used relics of a former phloem-*un*loading pathway. Apart from instances of complete obliteration (see Chapter 4), electron microscopists have not yet noted structural features interpretable as one-way valves, though other evidence for partial closure exists (Chapter 9; Spanswick and Costerton 1967; Tyree and Tammes, 1975). The very fact that two cell types with such very different osmotic properties as companion cells and phloem parenchyma can be interconnected also speaks of some form of plasmodesmatal control, otherwise the pressure difference would not easily be maintained.

11.3. UNLOADING THE PHLOEM

The reminder in the preceding paragraph that sieve tubes are unloaded as well as loaded leads on to the next topic for discussion. To some extent the structures that we have been considering are relevant because the source tissue that was described develops *via* a stage when it is a sink. The same change in physiological activity applies over a wide range of sources, including storage tissue in vegetative organs like tubers, and reproductive structures like seeds, as well as to leaves. Other parts of the plant, such as growing root or stem apices are, by contrast, perpetual sinks.

A fundamental rule of translocation between sources and sinks seems to be emerging (Canny, 1973; Wardlaw, 1974). It is that while the sources provide the impetus for translocation, the sinks govern the direction of translocation. What, then, might be the pathway(s) of unloading at the sinks?

The immediate choice is between a symplastic route and a leakage or secretion into the apoplast, or some combination of the two. There is very clear evidence favouring a leakage process where the phloem is supplying a rather unusual sink, namely the feeding hyphae of the parasitic angiosperm *Cuscuta* (Jacob and Neumann, 1968; Wolswinkel, 1974b). Lengths of host stem with attached haustoria will continue to leak translocate when perfused much longer than do control, uninfected stem segments. The ultrastructure of the host-parasite interface in this case suggests strongly that it is the plasma membrane of the host sieve tubes that becomes leaky (Dörr, 1972). The phenomenon seems not to be restricted to sieve elements: it is not impossible that parasitic generations (embryo, endosperm etc.) might be similarly nourished by leakage from their parents, just as a number of parasites and symbionts are able to induce permeability changes in their hosts.

The other extreme is a completely symplastic unloading system. Indeed if a sink does no more than attract a translocation stream that is powered at its source, it is hard to see why a symplastic pathway should not operate. As we have seen, the distinctive sieve element companion cell plasmodesmata occur throughout the phloem, except in the protophloem, where they are often simpler. As long as the incoming molecules of translocate are removed by respiration, growth, or formation of storage macromolecules, a further supply should be attracted.

This may be what happens in root tips, where the protophloem approaches the extreme tip to within a few hundreds of micrometres (see tabulated examples in Esau, 1969), where plasmodesmata lead from the protophloem and as far as is known, from cell to cell throughout the apex, including the cells of the quiescent centre (Phillips and Torrey, 1974a) and across the root cap junction (Clowes, 1970; Juniper, 1972; Phillips and Torrey, 1974b) and where the dependency of the sink upon the incoming supplies is obvious, for example in the mitotic arrest that is brought about by carbohydrate deprivation (Van't Hof, 1973). Dick and ap Rees (1975) have recently shown that flooding the free space of root cortex and root apex of pea seedlings with unlabelled sugars does not impede the entry into the cortex and apex of labelled sugars derived from sugar supplied to an older part of the root or from products of photosynthesis in labelled carbon dioxide. Asymmetric labelling is retained unaltered in the incoming sugars, yet there are invertases in the free space. The implication is that the cortex and apex are indeed supplied *via* a protected, presumably symplastic, route.

A wholly symplastic pathway also seems a reasonable interpretation of structural and physiological observations on transport in legume root nodules, where the phloem supplies carbon that is used in nodule growth and respiration, and for the manufacture of carbon skeletons destined to receive nitrogen atoms and be exported again *via* the xylem (Pate, Gunning and Briarty, 1969; Gunning *et al.*, 1974). In this case the sink is of a different nature from a growing root tip or an accumulating storage tissue. It takes the form of secretion of molecules into the xylem, molecules which, it is thought, originate in the phloem, pass symplastically through the endodermis to the *Rhizobium*-containing cells in the core of the nodule, become metabolised, and leave, again symplastically, back through the endodermis to the secretory pericycle cells. The magnitude of the fluxes involved and the necessity for bidirectional movement across the endodermis are features of this unusual sink which are referred to in 10.5.5.

When an aphid inserts its stylet into a sieve tube, a minute sink is created, and phloem sap is extruded. The stylet is larger, of course, but in principle is it any different from a plasmodesma? The question highlights one of the difficulties in thinking about the role of plasmodesmata in unloading the phloem. They are there, all up and down the phloem, so why should phloem sap not flow out through them? The facile answer is that no unloading will occur unless the translocate can be consumed, but an operational explanation is elusive.

One possibility is that the plasmodesmata in the side walls of the sieve element might be sites of control. Thus Webster and Currier (1968) show that induction of callose formation by heat treatment reduces lateral movement of translocate in cotton cotyledonary petioles. Without making any claims from their own results, they recognise that phloem callose could limit or regulate lateral movement under normal conditions, and that callose deposition in the parenchyma of the petiole could be responsible for the reduction in sink activity that is observed as the leaf matures.

Another possibility is that lateral transport does occur, but only within the confines of a membrane-limited compartment: only if the membrane possesses the necessary vectorial trans-membrane transport systems will onward lateral movement occur. Fellows and Geiger (1974) support this concept, showing that in sugar beet leaves that are young enough to be sinks, the companion cells are at approximately the same osmotic pressure as the sieve elements. By contrast the mesophyll cells have a much lower osmotic pressure (just as in leaves that are mature enough to be sources) - so it is clear that lateral transport from the sieve elements has not simply flooded the entire leaf symplast. The veins of those parts of leaves that are still importing have *some* capacity to accumulate radioactivity derived from labelled carbon dioxide, and Fellows and Geiger suggest that the progression from sink to source in the leaf is in fact the progression in the accumulative powers of the sieve element-companion cell complex - or presumably the plasma membrane that bounds the complex. If that membrane can accumulate solutes and raise the osmotic pressure within the compartment it bounds sufficiently to prevent a pressure driven inflow of translocate from neighbouring sources, then unloading will cease. If the membrane can do more than this, then phloem loading and export will occur.

The idea that a membrane can prevent unloading, not so much by being a permeability barrier as by working in the opposite direction to create an unfavourable pressure balance, suffers from the same problem that was raised in the concluding paragraph of the section on phloem loading in leaves. There are plasmodesmata leading from the companion cell not only to the sieve elements, but also to either the phloem parenchyma, or to the border parenchyma (especially in *Cucurbita* - Turgeon *et al.*, 1975), or to both. Once again it seems inevitable that the latter plasmodesmata must be sealed in some way if the pressure is not to be dissipated, or else there is some entirely different method of controlling unloading.

Quantitative data on the mass transfer of translocate through sieve tubes and into sinks are available for several storage organs (Canny, 1973). Estimation of flow rate through plasmodesmata in growing potato tubers has been attempted (Crafts, 1933), but without any knowledge of plasmodesmatal frequencies. Recent observations on a nectary demonstrate how extremely intensive phloem unloading processes can be and what high fluxes plasmodesmata may be able to carry. The following summarises recent work on *Abutilon* nectaries (Gunning and Hughes, in prep.).

Abutilon nectaries have been studied by several workers, especially Mercer and Rathgeber (1962), Findlay and Mercer (1971a,b), Findlay, Reed and Mercer (1971), and Reed, Findlay, and Mercer (1971). The nectar is secreted from the apical cells of trichomes, each trichome surmounting a stalk cell that in turn is connected to an eminence on the outer face of an epidermal cell. *Abutilon* secretes extremely concentrated nectar and, as would be expected, the underlying tissue contains a dense web of phloem, approaching to as close as two cells below the epidermis. A permeability barrier in the anticlinal wall of the stalk cells prevents movement of an apoplastic tracer (the fluorescent brightener Calcofluor white M2R new (Hughes and McCully, 1975) from the subjacent tissue into the trichomes. It is inferred that everything that enters the trichome must pass through the protoplast of the stalk cell.

Attention was focussed upon the first periclinal wall on the distal side of the permeability barrier, i.e. the distal wall of the stalk cell. It proved to have a surface area of 126 μm^2, and averages derived from ultrathin sections cut so as to reveal face views indicate the presence of 12.6 plasmodesmata μm^{-2}, i.e. a total of 1586 per trichome. The plasmodesmata are illustrated in Fig. 1.1.; they are simple, with no marked constrictions limiting the cytoplasmic annulus (inner radius 8 nm, outer radius 14.5 nm), and with a desmotubule that is quite clearly attached to the endoplasmic reticulum, cisternae of which are abundant close to the plasmalemma. When sectioned transversely the desmotubules do not always appear open, but some are seen with an electron-lucent annulus (outer radius 5 nm) surrounding a central rod (radius 1.5 nm). The average length of the plasmodesmata in this very thin cell wall is 87 nm.

The pre-nectar passing between the phloem and the site of secretion must pass through the plasmodesmata in this wall or across the plasma membranes that line its opposed faces. Secretion rates ranging from 8 to more than 50 μm^3 s^{-1} trichome $^{-1}$ were observed using the method of Findlay and Mercer (1971a). Taking a mid-range rate of 33, the flux per plasmodesma is 2.1×10^{-2} μm^3 s^{-1}, this representing 580 volume changes s^{-1} (velocity of flow 50 μm s^{-1}) through the cytoplasmic annulus, or 3400 volume changes s^{-1} (velocity nearly 300 μm s^{-1}) if the flux is confined to the desmotubule (excluding the central rod). Assuming laminar flow through an annulus (equation 3 in Chapter 1) the hydrostatic pressures needed to generate such flows are 0.015 bar (cytoplasmic annulus) or 0.3 bar (desmotubular annulus). These pressures are 1-2 orders of magnitude smaller than would be required for the trans-membrane alternative to the plasmodesmatal pathway, if the hydraulic conductivity of each membrane is 5×10^{-8} m s^{-1} bar^{-1}.

Here, then, is an example of phloem unloading with transport rates very much higher than in any other well documented case. It cannot, however, be as simple as a pressure driven release of phloem sap. Certain substances, notably nitrogenous compounds and potassium ions, are retrieved from most nectars (Ziegler, 1968), and there are other metabolic modifications to the sugars (Lüttge, 1971). It is also reported that one sample of phloem sap obtained from an aphid on an *Abutilon* petiole had a lower osmotic pressure than the nectar (Reed, Findlay and Mercer, 1971). These aspects await a satisfactory explanation, but meanwhile the flux estimates have a bearing on the mechanism of transport through plasmodesmata.

Mercer and Rathgeber (1962) discussed the possibility that the pre-nectar might travel from the phloem to the exterior in the internal space of the endoplasmic reticulum by a sequence of pinocytotic and reverse pinocytotic absorption and secretion events at each successive plasma membrane, but were worried by the small volume of the endoplasmic reticulum compartment. If the endoplasmic reticulum systems of adjacent cells are connected by desmotubules, the need for pinocytosis and its reverse is avoided, and although the compartment is small, it may validly be suggested that transport through it is (in at least this case) a necessity. The total volume of the stalk cell is 600-700 μm^3, yet it is passing 2000 μm^3 or more of pre-nectar per minute. If this flow were to pass through the cytoplasm and the cytoplasmic annuli of the plasmodesmata there would be an inevitable wash through of low molecular weight substances of all kinds. It seems most unlikely that any cell could for long survive such perfusion unless back diffusion replenishes the system or unless the flow is compartmentalised. Incredible though the velocity of flow through the desmotubule may appear, the endoplasmic reticulum-desmotubule system is, at the moment, the only likely compartment.

We are thus led back to a general point on which to conclude: the idea that the symplast is, in effect, intercellularly connected endoplasmic reticulum. Here may be the *raison d'être* of plasmodesmatal ultrastructure: the plasma membrane canal is needed only to provide a route for the endoplasmic reticulum, and the cytoplasmic annulus, being undesirable but necessary is in the majority of cases sealed off as far as possible by constrictions. In some cases the annulus may be usefully employed, e.g. it could provide a route for the retrieval of amino acids and ions from pre-nectar; certainly in the distal wall of the stalk cell of mature *Abutilon* nectaries is not constricted (though it *is* constricted in immature nectaries) and if, as envisaged above, the outward flow of pre-nectar is compartmentalised, back diffusion of the retrieved molecules would be operationally feasible. The idea that the annulus and the desmotubule may both be used is not so easily applied in other instances where bidirectional transport through plasmodesmata has been proposed, as for example in the tangential walls of endodermal cells in roots and legume nodules (Chaper 10) and in the bundle sheath of C_4 plants (Chapter 12), for in these sites the annulus is constricted twice if not thrice.

11.4. OPEN DISCUSSION

Vallisneria and *Abutilon* provided the focal points for the discussion.

LUMLEY reported his anatomical and autoradiographic observations on *Vallisneria*. There are two sorts of leaf in this plant (as in many water plants) and the interpretation of the long series of experiments

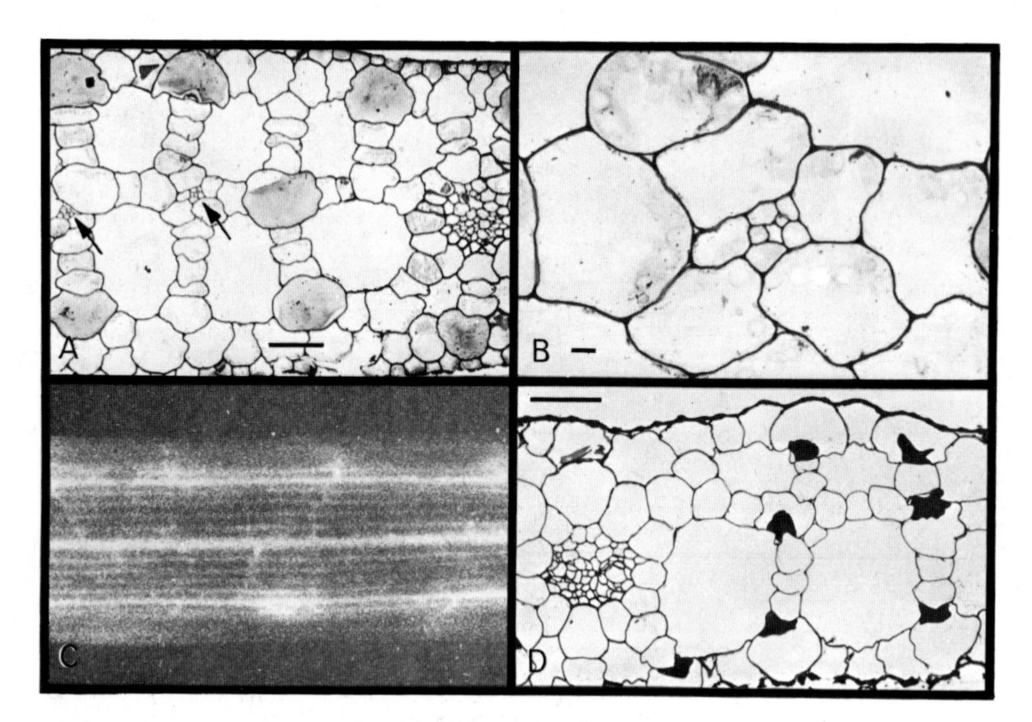

Fig. 11.3. *Vallisneria* leaves: (A) shows the type of leaf which in addition to major longitudinal vascular bundles carries minor strands of phloem tissue in some of the parenchyma diaphragms; two such bundles are included in (A) (arrows) and one is enlarged in (B). When a transporting zone of this type of leaf is autoradiographed after $H^{14}CO_3^-$ has been supplied to an absorbing zone silver grains appear (in reversed contrast in (C)) over all types of bundle - major, minor and cross veins. The other type of leaf is shown in (D), there being no small bundles in the parenchyma diaphragms between the major veins. Scale markers 100 μm in (A) and (D), 10 μm in (B)

by Arisz and co-workers depends crucially on which one they used. The best evidence for symplastic transport in *Vallisneria* parenchyma comes from work on leaves that were reduced to parenchyma bridges. But in one type of leaf, *Vallisneria* possesses very small bundles of phloem running along most, but not all, of the longitudinal parenchymatous diaphragms (Fig. 11.3a,b.). This fact is not mentioned in any of the papers from Groningen. Furthermore, if the apical part of one of these leaves is immersed in radioactive bicarbonate solution, then autoradiograms of neighbouring transporting zones show longitudinal labelling patterns strongly suggestive (but not proof) of vascular transport in the small as well as the large longitudinal veins (Fig. 11.3c.), not to mention the cross veins which Arisz *et al.* do include in their diagrams. In the other, thin, narrow, leaves of *Vallisneria*, the minor bundles are absent (Fig. 11.3d.). Hence if the parenchyma bridge experiments employed the former type, phloem translocation could have contributed to the observed transport; if the latter type was used, then all longitudinal bundles could indeed have been removed, leaving cell-to-cell transport through non-vascular tissue as the only available symplastic pathway.

The general reaction to the above information was threefold: that regardless of the type of leaf that had been used, Arisz and co-workers had still demonstrated retention of mobile solutes within the symplast; that it seemed more than ever unnecessary to ask symplastic transport through *parenchyma* to account for the relatively high rates of movement that had been observed; and that once again the desirability of physiological and structural investigations proceeding hand in hand had been demonstrated.

GUNNING had suggested that because the volume flow of pre-nectar through the stalk cell of the *Abutilon* nectary trichomes amounts to several cell volumes per minute, it would be reasonable to expect the flow to be compartmentalised in order to avoid wash out of essential solutes. This was criticised by WALKER, who analysed an equilibrium state in which the upward wash through effect (equal to concentration (c) times velocity (v) of flow) is balanced by back-diffusion (equal to the diffusion coefficient (D) times the standing concentration gradient ($\Delta c/L$)). Re-arranging, $\Delta c/c = vL/D$, and assuming $v = 50\ \mu m\ s^{-1}$ through plasmodesmata of length 87 nm and $D = 2 \times 10^{-9}\ m^2\ s^{-1}$, the concentration drop at the distal wall of the stalk cell at the steady state is about 0.2% of the total. Substituting values appropriate to the cell as a whole, the steady state drop is about 0.1% of the total. Hence for one cell plus one wall, multiplied up to a file of 12 cells, a gradient of $\approx$3% from tip to base suffices to balance the unidirectional upward flow. This seemed an entirely feasible situation and it also shows that if solutes such as potassium or amino acids accumulate in the trichome due to selective secretion of sugars from the pre-nectar to produce nectar, then they too could be retrieved by simple diffusion operating against the mass flow. Transport could, in other words, be bidirectional within the same compartment and the stalk cell would not become depleted of its essential metabolites.

It was pointed out in reply that if the mass flow is confined to the desmotubule the six-fold increase in velocity makes the overall gradient about 15%, but WALKER said it seemed to him reasonable to think of the flow as being in the cytoplasmic annulus since that pathway is shown by the electron microscope to be open, also he did not understand how there could be bulk flow through the desmotubule if the annulus surrounding it is open. At this point other aspects of the system, notably the mechanism of selective secretion of sugars and the location of the 'pump' became relevant to the discusision.

GUNNING described nectar secretion in *Abutilon*, where (as shown in the earlier papers - see text) discrete droplets emerge very rapidly at intervals of about 10 seconds through presumed cuticular pores at the apex of the trichome, the mechanical hysteresis of the cuticle and the sequential slow build up of pressure in the sub-cuticular space and rapid release of pressure in the discharge phase giving rise to the quantised secretion process. QUAIL asked what happens at the final secretion step if the pre-nectar passes from cell to cell *via* endoplasmic reticulum; does it have to enter the cytoplasm and then cross the plasmalemma? GUNNING said the details are not known. The ultrastructural observations that he and HUGHES were making were suggestive of an 'explosion' process, in which endoplasmic reticulum cisternae distend and liberate their contents through their own limiting membrane and the plasmalemma, both of these membranes self-sealing again afterwards. Given such a system, the selectivity of secretion could be achieved by retrieval of solutes from the cisternae into the cytosol, alternatively if the pre-nectar is in the cytosol, the selectivity should be governed at the plasmalemma or by retrieval *into* the endoplasmic reticulum. In reply to CARR it was said that the apical cell of the trichome does have a considerable volume fraction of mitochondria, comparable with, say, transfer cell cytoplasm, but that the endoplasmic reticulum is the most conspicuous cytoplasmic element; there are numerous extensive cisternae, the degree of swelling of which depends upon the osmolarity of the fixative that is used.

FINDLAY recalled that in the original papers (see text) a 'Curran-MacIntosh' model had been proposed, with a pump somewhere in the trichome operating across a membrane to increase the internal pressure in a compartment connected to the exterior by a perforated wall (the cuticle), the nearest likely membrane being the plasmalemma of the apical cell, with the compartment in question then being the sub-cuticular space. GUNNING did not know how much energy would be needed for this but doubted whether the mitochondria or the plasmalemma present in the apical cell could cope with the very large trans-plasmalemma fluxes that occur (of the order of 10^{-2}-10^{-4} mole sugar m^{-2} s^{-1}); he suggested that if the 'Curran-MacIntosh' model is extended, then a much more extensive loading membrane can be envisaged, namely the plasma membranes of the companion cells that load the phloem strands supplying the nectary. The compartment loaded by this 'pump' would then be the symplast extending from sieve element right up to and along the trichomes (plasmodesmata had been observed all along this pathway) and the perforated wall would in effect be the final plasmalemma through which the nectar 'explodes' (see above) plus the perforated cuticle which, as before, brings about the pulsation type of secretion. The loading membrane on this extended model could not only be the plasmalemma of the companion cells near the nectary (it has to be proposed that these cells *can* load the system, in that isolated portions of sepal that carry nectaries will continue for long periods to pump sugar) but also the companion cells in the ultimate source of sugar for the nectary, i.e. in the source leaf or leaves. Much depends upon whether the nectary is supplied from concomitant photosynthesis (involving a loading process) or from stored polysaccharide which is already in the symplast.

WALKER, FINDLAY and GUNNING then discussed another possible location for a 'pump', namely the base of the trichome. The cytology of the stalk cell did not make it a likely candidate, said GUNNING - it has an even smaller area of plasmalemma than the apical cell, and less total mitochondrial volume; he saw the main function of this cell as providing the seal which prevents intermingling of the contents of the trichome and the subtending apoplasts. FINDLAY then proposed a two-stage pumping system, with a remote driving force centred in the phloem and a final process in the apical cell. LUMLEY and GRESSEL suggested experiments aimed at locating the pump(s) by progressively damaging trichomes and their underlying tissue while watching to see when secretion stops, and GUNNING predicted that *if* the experiments could be done, it would be found that the phloem drives the system but that the trichomes govern the composition of the nectar.

Since none of the models was throwing new light on the problems of bidirectional transport or the pathway through plasmodesmata, the discussion closed.

12

SYMPLASTIC TRANSPORT OF METABOLITES DURING C_4-PHOTOSYNTHESIS

C.B. OSMOND[1] AND F.A. SMITH[2]

[1]Department of Environmental Biology, Research School of Biological Sciences, The Australian National University, Box 475, P.O., Canberra City, A.C.T. 2601, Australia

[2]Department of Botany, University of Adelaide, Adelaide 5001

12.1. INTRODUCTION

The C_4 dicarboxylic acid pathway of photosynthetic carbon metabolism is a major elaboration of the photosynthetic carbon reduction cycle (PCR-cycle) which has stimulated a great deal of new research into all levels of carbon metabolism in higher plants in recent years (Hatch and Slack, 1970; Hatch, Osmond and Slatyer, 1971; Black, 1973; Hatch, 1975). The unique features of the C_4 pathway centre around leaf anatomy and the location and activity of enzymes involved in the carboxylation and decarboxylation of C_4 acids. There is now good evidence that these acids are synthesised in one cell layer (the mesophyll) and decarboxylated in another (the bundle sheath). This seemingly pointless exercise is believed to utilise C_4 acids as CO_2 carriers to bundle sheath cells and to result in an increase in CO_2 concentration in the vicinity of RuDP[3] carboxylase in these cells. The higher intracellular concentration of CO_2 is believed to optimise the carboxylation activity of this enzyme and to effectively abolish the inhibition of the carboxylase due to atmospheric levels of O_2.

This review briefly restates the evidence for compartmentation of carboxylation and decarboxylation enzymes in leaves of C_4 plants (reviewed in detail by Hatch and Osmond, 1976) as well as the structural basis of this compartmentation (reviewed recently by Laetsch, 1971, 1974; Björkman, Troughton and Nobs, 1973). We will, however, place particular emphasis on the structure of the mesophyll cell - bundle sheath cell interface and will attempt to extend earlier conceptions of the role of plasmodesmata in the rapid symplastic transport implicated in C_4 photosynthesis (Osmond, 1971; Smith, 1971).

[3]Abbreviations: PEP = phosphoenolpyruvate; RuDP = ribulose-1,5-diphosphate; 3-PGA = 3-phosphoglycerate; ATP = adenosine triphosphate

12.2. COMPARTMENTATION IN C_4 PHOTOSYNTHESIS

The involvement of symplastic transport in C_4 photosynthesis is based on the kinetics of $^{14}CO_2$ fixation, the strict compartmentation of key enzymes between mesophyll and bundle sheath cells and on the metabolic capacities of the two cell types upon isolation. The progressive development of more satisfactory techniques for isolation of mesophyll and bundle sheath cells has introduced some confusion during the evaluation of compartmentation of these reactions in leaves of C_4 plants (Coombs and Baldry, 1972; Black, 1973; Laetsch, 1974). For example, the early differential grinding techniques (Osmond and Harris, 1971) show a degree of cross contamination of photosynthetic carboxylase enzymes between cells which is not apparent in the subsequent more satisfactory cell separation techniques.

There is now abundant evidence for the generalised model shown in Fig. 12.1. (Gutierrez, Gracen and Edwards, 1974; Hatch and Osmond, 1976). The principal features of this model are:

(1) The reactions leading to the synthesis of the C_4 acids malate and aspartate, and to the regeneration of the C_3 precursor PEP, are confined to mesophyll cells.

(2) The carboxylation of PEP takes place in the cytoplasm of mesophyll cells, but the reduction of oxalacetate to malate and the phosphorylation of pyruvate are light-dependent events confined to mesophyll chloroplasts.

(3) Mesophyll chloroplasts are unique in that they do not contain the enzymes involved in the regeneration or the carboxylation of RuDP (i.e. they lack fraction-1 protein). These chloroplasts do, however, contain some PCR-cycle enzymes.

(4) The reactions involved in the decarboxylation of C_4 acids, the carboxylation of RuDP and the reduction of carbon to the level of carbohydrate (the PCR-cycle) are confined to bundle sheath cells.

(5) In some species the C_4 acid decarboxylase is found in bundle sheath chloroplasts, and in others it is found in bundle sheath mitochondria. The PCR-cycle is confined to bundle sheath chloroplasts.

This differential localization of enzymes implies that there is movement of C_4 acids from the vicinity of mesophyll cell chloroplasts to the cytoplasm of bundle sheath cells. There is no radiotracer kinetic evidence to suggest that significant amounts of externally applied $^{14}CO_2$ have direct access to bundle sheath cells during C_4 photosynthesis (Hatch and Osmond, 1976). Thus we believe that the symplastic transport of C_4 acids, as CO_2 carriers, to bundle sheath cells and the return of C_3 acid precursors to mesophyll cells (Fig. 12.1.) is an essential feature of the C_4 photosynthetic pathway.

The properties of symplastic transport in this system are difficult to approach by direct experimentation. The process is evidently very rapid and the transport path length is only about 50 μm between the two cell layers, which are not readily accessible in intact leaves. It is difficult to derive quantitative data from autoradiographic studies (see Osmond, 1971), and this approach seems unlikely to be very useful. However, it is feasible that radiotracer kinetic experiments, coupled with non-aqueous separation of cell layers and contents may yield useful data. For example, early studies suggested that a large part of the ^{14}C was in the soluble, cytoplasmic phase rather than associated with organelles (Berry, Downton and Tregunna, 1970).

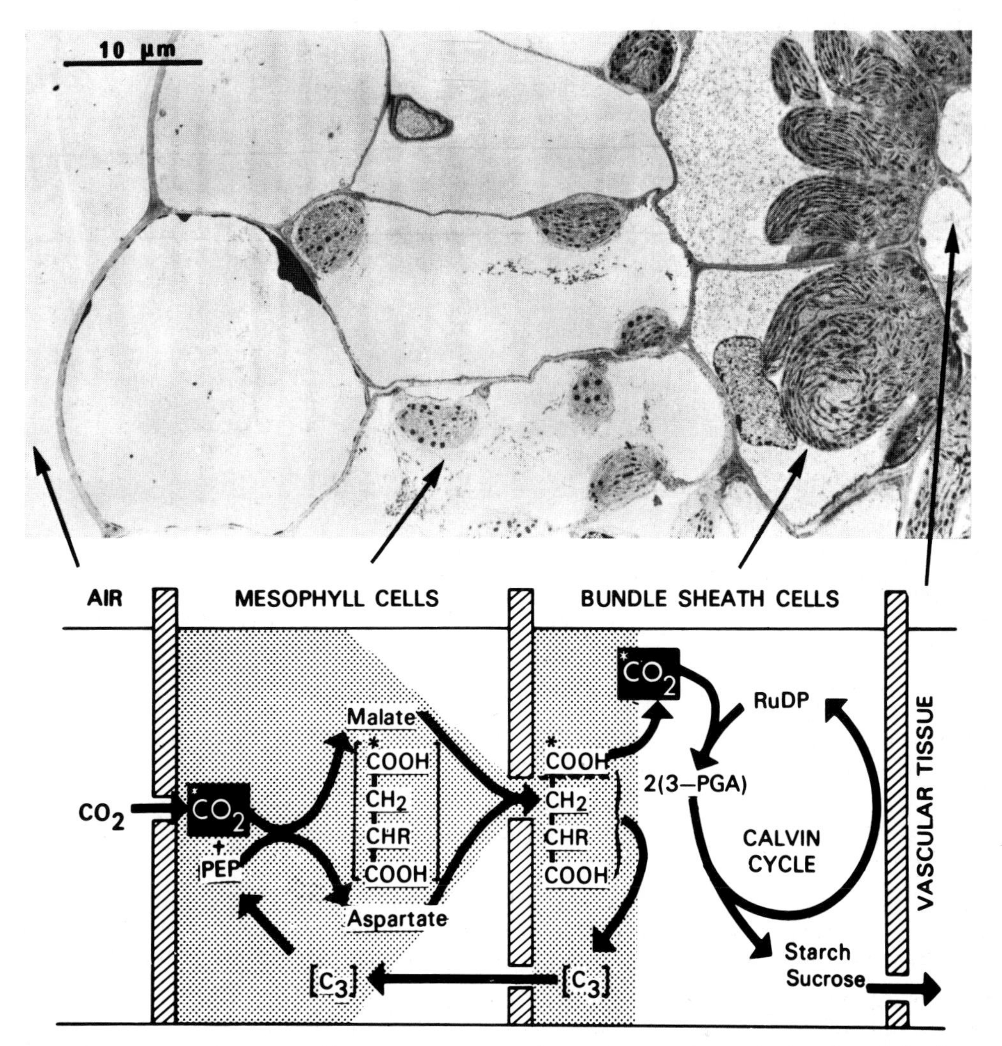

Fig. 12.1. A simplified scheme showing the compartmentation of the basic reactions of the C_4 pathway in mesophyll and bundle sheath cells. This scheme is shown in relation to a micrograph of *Panicum miliaceum* (from Hatch and Osmond, 1976). The arrows depicting the entry of CO_2 into the system and the departure of sucrose do not imply symplastic movement

Hatch (1971a) allowed the C_4 acid pools of *Zea* and *Amaranthus* to attain isotopic equilibrium with $^{14}CO_2$ prior to transfer to $^{12}CO_2$. During the 'chase' semi-logarithmic plots of the loss of ^{14}C from the C_4 acids are distinctly biphasic (Fig. 12.2.) whereas the loss of ^{14}C from the free $^{14}CO_2$ pool in these leaves and that from 3-PGA show a single phase with a decay constant identical to that of the slow phase of the C_4 acids. (These parameters describe events which take place after the C_4 acids have entered bundle sheath chloroplasts). The same analysis applied to a 'chase' experiment in which the C_4 acid pool had not attained isotope equilibrium, failed to reveal biphasic kinetics for loss of label from C_4 acids (Galmiche, 1973).

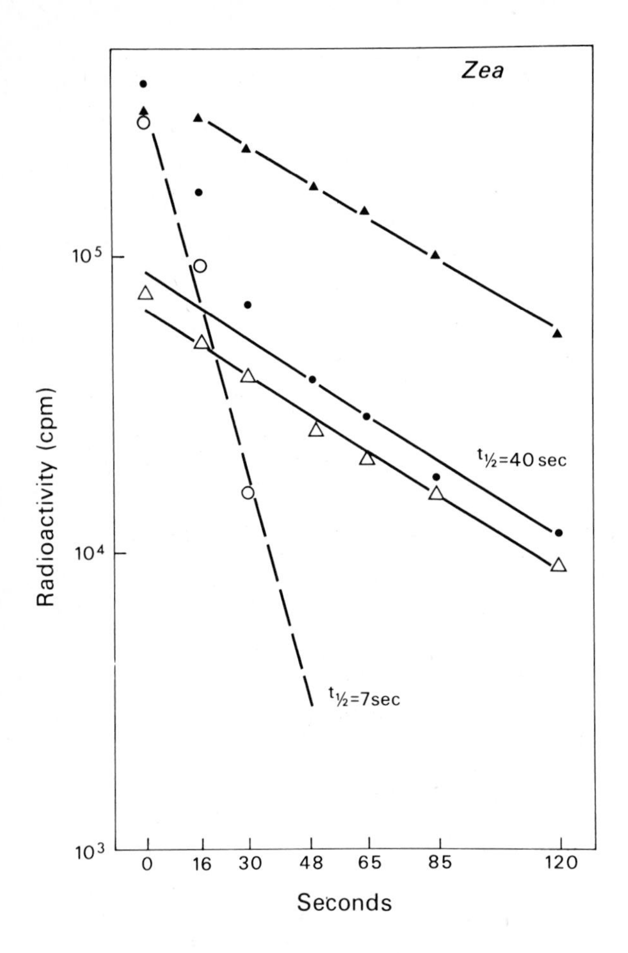

Fig. 12.2. The kinetics of ^{14}C loss from the C-4 carboxyl of malate + aspartate (●) the C-1 carboxyl of 3-phosphoglycerate (▲) and the free $^{14}CO_2$ pool (△) during a chase in $^{14}CO_2$ following exposure of *Zea mays* leaves to $^{14}CO_2$ for 35 seconds (data of Hatch, 1971a). The semi-log plot shows an initially rapid phase of C_4 acid decarboxylation (○), indicating two pools of C_4 acids. Further work is required to identify these pools which may reflect those of the cytosol and chloroplast. Aside from this, and an initial lag in the case of 3-phosphoglycerate, the kinetics of turnover of these carbon atoms are very closely related, as indicated in the simplified scheme of Fig. 12.1.

The analysis of these experiments is complex and the equations used for flux analysis at flux equilibrium are not appropriate in this case (MacRobbie, 1971b). The biphasic nature of the malate decarboxylation shown in Fig. 12.2. may reflect two compartments for this metabolite in bundle sheath cells.

It is important to stress that the carboxylation-decarboxylation sequence of C_4 photosynthesis is in effect a closed system in which charge (and pH) balance must be maintained without large-scale ionic movements into and out of the symplast. In this respect, the system differs from the somewhat analogous situation in Crassulacean acid metabolism, and from the carboxylation or decarboxylation reactions associated with excess uptake of inorganic cations or anions into plant cells (Raven and Smith,1974; Osmond, 1975). It appears that the subsidiary phosphoglycerate/triose phosphate shuttle (not shown in Fig. 12.1.) must be involved in correcting the charge imbalance which would otherwise result from the synthesis and transfer of malate or aspartate (with two carboxyl groups) inwards and pyruvate or alanine (with one carboxyl group) outwards (Hatch and Osmond, 1976). This aspect of C_4 photosynthesis has received surprisingly little attention, when compared with the emphasis on charge-balance in studies of ion transport (i.e. salt accumulation) and its control.

In the simple treatment in 12.4. we have followed the 'CO_2 carrier' concept by assuming that the minimum flux of C_4 acids into the bundle sheath cells is equivalent to the rate of net photosynthesis. These fluxes are examined in terms of the geometry of the adjacent cell layers, and the role of plasmodesmata is considered.

12.3 STRUCTURAL ASPECTS

In the majority of C_4 plants the photosynthetic tissues are organized in a double hollow cylinder of cells which is wreath-like or 'Kranz' in cross section. In the generalised Kranz complex the outer cylinder of relatively thin-walled mesophyll cells closely adjoins the inner cylinder of thick-walled bundle sheath cells which surround the solid cylinder of vascular tissue. This generalised arrangement is found in *Zea* and *Amaranthus*, the two C_4 plants discussed in detail below, but significant variations (Brown, 1975) are known which may be particularly valuable for testing some quantitative aspects of symplastic transport in this system. In the Cyperaceae many species have a layer of chloroplast-free cells between the mesophyll and bundle sheath cells (Laetsch, 1971); in *Arundinella* the counterparts of the bundle sheath cells form a solid cylinder without vascular tissue in the interveinal Kranz complex (Cookston and Moss, 1973) and in *Aristida*, the bundle sheath contains two concentric cylinders of cells (Laetsch, 1971; Carolin, Jacobs and Vesk, 1973). More complex arrangements with incomplete cylinders are common in the Chenopodiaceae (Björkman, Troughton and Nobs, 1973; Carolin, Jacobs and Vesk, 1975).

Mesophyll cells are more or less radiately arranged about the bundle sheath in the generalised Kranz complex and tend to be separated by normal intercellular spaces. Scanning electron micrographs provide a particularly dramatic contrast between the exposure of mesophyll cell walls and those of the bundle sheath to the leaf air spaces (Björkman, Troughton and Nobs, 1973; Chen *et al.*, 1974). The bundle sheath cell walls are always extensively thickened, may sometimes stain for lignin (Hattersley, unpublished) and in the monocotyledons, transmission electron microscopy shows a darkly staining band of material believed to be suberin (O'Brien and Carr, 1970; Laetsch, 1971).

Figure 12.3. is a reproduction of electron micrographs published by Laetsch (1971) which show two characteristics of the suberin lamella in the cell walls between mesophyll and bundle sheath cells. In *Saccharum* the suberin lamella surrounds the bundle sheath cell cylinder and penetrates a short way into the walls of adjacent bundle sheath cells (Fig. 12.3A,B.). Laetsch (1971) indicated that the development of the suberin lamella may depend on age and growth conditions. A second feature of the suberin lamella is the way in which it is greatly thickened in the vicinity of the plasmodesmata (Fig. 12.3C.). The plasmodesmata appear to be constricted as they cross the thickened suberin lamella (Laetsch, 1971). The suberin lamella appears to inhibit the outward movement of dyes from bundle sheath cells (O'Brien and Carr, 1970) and presumably further restricts gas and solution phase transfer of materials across an already formidable cell wall barrier. The absence of the suberin lamella in dicotyledonous C_4 plants however, precludes the assignation of any special role to this structure in the transport events of C_4 photosynthesis.

In all C_4 plants, the mesophyll and bundle sheath cells are connected by numerous plasmodesmata (O'Brien and Carr, 1970; Laetsch, 1971, 1974; Carolin, Jacobs and Vesk, 1973). Figs. 12.3C,D, show the plasmodesmata of *Saccharum* in longitudinal and transverse section;

Fig. 12.3E. shows the plasmodesmata connecting adjacent bundle sheath cells of *Aristida*. Olesen (1975) has described the ultrastructure and frequency of the plasmodesmata in the mesophyll-bundle sheath interface of *Salsola kali*. Structurally, the plasmodesmata conform to the

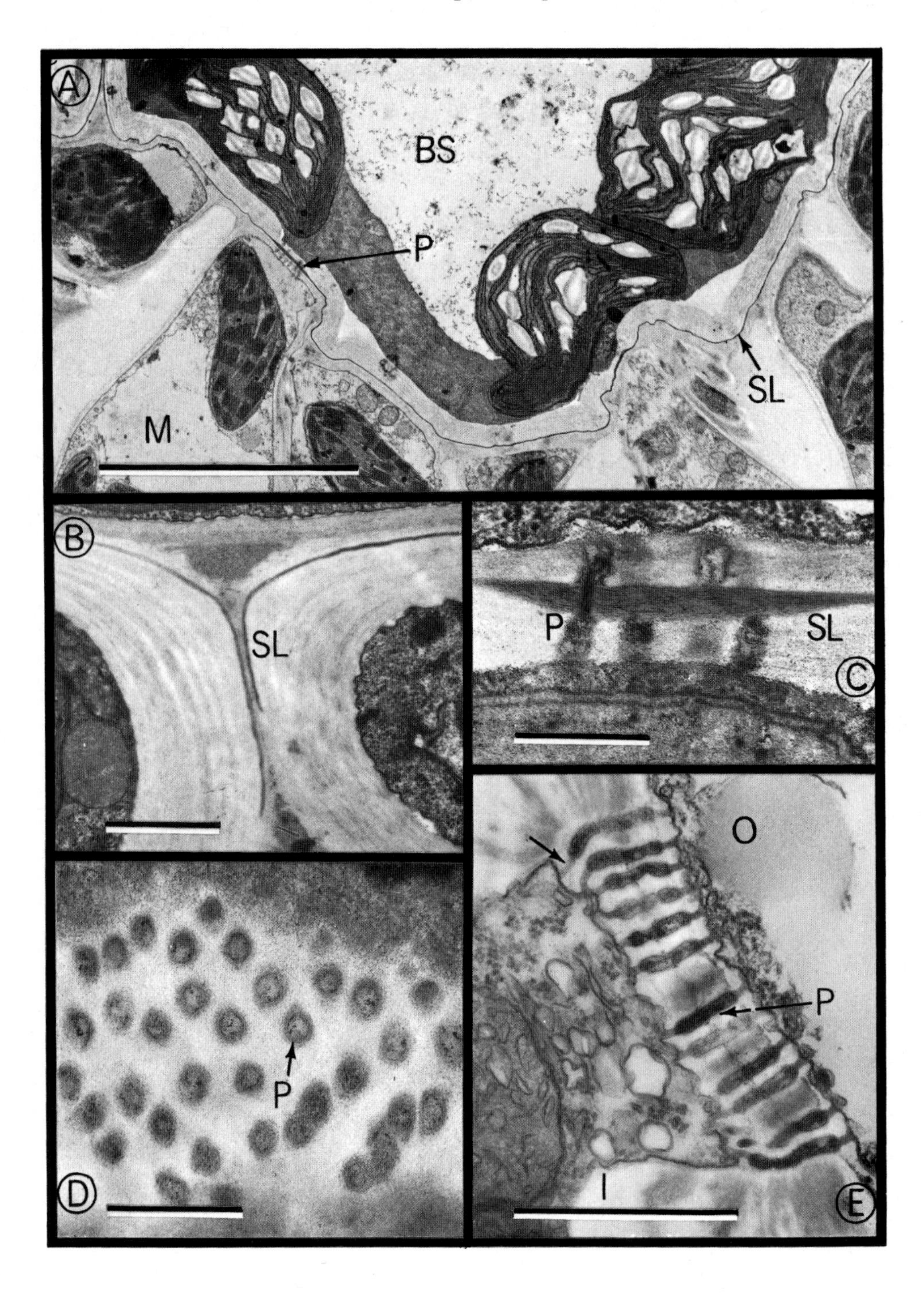

description by Robards (2.4.). They are about 450 nm long, have a pore radius of 22±3 nm, and the radius of the desmotubule is 7.5±1 nm. The plasmodesmatal frequency in the cylindrical sheath interface is about 15 μm^{-2}, a comparatively high value (cf. Table 2.1.). Quantitative studies of plasmodesmatal frequency in bundle sheath cell walls of *Zea mays* are only now being made. O'Brien (personal communication) has indicated that they resemble those of the 'intermediate' bundle type of wheat (Kuo, O'Brien and Canny, 1974).

Further studies will be required to resolve uncertainties as to the frequency and ultrastructure of plasmodesmata in different C_4 plants, as well as the nature of the connections between plasmodesma membranes and other cell membranes. Slack, Hatch and Goodchild (1969) proposed that the chloroplast peripheral reticulum may be associated with the plasmodesma membranes; a view recently re-iterated by Chapman, Bain and Gove (1975). Irrespective of their role in symplastic transport implied below, the plasmodesmata are evidently effective in maintaining the strict compartmentation of soluble enzymes. For example, PEP carboxylase, a protein of molecular weight about 300,000, is present only in the cytosol of mesophyll cells, and several amino-transferases show similar compartmentation of soluble isoenzymes.

12.4. THE SYMPLAST FLUX DURING C_4 PHOTOSYNTHESIS

Some properties of symplast transport during C_4 photosynthesis may be derived from net photosynthesis rates, from the pool sizes of photosynthetic intermediates measured in steady state isotope equilibrium experiments, and from the geometry of the generalized Kranz complex. The following treatment, based on that proposed by Osmond (1971) and recently amplified on the basis of more accurate estimates of several parameters (Hatch and Osmond, 1976) provides only a preliminary assessment of the role of plasmodesmata in symplast transport in this system.

◀ Fig. 12.3. The plasmodesmata (P) and suberin lamella (SL) of the mesophyll - bundle sheath interface in C_4 plants (electron micrographs reproduced from Laetsch, 1971; with permission)

A. Bundle sheath (BS) and mesophyll cells (M) of sugarcane. Note suberized layer (SL) in the primary wall of bundle sheath cells and pit field with plasmodesmata (P) traversing this layer. Scale mark = 10 μm

B. Walls of adjacent bundle sheath cells showing termination of suberized layer. Scale mark = 1.0 μm

C. Plasmodesma (P) traversing the thickened suberized layer (SL) in pit field. Note constricted plasmodesma in the region where it traverses suberized layer. The constricted region is filled by an electron-opaque core. Scale mark = 0.5 μm

D. Pit field with transverse section of plasmodesmata (P). Electron-opaque core is not observed in most profiles of plasmodesmata. Scale mark = 0.5 μm

E. Pit field with plasmodesmata (P) in wall between cells of inner (I) and outer (O) bundle sheath in *Aristida ascensionis*. A suberized layer is not visible in these walls. Note constricted ends of plasmodesmata (arrow), thin wall, and (as in C) the presence of desmotubules. Scale mark = 1.0 μm

Primary estimates of C_4 acid flux across the mesophyll bundle sheath cell interface may be obtained by assuming that all CO_2 fixed by C_4 plants crosses this interface as malate or aspartate. The net flux of CO_2 in C_4 plants is unidirectional (i.e. there is no significant exchange of CO_2 in photorespiration as in C_3 plants) and ranges from 0.05 - 0.1 μmole s^{-1} (mg chlorophyll)$^{-1}$. The surface area of the bundle sheath cylinder may be estimated by microscopic examination and both surface area and rate converted to a common leaf volume basis as given in Table 12.1. The minimum flux (J_M) obtained by dividing rate of photosynthesis by area is about 10^{-2} nmole $mm^{-2}s^{-1}$, across the whole surface of the bundle sheath cylinder. It is unlikely that movement of C_4 acids across two cell membranes (plasmalemma) and a thick, often suberised, cell wall could proceed at anything like this rate (cell membrane fluxes for charged molecules are commonly 10^{-5} to 10^{-4} nmole mm^{-2} s^{-1}). Tyree (1970) estimated that the plasmodesmata may be 10^2 to 10^3 more permeable than the cell membrane - cell wall path. On this basis alone it seems that plasmodesmata provide a low resistance pathway approximately adequate to sustain the estimated fluxes. It should be noted that the rates (J_M) across the mesophyll - bundle sheath interface are at least 10 times greater than those for inorganic ions moving across the root endodermis (see 10.5.3.).

The reality of these plasmodesmatal fluxes may be assessed by comparison with independent flux estimates based on concentration gradients and probable diffusion coefficients. The pools of C_4 acids involved in photosynthesis may be estimated from steady state radiotracer experiments (Hatch, 1971a; Osmond, 1971) and those of other metabolites such as pyruvate and alanine, may also be estimated but with less certainty (Hatch and Osmond, 1976). It seems likely that these pools of metabolites are restricted to the cytoplasm of mesophyll cells in the case of C_4 acids. An approximate minimal concentration in the mesophyll cells (C_m) may be estimated on the basis of area and volume measurements taken from transmission electron micrographs, as shown in Table 12.1. Osmond (1971) estimated the possible values for such diffusive fluxes (J_C) from the equation:-

$$J_C = \frac{\alpha D\ (C_m - C_b)}{r_m (\log_e \frac{r_m}{r_b})} \qquad \ldots.(1)$$

where α is that fraction of the surface area across which diffusion occurs, D = the diffusion coefficient, C_m and C_b are the concentrations of C_4 acids in the mesophyll and bundle sheath respectively, and r_m and r_b are the average radii of the chloroplast layers in the two cell layers. This equation is a modification of Fick's Law and was used by Pitman (1965b) to describe the diffusion of inorganic ions in the root cortical cylinder. Assuming for convenience that C_b is zero, and that D = 8×10^{-4} mm^2 s^{-1} (a mean value for compounds of similar charge and molecular weight: Weast, 1963), then values of J_C can be calculated as shown in Table 12.1. The maximum values (i.e. where α = 1) are about 100 times greater than the corresponding values for J_M (Table 12.1.). As suggested by Osmond (1971), J_C becomes equivalent to J_M if the effective surface area across which diffusion is occurring is only about 1% of the total surface area (i.e. where α = 0.01). This effective surface area coincides with the area occupied by plasmodesmata, at least in *Salsola kali* (Olesen, 1975) and probably also in *Zea mays*, assuming that in the latter the pitfields resemble those in wheat (see above). There is, however, a major complication arising from the ultrastructure of the plasmodesmata. If, as suggested by Olesen (1975), and elsewhere in this book, movement (in this

TABLE 12.1.

CALCULATED C_4 ACID FLUXES IN THE KRANZ COMPLEX DURING C_4 PHOTOSYNTHESIS

Data are from Hatch and Osmond, 1976

Parameter	Symbol (where used in text)	Units	*Zea mays*	*Amaranthus edulis*
Net photosynthesis, per leaf volume	-	nmole mm^{-3} s^{-1}	0.185	0.180
Surface of bundle sheath cells	-	mm^2 mm^{-3}	12.5	22.0
C_4 acid flux, across bundle sheath	J_M	nmole mm^{-2} s^{-1}	1.48×10^{-2}	0.82×10^{-2}
Photosynthetic C_4 acids, pool size per leaf volume	-	nmole mm^{-3}	1.85	1.82
Estimated C_4 acid concentration in mesophyll *	C_m	nmole mm^{-3}	38	60
Radius, chloroplast layer, mesophyll cells	r_m	mm	60×10^{-3}	80×10^{-3}
Radius, chloroplast layer, bundle sheath	r_b	mm	45×10^{-3}	25×10^{-3}
Calculated maximum C_4 acid flux (α = 1) **	J_C	nmole mm^{-2} s^{-1}	1.76	0.52
Calculated C_4 acid flux (α = 0.01) **	J_C	nmole mm^{-2} s^{-1}	1.76×10^{-2}	0.52×10^{-2}
Calculated 'restricted' C_4 acid flux (α = 0.001)***	J_C	nmole mm^{-2} s^{-1}	3.04×10^{-2}	4.8×10^{-2}

* assuming that C_4 acids are restricted to cytoplasm (10% of mesophyll cells).

** using equation (1) (see text); assuming that $D = 8 \times 10^{-4}$ mm^2 s^{-1} (Weast, 1963), and that C_b = zero.

*** using equation (2) (see text); other assumptions as for **.

case of C_4 acids) is restricted to the desmotubules, then α decreases by a factor of ten and J_M then apparently greatly exceeds J_C. The solution to this apparent conflict, as pointed out by Olesen (1975), is that the restriction imposed by the plasmodesmata only occurs in a very small part of the total diffusion path length. The shorter this restricted length, when compared with the total path length, then the further the system departs from the uniform system for which equation (1) applies. Accepting that for most of the path length, which is intracellular cytoplasm, $\alpha \approx 0.1$, it seems valid to apply the simple Fick's equation:-

$$J_C = \frac{\alpha D \, (C_m - C_b)}{L} \qquad \ldots.(2)$$

where L is now an effective diffusion path length largely determined by the plasmodesmatal length, and may be taken as 1 μm. This assumes that the mesophyll layer is only one cell layer, i.e. solutes pass through only one layer of plasmodesmata. As shown in Table 12.1., J_C would then be greater than J_M even when α (within the plasmodesmata) = 0.001. It may be noted that the assumption that the intracellular diffusion can be neglected strictly only applies if the cytoplasmic contents are well mixed, and does not apply if the mesophyll has more than one layer of cells. It may not be necessary to invoke cytoplasmic streaming as the means of ensuring mixing, as Tyree (1970) calculated that over distances less than 40 μm diffusion ought to account for movement as efficiently as streaming (but see Chapter 5). It may be concluded that, provided sufficiently large concentration gradients are maintained by carboxylation and decarboxylation processes, transport of C_4 acids (and the other shuttling metabolites) can occur efficiently by diffusion. In this case at least, the uncertainty about the size of the plasmodesmatal pore (or the 'open' portion thereof) does not seem important. (Of course, this may not be the case where concentration gradients are small). Nevertheless, it should be pointed out that if the desmotubule (Figs. 12.3C,E.) provides the pathway for solute movement, and is attached to the endoplasmic reticulum then it is necessary to postulate solute transport across the endoplasmic reticulum membrane. This would further remove the kinetics of solute movement from a simple diffusion system.

A major uncertainty not considered so far relates to the direction of water movement and its effect on solute movements in the symplast. It is assumed that the above calculations apply to the movement of C_4 acids inwards to the bundle sheath and to the movement of C_3 metabolites back to the mesophyll cells. Water flux in transpiring leaves is opposed to the former but in the same direction as the latter. In wheat Kuo *et al.* (1974) showed that the very much higher water fluxes were principally confined to large vascular elements which carried little sugar. Whether a similar specialization applies in leaves of C_4 plants or whether the water movement is restricted to the apoplastic pathway remains to be assessed.

12.5. THE 'CO_2 CONCENTRATING MECHANISM': FUNCTIONAL ASPECTS OF SYMPLASTIC TRANSPORT IN C_4 PHOTOSYNTHESIS

As indicated at the outset, the intricacies of C_4 acid carboxylase and decarboxylase enzyme systems and the structural organization of the Kranz complex evidently form the basis of a CO_2 concentrating mechanism in bundle sheath cells. Leaves of C_4 plants exposed to $^{14}CO_2$ in the light show a substantial internal pool of free $^{14}CO_2$ (1-2 mM CO_2 + HCO_3^-, assuming that the pool is confined to the cytoplasm plus chloroplasts). This is at least 10 times larger than found in the same leaves in the dark or in leaves of C_3 plants in the light (Hatch, 1971a). The kinetics of labelling of this free CO_2 pool are intermediate between the C-4 carboxyl of C_4 acids and the C-1 carboxyl of 3-PGA and it probably represents the CO_2 released by the decarboxylase systems in bundle sheath cells. The minimum estimate of the free CO_2 concentration in the cytoplasm plus chloroplasts of bundle sheath cells (equilibrium at pH 8.0) is approximately 25 μM (Hatch and Osmond, 1976). This is about 3 times the free CO_2 concentration in air-saturated water. Because the system is unlikely to be in equilibrium, CO_2 concentrations would be rather higher than this. The maintenance of high intracellular CO_2 concentrations in bundle sheath cells must be carefully assessed, for as pointed out by Smith (1971) the C_4 acids and CO_2 or HCO_3^- presumably share the same diffusion path.

The CO_2 concentrating mechanism could function in two ways. First, if the rate of CO_2 transport and release *via* the C_4 acids vastly exceeds the rate of fixation in bundle sheath cells, CO_2 would be concentrated in these cells. This mechanism presents grave problems of nucleotide and ATP balance, however, and would imply that C_4 plants spend vast amounts of energy simply to accumulate CO_2 in the bundle sheath. Quantum efficiency measurements rule out this explanation of the CO_2 concentrating mechanism. The second and most probable explanation of the CO_2 concentrating mechanism is that quite low concentrations of CO_2 plus HCO_3^- (1-2 mM) are required to saturate RuDP carboxylase in bundle sheath cells. Even if the CO_2 plus HCO_3^- concentration in mesophyll cells is zero, this gradient is only about 10% of the C_4 acid gradient (Table 12.1.). Any back flux of HCO_3^- would be restricted to the plasmodesmata, and though D is greater than for C_4 acids ($D \approx 15 \times 10^{-4}$ mm^2 s^{-1}) substitution in equation (2) gives a back flux of at most 10% of the C_4 acid flux. The pathway for CO_2 diffusion is more problematical, since for this uncharged molecule the plasmalemma plus cell wall pathway is a possibility (i.e. the value for α is uncertain). In this case, returning to equation (1), and assuming a 25 μM concentration difference; the maximum back flux of CO_2 ($\alpha = 1$) would still be less than 10% of the C_4 acid flux. Specific anatomical features (e.g. the suberin lamella) might tend to reduce this value, essentially by reduction of α or by reducing D for CO_2. It may be concluded that reasonably strict stoichiometry between the C_4 acid carboxylase - decarboxylase system and the PCR cycle is likely.

Although we have no way of estimating the status of the HCO_3^- + CO_2 equilibrium in bundle sheath cells, the maintenance of CO_2 concentrations in excess of 25 μM in these cells seems very probable. Concentrations of this order are sufficient to abolish the inhibition of C_3 photosynthesis *in vivo* due to 21% O_2, and to stimulate the rate of CO_2 fixation in C_3 species to that of C_4 species (Osmond and Björkman, 1972). Recently, *in vitro* studies with RuDP carboxylase/oxygenase from both C_3 and C_4 leaves have shown that CO_2 concentrations between 25 and 50 μM largely eliminate the competition due to 250 μM (21%) O_2 (Badger, Andrews and Osmond, 1974). Thus the symplast transport of C_4 acids as CO_2 carriers to bundle sheath cells initiates a CO_2 concentrating mechanism which reasonably accounts for the proposed optimisation of RuDP carboxylase activity in bundle sheath cells (Hatch, 1971b) and the elimination of O_2 inhibition of this enzyme in C_4 plants in 21% O_2 (Bowes, Ogren and Hageman, 1971).

12.6. ASSESSMENT

This armchair treatment of the role of plasmodesmata in symplastic transport during C_4 photosynthesis does not at present permit any firmly based conclusions. Sufficient data are available, however, to support the following possibilities.

(1) The fluxes of metabolites between the mesophyll and bundle sheath cells of C_4 plants are evidently too rapid to involve transport across cell membranes and the cell wall.

(2) Even if these rapid fluxes are restricted to plasmodesmata occupying between 0.1 and 1% of the interface between mesophyll and bundle sheath cells, the likely concentration gradients of metabolites are probably sufficient to permit these fluxes on the basis of diffusion alone.

(3) The maximum gradients of CO_2 and HCO_3^- concentration between bundle sheath and mesophyll cells could result in a leakage of CO_2 from the bundle sheath cell at approximately 10% of the rate of C_4 acid exchange. It is possible that the CO_2 concentrating mechanism in bundle sheath cells of C_4 plants may be described in these simple terms.

Thus we conclude that the plasmodesmata connecting the adjacent cells of the Kranz complex are capable of carrying a rapid diffusion of organic metabolites during photosynthesis in leaves of C_4 plants. It is difficult to see a role for the suberin lamella in these processes, but clearly this structure may have important implications for apoplastic transport in the region of the plasmodesmata of monocotyledonous C_4 plants. The above discussion does not preclude the involvement of other specialized structures, such as the chloroplast peripheral reticulum, in symplastic transport of metabolites. Hopefully, it may stimulate further quantitative ultrastructural and microphysiological research into this complex symplastic transport system which is fundamental to photosynthesis in C_4 plants.

12.7. OPEN DISCUSSION

The major problems of delivering C_4 acids to the bundle sheath and minimising CO_2 loss from sheath to mesophyll having been dealt with, several people thought that there were still areas of uncertainty regarding the system. In LORIMER's view, the passage of transpirational water might set up a mass flow from veins to mesophyll that would override the effect of the concentration gradients that are supposed to drive the flux of C_4 acids into the bundle sheath. After all, 100-200 g H_2O is lost for every g of carbon dioxide fixed in the typical C_4 situation. WARDLAW too believed that for the system to work the transpiration stream would have to be dissociated from the photosynthetic fluxes. The latter are delicately poised and bidirectional. CARR pointed out that we know very little about the permeability of the suberin lamella - it might allow an outward apoplastic movement of water that would not hinder the symplastic fluxes. SMITH agreed that there is a potential apoplastic pathway for the water. He suggested that it would be useful to subject some C_4 plants to the type of study that Weatherley has employed for locating relative resistances to flow through the leaf. He also pointed out that not all C_4 plants (especially dicotyledonous ones) possess a suberin lamella, and that suberin lamellae (even in the monocotyledons) are not confined to C_4 plants. KUO added that in wheat there is no suberin lamella around the cross veins, which therefore might provide a route for water to enter the mesophyll without jeopardising trans-bundle sheath fluxes in the longitudinal veins. KUO said that he was attempting to map the distribution of the suberin lamella, and the plasmodesmata that pierce it, in some C_4 representatives.

Another area of uncertainty that was mentioned is the fate of the final photosynthate. In what form is it translocated, and what path does it follow? WARDLAW wondered whether a mass flow 'sucking' the photosynthate might occur and might influence the overall kinetics. The point made in 11.2.4.2. was reiterated: that the path from the bundle sheath to the sieve elements had been sadly neglected compared with the path from mesophyll to bundle sheath. The export path of photosynthate from the tracts of Kranz-type cells that occur remote from veins in certain C_4 grasses is especially obscure.

The mesophyll-bundle sheath junction provides an excellent example of adjacent cells showing pronounced morphological and biochemical differentiation despite the presence of numerous plasmodesmata. GUNNING suspected that this, together with the observed constrictions of the plasmodesmatal plasmalemma, and the need to keep cytosol enzymes in their respective cells, added up to a good case for the cytoplasmic annulus being of little function, with the desmotubule providing the operational pathway. In terms of the flux through the plasmodesmata, said SMITH, the figures do not rule this out - but there remains the unknown quantity of transport across the endoplasmic reticulum membrane. In the case of PEP carboxylase, the enzyme mentioned in the text, however, a molecular weight of 300,000 implies a radius of about 4.5 nm, and such a large molecule would not readily pass into or through a plasmodesma, so maintenance of the biochemical differential is, for this enzyme, not difficult to envisage.

13

PLASMODESMATA IN GROWTH AND DEVELOPMENT

D.J. CARR

Department of Developmental Biology, Research School of Biological Sciences, The Australian National University, Box 475, P.O., Canberra City, A.C.T. 2601, Australia

13.1. HYPOTHESES AND ASSUMPTIONS

The evidence of a role for plasmodesmata in differentiation and development is as yet purely circumstantial. However, the amount and quality of the evidence, supported by the recent, although equally circumstantial, evidence from equivalent animal systems, is now becoming difficult to explain away. I shall begin by stating three hypotheses, fundamental to this Chapter, and to deal with certain assumptions. The first hypothesis was stated by Strasburger (1901) and repeated by Lundegårdh (1922) and by Münch (1930): it is that the plasmodesmata are constructed and function not only so as to facilitate transport of solutes between cells but also to guard the genetic individuality of those cells. Similar ideas were expressed by Bennett (1973a,b) concerning the construction and functioning of junctions between embryonic animal cells. The second hypothesis is also an old one, and can be traced back at least as far as Goebel (1897). It states that cells isolated from neighbouring cells will be free to undergo some form of special development. In lower plants they may even regenerate whole new individuals. The converse is not true. Cells in communication with other cells remain free to develop or differentiate in special ways. This is a corollary of the first hypothesis. Again, it is true also of animals, that cells not electrotonically coupled have either ceased to divide, such as voluntary muscle fibres or some neurons, or they are malignant. On the other hand, absence of coupling is not a necessary condition for uncontrolled growth in animal cells, and some tumour cells are electrotonically coupled (Loewenstein, 1968). The third hypothesis really depends on the first two. It is, that tissues and organs in which function is dependent on cell communication or metabolic coupling by way of plasmodesmata, will fail to carry out that function if the coupling is broken. So far, we have only uncertain evidence in favour of this hypothesis, but it brings the plasmodesmatal role into line with the great deal that is known of the co-ordination of activities in the plant and its internal correlations.

The assumptions concern the plasmodesmata themselves. The mere morphological demonstration of existence of plasmodesmata between cells tells us nothing of their functioning. We have no means - as yet - of demonstrating that individually they are functioning at a given time, and we have no means of knowing whether the function switches on or off or is even controlled by the protoplasts themselves. As we shall see, there is evidence of such control in animal cells and some models of possible regulatory systems have been suggested. Evidence of plasmodesmatal connection between cells is only one side of the story: we need to know also that the plasmodesmata are functional, and this can come from measurements of electrotonic coupling and also from demonstrations of the movements of substances between the cells. Even then, we are making assumptions linking these processes with the plasmodesmata, although direct visual demonstration of the movement of tracer substances within the plasmodesmata should now be possible, and electrotonic coupling would now be difficult to explain except in terms of plasmodesmatal continuity. The possibilities of both rapid establishment and dismantling of cell junctions have been adequately demonstrated for animal cells (see Loewenstein, 1973). We have as yet no information on the time taken to establish a secondary plasmodesma (see Chapter 4 for a definition of this term) across a cell wall, nor how long it takes to occlude plasmodesmata and obliterate them. It may take hours or days to make or seal the necessary holes in the cell walls. However, there is evidence of both the neoformation and the obliteration of plasmodesmata (see below and Chapter 4), and of correlations between these processes and aspects of growth and development. One can agree that there is a certain truth in the teleological statement that plasmodesmata are formed in the numbers and in the walls where they appear to be needed, and with Pfeffer's (1897) statement, given in full at the beginning of this Volume, that if plasmodesmata were not known to exist, it would be necessary to propose them. However, until we know more about the quantitative aspects of plasmodesmatal transport and its regulation, statements on the degree of coupling between cells in terms of frequency or number of plasmodesmata must be treated with caution.

13.2. CELL COMMUNICATION IN ANIMALS

"*It is almost self-evident that the development of an embryo from a fertilized egg and the regeneration and repair of adult tissues involves communication and control processes between the cells of the organism. Information must flow from cell to cell, allowing the organism to compare what is, with what ought to be and to correct the difference*" (Goldacre, 1958). Goldacre and Bean (1960) went on to develop an electronic model of coupling and feedback between cells, although Goldacre regarded his model as merely an analogue. It was indeed ahead of its time.

Loewenstein (1968) has succinctly summarised the possibilities of communication between cells: "*(1) they may have information signals and receptor processes for these signals on their surface membrane; (2) they may release signals to the exterior and these may reach receptor processes of other cells* via *the intercellular spaces (hormonal systems); (3) the signals may pass directly from one cell interior to another*". He cites plasmodesmata in plants as likely media for such direct signalling. His own discovery of low resistance membrane contacts or junctions between animal cells has led to considerable speculation on the possibility that such junctions might play an important role in the control of growth and differentiation. Work on these junctions constitutes a model for what might be done with plant systems and it will therefore be discussed here briefly. Space does not allow an exhaustive treatment (see Cox, 1974 for reviews) and, in any case, the field is expanding so rapidly that it would be inappropriate to do more than mention salient facts and the more relevant hypotheses. The morphological channels for intercellular communication have been known so much longer for plants than for animals, that it is a matter of regret that plant physiologists have failed to seize the advantage which this gave them and have fallen so far behind the animal physiologists in their investigation.

The animal work began with the discovery that when current is passed between a cell interior and its exterior, the voltage resulting inside a contiguous cell is nearly as high as that in the cell containing the source of the current. The electrical coupling behaves linearly over a wide range of voltage and is linearly attenuated over several cells. This discovery led to the hypothesis of junctional elements in which, according to Loewenstein, areas of membrane of low resistance are apposed at a junction and are in turn sealed off from the exterior by some sort of 'perijunctional insulation'. The evidence that the morphological correlates of this model are the *gap junctions*, while *tight* or *occluding junctions* or *desmosomes* are areas of low cell-cell permeability, has been assembled by Gilula (1974). Gap junctions have been found in all multicellular animals which have been investigated. Some types of cell rarely form gap junctions (e.g. mature neurons) and some (e.g. skeletal muscle cells, circulating blood cells) normally lack them. Revel and Karnovsky (1967) showed that, in face view, the gap junction comprises a polygonal lattice. According to Bennett (1973a,b), the gap junction consists of arrays of hexagons, each 10 nm diameter, and each surrounding a bridging structure joining the apposed membranes. Each of these bridging structures is believed to have a 1 nm channel connecting the two cells. Either these are the sole sites of the low electrical resistance or else, as Bennett (1973a) states, coupling might merely involve close apposition, to within 10-20 nm, of low resistance membrane. This is an unlikely model, for substances in solution would presumably leak from such clefts into intercellular spaces, rather than pass directly to the coupled cell. Indeed, coupled cells do behave as though they have 'private pathways' for solute exchange (Bennett, 1973b).

Cells are impermeable to the dyestuff, Procion yellow (molecular weight about 550), yet if this is injected into a cell it will also move into one which is electrotonically coupled to it, evidently by a pathway not involving intercellular space (see also 6.5.). Other tracers - dansylated amino acids, sucrose and small proteins - have been shown to move, presumably through gap junctions, in the same way. The gap junction thus appears to behave as a molecular sieve, and the upper limit of size of molecule which can pass through may be of the order of 10^3 (Bennett, 1973a,b) or even (Loewenstein, 1973) 10^4 daltons. A polypeptide of molecular weight c. 1800 fails to move across junctions except *after* fixation (Bennett, 1973a). In ingenious experiments in which 'donor' cells made heavy by ingesting tantalum particles could easily be separated from recipient cells after being allowed to mix (and presumably to form junctions) with them, Kolodny (1974) has shown that RNA and some small proteins (especially histones) but not DNA can be exchanged between cells. By careful controls, most of the possibilities of exchange merely through the medium, or of labelled degradation products, or by pinocytosis of cell fragments or by transfer by contaminating micro-organisms were eliminated. Kolodny believes that both structural and informational macromolecules may be exchanged between animal cells through junctions. However, most investigators believe that junctional exchange is restricted to molecules too small to be informational macromolecules (Sheridan, 1974).

Unlike plant cells, animal cells can migrate and so must be able to make and unmake gap junctions with ease. Nevertheless, during the growth and development of plants, plasmodesmata can be formed secondarily and can also be occluded, as will be discussed later in this Chapter. Interest therefore attaches to the formation and dissolution of gap junctions in animal cells, as model systems. When animal cells are manipulated together the establishment of gap junctions can be monitored electrically. Junctions are then formed within minutes (sponge cells - Loewenstein, 1973; newt embryo cells - Ito, Sato and Loewenstein, 1974) or hours (Novikoff hepatoma cells, see Sheridan, 1974). Junctions can be made between cells of widely differing species of mammals and apparently at any locality on their membranes.

In searching for biological roles of gap junctions, Loewenstein (1968) showed that certain lines of tumour cells (e.g. L cells) are unable to form gap junctions. However, cells of other types of cancers are electrotonically coupled (see Loewenstein, 1973; Furshpan and Potter, 1968 for reviews). Cells taking different pathways of development may lose junctions (see, for example, Potter *et al.*, 1966). Cases in which there is likely to be exchange of cyclic AMP, cyclic GMP and calcium ions through junctions between cells in normal hormone-regulated animal development are discussed by Sheridan, 1974. Loewenstein and his colleagues (Loewenstein, 1973) have shown that non-coupling tumour cells can be genetically corrected by fusion with normal cells and that the hybrids are then not only non-cancerous but also form junctions with normal cells. In other studies (Gilula *et al.*, 1972; Loewenstein, 1973) it has been shown that inability to transmit nucleotides or nucleotide derivatives to other cells, which lack ability to synthesise a compound of this type essential to nucleic acid synthesis, correlates with the lack of electrical coupling and of gap junctions. Conversely, metabolic co-operation was shown to correlate with coupling and with the formation of gap junctions. While there is no proof that the substances exchanged in metabolic co-operation move through gap junctions, that they do so seems highly probable (Cox *et al.*, 1974). Moreover, it seems likely that it is a process which takes place also *in vivo* and which is involved in the normal regulation of growth, differentiation and embryonic development.

The extension of these ideas to developmental problems in animals (and later, to plants) leads us to consider briefly a recent hypothesis on the control of patterning by 'positional information' (Wolpert, 1969). To establish and periodically regulate pattern in, for instance, the developing insect epidermis, the hypothesis supposes that there is intercellular movement of information, perhaps in the form of specific molecules, although of that there is as yet no definite evidence (Wolpert, 1971). Although it is possible to show that there is electrotonic coupling between the cells of the epidermis (Caveney, 1974), the boundaries which one would expect to find (e.g. intersegmental boundaries) are not apparent, nor is there an axial gradient in coupling. Caveney (1974) states that "*if an intercellular gradient in bioelectric potential does exist, its range must be less than 1 mV and masked by the experimental error of the recordings*". On the other hand, clones of mutant cells belonging to a developing region or 'compartment' of *Drosophila* do not migrate across the boundaries of the compartment - they 'know' they belong there (Lawrence, 1975). Discontinuities such as those at segment borders are sharply bounded by cells which appear to be able to change shape and may suffer abrupt changes in adhesiveness. Moreover, the segment border is an important discontinuity between neurons belonging to one segment and those belonging to another. Since Lawrence agrees with Caveney that the epidermal cells are coupled between segments, individuality of the segments appears not yet explicable in terms of electrotonic coupling

pale areas in the middle of the lamina of variegated ('*medio-picturata*' variegations) of *Pelargonium zonale* 'Happy Thought', *Euonymus japonica* var. *aurimaculata* and *Hedera helix* 'Goldheart'. The leaves of these chimaeras (green-green-white, or green-green-yellow) have deep green edges but pale centres where the 'green' L 2 layers may lie a single layer over 'chlorophyll defective' L 3 cells. Bergann explains the pale centres as due to some substance, formed by the chlorophyll-defective partner, diffusing to the cells of the overlying layer where it destroys the chlorophyll, or inhibits its production.

13.3.2.2. Anthocyanin formation. Similar examples are known, involving anthocyanin production. In pal_{rec}/pal_{rec} (anthocyaninless) plants of *Antirrhinum* the rate of back-mutation of some cells to Pal/pal_{rec} is unusually high. Larger or smaller groups of red cells thus arise, the size of the groups depending on how many divisions the reverse-mutant cells undergo, subsequent to mutation. Whole red shoots, sectors or areas ranging in size down to small red spots may thus appear (Fig. 13.2.). The whole plant is thus often red spotted on a non-red ground (Bauer, 1930). Bauer showed that close inspection of the red spots on petals shows *"that the reverse-mutant factor also affects neighbouring cells. The genetically red cells are surrounded by a halo of pale red cells, themselves not reverse mutants"*.

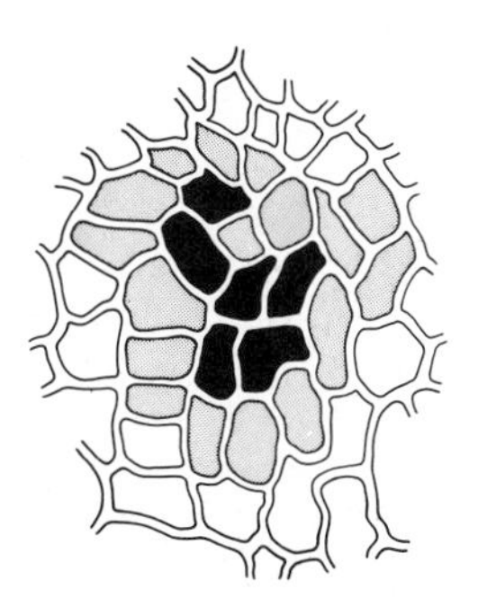

Fig. 13.2. Epidermal cells of petal of anthocyanin-less mutant of *Antirrhinum*. In the centre, a group of revertant anthocyanin-forming cells surrounded by non-revertant (mutant) cells which have nevertheless formed some anthocyanin. (Redrawn from Bauer, 1930)

The most intensive investigation of this sort of phenomenon is that of Bergann (1962) on the periclinal chimaera, *Euphorbia pulcherrima*, Ecke's Pink (Fig. 13.3.). In this species, the bracts are normally deep red due to anthocyanin formed mostly in the upper epidermis in the presence of the gene Wh (i.e. Wh/Wh or Wh/wh). In the homozygous recessive Ecke's White wh/wh, the bracts are white. Pink bracts could, in principle, be due to changes at other gene loci, reducing the amount of anthocyanin formed. Bergann showed however, that Ecke's Pink, which sports relatively frequently to red and rarely to white, is a periclinal chimaera, in which L 1 is wh/wh and L 2 and L 3 Wh/Wh. Because of the reduced light screening of the upper epidermis the hypodermis, which is Wh/Wh, forms more pigment than is formed in the hypodermis of the red parent plant. There is even some pigment in the layer beneath the hypodermis (Fig. 13.3.). The pigments are the same and in the same relative proportions in the pink chimaeral plant as in red plants. Analysis showed that in Ecke's White, which forms no anthocyanin at all, the bracts contained much higher levels of flavonols and leucoanthocyanins than those of a red plant, and the three flavonols were the same in both. Moreover, in the white plant the upper epidermis had three times as much flavonol as the lower. Evidently flavonols are formed from precursors common to them and to the anthocyanins. Because of the formation of the flavonols lack of anthocyanin formation by failure of glycosidation or absence of sugars in the epidermis of the bracts of white plants could be eliminated. Thus Bergann was led to the conclusion that pigmentation of the upper epidermis of Ecke's Pink must be due to diffusion from the underlying 'normal' cells of a direct or indirect product of the gene Wh into the defective wh/wh cells, enabling them to complete the synthesis of anthocyanin. The biosynthetic step involved might be the reduction to form the oxonium ring. Bergann himself believed that the diffusing agent must be in the nature of a co-enzyme or co-factor of the enzyme responsible for this step in biosynthesis.

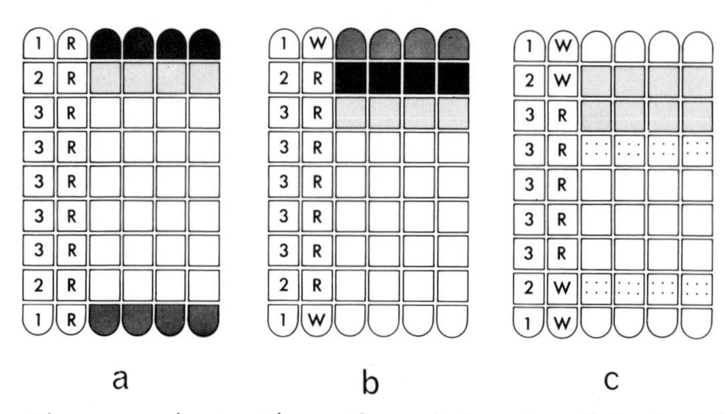

Fig. 13.3. Diagrammatic sections through bracts of *Euphorbia pulcherrima* to show location and intensity of anthocyanin formation. Letters on left hand side of each figure show potential phenotype on the basis of the genotype. a) 'wild type' = red bracts. b) Ecke's Pink. c) "*Trebstii alba*". (Redrawn from Bergann, 1962)

It seems not unlikely that similar phenomena are to be found in other cases of pink varieties of red-flowering and mutant white-flowering plants (e.g. *Bouvardia*, Bateson, 1921). Since a very late step in the biosynthesis of anthocyanin appears to be that involved in the partner-induction in *Euphorbia* the explanation may be free from criticism that the postulated inducing substance might be freely diffusible in the apoplast, rather than pass solely through the symplast. Apoplastic transport appears to be involved in a similar case in maize, where McClintock observed that complementary pigment synthesis occurs over a zone of two or three cells at the line of contact between genetically colourless (inhibited) C^IBz cells and nearly colourless (bronze) C bz cells (see Rhoades, 1952). The possibility that this phenomenon might be due to freely diffusible substrates was tested by Reddy and Coe (1962). They achieved complementary interaction by merely pressing together pieces of genetically competent aleurone tissue, peeled from fresh kernels, and placing them on agar at 25°. Under these circumstances pigment develops in only one of the partners - the one with its genetic block earlier in the chain of synthesis. It seems highly probable that the substances in this example would have traversed extracellular spaces between the apposed pieces of aleurone tissue, which are unlikely to have formed cell-to-cell connections in the 1 or 2 days of the experiments. Since pigment was found exclusively in only one of the partner tissues it is argued that neither enzymes nor catalysts could have diffused from one tissue to the other.

13.3.2.3. Mutual morphogenetic influences in chimaeras. Further possible evidence of informational transfer from one kind of tissue to another comes from studies of the progeny of interspecific periclinal chimaeras. There have been many investigations of the possibility of reciprocal influence in such chimaeras (e.g. Krenke, 1933; Bergann, 1956). Such influences were expected, indeed believed, inevitable but it was not believed that they could be of a specific, inheritable nature. Winkler (1910) obtained large numbers of seeds from his monectochimaeras, *Solanum tübingense* and *S. gaertnerianum* but all seedlings were typical of the L 2 parent, *S. nigrum*. Bergann (1956) found large effects on the size, specific gravity and numbers of seeds in fruits of the *Crataegomespilus* chimaeras and cites Buder's observations of effects on the size and numbers of stomata per unit area in *Cytisus adami* as examples of such interactions. Günther (1956-1957) reinvestigated the matter critically and admitted that divergences from parental behaviour in the chimaeras themselves (affecting characters of the epidermis or mesophyll) are difficult to attribute to exchange of formative substances, since they may be due merely to mechanical adjustments which the chimaeral layers must make to each other. She substantiated Winkler's conclusions concerning the characters of the offspring - no divergences from the parental norms were both reproducible and heritable.

Nevertheless, reproducible divergences did appear, for instance in the seedling leaves (up to leaf 5) of progeny of the dichlamydeous chimaera, *S. lycopersicon* (4n) + *S. nigrum*. Abnormalities in leaf shape were also found in progeny of a plant of *S. tübingense* compared with plants of parental *S. nigrum*. It is difficult to assess the possibility that these temporary modifications of the seedlings might have been due to a sort of 'partner induction' of the type discussed above. Nevertheless, they remind one of the similar abnormalities which appear in the progeny of plants raised from seeds treated with heterologous DNA (Hess, 1972; Ledoux, 1971). It is very unlikely that DNA is exchanged between cells of partners in the chimaera, and it is clear that the modifications which appear in progeny are not heritable. However, there are grounds for maintaining the hypothesis of some sort of mutual chemical interaction in the parent chimaeral plant with effects transmitted, perhaps, to offspring. It is well attested for instance that growth hormones applied to parent plants can be accumulated in seeds and affect their performance in germination and early seedling growth (Zeevaart, 1966) and there is evidence to show that germination is affected by climatic conditions under which seeds are formed (Schwemmle, 1969; Evenari *et al.*, 1966). More specific effects on other seed and seedling properties brought about by substances transmitted by the non-generative chimaeral partner seem therefore not improbable. However, we are no nearer than in the examples of metabolic co-operation discussed earlier to being able to claim transmission *via* plasmodesmata.

13.3.3. The Transport of Hormones

13.3.3.1. Auxins, gibberellins and cytokinins Proponents of the symplastic transport of regulatory substances in plants must have been severely discouraged by the demonstration from the twenties onwards by Paal, Boysen-Jensen, Went and others that growth substances such as auxin could be transported apoplastically. Although from time to time theories of polar transport have been put forward involving plasmodesmata (e.g. O'Brien and Thimann, 1967b) according to a recent view *"no experimental evidence is available as to whether or not plasmodesmata function in auxin transport"* (Goldsmith, 1969). Plant cells are permeable to all the main classes of hormones (cytokinins, gibberellins, auxins and various types of inhibitors) as well as to synthetic growth regulators . Cytokinins and gibberellins have been found in xylem sap. There seems no reason to suppose that any of these substances move preferentially *via* plasmodesmata. Possible models of auxin transport are considered by Goldsmith and Ray (1973) but none of them include a role for plasmodesmata. They remark that ....*"at the cellular level polar transport consists of uptake from the free space at the apical end of the cell, transport through the cell to its basal end by some means other than cytoplasmic streaming and release into the free space at the basal end of the cell, rather than cell-to-cell transport* via *plasmodesmata."* Rubery and Sheldrake (1974) assume the existence in the plasmalemma of an auxin carrier and explain polar transport as due to the localisation of an *"auxin secreting system"* (involving the carrier) in the plasmalemma at the basal end of the cells.

13.3.3.2. An inhibitor in blue-green algae. It is highly probable that the hypothetical inhibitor which prevents heterocysts arising too close together in filaments of *Anabaena* is transmitted in the blue-green algal version of the symplast *via* the 'plasmodesmata' which Wilcox *et al.*(1973; their Fig. 2.) show between the cells. As pointed out in 3.2. the structure of blue-green algal intercellular connections is still unclear. This inhibitor is held to be responsible for the non-random pattern of distribution of heterocysts along the filaments. Other phenomena - movement and nitrogen fixation - require co-ordination of the activities of cells of the filaments, *via* the plasmodesmata.

13.3.3.3. Flower hormones. Of the plant hormones whose transport has been studied, the flowering hormone(s), still hypothetical (see Carr, 1967) are the most likely to be restricted to the symplast. The view that flower hormones are constellations of hormones of known type (e.g. gibberellins) receives its most salient criticism in the fact that the known plant hormones (not excluding the still questionable 'florigenic acid' - Hodson and Hamner, 1970) can be transmitted across aqueous gaps and so between cut surfaces of tissues merely apposed, whereas we have no certain evidence that the floral stimulus can move otherwise than in cell to cell connections, such as are presumably established in grafting. In order to assess the significance of this distinction, we need to enquire into the nature of graft unions, a topic not yet surveyed ultrastructurally, and in particular into the establishment of symplastic continuity across them.

In his classic book on grafting, Vöchting (1892) observed pitting in walls common to stock and scion and assumed (p. 119) that plasmodesmata are formed, but also raised the difficulty of deciding whether a particular cell wall really lies between the two partners or belongs to one or the other. Bearing in mind this caution, Strasburger (1901) carefully investigated grafts between pear varieties and *Cydonia*, *Syringa* grafted on *Ligustrum*, and *Datura* on potato. Particularly in the last, but also in the others, opposite pitting was observed in what were believed to be fusion walls. Plasmodesmata were actually observed in walls believed to be fusion walls, in grafts of species of *Abies* and *Picea*, e.g. *Abies nobilis* on *A. pectinata*. Careful observations enabled him to distinguish the fusion zone by its relatively larger intercellular spaces, differences in the cortical cells of the partners, etc. and to identify common walls, equipped with secondarily-formed plasmodesmata. These observations did not go unchallenged. A. Meyer (1914) who had endeavoured, and failed, to observe plasmodesmatal connections between scion and stock (either in homoplastic or in heteroplastic grafts) cited Herse (1908) on the difficulties of determining the boundary between scion and stock in woody species. Herse had looked especially for the expected phloem fusions and found that phloem bridges were formed from a common callus tissue proceeding mostly from the cambial zone. One could therefore not decide which sieve tubes formed from cells of the stock, which from the scion.

According to Herse (loc.cit) "*the more one observes such unions, the more one recognises that criteria such as the course of larger intercellular spaces, the occurrence of thicker cell walls etc, are insufficient to determine precisely the boundary between otherwise similar callus cells of different origin*". These and similar criticisms were also raised in Meyer and Schmidt (1910).

In some of Vöchting's grafts of red beet, a good union was established without vascular connection, and even in autoplastic grafts, plasmodesmata would not be formed (he believed) if the cells were oriented in opposite directions of polarity. A form of autoplastic graft is to be found in the post-genital fusions (Baum 1948a,b) common in the floral development of many Angiosperm families (Asclepiadaceae, Ranunculaceae Papilionaceae, Campanulales). In some cases (see 4.3.1.) such fusions have been shown to involve the secondary formation of plasmodesmata. Nemeč (1924) was of the opinion that one can only speak of a real growth union of heterologous tissues if the walls between cells of the two partners are traversed by plasmodesmata. It is evident that in cases where no vascular connection is formed, symplastic exchanges between the partners must proceed *via* plasmodesmata. However, Ball (1969) found evidence of movement of labelled substances between callus tissues of different genera cocultured on agar media. In mixed cultures certain isozymes, present in pure cultures were replaced by others. These effects seemed not to depend on intimate fusion or the formation of plasmodesmata.

In Monocotyledons successful grafts have been established but only in a few has convincing evidence been produced of the establishment of vascular connection between the partners. Calderini (1846) claimed that grasses could be grafted at the nodes by pulling out the culm and re-inserting another (or the same) culm, similarly prepared. Thus a rice culm grown on a stock of *Panicum crus-galli* was said to have done better than on its own root system and fruited well. The technique has been successfully imitated by Muzik and LaRue (1954) with some large tropical grasses. A relatively high percentage of successful autografts of *Pennisetum purpureum* (Merker grass) were made which lived and grew for 6 months and established excellent vascular connections. Autografts of *Bambusa longispiculata* also succeeded. Lianas (*Scindapsis*, *Philodendron*, etc.) were relatively easy to graft and, peculiarly among Monocotyledons, readily formed masses of callus tissue. Muzik and LaRue also grafted sugar cane and some members of Commelinaceae (1952) and Muzik (1958) grafted orchids, with resultant growth of the scion (vanilla) for at least two years and to several centimeters, but without any vascular connection to the stocks. Presumably the capability of the scion to produce its own roots made it independent of the stock (which died if defoliated) and, as far as we are here concerned, the achievement may be trivial. Similarly, the graft of *Tradescantia fluminensis* on a leafy and vigorous *T. zebrina* to which Krenke (1933) devotes 21 pages and which lived for over a year, may be regarded as one in which there were exchanges only of apoplastically-transported materials. Vascular connection across the graft union was - even in Krenke's own words - not satisfactorily demonstrated. Schubert (1913) found no evidence for the establishment of vascular connections in his Monocotyledonous grafts. A graft of *Campelia zonata* (Commelinaceae) on its own defoliated stock lived for over a year. The scion grew and formed new leaves. An intermediate 'callus' was formed but no vascular tissue regenerated. According to Krenke (1933) successful grafts of ferns have never been reported and this statement is repeated by Brabec (1965). The latter briefly reviews the literature on the grafting of algae, Phycomycetes and higher fungi. In most of these plasmodesmata are not involved, nor are they in the intergeneric transplantations of sporophytes of mosses on gametophytes

(Arnaudow, 1925), since plasmodesmatal connections appear not to be present between the two generations.

In Dicotyledons the process of grafting itself has been investigated histologically by a number of workers, including some recent ones (Krenke, 1933; Homès (cotton), 1958; Buck (roses), 1954). (For a discussion of the fate of non-articulated laticifers in grafting see 4.9.). According to Krenke's extensive account, at the cut surfaces the cells die and brown pigments are deposited in an 'isolation layer' ('contact layer' of Muzik and LaRue, 1952). Below this layer intensive cell division may take place in either or both partners. The isolation layer then begins to break up either through resorption by the dividing tissues or (and especially in heteroplastic grafts, according to Simon (1930), who grafted *Solanum melongena* on *Iresine lindeni* (Amaranthaceae)), by mechanical pressure from either side of the graft. Resorption of the isolation layer takes place even in Monocotyledonous grafts (Muzik and LaRue, 1954; Schubert, 19]3). The callus-like tissue which is formed can be contributed to by parenchyma of all kinds, but the cambium is the most important donor. The phloem of the two partners becomes joined by newly-formed anastomosing sieve-tubes differentiated from parenchyma cells. According to Krenke (1933) the new sieve elements appear to grow, at least terminally, intrusively between parenchyma (his Figs. 121, 123) to reach a sieve-tube of one of the partners. This is not substantiated in work on phloem regeneration across wounds (Eschrich, 1953). Vascular regeneration may take place, independently of the cut vascular tissue of the partner, in parenchyma lying between them, and connections will then be made to the vascular tissue of both. Alternatively, cells of the cambial tissue of one or both partners may differentiate and form a bridge of phloem to the other. Formation of the companion cells to the new sieve elements takes place after the formation of the sieve tube, according to Krenke. However, he points out that there is a paucity of information on this and he himself investigated only two cases. Eschrich showed that eosin will travel across the regenerated phloem bridge even before sieve plates have been formed, i.e. presumably through the plasmodesmatal progenitors of the sieve plate pores.

Having digressed to provide a necessary background, we can now return to the main point of this discussion. Fusion between graft partners can take place remarkably rapidly, under ideal conditions. Buck (1954) reports that rose bud grafts become 'firm' in four days and fully established in fourteen. Information on the rapidity with which physiological union takes place can be obtained from experiments on the transport of flower hormone or labelled assimilates from one partner to the other. In Zeevaart's (1958) anatomical study of graft union in *Perilla ocymoides*, xylem and possibly phloem connections were established after, at the most, eight days. In Moskov's (1935) experiments on the same species union took nine to twelve days. However, Zeevaart's experiments in which grafts were allowed to establish between a donor of flower hormone and a receptor for only four, five or six days resulted in flowering of 10, 70 and 100% of the receptor plants respectively. This suggests that some plasmodesmatal connections, at least, can be established within four days and probably that the flower hormone can cross the graft union by means of secondary plasmodesmata between cells of the partners. In other experiments of Zeevaart with *Perilla*, the minimum period of contact for both the floral stimulus and C^{14}-labelled assimilate to move from the donor leaves across the graft to the receptor was six to seven days. In Chailakyan and Butenko's (1957) study, the movement of florigen correlated well with the movement of C^{14}-labelled assimilate. Photosynthate was transmitted across a graft union of *Silene armeria* in seven, but not in five, days after grafting, and well before sufficient xylem connection had established to prevent wilting of the scion. Although phloem usually follows xylem in regeneration de Stigter (1966) hypothesized that the transmission occurred in regenerated sieve tubes in a prefunctional stage of differentiation (as in Eschrich, 1953). Against the evidence for an obligatory association between flower hormone and assimilate in transport can be put that of Evans and Wardlaw (1966): in *Lolium temulentum* (and *Xanthium*) *"the floral stimulus can be exported by leaves still too young to export assimilates in significant amounts"*. From experiments on rates of transport the authors conclude that the floral stimulus and assimilates may both move in the phloem but they can move independently and by different mechanisms.

Vascular tissue appears not essential for the transport of florigen. Chailakyan (1940a,b) found that florigen will move through parenchyma of the stem and root, although much more slowly than in the phloem. By cutting the veins at the base of the lamina of *Perilla*, he was able to show (1940c) that transport will take place across a bridge of leaf parenchyma: indeed, if flower hormone is made in leaf parenchyma cells, it must move across them to reach the phloem. Selim (1947) showed that the flower hormone can be transported across a very thin bridge of bark of the stem of *Perilla* - perhaps in sieve-elements newly formed from a callus tissue.

Withrow and Withrow (1943) re-examined critically the statement of Hamner and Bonner (1938) that flowering in *Xanthium pennsylvanicum* could be transmitted across a 'pseudograft' in which the partners were separated by a membrane of Japanese lens paper - alleged to prevent graft union. In their experiments, Withrow and Withrow found that tissue union occurred in more than four and less than eight days (only hand sections were examined) and lens paper did not prevent either tissue union or the transmission of the flowering stimulus. In odd cases, the same was true of rice-paper membranes, even when impregnated with lanolin. This was said to be due to breakdown of the paper, allowing tissue union (on a small scale) to take place. Similar results were reported by Melchers and Lang (1948) for grafts of *Hyoscyamus niger*. Filters of different pore sizes were found to prevent transmission of 'vernalin', the hypothetical precursor of 'florigen', in grafts of non-induced annual and unvernalized biennial plants. *"Contact periods of a few days, during which vascular union was probably not established"* were sufficient for ensuring a certain degree of flowering response (Melchers, 1937) so that at least some transmission can apparently occur through parenchyma, possibly secondarily connected by plasmodesmata.

The fact that *Silene armeria* (Caryophyllaceae) can be made to flower by grafting to flowering *Perilla* (Labiatae) (van der Pol, 1972) suggests that symplastic connections, involving the formation of secondary plasmodesmata, can be made between species from taxonomically different groups. The same conclusion can be reached from the successful grafts of *Zinnia* (Compositae) on tobacco (Solanaceae) (Mothes and Romeike, 1955). Other examples of taxonomically 'wide' heteroplastic grafts are to be found in Brabec (1965, p. 424) and Mothes and Romeike (1958).

It adds to the difficulties of isolation of a flower hormone that cells may be impermeable to this class of substance, which may be able to move only in the symplast. Indeed it is one of the ironies of the search for plant hormones that substances to which the plant cell is quite impermeable could not be reapplied to the plant cell, (at least not by the usual methods) and could therefore not be discovered to have regulatory properties. This limitation may have restricted discovery to such substances to which the plasmalemma is permeable, leaving the more interesting, symplastically confined, regulatory substances whose existence has been inferred still to be identified!

13.3.4. Conduction of Stimuli

13.3.4.1. Thigmotropic responses. Gardiner (1882a,b) observed abundant plasmodesmata equally in the upper and lower halves of the pulvinus of *Mimosa pudica* as well as in the leaves of the insectivorous plant, *Dionaea muscipula*, and was the first to express the view that plasmodesmata were involved in the perception and transmission of stimuli. According to Pfeffer only the lower half of the *Mimosa* pulvinus is sensitive to stimulation. Nevertheless from his studies of contact stimuli he reached the conclusion that "*plasmodesmata, like the nerves of higher animals are the pathways of conduction of stimuli*" (Pfeffer, 1885). Plasmodesmata were also invoked by Oliver (1887) to explain the transmission of thigmotropic stimuli from one stigma lobe to the other in *Mimulus* and *Martynia*. His explanation is rendered the more plausible in that there are no vascular connections between the lobes.

In *Mimosa*, stimulus conduction is possible over exceptionally long distances - wounding the tap root may cause the leaves to fold. Haberlandt (1890) claimed to have recognised specific conducting systems involving long cells, resembling sieve elements, connected by plasmodesmata, and responsible for transmission to the pulvini of the leaves, of a hypothetical product of wounding or of other means of stimulation. This substance - 'excitation substance' ('Erregungssubstanz') - has not yet been isolated. The work done on it was summarised by Umrath (1959) and more recently by Pickard (1973). It seems probable from the rates of movement in the plant of this analogue in plants of a neuro-transmitter that it is conducted symplastically, although perhaps not in the cell-system identified by Haberlandt which although longitudinally interconnected by pits (Esau, 1973) may be secretory in function (Fitting, 1907; Molisch, 1915). In grafts between different *Mimosa* species, Lieske (1921) showed that the stimulus is transmitted across the graft union, retaining characteristics of the partner species with regard to the rate of transport of the stimulus and the specificity of the reaction induced.

As an example of a multidisciplinary approach to the problem of information transfer through plasmodesmata, one cannot do better than to cite the work of Barbara Pickard (Williams and Pickard, 1972a,b; 1974) on *Drosera* (see also Chapter 6). The tentacle of this insectivorous plant has a complex structure in which all the living cells appear to be electrotonically interconnected and all the living cells of the stalk are excitable. When an insect alights on the mucilaginous head of a tentacle it causes a lowering of the 'receptor potential' there (there are small pores in the cuticle through which the walls of the outer cells are in direct contact with the mucilage). As the receptor potential falls to a critical value, action potentials are generated, which travel down the tentacle to its base, where they cause the tentacle to inflect inward to the leaf surface. Pickard (1973) develops the suggestion that action potentials themselves might control the formation and release of various substances in the plant, such as ethylene or abscisic acid, or 'short-range' control substances, and that, in turn, such substances might trigger action potentials as a form of 'electrical signalling' in plants.

13.3.4.2. Geotropic responses. Pfeffer's concept of symplastic conduction of stimuli was strongly supported by the finding by Gardiner and Hill (1901) that plasmodesmata were especially abundant in the extremely sensitive cells of root tips. This has since been confirmed, and the distribution of plasmodesmata has been particularly well studied in root tips (Juniper and Barlow, 1969). It is particularly the root cap columella - its central region - which appears to be sensitive to geotropic stimuli and in this tissue in *Vicia faba* Griffiths and Audus (1964) reported larger numbers of plasmodesmata on transverse walls than on longitudinal walls, while in *Zea* (Juniper and Barlow, 1969) there is a similar, tenfold difference (see Table 2.1.). in *Zea* (Juniper and Barlow, 1969) and *Convolvulus arvensis* (Phillips and Torrey, 1974a,b) the ratio of the number of plasmodesmata on transverse to that on longitudinal walls increases in cells of the columella of the root cap as they are progressively displaced from the root cap initials. According to the experiments of Juniper *et al.* (1966), Gibbons and Wilkins (1970), Shaw and Wilkins (1973), Pilet (1972) and Wilkins and Wain (1975) the root cap is the site of perception of gravity by the root. The root cap also produces an inhibitor which moves into and reduces the rate of growth of the root. Horizontal displacement of the root results in downward lateral transport of the inhibitor and consequent unequal growth of the upper and lower longitudinal halves of the geostimulated root. According to Juniper (1972) experiments on removal (decapping) of the root cap of maize and replacing it on the root tip failed to restore the gravi-perceptive system and *"it looks as though, at least within the cap, plasmodesmatal continuity is necessary for the transmission of the stimulus"*. In Pilet's (1971) experiments, however, re-attaching the cap with dilute Ringer's solution (but not with an oleate solution) restored geosensitivity, albeit with some delay. Pilet claims therefore that the inhibitor will cross an aqueous but not a lipoidal gap and is likely to be water-soluble.

The discrepancy between the concept of Juniper of an essentially symplastic movement of an inhibitor produced in the root cap and the possibly apoplastic movement reported by others, including Pilet, remains to be resolved. Similarly, there are disagreements concerning the ability of the root to execute geotropic curvatures without prior irradiation. Most experiments have been carried out with irradiated roots. Pratt and Coleman (1974) have shown that the root cap of the dark-grown root is rich in phytochrome and likely to be very sensitive to light. Wilkins and Wain (1975) confirm this sensitivity and show that a powerful inhibitor is produced subsequent to irradiation. The production of the inhibitor following irradiation is confirmed by Pilet (1974). According to Scott and Wilkins (1969) dark-grown *Zea* roots do not execute geotropic curvatures but Pilet (1974) claims that they do, but with some delay. Whether the 'dark' inhibitor responsible for the 'dark' georeactions is the same as that produced after irradiation and whether both move apoplastically remains to be seen. It may be that the abundance of plasmodesmatal connections in the transverse walls of the root cap columella does play some role in the geotropic response of the root, perhaps in the transmission of the geoperceptive stimulus. For other effects of root cap removal, see 13.4.

Strasburger (1901) spent some time in trying to find out whether or not plasmodesmata were re-formed following plasmolysis and came to the conclusion that they were not. He therefore decided to carry out experiments to test their involvement in certain plant functions. The experiments involved geotropic stimulation of roots on intact plants (*Vicia faba*) and of cut shoots subsequent to a temporary plasmolysis. 'Saltpeter' was used as the plasmolyticum and the salt was removed by washing in water until the plant parts regained their former turgor. ('Saltpeter' could be either $NaNO_3$ or KNO_3 but elsewhere Strasburger refers to 'Kalisaltpeter' for the latter). The plants were placed horizontally. Two days later the roots were still alive (some died later) and some had executed random curvatures. Cut shoots of *Hippuris*, *Impatiens*, *Tradescantia*, culms of *Alopecurus* and flower-stalks of *Taraxacum* were also plasmolysed and deplasmolysed, then laid horizontally. It was not possible to eliminate the geotropic curvature at the nodes of *Alopecurus* even following 25 hours treatment with 20% 'Saltpeter' solution. This, and the occasional failure to eliminate curvature in the other shoots, was attributed by Strasburger to failure to plasmolyse all the interior cells. Whether this explanation is likely is not clear. The *Hippuris* shoots and the *Taraxacum* flower-stalks died as a result of the treatment. However, Strasburger does not explain whether or not any of the other shoots or the control *Hippuris* died following the experimental treatments.

Strasburger was himself aware of the unsatisfactory nature of these experiments but went on to say that they showed that plants which had once been thoroughly plasmolysed no longer had the capacity to carry out tropistic curvatures. One explanation of the failure of the plant parts to respond to stimuli might well be that symplastic transport of nutrients, e.g. to root cells, might well be necessary to keep them alive (c.f. Meeuse, 1957). The rapid death of the plasmolysed cells - in roots within four days - suggests this as an explanation. This suggestion receives support from the experiments of Geiger *et al.* (1974) on the inhibition of translocation in mesophyll cells by plasmolysis. Still more direct evidence that plasmolysis impedes cell-to-cell transport is provided by Smith (1972) in his work on the movement of neutral red along the filamentous young gametophyte of the fern, *Polypodium vulgare*. The dyestuff moved only along protoplasmic connections connecting one plasmolysed protoplast to another. Another possibility is that 'saltpeter' might not have been a wise choice as a plasmolyticum. Karzel (1926) found that 0.5 M KNO_3 for as little as half an hour kills cells of *Lunularia*.

Certainly it cannot be taken as inevitable that plasmolysis is irreversibly lethal to cells, since Klebs (1888) was able to keep cells of many filamentous algae, fern prothalli, *Funaria*, *Elodea*, *Vallisneria*, *Lemna* and the flesh of *Symphoricarpus* fruits alive in the plasmolysed state in sucrose solutions for many weeks, during which many of them (including those of *Elodea*) formed new cell walls. Klebs' experiments were substantiated and extended by Küster (1939) and Küster (1956) devotes a section of his book (p. 870) to a review of the work in this field. Moreover, Klebs and many others (see 4.5.5.) have obtained regeneration from cells isolated by plasmolysis from symplastic connection with other cells: a topic considered in detail in 13.4.1., where it will be necessary to return to other effects of plasmolysis upon plasmodesmata. In view of this, further observations on plasmolysis itself are relevant.

Strasburger (1901) found that in *Mnium* the plasmodesmata were pulled out of the wall and he had the impression that each cell contributed half to each plasmodesma, so that the halves merely abutted one another in the wall. Following plasmolysis with 'Saltpeter' solutions and subsequent deplasmolysis *Mnium* plants seemed normal for at least two weeks, during which he sampled them to see whether they had re-formed their plasmodesmata. Occasional plasmodesmata were found but they seemed to be *"where the protoplast in plasmolysis had not retracted from the wall"*. In any case, he says, the fixations were poor, the protoplasts tended to shrink from the cell walls. Strasburger points out that the fact that the plants appeared to have suffered no damage from the plasmolysis treatment is not surprising since Klebs had kept even enucleate protoplasts of *Funaria* alive for six weeks. The treated *Mnium* plants did appear to dry more rapidly than untreated plants when exposed to the air. Strasburger had expected that following a first plasmolysis, a second would take place more rapidly owing to the absence of plasmodesmata; this proved not to be the case. The only recent work on this topic appears to be a brief report of Burgess (1971) in which wheat root meristem cells were fixed after plasmolysis in mannitol for an hour. According to Strasburger (1901) meristem cells withstand plasmolysis better than highly vacuolate cells, partly because of their thin cell walls which can permit of a degree of 'negative turgor'. The whole topic of the effects of plasmolysis on plasmodesmata and the possibility of their re-establishment following deplasmolysis needs re-investigation.

Sachs (1972) has suggested that the patterns of plasmolysis can be used as an indication of the relative importance of the interactions, through plasmodesmata, between a cell and its neighbours. In hypocotyls of *Helianthus* the patterns changed predictably following wounding, taking at least sixteen hours to do so. In part, Sachs was able to repeat older work which suggested a re-orientation of the patterns of plasmolysis in response to lateral treatment with auxin. According to Strasburger (1901) plasmodesmata are withdrawn from the walls of damaged cells so that one might expect a change in plasmolysis patterns to result from wounding. However, there are many situations in which conclusions concerning cell communication cannot be drawn from the pattern of plasmolysis (see also Stadelmann, 1956). It was pointed out long ago (Bower, 1883) that plasmolysed epidermal cells have strands of cytoplasm running to the outer walls where there are no plasmodesmata. Weber (1925) demonstrated that stomatal guard cells can show either convex or concave plasmolysis figures, depending upon whether the stomata were shut or open (respectively) at the time of plasmolysis. However, in general, mature guard cells have few or no plasmodesmatal connections to their subsidiary cells (see 2.3.). In his work on neutral red transport in plasmolysed cells of filamentous fern gametophytes, Smith (1972) distinguished between cytoplasmic strands which interconnect protoplasts, which conduct the dyestuff, and strands which attach the plasmolysed protoplast to the cell wall, which do not conduct.

13.4. DEVELOPMENTAL CONSEQUENCES OF ISOLATION OF CELLS FROM TISSUES

"It is a familiar fact that the physical isolation of parts of the body of organisms is in general followed by changes in behaviour. In the higher animals the specialization of structure and function is so great that such physically isolated parts soon die, though under properly controlled conditions, e.g. in tissue culture media, even small groups of cells may be kept alive for long periods and may grow and divide. In the simpler animals such physical isolation of parts, if not carried too far, is followed by differentiation of some or all of the cells of the isolated piece and this is in turn followed by a new developmental process which gives rise either to a complete new individual or to the more apical or anterior portions of such an individual....Among the lower plants every cell of the body may be capable, when physically isolated, of transforming into a new growing tip and so into a new individual..." (Child, 1924).

Child had introduced the term "*physiological isolation*" over a decade earlier (Child, 1910, 1915). Already in 1892 Vöchting had recognised that removal of dominating apical regions of lower plants leads to the formation of numerous adventitious buds, each capable of regenerating a new individual. Klebs (1887) had shown that reducing the sap volume by plasmolysis could stimulate further growth and cell division and even lead to regeneration of new individuals in a number of green algae (e.g. *Zygnema, Cladophora*). However it appears to have been Goebel (1897) who first suggested that regeneration and rejuvenation brought about by isolation of cells might be due to the severance of protoplasmic connections - i.e. by physiological isolation from other cells. This concept has passed into the mythology of plant development, having been invoked from time to time to explain a wide range of phenomena, extending from full regeneration of new individuals to special cases of cell differentiation, but has never achieved the status of respected hypothesis. For instance, it is not mentioned at all by Muller-Stoll (1965) in his otherwise comprehensive review of regeneration in lower organisms. Underlying the concept is the assumption that the performance of individual cells is kept short of a full expression of totipotency by 'information' reaching them from surrounding cells. It used to be commonly assumed that this 'information' consists of auxin produced in a centre of dominance, although why the auxin should not be able to cross a zone of damaged or dead cells surrounding the isolated cell or cell group is not immediately apparent. However, it is clear that in many cases, regeneration is dependent on a resumption of expansion growth by the isolated cell or group of cells and insofar as an inhibition of growth can be attributed to auxin (or other growth substances involved in correlative inhibition) the first steps towards regeneration might thus be prevented. It seems more probable that the information which is exchanged between cells of a stable tissue has a wider connotation and that in addition to inhibiting growth, it is able to inhibit cell division and may be responsible for the establishment and maintenance of steady states of genetic expression.

In both normal and pathological development, there are many instances in which removal of cell-to-cell contact in tissues leads to repair or restitution, during which cells reveal potentialities which are suppressed in the intact tissue. A good example is that of the mesophyll cells which border the holes which appear during normal development in the leaf of *Monstera* (Bloch, 1944). The mesophyll cells de-differentiate and then differentiate to form a new epidermis, bounding the holes. Recent work on the effects of removal of the root cap of *Zea mays* (Clowes, 1972) shows how complex the interactions of cells on each other and on adjoining tissues can be. Removal of the whole cap results in a sharp increase in the rate of mitosis in cells of the quiescent centre and a decrease in that of the stele. Removal of the distal half of the cap also results in a sharp increase in the rate of mitosis in the quiescent centre, but in the cap initials a fall to less than half the normal rate. After 24 hours the rate of mitosis in the cap initials increases again due to the origin of a new set of cap initials from cells of the quiescent centre. Whereas other parts of the root form a callus after wounding, the cells of the cap do not. The simplest explanation of the stability of the differentiated state and of normal pathways of development is that information is exchanged reciprocally between cells *via* cytoplasmic connections and that the information flow ceases when these connections are broken. It must be borne in mind, however, that non-photosynthetic cells will be dependent on supplies of organic nutrients and may fail to display totipotency following isolation simply because of starvation.

13.4.1. Consequences of Isolation Induced Plasmolytically or Mechanically

Plasmolysis was introduced by Klebs (1888) as the most convenient method of producing cell isolation. The method was used by Miehe (1905) on a marine *Cladophora* species. Following plasmolysis in concentrated sea water, the individual protoplasts formed new cell walls. Then on transference to normal medium rhizoids and eventually deep-green apical photosynthetic filaments were formed. A similar regeneration in *Cladophora* following protoplast isolation by repeated centrifugation was described by Czaja (1930). The phenomenon was re-investigated in great detail by Schoser (1956).

In experiments designed to test whether multicellular Volvocaceae are to be regarded as colonies or as multicellular organisms, Bock (1926) experimentally destroyed all except one or a few of their cells and studied the subsequent regeneration. In *Gonium*, *Eudorina* and *Pandorina* the fewer the number of cells left in the colony the faster they divided to restore the colony. In *Gonium* extirpation of all but one cell led to rapid divisions restoring the full complement of cells. In *Pandorina*, and even less so in *Eudorina*, the restored colonies had fewer cells than normal. If, however, 'old' *Eudorina* colonies were operated upon just before the onset of normal asexual reproduction, in which each cell gives rise to 32 daughters, a full complement of 32 cells was produced from a single unextirpated cell. On the other hand extirpation of a single cell from a colony provoked no restoration. Attempts to carry out the same sort of experiments with *Volvox* failed.

These results indicated to Bock that the colonies behaved as integrated multicellular organisms and he sought an explanation in terms of cellular interconnection. Conrad (1913 - cited by Bock) had already shown intercellular connections in *Eudorina*. Bock was able to find them in *Gonium*. (Stein, 1965, also finds them but is more concerned to show that they are not plasmodesmata *sensu strictu*). Bock also found intercellular connections between at least the inner cells of *Pandorina* (and Lefort, 1963, also reports them). As the colonies enlarge and the cells separate, the cell-to-cell connections become tenuous (c.f. Stein, 1965) and disappear. This is held by Bock to account for the lack of exact synchrony in mitosis in (e.g.) *Gonium*.

Regeneration from individual cells of red algae was described in the extensive investigations of Tobler (1902; 1903; 1906) and Child (1917). The latter found that the thallus of *Griffithsia bornetiana*, kept in standing sea water, falls into its individual cells or small groups of cells, which thus become physiologically isolated. *"The fact that physiological disintegration of the individual is soon followed by physical separation of the cells is of interest, as indicating the very direct dependence in this simple individual, of the gross morphological order upon the dynamic integrating factor"* (Child, 1917, p. 221). The isolated cells or cell-groups eventually regenerated new rhizoids and new photosynthetic filaments. A remarkable feature of regeneration in filamentous red algae is the capacity to restore the continuity of the thallus across dead cells by outgrowths from cells bordering on them (Lewis, 1909; for *Griffithsia bornetiana*; Höfler, 1933, for *G. schousboei*; Weide, 1938, for *Callithamnion*; Funk, 1955 for *Ceramium*, *Sphondylium*; and more recently Waaland and Cleland, 1974, for *G. pacifica*). Waaland and Cleland show that the downward-growing cell ('repair rhizoid') has rhizoidal characters, while the upward-growing cell has shoot characters. They believe the repair-rhizoid to be the source of factors which may regulate cell division, morphogenesis and the direction of growth.

Goebel (1897) was able to obtain regeneration of prothalli from cut stems of *Sphagnum* but not from leaves. This was, however, achieved by Woesler (1934). In a thorough investigation of the development of the leaf of *S. cymbifolium*, Zepf (1952) studied the regeneration from both chlorophyllose and young hyaline cells of the leaf (Fig. 13.4.). Leaves were detached and placed in dilute Knop's solution to which was added a decoction of peat. Within a few days regeneration took place from individual cells, including the youngest chlorophyllose and hyaline cells in the basal part of the leaf. Even hyaline cells which had begun to form the spiral wall thickenings or even advanced to form one of the wall pores were able to regenerate, but slowly. In the latter case the pore was first closed over by a new wall in fourteen to eighteen days. The number of chloroplasts per cell rose from about 10 to about 40 during the course of regeneration of hyaline cells (c.f. other cases in mosses reviewed by Giles, 1971). Zepf found that the middle lamella between older cells of the tissue was solubilized wholly or in part during the first few days in nutrient solution (or in water). He was thus able to obtain single cells which could be picked out and cultivated further in hanging drops. Under these conditions hyaline cells died, presumably of starvation, after six to eight days. Single chlorophyllose cells regenerated as well as similar cells within the whole leaf and after four to six weeks most of the protonemata thus formed had initiated a marginal bud. Since the middle lamella appears to be rapidly solubilized when the whole leaf is placed in nutrient solutions, we can surmise that the plasmodesmata (which Zepf does not mention) were soon broken or withdrawn thus isolating cells and permitting their regeneration.

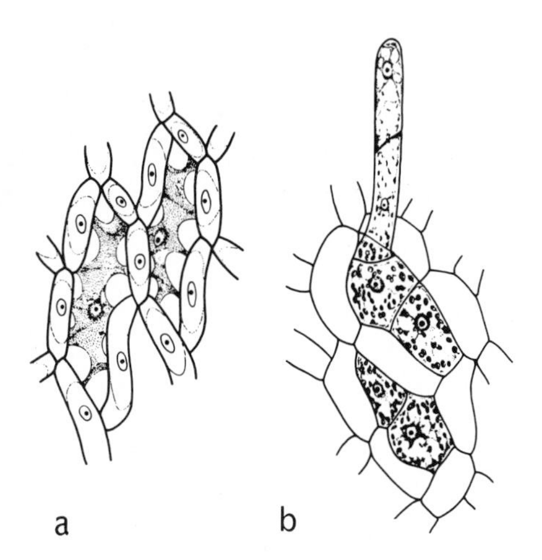

Fig. 13.4. *Sphagnum cymbifolium*
a) Portion of differentiating leaf, to show chlorophyllose and future hyaline cells, at this stage still with protoplasts.
b) Regeneration of protonema from de-differentiated hyaline cell.
(Redrawn from Zepf, 1952)

Stange (1957) was able to obtain formation of 'cauloid regenerates' (protonemata) from single cells of the wing of the liverwort, *Riella*, either by cutting it into small pieces or following plasmolysis with KNO_3 for 20 minutes. Certain differences in the rate and manner of formation of the regenerates, depending on the method of isolation, could be attributed to assistance by the non-regenerating cells of a cut fragment to the development of the regenerating protonemata. Following isolation, every cell of the wing tissue may undergo division. A few days after plasmolysis, the regenerates occupied a marginal zone of the wing, (the zone increases in width and the regenerates increase in density towards the apex). Plasmolysis was at least ten times as effective as cutting in bringing about regeneration. Although plasmolysis killed some cells, cells which gave rise to regenerates could sometimes be seen to have living neighbours which did not regenerate and which were distinguished by their high content of starch. Cell isolation leading to regeneration could therefore not be attributed solely to killing of some cells by plasmolysis. The marginal zone, in which regenerates appeared, consists of cells which, by their rich starch content, differ sharply from the cells of the inner part of the wing. Stange made a detailed investigation of the cytology of the regenerating cells but failed to investigate their plasmodesmata.

In his investigation of regeneration following plasmolysis of gemmae of *Marchantia* for one hour with potassium nitrate or sucrose, followed by slow deplasmolysis, Förster (1927) found that groups of cells co-operated to produce adventive thalli. However, he was able to observe regeneration from 'physiologically isolated' single cells in gemmae in which most of the cells had died, and only a few marginal cells remained alive. He cites a number of previous observations of regeneration from physiologically isolated cells in Marchantiaceae. Regeneration has been studied much more intensively in fern gametophytes and the subject has been reviewed - somewhat uncritically- by Miller (1968). Under suitable conditions, any single cell is capable of regenerating a complete prothallus. *"The basic stimulus which seems to underlie all examples of regeneration is the mechanical or physiological isolation of cells from the rest of the prothallus. A cell which is cut off from communication with its neighbours assumes a sporelike behaviour, resumes growth, and often recapitulates the pattern of growth by a germinating spore"* (Miller, 1968).

Isaburo-Nagai (1914) found no differences due to the chemical nature of the plasmolytica used to isolate the cells, so long as they were not toxic. Nor could the effects of plasmolysis (usually for 20 minutes followed by transfer to Knop's solution) be initiated by losses of turgor, produced by temporary drying. Since some cells die after plasmolysis, Isaburo-Nagai tested a hypothesis that isolation occurred because of cell deaths, by repeatedly stabbing a prothallus with a fine needle, but this did not cause regeneration. Even cells of prothalli kept plasmolysed as long as 187 hours in molar sucrose solution were able to regenerate after transfer to 0.5% Knop's solution. Only such cells as had been plasmolysed could succeed in regenerating - weakly hypertonic solutions produced no results. Linsbauer (1926) also endeavoured to isolate single cells mechanically, by killing the surrounding cells. He was able to cause regeneration but could not be certain that this was from single cells only. He was able to observe that prothalli, most of the cells of which had been killed by fungal attack, contained single cells which had subsequently regenerated. Ito (1962) repeated the 'isolation by killing' experiment successfully and Meyer (1953) isolated and cultured single cells from gametophytes partly destroyed by a fungus which attacked the middle lamella. Linsbauer used X-rays and Kato (1964) ultra-violet radiation to induce regeneration from single cells. Nakazawa (1963) plasmolysed young *Pteris vittata* prothalli in 0.2 M $CaCl_2$ for 30 minutes then transferred them to dilute Knop's solution. 85% of the thalli remained alive after the treatment in which plasmolysis was clearly observed in all the cells. Unlike previous authors, Nakazawa examined

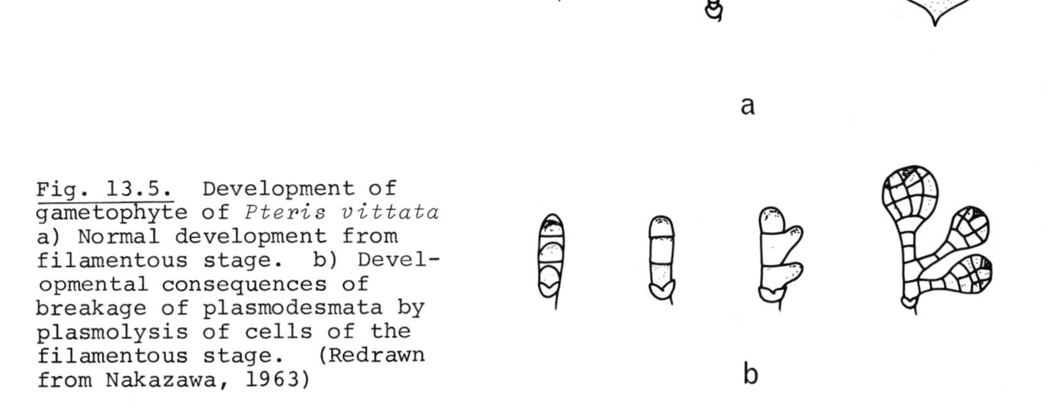

Fig. 13.5. Development of gametophyte of *Pteris vittata* a) Normal development from filamentous stage. b) Developmental consequences of breakage of plasmodesmata by plasmolysis of cells of the filamentous stage. (Redrawn from Nakazawa, 1963)

the cells for plasmodesmata. He was able to stain and observe them and then subsequently plasmolyse the cells, to see the plasmodesmata left in the cell walls by the withdrawing protoplasts. (This latter is corroborated by Burgess, 1971). Many of the cells isolated by plasmolysis went on to develop new protonemata and prothalli (Fig. 13.5.) - i.e. following cell division, the previously plasmolysed cells must have been able to make plasmodesmata across the new division walls. He did not re-examine the prothalli to see whether the 'old' plasmodesmata were re-formed, but drew the conclusion from the results of regeneration that they were not. Fraser and Smith (1974) have confirmed the existence of plasmodesmata between cells of the filamentous gametophytes of ferns.

Cell isolation need not result only in complete regeneration - the retracing, in Goebel's terms, of the ontogenetic pathway from juvenility onwards. If fern prothalli are sufficiently old, temporary plasmolysis may promote the development of antheridia (Näf, 1961).

13.4.2. Plasmodesmata and Cell Differentiation

In the chapters of his book devoted to "*The internal factors of differentiation*" Bünning (1951) introduced some new concepts into the topic of cell differentiation. In this section we shall deal with some of these concepts insofar as they involve plasmodesmatal connections between cells. Bünning laid stress on the role of 'physiological isolation' in normal cell differentiation. By this he meant presumably not *total* isolation (such as will be considered in 13.4.4.) but *relative* isolation of individual cells due to their positional relationships to adjacent cells. Positioning of cells over anticlinal walls was held to be an isolating factor of considerable importance in normal development. In *Sinapis* roots, for instance, the protophloem cells lie in such a position. Bünning (1951) cites an interesting example from the algae: 90% of the tetraspores formed in *Dictyota* originate from cells each of which lies over an anticlinal wall of the underlying layer of cells. The plates of tracheidal cells in pine needles originate from cells which abut radial walls of the endodermis, the alternating plates of parenchyma from cells which abut tangential walls of the endodermis (Huber, 1948). It would be interesting to have more information on the electrotonic and plasmodesmatal reality underlying these assumptions of 'physiological isolation' leading to differentiation.

At about 270 µm from the root tip of *Sinapis alba* certain cells of the rhizodermis can be distinguished as future root hair initials by their high basiphilicity and their position in relation to the cells of the hypodermis. Each root hair initial abuts two hypodermal cells, and lies over a radial wall separating the two (Cormack, 1947). Other cells which are not determined as root-hair initials each abut only one hypodermis cell and each lies on the outer tangential wall of the latter. Bünning (1951) asked why it is that a position over the anticlinal wall could thus determine the fate of the rhizodermis cell as a root-hair initial. He refuted Cormack's suggestion that the wall might (as part of the apoplast) increase the nutritional status of the initial, since there is an air-filled intercellular space at the junction of the anticlinal hypodermis wall and the initial. He believed that this gives the initial an isolated position early in development. Experiments on detached seedling roots to try to increase the number of root hair initials by feeding with glucose failed. Bünning therefore advanced the hypothesis that the inner part of the root exerts some influence which depresses the possibility of root hair initials being formed in the rhizodermis and that only the 'physiologically isolated' cells lying over the anticlinal walls escape this inhibition.

Short longitudinal cuts into the cortex of the root, separating the rhizodermis locally from the inner tissues resulted in virtually every cell of the rhizodermis taking on characteristics of root hair initials. This certainly supports the hypothesis. As testimony to the reality of the 'physiological isolation' of the normal root hair initial, Bünning advances the concept that regulatory substances such as hormones do not move through the apoplast but by way of plasmodesmata and that *"pits and plasmodesmata are usually not found on the cell edges but in the middle of common walls"*. No investigation was made to see whether plasmodesmata were present in the walls of the rhizodermis or hypodermis or whether or not they were distributed in the assumed pattern. Instead, the fact that wounding causes cell division in the hypodermis, even at some distance from the wound, and results in an increased number of cells becoming root hair initials is drawn upon to support the view that the pathway of the 'normal' inhibitor of root hair initial formation is the same as that of the supposed 'wound hormone', i.e. that it moves along pathways in the hypodermis which in the experiment were traced by cell divisions induced at right angles to the gradient of the wound hormone, and thence into the rhizodermis *via* its cell contacts with the hypodermis.

As yet, no investigation appears to have been made of the actual degree of plasmodesmatal connection between the root-hair initial of *Sinapis* and adjacent cells. We can however, cite information from another slightly different case of root-hair initiation, that of *Hydrocharis morsus-ranae* (Cutter and Hung, 1972). In Bünning's analysis this comes under the heading of differentiation resulting from unequal division. In the Monocotyledons and many ferns, the root epidermis consists of a mosaic of trichoblasts (root-hair initials) and atrichoblasts. Unequal cell divisions take place in the cells of the rhizodermis near the root apex, and each results in a small, densely cytoplasmic daughter cell (the trichoblast) proximal to the root apex and a larger more vacuolate daughter cell, distal to the root apex. Plasmodesmata are present in the wall between the two cells, and are more numerous in the transverse wall than in the longitudinal walls. *"Cytoplasmic communication thus remains possible between cells which subsequently differentiate very dissimilarly indeed"* (Cutter and Hung, 1972).

Despite the lack of information concerning the degree of physiological isolation involved in root hair initiation in *Sinapis*, one can point to other examples in which some degree of isolation certainly does precede cell differentiation. Fibre idioblasts in the air roots of *Monstera deliciosa* (Bloch, 1946) arise from unequal divisions and on at least one longitudinal face they abut an intercellular space, i.e. lose the plasmodesmata on at least one wall. Later on, these idioblasts grow out into this intercellular space and intrusively between the hypodermal cells. Intrusive growth is characteristic of other sclerenchymatous idioblasts, which, lacking or freed from cell communication, commence a special development usually late in the growth of the organ which contains them. Cells in the epidermis also have no plasmodesmata on their outer walls, and there are many kinds of epidermal idioblast which in principle might owe their differentiation to lack of, or a reduced, cell communication. In the epidermis of grasses, unequal division give rise to cells each of which subsequently divides to give rise to a pair of idioblasts, the silica-cork cell pair. Kaufman *et al.* (1970b) have made a study of the development of the silica-cork pairs in *Avena*, and from their micrographs it can be deduced that early in its formation the silica-cork pair initial still retains plasmodesmatal connections both with cells in its own row and with cells in lateral rows. However, the plasmodesmata appear to be lost before the initial divides to give rise to the two idioblast cells. Thus before they proceed to their special pathways of differentiation, the silica cell and the cork cell are plasmodesmatically isolated, from each other and from cells in their own and lateral cell rows. Cells of the epidermal trichomes of a number of species of Asclepiadaceae have been shown to be well connected by plasmodesmata (Inamdar *et al.*, 1973).

Stomatal initials arise by unequal division of epidermal cells in which the smaller, more densely cytoplasmic daughter cell subsequently divides equally to give rise to the stomatal guard cells. According to Bünning stomata and trichomes belong to a class of structures called 'meristemoids' the cells of which not only resume their own brief pseudo-meristematic activity but also exert some sort of influence preventing initiation of other 'meristemoids' within their vicinity. By this means they can give rise to patterns in development (Bünning, 1953; 1955). Bünning and Sagromsky (1948) showed that the origin of the non-random pattern of stomata and trichomes on the leaf epidermis of many Dicotyledonous plants can be traced to these properties of meristemoids. What concerns us here is the degree of plasmodesmatal connection maintained during the differentiation of these structures. In Dicotyledons, stomatal mother cells arise irregularly in the epidermis, and so provide difficulties of identification for the electron microscopist. Landré (1972) illustrates a guard mother cell in the cotyledonary epidermis of *Sinapis alba* with some plasmodesmatal connections to adjoining epidermal cells. No plasmodesmata are evident on the walls of the guard cells of the fully-formed stoma. In the pea plant also (Singh and Srivastava, 1973), the guard mother cell is connected to surrounding epidermal cells by plasmodesmata. Plasmodesmata appear also in the wall between the two guard cell initials. They are lost, however, as soon as the stomatal pore forms. The plasmodesmata between the guard cells and neighbouring epidermal and subsidiary cells are also lost at maturity. Plasmodesmata remain, however, between subsidiary cells (Burgess *et al.*, 1973). Plasmodesmata are abundant between immature sister-guard cells in *Vicia faba* (Pallas and Mollenhauer, 1972a,b) and there may be some between immature guard cells and epidermal cells. The evidence produced by Pallas and Mollenhauer (1972a,b) for complete plasmodesmatal connections between *mature* guard cells and epidermal cells is not convincing. No plasmodesmata are shown in the mature guard cell walls of *Nicotiana* in their (1972a) paper. In a study by light microscopy, Litz and Kimmins (1968) also claim to show plasmodesmatal connections between mature guard cells of *Nicotiana tabacum* (as well as those of *Phaseolus vulgaris* and *Datura stramonium*) and subsidiary cells. In a study by electron microscopy of the stomata of *Opuntia ficus-indica*, Thomson and de Journett (1970) show that there are no plasmodesmata between mature guard cells and subsidiary cells.

In Monocotyledons, the regularity of formation of stomatal initials along parallel cell rows makes it more convenient to study their progressive development. The young guard mother cell of *Commelina cyanea* is certainly connected to adjacent cells by plasmodesmata, judging by the photographs published by Pickett-Heaps (1969). The cell walls of the initial appear to thicken slightly (Fig. 11 of the publication) and to lose their plasmodesmata before the final division to form the guard cells. The fully-formed guard cells appear to lack plasmodesmata but the surrounding cells retain them. Similarly in *Zea mays* (Srivastava and Singh, 1972b) and *Avena sativa* (Kaufman *et al.*, 1970a) the guard mother cell is not isolated from adjacent cells. At all stages up to the early quartet stage (guard cells with subsidiary cells) plasmodesmata can be found between all cells in the developing stomatal apparatus in *Avena*, and Kaufman *et al.* believe on dubious evidence that they remain present between the guard cells even in the mature condition. In *Zea*, plasmodesmata are present throughout development but are lost from the walls of the guard cells at maturity (Ziegler *et al.*, 1974). However, in both cases plasmodesmatal connection between the subsidiary cells and adjoining epidermal cells is retained at maturity. In *Zea*, contrary to the report of Brown and Johnson (1962) the subsidiary cells are also connected to the epidermal cells, which in turn are connected to the mesophyll (Srivastava and Singh, 1972b).

It is evident that earlier reports of plasmodesmatal isolation of mature guard cells are in the main substantiated by the recent studies using electron microscopy. Thus Kienitz-Gerloff (1891) was unable to find such connections. Sheffield (1936) could not find plasmodesmata in *Hyoscyamus niger*, *Solanum nodiflorum*, tomato or tobacco (c.f. Livingston, 1935). Schumacher (1942) failed to find plasmodesmatal connections between guard cells and subsidiary cells in *Passiflora triloba* and Brown and Johnson (1962) failed to find them in a number of grasses. In a number of other cases (e.g. Gardiner and Hill, 1901, for *Pinus pinea;* Kuhla, 1900 for *Viscum*) claims to the contrary have been made on the basis of light microscopy. Plasmodesmata in the guard cell-subsidiary cell walls were demonstrated in several ferns by Poirault (1893) but only after 'violent attack' by the swelling reagent. Tucker (1974)(c.f. Haberlandt 1922) has found that the guard cells of injured leaves of certain Magnoliaceous plants can divide, taking part in the regeneration. Having seen 'prominent pits' between guard cells and other cells in one of these plants, she proceeded to the assumption that this means that *"there is no loss of plasmodesmatal connections to adjacent cells as the epidermal cells mature"* in Magnoliaceous plants. She thus attributes the *"usual loss of regenerative potentiality of guard cells"* in other plants to ....*loss of plasmodesmatal connections to adjacent cells as the epidermal cells mature"*. We have here several unproved assumptions, but nevertheless the phenomenon is certainly worth further investigation.

Early observations of plasmodesmatal connection between cells of different states of differentiation (e.g. anthocyanin cells, tannin cells, and surrounding parenchyma cells) are discussed by Münch (1930) who remarks that since the products of these cells are retained in vacuoles it is not surprising that they do not leak out into adjoining cells. Esau (1963) shows plasmodesmata between tannin cells of *Vitis* and neighbouring tannin-free cells. So too do Ledbetter and Porter (1970), but Van Steveninck (Chapter 7) notes that there seem to be fewer plasmodesmata leading to tannin-rich cells than to others. Plasmodesmata are also found between myrosin cells and neighbouring normal cells in the cotyledons of *Sinapis alba* early in germination (Werker and Vaughan, 1974). Later in development the cell contents of the myrosin cell become detached from the cell wall and the plasmodesmata are ruptured.

From this examination of the literature on cell differentiation we may conclude that physiological isolation is not essential to the early initiation and differentiation of many kinds of plant cells. Plasmodesmatal connections may be retained until a late stage in the differentiation of idioblasts and meristemoids of many different kinds, although it is true that in some cases, the mature differentiated cells do appear to lose plasmodesmatal connections with neighbours. However this loss is to be regarded as part of the differentiation process, not its cause. It may be necessary to their functioning, for instance, for guard cells to be isolated plasmodesmatically from adjacent cells. It is true that in some aspects of their physiological behaviour they behave as though they were isolated from the rest of the epidermis. For instance, Leitgeb (cited in Strasburger, 1901) found that in cut flowers or detached epidermal strips kept in a moist atmosphere the guard cells outlived the remaining epidermal cells. This was confirmed by Hagen (1919). Kienitz-Gerloff, 1891 (p.57), finding that the chloroplasts in guard cells retained their chlorophyll in yellowed leaves, attributed it to lack of plasmodesmatal connection to the epidermal cells. The observation was repeated by Weber *et al.* (1953). Sachs (cited by Strasburger, 1901) had already remarked that the only starch to be found in abscised, yellowed leaves was in the guard cells.

13.4.3. Plasmodesmata and Differentiation in Tissue Cultures

The technique of tissue culture and the possibilities of obtaining compact or friable masses of aseptic proliferating tissue from haploid or diploid cells, of culturing suspensions of single cells or of obtaining quantities of free cell protoplasts, capable of fusing together or developing cell walls, and finally the induction of differentiation of cell types or organs, even whole plants from these cultures, must surely offer one of the most important experimental approaches to the role of plasmodesmata in development. As Street (1973) has put it "... *in those cases where organogenesis can be readily induced there is a need for much more intensive anatomical and histochemical examination by both light and electron microscopy to trace the emergence of organization within the initially relatively uniform cell mass of the callus or suspension culture aggregate. One aspect of possible importance here is that of* the degree of cellular contact and metabolic interchange" (my emphasis) "*within the culture. In individual cases it is often possible to predict morphogenetic potential from observation of the friability or compactness of the tissue but we understand the nature and significance of such intercellular contact neither in terms of middle lamella and wall-composition, formation and functioning of plasmodesmata nor of development of tensions and compressions*".

Early reports of a lack of plasmodesmata in tobacco callus tissue were later discounted by the suggestion that the techniques used were inadequate (Kassanis, 1967). Using light microscopy and electron microscopy, Spencer and Kimmins (1969) showed conclusively that cells of normal and crown-gall callus isolated from tobacco do indeed have plasmodesmata. They have been observed also in tissue cultures of Jerusalem artichoke (Bagshaw), *Atropa belladonna* (Yeoman and Street) and *Ranunculus sceleratus* (Thomas, Konar and Street) (references in Street 1973).

On the other hand, many authors have described friable callus tissues in which there is at least some probability of the absence of plasmodesmatal connections, at least between cells on the outer surface of the callus. Some tissue cultures regularly undergo dissociation into single cells or small groups of cells and this disaggregation is accentuated by growing the tissues on a shaker. Thus so-called 'single-cell' suspensions can be obtained. Details of the sequence of events leading to dissociation of the cells have been described. "*Cell separation results from disruption of the outer layers of the cell wall. Following formation of a new cell plate after mitosis, a thin wall is formed ....Cell wall material continues to be deposited during several cell generations, leading to the formation of walls of progressively greater thickness around the older cells of an aggregate ...Separation of the cells involves both the separation of the walls in the region of the middle lamella and the loosening of material in the outer wall layers*" (Sussex and Clutter, 1968). Small pieces of fibrillar wall material are seen extending outwards from the walls; this material eventually separates from the cells and accumulates in the medium as an amorphous substance.

What part, if any, the withdrawal or even non-formation of plasmodesmata plays in all this is not known. The original callus (*Euc-*

alyptus hypocotyl) used by Sussex and Clutter (1968) could form a coherent nodular tissue under suitable circumstances: the cells within these nodules were interconnected by plasmodesmata. By changing the medium a peripheral zone of tracheary elements could be caused to differentiate in the nodular tissue. Sussex and Clutter therefore carried out experiments to see whether compressing free cells together to form nodular clumps of 'pseudotissue' of about the same size as the differentiating compact tissue nodules would also result in the formation of tracheary elements. Cells were put into dialysis tubing which was tied at intervals so that the cells were tightly packed into nodules of varying size and the whole re-immersed in the inductive culture medium. After 21 days, no tracheary elements had been formed in the reaggregated pseudotissue nodules, despite very intimate contact between the separate cells. If corroborated, this result would lend considerable support to the hypothesis that information for cell differentiation is transported only through plasmodesmata, and that a non-interconnected pseudotissue is not capable of sustaining the informational exchange or diversity required for even such a relatively simple step in differentiation. Similar experiments have been done by Wilbur and Riopel (1971a,b) using *Pelargonium* cells and tissues. However, the emphasis in their investigation was placed on the fact that, despite the capacity of all cells of a culture for growth, the growth displayed depends on the cell density. Free cells from cell suspensions were pipetted into small nylon-mesh cones to achieve different initial cell numbers per cone. 'Sclereids' (equivalent to Sussex and Clutter's tracheary elements) were formed by cells in the cones and their formation depended on a critical cell number being present. The number formed also bore a quantitative relationship to the cell number, the more cells the more sclereids formed. Unfortunately no examination was made in this instance to see whether or not the sclereids were formed in groups of cells, originating from the 'free cell' inocula, interconnected by plasmodesmata (as are found in suspension culture aggregates of carrot cells, Halperin and Jensen, 1967). The authors presume that "*it is not unreasonable to assume that such connections did form prior to differentiation as the cells divided in the cone*". They thus support the contention of Sussex and Clutter that plasmodesmata are essential for cell differentiation, but fail to provide crucial evidence.

13.4.4. Isolation in the Development of Alternating Generations and during Reproduction

"*The only plant cells that are normally isolated are the initiating cells in alternate generations, gametes and spores. This is an obvious fact in non-seed plants, where the male gametes and spores are free-living*" (Halperin, 1970). Plasmodesmatal connections between cells involved in sexual and asexual reproduction in the algae are dealt with in Chapter 3. The general cytology of differentiation of gametes and gametangia in the cryptogams is well reviewed by Vazart (1963). According to Górska-Brylass, 1968; 1969, individualisation is always effected by deposition of a callose wall. Callose may serve to isolate archegonial initial cells (e.g. in *Selaginella* and ferns), spermatids in antheridia, as well as the generative cells of pollen grains and microspores (e.g. *Selaginella*).

13.4.4.1. Sporophytes and sporogenous cells of bryophytes and ferns. As mentioned earlier and in Chapter 14 Kienitz-Gerloff (1902) could find no plasmodesmata in the foot cells of the sporophytes of *Polytrichum* and *Lepidozia*, connecting them to the gametophytes. He had already stated (1891) that he believed each individual plant to be a plasmodesmatically interconnected but closed system, and the new observations supported this view. However, he had then to concede that transport of solutes in the plant must be able to take place without the intervention of plasmodesmata, across the apoplastic gap between the generations. In many mosses (*Hypnum*, Genevès, 1966; *Polytrichum*, Paolillo, 1969; *Mnium* etc., Eymé, 1969; *Fissidens*, Mueller, 1974) the pre-meiotic sporogenous cells are interconnected by plasmodesmata but there are none between sporogenous and vegetative cells. In preparation for meiosis, the protoplast of the spore mother cell withdraws from the cell walls (*Hypnum*, Genevès 1967a; *Fissidens*, Mueller, 1974) and the plasmodesmata are consequently broken. Within each spore mother cell, the sporocytes are initially interconnected by cytomictic channels through the infurrowing walls (*Riccardia pinguis*, Horner *et al.*, 1966). The walls and cytomictic channels (*Hypnum*, Genevès, 1972a,b) are temporary - the sporocytes become separated (in *Hypnum* by secretion of a copious mucilage) and each forms its own thick spore wall. Meanwhile, the parent wall disintegrates.

Most of the plasmodesmatal connections between sporogenous cells of the fern, *Blechnum* are lost at a pre-meiotic stage but some persist and enlarge to form cytomictic channels (Lugardon, 1968). Such channels are also formed in *Osmunda* (Lugardon, 1968). At the end of Prophase I of meiosis the spore mother cells become isolated.

13.4.4.2. Archegonial development Kienitz-Gerloff (1891) was unable to demonstrate plasmodesmata in the axial cells in the venter of the archegonia of mosses. These cells behave as though they are physiologically isolated. Wettstein showed that they can regenerate separate protonemata (see Bünning, 1955).

Diers has provided a well-documented study of archegonial development in *Sphaerocarpus donnellii* (Diers, 1965a,b). At all early stages plasmodesmata exist between all the cells. After the division of the neck canal mother cell, plasmodesmata are no longer seen between the cells of the axial row (i.e. the central cell and neck canal cells) and the cells of the archegonial venter and neck. Plasmodesmata remain between the cells of the axial row themselves. At this stage the plasmalemma of these cells begins to retreat from the cell wall. The central cell divides giving rise to the egg cell and the central canal cell and plasmodesmata are formed in the cross wall. Finally the plasmodesmata between the cells of the axial row are lost and all, with the exception of the egg cell itself, degenerate having produced a 'slime' (Fig. 13.6.). The early stages of archegonial development in the moss, *Mnium undulatum* (Barbièr, 1972) and in *Marchantia* are very similar to

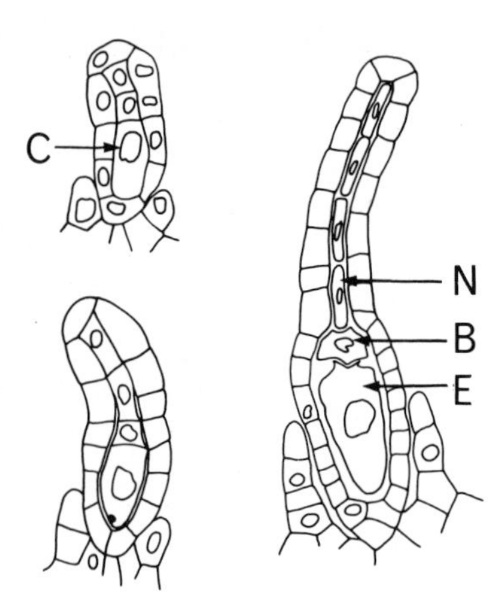

Fig. 13.6. Median sections through developing archegonium of *Spherocarpus donnellii* to show the stage at which plasmodesmatal connections are lost between cells of the axial row and the neck and venter cells. C = central cell; B = basal canal cell; N = neck canal cell; E = egg cell. (Redrawn from Diers, 1965a)

those of *Sphaerocarpus*. Zinsmeister and Carothers (1974) give details of archegonial development in *Marchantia polymorpha*. As the neck elongates, so do the neck canal cells and they lose some of their plasmodesmata. It is at this stage that the asymmetric division of the central cell gives rise to the egg cell and the ventral canal cell, with numerous plasmodesmata in the cell plate, in sharp contrast to their complete absence between the cells of the axial row and the archegonial wall. Zinsmeister and Carothers find an unusual association between plastids and mitochondria in the cells of the axial row, and they attribute it to an altered cellular environment consequent on the loss of plasmodesmata.

The archegonia of the leptosporangiate ferns are much simpler in construction than those of the bryophytes and are partially sunk in the tissues of the gametophyte. Plasmodesmata are not found in the cells of the wall of the archegonium in *Pteridium aquilinum* after the primary cell - the cell which gives rise to the axial row - has begun to develop (Bell and Mühlethaler, 1962). Their absence is attributed by the authors to the rapid expansion of the archegonium. The authors discuss the effects of this isolation on the nutrition of the axial cells. In a fine study of the archegonium of *Dryopteris filix-mas*, Menke and Fricke (1964) show the extremely thick walls, without plasmodesmata, of the cells of the venter. In this case, the neck cells retain plasmodesmatal connections with each other but the neck-canal cell, central cell and basal cell (which forms the base of the venter) are plasmodesmatically isolated from the rest. Moreover the basal cell appears not to be connected to the central cell, although the central cell is connected by plasmodesmata to the neck canal cell. The wall between the basal cell and the central cell later becomes very much thickened. The central cell divides to give rise to the egg cell and ventral canal cell. The cell plate which forms contains numerous plasmodesmata. However, they are rapidly lost so that the egg cell becomes completely isolated. The egg cell of *Selaginella* follows a similar pathway of development to isolation (Robert, 1969).

13.4.4.3. Antheridial development In the antheridium of the liverwort *Fossombronia* (Vian, 1970) and the mosses *Pogonatum aloides* (Bonnott, 1967b), and *Polytrichum formosum* (Genevès, 1967a,b) the spermatogenous cells are interconnected by plasmodesmata but are isolated from the cells of the wall. Owing to the unusual elongation of the plastids, occasionally a plastid may become trapped across a cell plate of the spermatogenous tissue, as shown in *Pogonatum* by Bonnot, 1967b. (A similar phenomenon is recorded for the sporogenous cells of *Hypnum* by Genevès, 1966). At spermatid formation (also in *Bryum*, Bonnot, 1967c; and *Blasia*, Carothers, 1972, 1973) the protoplast rounds off and the plasmodesmata are broken. In *Polytrichum* (Genevès 1967b) the plasmodesmata are retained until the differentiation of the spermatids. In the antheridia of *Marchantia* and *Riccia* strict synchrony is restricted to the derivatives of each primary spermatogenic cell (Górska-Brylass, 1969). At the final division to give rise to spermatids a callose wall is formed around each mother cell and between the spermatids. This wall disintegrates after maturation of the spermatozoid.

In *Equisetum* synchrony is maintained in the antheridium up to early stages of spermatogenesis. Plasmodesmata are lost as spermatid protoplasts retract from their walls, but synchrony of maturation is maintained (Duckett, 1973).

13.4.4.4. Megaspores and embryo sacs A number of studies has been made by French authors of the ultrastructure of development of the oosphere in conifers, but little attention has been paid to plasmodesmatal isolation. Plasmodesmata between the young central cell (which after the second meiotic division becomes the oosphere) and a jacket cell (or 'follicular cell') can be seen in abundance in Plate 12 of the study of oogenesis in *Pinus laricio* by Camefort (1962). It would be interesting to know whether the oosphere itself becomes plasmodesmatically isolated at maturity.

In the Angiosperms, the megaspore mother cell of *Crepis tectorum* is connected by plasmodesmata to the cells of the nucellus (Godineau, 1968) but the plasmodesmata are less numerous than they are between the tegumentary cells. In *Dendrobium* the megaspore mother cell is well connected by plasmodesmata to the adjacent nucellar and epidermal cells (Israel and Sagawa, 1964a). In the 'archesporial stage' (i.e. after enlargement of the megaspore mother cell) these plasmodesmata are lost, while the other cells of the ovule retain theirs. Israel and Sagawa (1964b) attribute this loss to the rapid growth of the megaspore mother cell, which involves pinocytosis (also described by Godineau) and resorption of materials from the surrounding cells. The whole of the megagametophyte (embryo sac) of *Zea* is separated from the nucellus by walls without plasmodesmata (Diboll and Larson, 1966), but plasmodesmata occur between all cells within the megagametophyte and, it is assumed, between all the cells of the nucellus. The same is true of *Capsella* (Schulz and Jensen, 1968b). No plasmodesmata connect the nucellus with the embryo sac of *Myosurus* (Woodcock and Bell, 1968) or the zygote with adjacent cells. In *Capsella* (and *Nothoscordum* - see 13.4.5.) occasional plasmodesmata are found in the relatively thick walls which separate the antipodal cells of the embryo sac from the chalazal region of the nucellus (Schulz and Jensen, 1971) - an exception which might almost be said to prove the rule of 'no connections between generations' in that the antipodal cells in the chalazal proliferating zone do not contribute to the formation of the embryo, but are merely transitory.

After formation of the zygote and before its first division in *Capsella* (and in barley - Norstog, 1972, and *Quercus gambelii*, Singh and Mogensen, 1975) the plasmodesmata in the walls separating it from the synergids and central cell disappear. Nevertheless, the synergids continue their supposed nurse function and the principal absorption of nutrients now appears to take place through a modification of the micropylar wall of the basal cell of the embryo. At the two-celled stage

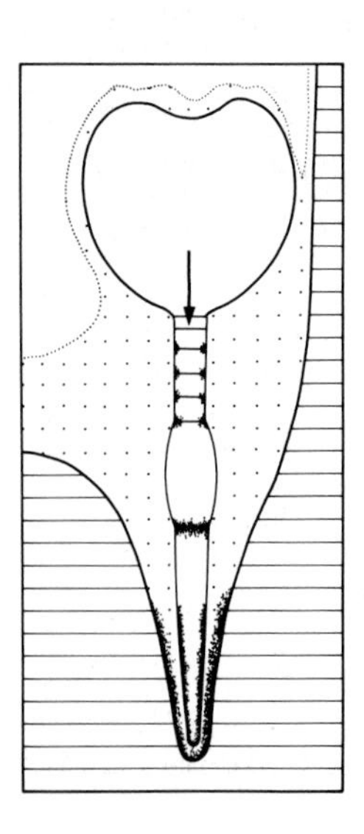

Fig. 13.7. Heart-shaped embryo and suspensor of *Diplotaxis erucoides* showing (arrow) the only wall said to lack plasmodesmata. (Redrawn from Simoncioli, 1974)

the cells of the embryo are well connected by plasmodesmata and this is true also of all the cells of the octant stage (Schulz and Jensen, 1968a). The end walls of the cells of the suspensor are also well-equipped with plasmodesmata, but there is no connection between these cells and the embryo sac itself. In neither *Capsella* nor barley are there plasmodesmata between the cells of the embryo and those of the endosperm. In general, what is true of *Capsella* is true of another member of Cruciferae, *Diplotaxis erucoides* (Simoncioli, 1974). However *Diplotaxis* has the peculiarity of showing no plasmodesmata on the wall separating the last cell of the suspensor and the hypophysis of the embryo itself. Thus symplastic connection between the suspensor and the embryo (the cells of which are well interconnected by plasmodesmata) is absent at their junction (Fig. 13.7.).

13.4.4.5. Microsporangia and microspores In *Ceratozamia mexicana* (Audran, 1974) four protoplasmic 'territories' are delimited in the sporocyte subsequent to meiosis. The cell plates which form at first contain cytomictic channels and plasmodesmata but these are occluded later by the deposition of material resembling callose. In *Podocarpus* (Vasil and Aldrich, 1970) the archesporial cells begin with numerous plasmodesmatal connections but the microspore mother cells are not connected to the tapetal cells or to each other, unlike the meiocytes of Angiosperms (see below). Moreover, they also lack the synchrony of meiosis found in Angiosperms. The microspores within a tetrad also become separated from each other by walls without plasmodesmata. However, within each microspore, the cells resulting from repeated mitoses are well connected by plasmodesmata. In a paper by Chesnoy (1969) on *Biota* (Cupressaceae) the cell plate, with plasmodesmata, is shown between the two sperm cells resulting from division of the spermatogenous cell. The cell wall between the two gametes of *Zamia* (Norstog, 1967) is shown *without* plasmodesmata. Later these cells separate, prior to fertilisation of the oosphere. In the pollen grain of *Ginkgo* (Dexheimer, 1970) there are numerous plasmodesmata between the spermatogenous cell and the tube cell.

In the angiospermous anther, plasmodesmata are absent from walls separating male meiocytes from the tapetum. Massive cytoplasmic channels interconnect the meiocytes in the anthers of many angiosperm species (Whelan 1974; Whelan *et al.* 1974) but they disappear before meiosis II, after which the pollen grains are separated from each other by callose walls. The interconnections are held to account for the synchrony of division and the all-or-nothing response of the developing anther to injury. Injury to the young anther affects the development of all the pollen grains within it (Heslop-Harrison, 1966a; 1968). The first gametophytic division, giving rise to the vegetative and generative cells, is unusual. Burgess (1970) gained "*the impression of the boundary (between the two cells) as constructed to allow a free passage of materials from the tube cell cytoplasm into the generative cell*". This contradicts the earlier conclusions of Angold (1968) and Heslop-Harrison (1968) that the two cells are isolated from each other. Isolation is supported by the micrographs of Sassen (1964), Jensen, Fisher and Ashton (1968), Lombardo and Gerola (1968), Vazart (1970), Giménez-Martín *et al.* (1970), Sanger and Jackson (1971), Lutz and Sjolund (1973) and Dunwell and Sunderland (1974a), whereas small connections may be seen in micrographs of Hoefert (1969a), Vazart (1969, 1971) and Dunwell and Sunderland (1974b). It seems likely that no real contradiction is involved, but that transitory connections are present in the early cell plate between the two cells, to be occluded during the stage when a callose wall is deposited, itself later to be removed (Gorska-Brylass, 1967; 1970), leaving the plasmalemmas of the two cells back to back, separated by a narrow space that is not, or not obviously, bridged. The vegetative cell of angiosperm pollen thus becomes isolated from the generative cell and from the sperm that arise from it (Hoefert, 1969b; Cass, 1973).

13.4.5. Adventitious Buds, Lateral Roots and Embryoids

As a result of wounding, plant organs may produce a callus tissue from which primordia of roots and shoots commonly arise. In some cases callus tissues may give rise to embryo-like structures. Barbara Haccius (Haccius and Lakshmanan, 1969; Haccius, 1971) has distinguished between 'embryoids', a term which she suggests should be used for all asexually-generated, embryo-like structures, including nucellar and integumentary embryos, and 'adventitious buds'. Embryoids, like embryos, have a bipolar construction i.e. a shoot pole and a root pole and are not connected by vascular strands to the parent tissue. The term adventitious bud refers to shoot structures which may eventually become detached and serve as propagules, which do not (at least initially) have a bipolar organization, and the development of which can or does involve a vascular connection to the mother tissue. Thus the buds formed adventitiously on detached leaves of various species (literature in Sinnott, 1960; Buvat, 1965; Haccius and Lakshmanan, 1969) or in the course of normal development in the notches of leaves such as those of *Bryophyllum* species are not to be regarded as 'foliar embryos' or embryoids.

Adventitious buds arise from cells of the epidermis (*Begonia rex*, *Linum* species, *Torenia fournieri*, *Daucus carota*, etc) or from cells of the hypodermis (*Euphorbia lathyrus* - literature in Haccius and Lakshmanan, 1969). Despite the fact that adventitious buds are often described as arising from single cells of e.g. the leaf epidermis, their exact origin is difficult to establish since divisions are initiated also in adjacent cells, and many such cells usually become involved in the development. No information appears to be available as to the state of the plasmodesmatal connections of the initiating cell or cells. In a study of adventitious bud formation from tobacco callus tissue, Ross *et al.* (1973) show plasmodesmata in the walls of a cell said to be 'postmeristemoid', i.e. likely to give rise to a bud.

Plasmodesmata appear to be lost from the walls of the endodermal cells which give rise to lateral root primordia in *Zea mays* at a stage when they begin to divide anticlinally to form the epidermis of the new lateral (Karas and McCully, 1973). As the new lateral root grows through the cortex, cortical parenchyma cells are attacked by hydrolases emanating from the lateral root tip. Before they are degraded, these cortical cells appear to lose their plasmodesmata (Bonnett, 1969).

It was Steward (Steward *et al.*, 1964) who first put forward the concept that cells in culture might have to be "*freed from organic connections with other cells*" as well as "*nourished by a medium competent to support their rapid growth and development*" if they are to exhibit their full totipotency, in particular, to give rise to new individual plants. We are here concerned only with the problem of cell isolation which was dealt with in 13.4.3. The difficulty of demonstrating initial plasmodesmatal isolation is compounded by the fact that embryoid formation can usually only be determined *post hoc*. Perhaps the most convincing demonstration that single cells can, in the right conditions, initiate embryos is that of Backs-Hüsemann and Reinert (1970). This was achieved from cells of a suspension culture originating from a carrot root callus tissue. Optimal conditions were first discovered as well as the fact that embryo formation usually involved small, isodiametric, densely cytoplasmic parenchymatous cells which generally formed part of small compact complexes. Individual cells of this type were isolated on nutrient agar and the subsequent division and eventual development of embryos followed by continuous observation. The first step appears to be an unequal division followed by repeated division of the more densely cytoplasmic cell to form an embryonal tissue complex which forms an axis of polarity and thus an embryo. Backs-Hüsemann and Reinert thus appear to have provided a formal demonstration that starting with a single somatic cell one can obtain an embryo by suitable treatment.

After pointing out (Halperin, 1969) that it would be enlightening to have electron microscope studies of the initiation of nucellar (i.e. somatic) embryos to compare the mode of isolation of a somatic embryo produced *in vivo* with those produced *in vitro*, Halperin attempted just such a study (1970). As in *Capsella* (see 13.4.4.4.) the chalazal end of the embryo sac of *Nothoscordum fragrans* has plasmodesmata between it and the cells of the nucellus, but the embryo sac is elsewhere lacking in plasmodesmata. All the cells of the nucellus are interconnected by plasmodesmata. The ovule thus appears to be plasmodesmatically as 'normal' as is that of *Capsella*. Yet in response to some unknown stimulus, nucellar cells proliferate into the embryo sac at the micropylar end to form, eventually, a somatic embryo. *"Whether this is the result of eventual physical isolation of one or more cells in the proliferating nucellus is still uncertain....Considerable histological disorganization is an obvious feature of the first stage of proliferation and the plasmodesmata which connect the cells involved may be incapable of serving the function which they have in organized tissues where polarized movement of hormones is important"*. In free cell cultures of the same plant *"embryos are clearly derived from the small multicellular units present in the inoculum"*...but...*"the first stages of development of embryonic cells in the explant have not been studied since the cells in question cannot be identified"* (Halperin, 1970). Nevertheless, it would appear that, as in the carrot cultures of Reinert, an unequal division to form a rapidly subdividing cluster of cells attached to a non-dividing vacuolate cell is the first recognisable stage. It is unfortunate, therefore, that we still do not have the critical information on which to make this extremely interesting comparison.

Button *et al.* (1974) have obtained a unique callus, said to be of nucellar origin, from unfertilized ovules of orange (Fig. 13.8.). The callus consists *"solely of numerous proembryoids ...adventive embryogenesis in the tissue is autonomous"*. Cells destined to form new proembryoids become surrounded by greatly thickened cell walls which lack plasmodesmata. It appears likely that the plasmodesmata become blocked and disappear prior to the onset of internal divisions in the embryogenic cells. It would be this tissue, *par excellence*, which would yield needed information on how a plasmodesma is obliterated. It is to be hoped that the South African group will provide us with this information in due course (Fig. 13.8.).

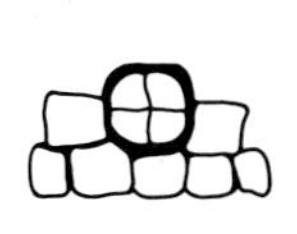

Fig. 13.8. Development of embryoid and plantling from ovular callus tissue of Shamouti orange. (Redrawn from Button *et al.*, 1974)

A callus tissue culture of *Ranunculus sceleratus* gives rise to embryoids which grow into plantlets. Single epidermal cells of the stems of these plantlets develop into embryoids (Konar *et al.*, 1972). These epidermal cells are said to be initially in plasmodesmatal connection with the underlying hypodermis and with adjacent epidermal cells. (Actually plasmodesmata are shown only in two micrographs and only in walls separating two 'proembryoid' cells. Walls separating such pairs of proembryoid cells are said to be rich in plasmodesmata. No proof is given of plasmodesmata elsewhere). Once embryogenesis has commenced the proembryo grows rapidly and ruptures the cuticle. Such an early proembryo can be detached easily from the stem. The isolation of the young proembryo is said (but not proved) to be associated with a change in wall composition. Embryos can also be formed directly from the callus and their development has been followed by Thomas *et al.* (1972). It is suggested, but not proved, that they originate from single cells. The earliest stages seen are groups of a few cells. It is surprising therefore that the embryogenic cells are said to be "*in continuity (through plasmodesmata) with surrounding cells when they embark on embryogenesis*". Only one micrograph shows plasmodesmata (two!) and they are in the wall separating the two cells of a two-celled proembryo! Moreover, this was from a 2,4-D culture in which, as the text reveals, embryo formation is retarded. It would be better to be realistic about the case of *Ranunculus sceleratus* and say that plasmodesmata have been seen but that no evidence for initiation of embryogenesis despite plasmodesmatal connection is presented. The good evidence for *Citrus sinensis* provided by Button *et al.* shows the opposite. On the whole, therefore, the critical evidence supports the view that resumption of totipotency leading to embryogenesis in cultured cells is preceded by cell isolation. Following that, there may be a stage of cell proliferation to create a mass capable of determining an axis of polarity. Subsequently, embryo formation follows somewhat (but by no means exactly) the pathway it takes in normal embryogenesis in the ovule.

It has been shown (13.4.4.5.) that the vegetative cell of the pollen of angiosperms is isolated from the generative cell. It is therefore of interest that an embryoid can develop from the vegetative cell (see, e.g. Sunderland and Wicks, 1971), or from the even more clearly isolated uninucleate precursor of the binucleate pollen grain (Rashid and Street, 1974). The first, and presumably also the subsequent, cell wall to be formed in the embryoid contains an abundance of plasmodesmata (Dunwell and Sunderland, 1975), showing that there is nothing about the haploid condition or the environment within the pollen grain which precludes formation of plasmodesmata. This is also well shown in the gametophytes of lower plants, in the multicellular pollen grains of *Ginkgo* and *Podocarpus* and the cell plate between the spermatozoids of *Biota* (13.4.4.5.).

13.5. PLASMODESMATA IN CELL WALL GROWTH

Little information is available on the role or fate of plasmodesmata during cell wall growth. Many plasmodesmata are grouped in the thinner areas of the primary cell wall called 'pit fields'. Evidently there is some association between the paucity of cellulose microfibril deposition in these areas and the presence of the plasmodesmata, and the relationship may be causal. But how the pattern of distribution of pit fields arises is not known. In derivatives of stem cambial cells, the pattern repeats that of the walls of the fusiform initials, since all such cells in a sequence (e.g. tracheids in pine wood) have pits (and hence pit fields) which are regularly aligned. In other cases the pattern of grouping may arise by early occlusion of plasmodesmata in areas which will not become pit fields (Chapter 4).

The first electron micrographs of preparation of cell walls, cleaned to show the cellulose microfibrils, led to the designation of areas (not initially recognised as pit fields) as centres of wall growth and to the concept of 'mosaic growth'. Wardrop (1955; 1956) then drew attention to the fact that the longitudinal separation of pit fields keeps pace with the growth of the entire wall, i.e. the wall grows over its whole surface. Moreover, labelled glucose was incorporated into microfibrils deposited over the whole growing wall. Support was thus found for Roelofsen's concept of 'multi-net growth', according to which microfibrils are deposited in a preferred transverse orientation, on the inner face of the growing wall but during later growth these microfibrils are pulled into an increasingly longitudinal orientation. Thus the outer layers of the wall have predominantly longitudinally arranged, sparsely distributed microfibrils, and each successive layer laid down during growth suffers the same fate of longitudinal dispersion. The arguments for and against this hypothesis are marshalled by Preston (1974). During the process, pit fields remain about the same in number but become increasingly subdivided by the deposition of groups of microfibrils across them (see 4.5.). In general, however, the microfibrils tend to take a course around the edges of the pit field or its subdivisions. The subdivision of the pit field may result in occlusion of some plasmodesmata and it may well be that the total number of plasmodesmata falls with cell expansion. The only figures available are for extreme states of two specialised cell types, endodermis and root cap cells (Table 2.1.). There is a possibility that secondary plasmodesmata may be formed during, or subsequent to, cell expansion (4.3.3.). Contrary to the expectations of the mosaic growth hypothesis, the plasmodesmata themselves appear to make no direct contribution by multiplication to the growth of the cell wall. However, direct investigation of the plasmodesmatal budget during growth of typical parenchymatous cells is very much needed.

In her investigation of the cortical parenchyma of *Viscum*, Reta Krull (1960) found that the area of primary pit field remained constant at about 30% of the total longitudinal wall area during the period of cell division, when the internode extended from 1 to 30 mm. During further growth to 5 cm. the specific area of pit field fell to about 20%, but the number of pit fields tended to remain constant (whereas it must increase during the period of cell division). During extension growth the pit fields enlarged and became subdivided, each subdivision retaining an area roughly the same as that in the dividing cells (5-60 μm^2). She believed, but gave no figures to show, that the number of plasmodesmata *increased* during the period of growth after cell division ceased. Subsequent 'secondary' thickening of the wall reduced the apparent size of each pit field, constraining each to a small circular patch; however, this brought about no change (she says) in the number of plasmodesmata. In the mature longitudinal walls, the plasmodesmata were in groups of up to nine, each group joined by an extended Mittelknoten. On the other hand, of the plasmodesmata in the cross walls, which grow relatively little, few had joined Mittelknoten, and of those that did, the Mittelknoten were small. Krull suggested that the striking correspondence between the size of the Mittelknoten, the degree of branching of the plasmodesmata and the extent of wall growth might be explained as due to increase in number of plasmodesmata by subdivision (a concept already rejected by Strasburger, 1901 as "*horrible*", for, in his time, and even up to 1957 (Meeuse) it was possible to say that "*branched plasmodesmata do not occur*"). Krull showed how branching could tie in with the multi-net growth hypothesis (Fig. 13.9.). She could find no evidence for a secondary origin of plasmodesmata during cell extension.

The persistence of plasmodesmata and the apparent difficulty of their replacement, if lost, is in sharp contrast with the lability of

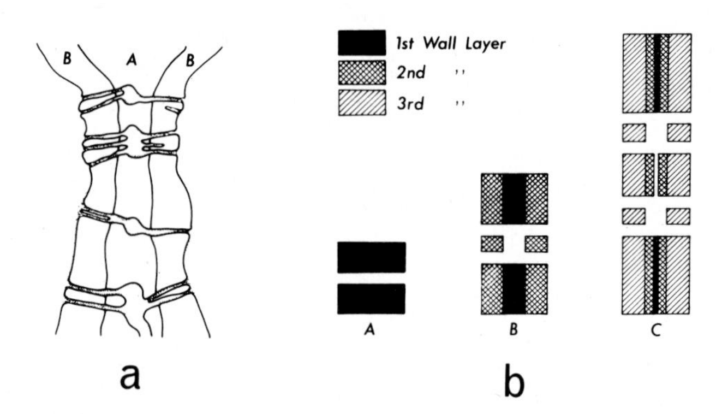

Fig. 13.9. *Viscum album* cortical cell.
a) Diagram to show the branched plasmodesmata in the longitudinal cell walls.
b) Diagram to illustrate the possible mode of involvement of multiplication of plasmodesmata in cell wall growth according to the multinet growth hypothesis. (Redrawn from Krull, 1960)

gap junctions in animal cells. The difference is to be attributed to the ability of the animal cell to migrate, and to the relative immobility of the plant cell with respect to surrounding tissue. Nevertheless, some plant cells have a pronounced ability to grow intrusively, usually by tip growth, between other cells. An example has been cited (4.9.) in the non-articulate laticifers which can form secondary plasmodesmata with adjacent parenchyma. No other example of intrusion followed by secondary plasmodesmatal connection is known. It seems likely that intrusive growth of fibres in internodes, or 'wandering fibres' of sclereids in leaves is not accompanied by the formation of secondary plasmodesmata. Intrusive growth of such cells may even, as suggested in Section 13.4.2., be initiated by partial or total plasmodesmatal isolation from surrounding tissue. The intrusive growth of foliar sclereids and fibres usually commences late in development of the leaf, thereby reducing the degree of disorganisation of the tissues. A certain amount of intrusive growth occurs in the maturation of longitudinal elements of the xylem, but the extreme examples are to be found in phloem fibres (e.g. flax).

There can be no question that the majority of tissue expansion takes place by symplastic growth of the constituent cells, during which the only losses of plasmodesmata need be those which abut on intercellular spaces or, probably, the angle thickenings of collenchyma. (It would be interesting to know whether the collenchymatous thickenings which form on the *side* walls of the palisade parenchyma of the poikilohydric fern, *Ceterach*, and *not* in the angles, as in normal collenchyma, involve loss of plasmodesmatal connection between the cells. Poirault (1893) illustrates the similar 'side wall collenchyma' of another fern, *Polypodium thyssanolepis* as having pits, but he does not say whether they are traversed by plasmodesmata). For a discussion of the now-discarded hypothesis of 'sliding growth' in relation to plasmodesmata, the reader is referred to Meeuse (1957) and for further references on intrusive growth to Sinnott (1960).

13.6. DORMANCY

The first investigations of the fate of plasmodesmata following the drying of plants were made by Poirault (1893) on herbarium specimens of some species of *Polypodium*. Following a few hours of soaking in water, plasmodesmata could be demonstrated. Since Poirault also reviewed previous work on the capacity of a number of species of ferns to revive after prolonged drying, and since these include some species of *Polypodium*, it is probable that he had in mind the problem of the state of the plasmodesmata in such 'resurrection plants'. In the cell walls of a completely dried but still viable moss (*Mnium affine*) Strasburger (1901) could demonstrate the plasmodesmata. When higher plants, not capable of surviving drying are dried, the protoplast adheres to the walls only where these are thin and flexible enough to collapse on the shrinking cytoplasm. In dried shoots of *Selaginella martensii* plasmodesmata were preserved as they were in cells of higher plants. Strasburger mentions *Selaginella lepidophylla*, a well-known poikilohydric ('resurrection') plant, but does not specifically state that he investigated the plasmodesmata in its dry cells. However, he remarks that in all instances where plants survive drying, the protoplast adheres firmly to the cell walls.

'Resurrection plants' are currently under investigation by Gaff and Hallam (1974). Gaff (pers. comm.) has clear evidence from electron micrographs of the survival of the plasmodesmata in dried but still viable specimens of *Borya nitida* (Liliaceae) and Hallam (unpublished) finds plasmodesmata in *Xerophyta villosa* eight hours after rehydration. In dried plants of that species and of *Talbotia elegans* plasmodesmatal fragments can be seen following fixation with aqueous fixatives in cell walls from which the cytoplasm has (artifactually?) retracted. Against these positive findings we must place the statement of Genkel and Pronina (1968) that poikilohydric plants lose their plasmodesmata on drying and that their protoplasts retract from the cell walls. This is said by them to be true of *Myrothamnus flabellifolia* (Rosales) (Genkel and Pronina, 1969). They also claim that the poikilohydric moss, *Neckera crispa* has no plasmodesmata at all. This claim, if true, would make it unique among the Embryophyta. To the contrary, the plasmodesmata are not lost from the stem cells of the poikilohydric liverwort, *Pleurozium schreberi*, after a prolonged drying (Noailles, 1974). Dry seeds of wheat and other cereals, and the endosperm cells of palms have been classical objects for the demonstration of plasmodesmata since Tangl and if electron microscopical confirmation were needed of their presence in dry seeds it is found in the study of the dry embryo of *Lactuca sativa* (Paulson and Srivastava, 1968), and in that of Perner (1965) on *Pisum sativum*. Drying to remove 75% of the water from a growing root of maize does not result in the disappearance of the plasmodesmata (Nir *et al.*, 1969). Nor does prolonged seed storage: plasmodesmata are still evident in ancient (1040±210 years) but viable seeds of *Nelumbo nucifera* (Hallam and Osborne, 1974).

Information on the state of the plasmodesmata in dormant shoot apices is scarce. Fabbri and Palandri (1968) kept cuttings of *Psilotum nudum* in moist sand over a period of three years. None rooted but all the apices remained alive but quiescent. The cell walls of the apical meristem, initially thin and equipped with plasmodesmata, gradually thickened and after two and a half years the plasmodesmata had apparently been resorbed. Individual cells separated at the middle lamella - a sort of encystment of the apical cells. This capacity to 'encyst' is held by the authors to account for the exceptional capacity for drought survival in this plant. A similar encystment of cells of overwintering plants such as winter wheat, cabbage etc. was described by Harvey (1933, cited in Vasil'yev, 1971), and his observations were repeated and developed by Genkel and Oknina (1948) and Oknina (1948). They described separation of the protoplast from the cell wall, involving retraction of the plasmodesmata. This was said to be observed in cells of leaves and apices of wheat and other grasses, needles of conifers and apices of fruit trees. It was not found in apices of walnuts and to a small extent only in *Vitis vinifera*. This 'isolation of the protoplast' is said to be accompanied by the formation of a thick superficial layer of lipoidal material on the protoplast. In a state of deep dormancy the plant is supposed to lose its 'connectedness' and to resolve itself into a mere collection of isolated cells. Plasmodesmata are supposed to be re-established at the end of the period of dormancy. Short-lived seeds are held to have cells unable to become isolated in this way (Genkel, 1950). Plasmodesmata are said to be absent from dormant currant buds (Genkel, Oknina and Bakanova, 1968 - referred to by Vasil'yev, 1971). The Russian authors also claim that Hechtian strands are not formed as a result of plasmolysis of dormant cells (although Levitt, in a discussion following Genkel (1950) holds to the contrary for frost-resistant cells). Kolomiec (1955) believes that a convex plasmolysis figure is diagnostic of a state of deep dormancy in red clover and many other Russian workers on dormancy have used plasmolysis tests as criteria of 'isolation of the protoplasts'.

Some of these claims were critically investigated by Krull (1960). She pointed out that they are certainly not true of seeds in general (as attested by Tangl's discovery!). Krull investigated mesophyll cells of 13 conifer species, five dicotyledons, seeds (cotyledons and endosperm) of seven dicotyledons and cortical parenchyma of stem or rhizome of 16 species as well as tubers, roots and bulbs. In no case could differences be observed in the state of the plasmodesmata between dormant and active tissues. Well-formed plasmodesmata were present in both winter and summer. Frozen sections or sections of rapidly fixed material of dormant wheat seeds (claimed by Oknina, 1948, to lose their plasmodesmata) revealed them intact. Onion bulbs of the same variety as used by Oknina and stored in the cold following the régime used by the Russian author, were tested and convex plasmolysis figures were obtained, fulfilling the conditions alleged to represent resorption of the plasmodesmata. Nevertheless, plasmodesmata were found in all the cell walls, remaining there after retraction of the protoplast in plasmolysis. Significantly, the Russian authors do not cite Krull's work.

An electron microscope study of cells of the wheat plant in winter shows plasmodesmata (Chien *et al.*, 1973) and in the dormant state the phloem of *Metasequoia* also has plasmodesmatal connections (Kollman and Schumacher, 1959). Genkel and Kurkova (1971) have reinvestigated the cells of dormant onions, using electron microscopy. Plasmodesmata are indeed readily seen in their preparations, although some of the plasmodesmata (labelled 'bifurcated') shown in their figures are clearly folds in the embedding plastic! The authors claim, however, a reduction in the number of plasmodesmata in the dormant state. In view of the difficulties attending the making of quantitative estimates of plasmodesmata (see Chapter 2) this claim - and indeed, probably all the claims of Genkel and Oknina, of the dissolution of plasmodesmata in dormancy - may be treated as very dubious. One must concede, however, that the plasmodesmatically-isolated stomatal guard cells (see 13.4.2.) have an unusually high degree of stress resistance (Döring, 1932).

13.7. PLASMODESMATA IN AGEING AND LONGEVITY

Plasmodesmata may be as long-lived as the cells which produce them or quite ephemeral (as they are in the cell plate between the generative and vegetative cells of the angiospermous pollen grain). Some plasmodesmata are relatively short-lived and are occluded during cell differentiation (e.g. those which abut future intercellular spaces). The record for longevity must surely be held by the plasmodesmata of *Nelumbo nucifera* seeds, mentioned in Section 13.6. However, quite commonly xylem parenchyma cells of trees may live as long as 100-150 years and since they will depend on their plasmodesmatal connections for organic nutrients, they must retain them while they remain alive. Indeed, since plasmodesmata are to be found even in yellowed abscised leaves (see 13.6. and 13.8.) they may persist structurally after the death of the protoplast.

Plasmodesmata are obliterated during or preceding morphogenetic cell death. Ziegler (1964) gives such an instance of isolation preceding death of a ray cell in *Taxodium distichum* and in 13.4.5. another is mentioned relating to death of cortical cells during outgrowth of a lateral root primordium. Plasmodesmata are occluded in the walls of cells abutting tracheids or vessel elements.

Little information is available on changes in plasmodesmata during ageing. In algae the plasmodesma-like interconnections between cells of some colonial Volvocaceae become attentuated during growth and enlargement of the colony and may eventually be lost (see 13.4.1. and 3.5.). The coarsening of plasmodesmata of *Chara* with age is recorded in 2.2.2.4.), and the opening of neck constrictions in *Abutilon* nectary stalk cells in 11.3..

13.8. ABSCISSION

Plasmodesmata might be involved in abscission in two possible ways: (1) they could be withdrawn or obliterated prior to degradation of the cell wall at the abscission zone or never formed there; (2) they might serve as channels for the pectinases and other lytic enzymes secreted to cause wall degradation (as in aleurone cells, where the plasmodesmata are foci for the liberation of phosphatase and esterase - see 4.7.). Studies have been made of the ultrastructure of cells at abscission zones in *Coleus* and *Gossypium* (Bornman, 1967) tobacco and tomato flower pedicels (Jensen and Valdovinos, 1967, 1968; Valdovinos and Jensen, 1968; Valdovinos, Jensen and Sicko, 1972, 1974) and *Phaseolus* primary leaves (Morré, 1968; Webster, 1968, 1973; Sexton and Hall, 1974 and Hall and Sexton, 1974). In no case have plasmodesmata been found (as far as can be ascertained from the published micrographs or from the texts of the publications) across cell walls actually undergoing breakdown during abscission. The evidence is therefore negative but supports the possibility that such walls either never have plasmodesmata (which seems unlikely) or that they are obliterated before abscission. On the other hand, ethylene treatment results in evident initiation of wall degradation within 2 or 3 hours (Valdovinos, Jensen and Sicko, 1972) in tobacco flower pedicels. Similarly, in *Phaseolus* explants fixed 8 hours after ethylene treatment, wall dissolution had proceeded to a considerable extent (Webster, 1973). If plasmodesmata are obliterated in these walls following ethylene treatment, an opportunity is provided to discover how quickly obliteration can take place. In the first two papers of their series, Jensen and Valdovinos suggest first, that plasmodesmata might be sites where wall dissolution occurs; second, that it might be initiated in the vicinity of plasmodesmata. However, no direct evidence of this is provided and the topic is not referred to in the later papers. There is no evidence of it in the papers on *Phaseolus*.

In *Phaseolus*, Webster (1968) shows that walls between cells may disappear during the natural abscission process, leaving the protoplasts more or less intact but plasmolysed. Such protoplasts do not fuse, from which one might suppose either that they were plasmolysed to begin with or else that plasmolysis occurred during wall dissolution or during fixation and preparation for microscopy. When isolated by enzymic degradation of the cell walls in a hypertonic solution protoplasts may still adhere by their plasmodesmata (Prat, 1972; Withers and Cocking, 1972; Fowke *et al.*, 1973). Highly vacuolate cells tend to produce separate protoplasts; less vacuolate cells tend to reaggregate following wall loss and can then undergo a process of cell fusion (Burgess and Fleming, 1974; Prat, 1972; Withers and Cocking, 1972). The lack of connection between the protoplasts during the abscission process supports the view that they were not interconnected at the beginning of the process otherwise the protoplasts might then have been drawn together. In *Coleus* and *Gossypium* there are marked changes in the appearance of the cytoplasm at a stage when abscission is about 50% completed. In ethylene-treated *Phaseolus* explants, the cytoplasm shows few signs of degradation at abscission.

In some cases (e.g. *Phaseolus* - Webster, 1968) cell divisions occur in cortical cells in the vicinity of the abscission zone. Such divisions might be due to assumption of 'meristemoid' activity following 'physiological isolation' due to occlusion of plasmodesmata across the future abscission zone. According to Webster (1973) groups of 4-5 plasmodesmata occur in the cell walls formed in such divisions. In yellowed, abscised leaves of *Fraxinus ornus* and other dicotyledons, Strasburger (1901) was able to find plasmodesmata apparently unchanged in the cell walls (and Poirault (1893) has a similar observation for a fern). Plasmodesmata were also present in abscised needles of *Abies*, gathered from the ground under the trees. Strasburger held as unlikely the hypothesis which had been put forward by Kienitz-Gerloff (1891) that the cytoplasm was withdrawn from such leaves before leaf-fall through the plasmodesmata.

13.9. PARASITES

In those cases in which parasites establish symplastic connections with their hosts the plasmodesmata which form must arise secondarily. The salient facts from recent electron microscopical observations on these secondary plasmodesmata are discussed in 4.3.2. A fuller discussion is given here of the symplastic relationship between host and parasite, in view of the great interest which attaches to this topic.

13.9.1. *Cuscuta*

Neither Kienitz-Gerloff (1891) nor Strasburger (1901) could find plasmodesmata (as such) between haustorial cells of the parasite and host cells. Pierce (1893) claimed to have seen sieve plates between the end of the contact hypha and the host sieve tube, forming a cytoplasmic union which Strasburger assumed to be unrelated to plasmodesmata and responsible for mass flow of assimilates to the parasite from the host. Thoday (1911) supported Pierce's claim but it is now known to be quite wrong. Schumacher (1934) distinguished between 'searching hyphae' which emanate from the initial haustorium and the 'contact hyphae' which enclose the sieve tubes of the host "*like fingers grasping a hand*". Using freeze-sectioning and Mühldorf's staining methods Schumacher and Halbsguth (1939) looked for plasmodesmata and immediately revealed that they are numerous in the cell walls of the searching hyphae. These plasmodesmata traverse the whole thickness of the parasite cell wall and they must (as the authors point out) be of secondary origin. However, despite diligent search no plasmodesmata could be found at the interface between the contact hypha and the host sieve tube, although plasmodesmata were readily evident in cross walls of the hyphae. Bennett (1944) confirmed the existence of the plasmodesmata in the searching hyphae. Schumacher's student, Dörr, re-investigated the problem, using electron microscopical methods. She was able to confirm (1967, 1968c) Schumacher's and Halbsguth's finding as far as the searching hyphae were concerned. For the difficult task of sectioning hyphal tips for electron microscopy special methods - growth of the dodder on a single bundle dissected from, but intact in, the petiole of the host, exploration of thick sections and re-embedment for electron microscopy - were used (Dörr, 1968a). Numerous plasmodesmata, 30-40 nm diameter were seen and they are believed to be surrounded each by a wall cylinder probably of callose (Dörr, 1969). Hyphal tips grew both inter- and intracellularly. The tip of the intracellular hypha has a cell wall, capped by a callose layer, laid down by the host cell. Between hyphal tip and host cytoplasm plasmodesmata extend, crossing both walls. These common plasmodesmata persist only until a sufficiently thick enclosing wall is formed by the host. The host half-plasmodesmata are then occluded. The parasite half-plasmodesmata remain, ending blindly in the middle of the wall. Dörr (1968b, 1972) and Kollman and Dörr (1969) were unable to find any plasmodesmata between the contact hypha and the host sieve tube. Indeed, the wall of the contact cell of the hypha becomes highly elaborated, i.e. the cell becomes a transfer cell. Kollman and Dörr believe that the cross-walls of the contact hypha become sieve plates, i.e. the hypha is transformed to a parasite sieve tube.

13.9.2. *Viscum*

Neither Kienitz-Gerloff (1891) nor Kuhla (1900) found plasmodesmata between parasite and host. Strasburger (1901) found pits between cells of the haustorium and of the host but no plasmodesmata traversed them. Both Kuhla and Strasburger found half-plasmodesmata extending from the parasite to the middle of walls common to both parasite and host. The pits in the cross-walls of the parasite cells are especially well-developed and Strasburger assumed that in the cambial zone, where the parasite tissue keeps pace in growth with the secondary growth of the host stem, the plasmodesmata are numerous to provide adequate transport to the parasite xylem which, of course, does not connect directly to that of the host. Haustoria which penetrate only to the cortex of the host contain sieve tubes but make no plasmodesmatal connection to cells of the host.

13.9.3. *Orobanche*

Schumacher and Halbsguth (1939) investigated several species of *Orobanche* in different hosts. The cells of the parasite are especially densely cytoplasmic and in *O. ramosa* can be multinucleate. The cells of the 'searching hyphae' attach to the side of sieve tubes of the host, especially in the region of the sieve plates, but the sieve tubes continue to function - callose plugs are not formed. As sieve tubes become empty they collapse and are resorbed, with the exception of the sieve plates themselves. Parasite sieve tubes appear in its tissues at some distance from the sieve tubes of the host. Plasmodesmata between host and parasite were not found either in the tubular 'searching hyphae' or in the cells in contact elsewhere with the host. Plasmodesmata were readily evident in internal walls of the parasite. However, plasmodesmata were believed to be seen between older cells of *O. speciosa* and parenchyma of its host, *Vicia faba*, but they were sparse and rare. The authors also investigated *Lathraea squamosa*. This parasite forms no connections to its host sieve tubes but appears to depend on resorption of collapsed regions of host tissue.

Dörr and Kollman (1974, 1975) re-investigated *Orobanche*, confirming that plasmodesmatal connections with the host are not established. Cells of the contact hypha slowly transform to sieve elements, each with a companion cell. The wall between the contact cell and the first 'transition cell' bears plasmodesmata and further back along the hypha such cross walls are transformed to sieve plates. Dörr and Kollman (1975) provide a useful review of the work of other authors on the structure and development of angiospermous parasites, especially in relation to sieve tube development.

13.9.4. *Arceuthobium*

Very close contact develops between cells of *Arceuthobium pusillum* and of its host, *Picea mariana*, according to Tainter (1971). Plasmodesmata between the two are restricted to small areas and to cells of the phloem parenchyma of the leaves of the host. Tainter did not observe the phloem-like connections of the parasite with the host phloem described by earlier authors, and believes the plasmodesmata he observed serve in transporting assimilates to the parasite.

13.9.5. *Phytophysa treubii*

The probability of plasmodesmatal connections between the green alga *Phytophysa treubii* (Phyllosiphonaceae) and cells of its host, *Pilea oreophilae* aff. (Urticaceae) which grows in the tropical rainforest of Java was suggested by Weber-van Bosse (1890); Zimmerman (1893 - cited in Strasburger) was convinced that such connections exist and Strasburger (1901) cites the instance as an extraordinary one of living connection between an alga and an angiosperm. Through the walls of the thick-walled, 2-2.5 mm wide bladders formed by the alga, enclosed in the parenchymatous tissue of the host, pass fine cytoplasmic strands which emerge through bulging 'bordered pits' (or areolae) formed by raised rings of 'cuticle' on their outer surface (Fig. 13.10.). Weber-van Bosse believed that these strands made connection, through plasmodesmata across the cell walls of the host, with host cytoplasm, but was not able to show this conclusively. The

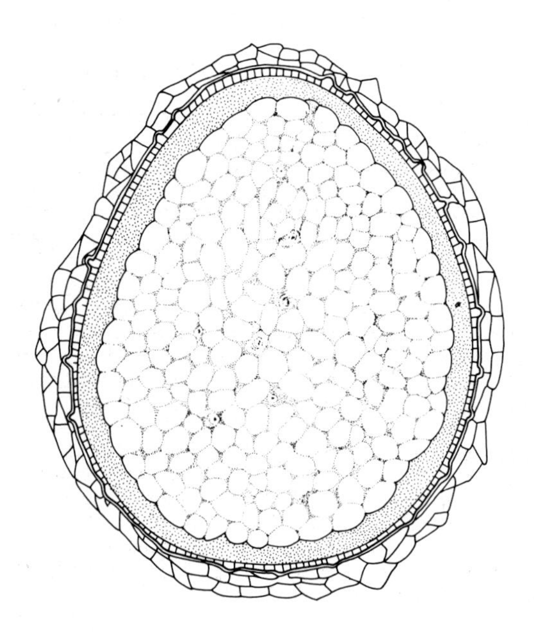

Fig. 13.10. *Phytophysa treubii* Longitudinal section of developing coenocyte in cortical tissue of host. (Redrawn from Weber-van Bosse, 1890)

areas of walls of parenchyma cells on which the algal areoles abutted were especially thin. The author used Russow's staining methods to slow plasmodesmata but the results were "*unsatisfactory*". But she did often see a fine dotted appearance of the thin *Pilea* cell walls apposed to the *Phytophysa* areoles and "*in all probability these simple pits were in contiguity with the pits in the areoles of the alga. It is evident that these canals would serve to transport materials which the* Phytophysa *extracts from the host which otherwise would hardly penetrate through the thick wall of the alga*". More areoles appear as the alga grows and fine strands of cytoplasm in the wall seem to indicate where future areoles will appear. Cytoplasmic connection with the host is lost at full size and before spore formation begins. The parasite induces the host to proliferate and form gall-like structures. The phenomenon is described here not so much because the claims for symplastic continuity between an alga and an angiosperm can be fully accepted, but because the matter certainly deserves re-investigation.

13.9.6. Physiological Implications of Host-Parasite Relations

Schumacher and Halbsguth (1939) drew the conclusion that the absence of penetration of parasites into the sieve tubes of their hosts gives proof that the unloading of the sieve tubes by sinks in the normal plant must proceed in the same manner as that in which parasites obtain assimilates from the phloem. Kollman and Dörr (1969) suggest that assimilates must move out of the sieve tube into the free space of the cell walls before they are absorbed by the parasite cell. In *Cuscuta* the absorption must be the function of the contact transfer cell. Jacob and Neumann (1968) have speculated that the phloem of the host is rendered leaky in the presence of *Cuscuta*. This has been examined experimentally by Wolswinkel (1974a,b). Using stem segments of *Vicia faba* parasitised by *Cuscuta* Wolswinkel was able to show that, in comparison with unparasitised segments, supplied assimilates are not retained by the phloem. He suggests that the leakiness might be induced by some hormonal influence of the parasite (c.f. Abou-Mandour *et al.*, 1968) (see also 11.3.).

13.10. PLASMODESMATA AND THE REGULATION OF GROWTH AND DEVELOPMENT

The evidence presented in this Chapter suggests that the plant has means to control the symplastic continuity of its parts and to regulate the degree of isolation of its constituent cells. This form of regulation is a counterpart to that which results from the release and action of hormones. One is led to speculate whether hormones themselves might act in part by opening or closing symplastic pathways. There is some indication of this in the control of abscission (see 13.8) by hormones such as ethylene, abscisic acid and auxin. Callose formation is one of the important, often temporary, steps in the occlusion of plasmodesmata and it is significant that application to plants of the synthetic growth regulator, tri-iodobenzoic acid (TIBA), can initiate callose formation causing cell degeneration (by isolation?) as well as blocking auxin transport (Bouck and Galston, 1967; Gorter, 1949).

A number of processes in morphogenesis which involve cell death involve severance of plasmodesmata (see 13.4.5.). On the other hand, plasmodesmata persist between living cells of quite different differentiation states, while in a uniform tissue only exceptional (idioblast) cells may differ in differentiation state. The suggestion is therefore at hand that the transmissibility of plasmodesmata for morphogenetic information (possibly also for genetic information in exceptional cases) may be under control by the plant.

When plasmodesmatically interconnected tissues in the plant are excised and cultured *in vitro*, the callus tissues which eventuate are often friable. In this state they presumably lack the plasmodesmatal continuity of the parent tissue (see 13.4.3.). One can infer then that part of the task of the whole plant in maintaining its integrity involves the maintenance of the symplast. It is interesting to compare the rapidity with which symplastic continuity can be established across graft unions between scion and stock (see 13.3.3.3.) and the reported absence of plasmodesmata between tissues of different origin cultured together *in vitro* (see 13.4.3.).

13.11 OPEN DISCUSSION

GUNNING asked about 'isolation'. Was it not possible that isolation from mechanical stresses might in some cases be as important as isolation from the symplast? He cited the work of Lintilhac and Jensen (1974) and Lintilhac (1974a,b) in respect of the isolation of the embryo sac. In cotton the embryo sac is isolated in both senses, for the cell walls in the integuments are disposed in such a way that it lies cushioned in a compression-free space. Lintilhac's concept of stress behaviour as a spontaneously arising morphogenetic trigger applies whether (as in the embryo sac example) or not (as in some of Bünning's meristemoids?) there is symplastic isolation in addition.

It is undeniable, said CARR, that tissue stresses affect the direction and rate of growth of cells and that applied stresses can affect the planes of cell divisions (e.g. classically in stem cambia). Epidermal cells may owe some of their capacity to differentiate in a wide variety of ways to the fact that they are unconstrained on their outer faces. Moreover such constraints are evident in the inner tissues of organs and must be involved in the conformation of 'inner' partners in periclinal chimaeras with 'outer' partners (see 13.3.2.). However, there is no evidence of mechanical stresses specifically determining the direction of differentiation. The unequal divisions, or divisions across axes of polarity, which according to Bünning (1953) precede so many cases of cellular differentiation are not explainable solely in mechanical terms and as Lintilhac admits, there are many exceptions to Errera's Law (of minimal cell plate area or of cell plates determined by minimal stress). More specifically, in the later growth of the cotton ovule (as a seed) there is a real symplastic isolation when the funiculus breaks. The biochemical and morphological changes which then ensue (as described by Ihle and Dure, 1972) are far more dramatic than the orientations of cell lines determined by effects of pressure of the inner integument on the nucellus as described by Lintilhac, or any attributable to 'mechanical isolation' in a system which retains symplastic connection.

The possibility that mechanical stress might be a factor in regulating development in tissue culture is referred to by Street (quoted in 13.4.3.) but the experiment of Sussex and Clutter (also discussed in 13.4.3.) shows that intimacy of cell contact achieved by compression cannot replace symplastic continuity in determining differentiation.

MARCHANT asked whether anything was known of the evolution of isolation in the genesis of megaspores and microspores, since (e.g.) the megaspore is always clearly free in the megasporangium in pteridophytes. CARR replied that the microspore is always clearly isolated, presumably because it is destined to be a dispersal unit. Martens (1966) had made a proposal, based on observations of callose formation, that the Cycadales (as typified by *Encephalartos*) among living gymnosperms are alone in preserving a presumed 'ancient mechanism' isolating the megaspores in the sporangium like the pollen in the anther. For instance he suggests that megasporophyte callose is absent in *Ginkgo*, *Larix* and probably the Gnetales. Unfortunately for this hypothesis there is abundant evidence for callose walls isolating the megasporocytes and megaspores of a wide range of Angiosperms (cf. Rodkiewicz, 1970). Evidently the 'ancient mechanism' is conserved because it is essential to enable the new generation to express its genetic individuality. As in the case of the microspore the callose wall disappears during differentiation of the megaspore.

14

HISTORICAL PERSPECTIVES ON PLASMODESMATA

D.J. CARR

Department of Developmental Biology, Research School of Biological Sciences, The Australian National University, Box 475, P.O., Canberra City, A.C.T. 2601, Australia

> *"We may ask ourselves if plant cells are in communication with each other and what is the nature of that communication. Many observers believe that they have seen in certain spots a complete absence of intervening cell wall and the cell contents able to pass from one cell to another... One must question whether the dark spots observed by Mirbel on the walls are such openings or whether the extreme tenuousness of the cell communications would probably not permit of their observation, even with the highest magnifications. The former hypothesis is the less probable"* Bernhardi, 1805.

14.1. DISCOVERY AND EXPLORATION

Immediately following publication of Tangl's (1879) paper describing 'open communications' between protoplasts of endosperm cells of *Strychnos*,*Phoenix* and *Areca* (Frontispiece) there was a spate of claims of prior discovery or at least of prior observation. These claims are discussed by Mühldorf (1937) who like Strasburger (1901) gives Tangl the credit for being the first to realise the full implications of his observations. In Tangl's own words: *"the observations made leave only one correct explanation of the matter, that the protoplasmic bodies of the inner cells of the endosperm are united by thin strands passing through the connecting ducts in the walls, which put the cells into connection with each other and so unite them to an entity of higher order"*. In the years following 1879 there was an astonishing output of work extending Tangl's findings to a wide variety of tissues and plants, virtually exhausting the possibilities of the microscopical methods of the times and enunciating most of the possible hypotheses on the origin, development and function of the structures, which Strasburger (1901) named 'plasmodesmata'.

Why did the Tangl paper arouse such immediate and intense interest? Why was his assertion so readily accepted that the 'connections' were thin strands of cytoplasm connecting the cells? Why were his conclusions on a very specialised tissue extrapolated to other kinds of cells? The reasons are not far to seek. The times demanded the replacement of the cell theory of Schleiden and Schwann (see Baker, 1948) by a new theory, the 'organismal theory'. Schleiden had stated the cell theory as: *"every plant is an aggregate of completely individualised entities, independent and separate, which are the cells themselves"*. Hofmeister (1867) and Sachs (1882) had come to the conclusion that the cell theory was insufficient to account for growth correlations, tropisms and the transport of substances, involving the whole plant, its organs and tissues. Already in 1873 Heitzmann had suggested that when cells divide, they remain connected by protoplasmic strands. Tangl's discovery was therefore timely: it was hailed with enthusiasm and any doubts postponed until later. In 1880 Pfeffer wrote: *"By means of the plasmatic threads .... a continuity of the living substance is maintained which is undoubtedly of the highest importance in ensuring the harmonious co-operation of the whole"* (Ewart's translation, 1900 - 1906); and Sachs added: *"the multicellular plant differs from the unicellular only in that, in the one case, the protoplasm is traversed by numerous sieve-like or lattice-like plates, while in the other these plates are absent"*. There was a rush to establish the existence of protoplasmic connections between all sorts of cells in both animals (references in Wilson, 1928) and plants. In his review of 1884, Klebs went overboard: *"the individuality of the cell is as good as abolished by this concept; the arguments, once so fierce, over the definition of the cell have no real meaning any more"*.

It should be borne in mind that, at the time, ideas about the protoplast - particularly its outer boundary - were still rather hazy. Nägeli and Cramer (1855) had conceived of a 'plasma membrane' as a permeability barrier and Pfeffer in 1877 postulated inner (tonoplast) and outer membranes to account for permeability and plasmolytic properties. (The ugly German word 'Hautschicht' (rather than the term 'ectoplast' of Hugo de Vries) was in general use (even in English) for the outer membrane and in 1901 Strasburger suggested that it should be replaced with'plasmoderma', but this did not catch on, even with German writers. In 1924, Mast gave us the modern term, 'plasmalemma'.) The plasmodesmatal canals were thought of as lined by 'plasmoderma'. It was also apparent that the cytoplasm could not be thought of as able to stream unhindered from cell to cell through the 'open communications'. It was necessary to suppose some kind of filtration. This was true even of the relatively thick cytoplasmic bridges of *Volvox* (Meyer 1896a). It had been the similarity of staining reaction of the plasmodesmata with that of the cortical layers of the protoplast which led Tangl to assume their equivalence and Strasburger (1901) held the plasmodesmata to consist only of cortical cytoplasm, bounded by 'plasmoderma'. The histochemical evidence for this can now be regarded as dubious - 'ectodesmata' (Franke, 1971)also stain with iodine and with 'pyoktanin'.

The intensity of research on plasmodesmata in the last two decades of the century can be gauged by that on algae alone. In Germany, Meyer worked on Volvocaceae, Kohl on Chlorophyceae. In Sweden, Wille (brown algae), in England, Massee (red algae), Moore and Hick were at work; in Italy, Borci studied the Cyanophyceae. The many discoveries of the early years are well documented in the reviews of Russow, 1883; Klebs, 1884; Meyer, 1897; Strasburger, 1901; Meyer, 1920; and Lundegårdh, 1922. By 1920 research was almost at a standstill, but a vast amount was known - much of it forgotten again until the late '50s . It was known that plasmodesmata are not distributed uniformly over all the walls of cells and that they are absent from some walls. Some counts had been attempted. In 1891, as Kienitz-Gerloff remarked, it was still not safe to argue from the then known facts what Gardiner (1883) and

Russow (1883) had argued - that the whole of the protoplasm of the plant was in 'open communication'. Certainly there were some plants or tissues in which such a demonstration could not be made or was very difficult. For instance, Strasburger and others were unable to find plasmodesmatal connections between host and parasite (but see 4.3.3. and 13.9). In 1902, Kienitz-Gerloff reported inability to demonstrate plasmodesmata between cells of the foot of the sporophyte and the gametophyte of *Lepidozia* (a liverwort) and *Polytrichum* (a moss) and remarked that the sporophyte "*thus appears to be an independent plant, merely living on the mother plant*". He repeated his assertion, made eleven years before that "*the individual plant is closed off from the environment completely and on all sides*".

As to the origin of plasmodesmata, the striking similarities of the endosperm figures to Strasburger's contemporary drawings of multipolar spindles were noticed by Tangl and led Russow specifically to advance the theory that spindle fibres might persist after telophase to become embedded in the newly-formed cell wall, thus giving rise to the plasmodesmata. Even so, he admitted the possibility of a secondary origin of some plasmodesmata. Kienitz-Gerloff (1891) closely followed the development of the cell plate in endosperms and came to the conclusion that there was no relationship between the spindle fibres, (which slowly gave way to the phragmoplast fibres and then disappeared, and the eventually-formed plasmodesmata, which could not be demonstrated in the newly-formed cell plate by light microscopy. He and others found secondary plasmodesmata between laticifers and parenchyma (see 4.9.). Gardiner (1897, 1900) however continued to subscribe strongly to the Russow hypothesis and it was left to Strasburger (1901) to demolish it finally by pointing out that in meristems, certain walls, *viz.* periclinal walls of the periderm and radial walls of the cambium, are formed, not as a result of mitosis, but by growth of existing walls, yet they do not differ fundamentally in numbers of plasmodesmata from walls which originate from cell plates, following mitosis. Plasmodesmata could also be demonstrated between cells of the partners of certain heterografts (see 13.3.3.3.).

Immediately following their discovery plasmodesmata were pressed into service to explain transport of non-permeating substances (e.g. sucrose) and the conduction of stimuli. They provided vehicles for the transmission of excitation in parenchymatous tissues, such as hairs and stigmas, as well as over long distances (e.g. in *Mimosa*, Haberlandt, 1890). Especially, they could be thought of as conducting the mysterious stimuli emanating from illuminated organs such as coleoptile tips, which, travelling to other parts of an organ, could there induce phototropic curvatures, an expectation soon, alas, to be disappointed.

14.2. ECLIPSE - THE TYRANNY OF FREE SPACE

After 1922 interest in plasmodesmata waned - the first flowering of research on them was over. Some interest was aroused in 1930 by Jungers who questioned their protoplasmic nature. Mühldorf in Rumania moved in with a crushingly long, exhaustively critical but not entirely convincing, reply. In America also, Livingston (1935) at least kept the concept alive by demonstrating the existence of plasmodesmata in various parts of the tobacco plant. But these and similar sporadic publications of the '30s were no more than rituals, designed after the fashion of displays of the liquefaction of a martyr's blood, to keep faith alive. One must take a little more seriously the work of Münch (1930) Crafts (1933) and Schumacher (1934) which showed that plasmodesmata could realistically be implicated in short-distance transport.

But Schumacher became more involved with the polemics concerning 'ectodesmata' (see Napp-Zinn, 1961 for a review of older work on these), and after demonstrating that the size and constitution of plasmodesmata provided no barrier to their participation in diffusive and even mass flow transport, Münch went on to point out that they were, in any case, inessential for his theory of transport! The credibility of plasmodesmata as specific pathways for the transmission of stimuli suffered a terrible blow from the discovery by Páal, Boysen-Jensen and Went in the '20s that the phototropic stimulus of coleoptiles could be transported in what Münch termed the apoplast.

In the period 1945-1960 plant cell physiologists were engrossed in work on ion uptake. Through studies beginning with those of Osterhout and continuing later with those of Stiles, Brooks, Steward and others on salt accumulation by plant cells, the classical concept of the plasmalemma began to lose ground. In the early days of these studies "*it was still believed that the protoplasmic lining of the plant cell constituted a semipermeable membrane, through which the dissolved substances diffused to the interior of the cell, but which exercised a certain degree of selective permeability, so that some ions were barred where others would enter*" (Robertson, 1950). For a while, most Continental European plant physiologists clung to the classical concept (Lundegårdh, 1941; Holm-Jensen *et al.*, 1957 - see Epstein, 1960) but many of them abandoned it in face of the pressure from the majority of English-speaking workers in the field. Not only was the classic concept of a plasmalemma as a barrier to diffusion disbelieved, even the existence of the plasmalemma itself was questioned. The ease with which electrolytes appeared to move into about 10% or more of the volume of plant tissue led to concepts such as 'free space', 'Donnan free space' (Briggs, 1957) and even 'outer space' (Epstein, 1955). Although Robertson (1950) warned about the prevailing state of ignorance of the exact location of the high electrical impedance and of the submicroscopic organization of the cytoplasm, he nevertheless cast doubts on the classical concept and made statements which less cautious followers were to accept as *ex-cathedra* dogma, implying the non-existence of the plasmalemma. "*It is perhaps necessary to warn against hypotheses which suggest the orientation of the functional constituents* (in salt accumulation) *in relation to hypothetical cell membranes whose existence is difficult to demonstrate .... That there is an external membrane consisting of a few layers of oriented molecules in a lipoprotein complex seems a very likely hypothesis <u>for some cells</u>*" (my emphasis) .... "*There is, however, no evidence that such an external membrane is of universal occurrence.*" Indeed, Briggs and Robertson (1957) referred to the concept of such a membrane as "*erroneous*".

With such authoritative statements dominating cellular plant physiology (one might with justification speak of the *tyranny of free space*) it comes as no surprise that this was an era when plasmodesmata were consigned to the oubliette. Apart from a brief reference to Collander's work, there is no mention of plasmodesmata in the first four volumes of Annual Reviews of Plant Physiology (1950-1953). It is hardly surprising that Meeuse's invaluable review (1957) was ignored. Nevertheless, there were some American workers, notably Levitt (1960), Epstein (1960 - which see for references) and Brouwer, who staunchly supported the classical view, as did Collander (1957) in Europe. In his publications in the '50s Arisz did not support the classical view, although he is often credited with keeping alive the idea of the symplast, first put forward by Münch as an inner space in which accumulated ions were not only retained but could pass - by means of the plasmodesmata - from cell to cell. Most authors maintained, as did Mercer, 1960 that "*the ectoplast has no existence as a physiological membrane .... The cytoplasm is freely exposed on its external surface but separated from an osmotic phase by a physiological membrane. The membrane could correspond to the tonoplast and the osmotic phase with the vacuole"*. Already, however, by 1957 Walker had shown (for *Nitella*) that the plasmalemma has a considerably greater electrical resistance than the tonoplast and this was confirmed in 1963 for *Chara* by Hope. A spate of results (e.g. Dainty and Hope, 1959 for *Chara;* Pitman, 1965a for beetroot) showed that rapid cation exchange with the cell wall accounted for most of the 'apparent free space' and the 'water free space' in beetroot is accounted for by cut cells and intercellular space.

14.3. REDISCOVERY AND REVIVAL OF INTEREST

Beginning about 1954 (references in Buvat, 1961) electron microscope studies began to reveal the reality of the existence of a plasmalemma in plant cells. Curiously enough, the plasmodesmata showed up on the first crude electron micrographs even earlier than the plasmalemma and could hardly be ignored. Thus in Mercer (1956) we find a diagram of plant cell ultrastructure showing clearly the tonoplast and (like a vestigial organ of a superfluous and unwanted symplast) a plasmodesma, but no plasmalemma. In his papers up to 1961 Arisz had assumed that ions taken up into the cytoplasm remained bound by Donnan forces and might pass from cell to cell, still bound, but as Robertson (1950) pointed out, the ions remain free in solution and readily liberated by rupturing the cell. But until 1963 Arisz remained cautious about accepting the newly-rediscovered plasmalemma as a membrane with sufficient ion retention capacity. Also, as late as 1964 - despite the accumulation of evidence from electron microscopy and electrophysiology to the contrary - Chambers and Mercer could write of *Chara australis* that: "*No definite structure has been identified at the external cytoplasmic boundary adjacent to the cell wall in osmium fixed material .... The absence of an external membrane is not likely to be due to poor fixation"*. In the same year, Livingston repeated the observations he had made in 1935 using a new staining technique to reassure himself and others that plasmodesmata really do exist. But surprisingly, in the following year (1965) Brabec remained, perhaps the last of the unconvinced: *"The existence of plasmodesmata .... is still problematical and requires renewed demonstration with modern methods"*.

It is of interest that Buvat (1961) was able to see a desmotubule ("*tractus axial dense*") towards which are directed the diverticula of the endoplasmic reticulum in wheat cells and that other early electron microscope studies (e.g. Krull, 1960; Porter and Machado, 1960b) also noted a relationship between the endoplasmic reticulum and the plasmodesmata. As section 2.4.2. shows, we are still exploring this and other aspects of the fine-structure of plasmodesmata. This brief history should serve to show how strongly the consensus of opinion affects discovery and the retention of the corpus of discovered knowledge. We are now at the beginning of a new phase in the history of plasmodesmata, a phase in which it is hoped to establish the details of their ultrastructure and their role in the functioning of the plant.

15

PLASMODESMATA: CURRENT KNOWLEDGE AND OUTSTANDING PROBLEMS

B.E.S. GUNNING AND A.W. ROBARDS

Department of Developmental Biology, Research School of Biological Sciences, Australian National University, Canberra, A.C.T. 2601, Australia.

This volume opens with a quotation from Pfeffer (1897) to the effect that if plasmodesmata had not already been discovered, what was known of plant physiology would have made it necessary to propose their existence. Can we now, 78 years later, and 96 years after Tangl's description, subscribe to his conclusion? We believe the correct answer to be a qualified 'yes, but'; 'yes' because various processes in plants require fluxes that could not be sustained were there no low resistance pathway of transport from cell to cell, and 'but' because we are still a very long way from being able to describe plasmodesmata in precise terms, let alone to define their functional attributes. We now attempt a selective summary of the current state of knowledge of some of the outstanding problems, as emerging from the preceding 14 Chapters[1].

The statement that plasmodesmata probably occur between all neighbouring living cells in plants cannot, we have seen, be accepted without three provisos (2.3.1.). One is that there are multicellular algae which do not have plasmodesmata (Chapter 3); another is that plasmodesmata very rarely bridge the generation gaps between gametophytes and sporophytes and either of these and the reproductive cells that they bear (13.4.4.); the third is that plasmodesmata can be present early in cell development but later become lost or occluded (4.5.; 4.6.).

A polyphyletic origin for plasmodesmata is probable in view of their occurrence in several algal and fungal groups (Chapter 3; 3.8.). Making due allowance for the problems of preserving ultrastructural detail in some of the lower plants, it would seem from the available micrographs, however, that evolution has not been strictly parallel. The desmotubule may or may not be present, or it may be replaced by

---

[1]Apart from literature not previously referred to, we will in this Chapter simply use cross-references to the earlier Chapters, rather than cite original sources.

other dense axial material; neck constrictions may or may not exist, whether or not there is a desmotubule. Nevertheless, neglecting the pits of red algae and many fungi, this variation in structure is relatively minor when set against the main conclusion, which is that the selection pressures favouring the development of a symplastic organisation must have been sufficiently similar and fundamental in a number of phylogenetic lines to lead to, as the structural agent of cell to cell connections, membrane bounded pores that are large enough to function adequately in carrying diffusive or convective fluxes, but small enough to prevent the intermingling of cytoplasmic constituents such as would obliterate any potential for biochemical and morphological individuality. This principle was recognised by Strasburger (14.1.) and it would appear to apply from chytrids to corn. Apart, however, from being able to talk semi-quantitatively about ultrafiltration, neck constrictions, and compartmentation as means of restraining the mixing of neighbouring cytosol components without precluding all transport (1.3.), we are very little closer than Strasburger was to being able to catalogue the molecules that are and are not permitted to pass from cell to cell (Chapter 6).

As to the nature of the selection pressures from which the symplastic system springs, surveys of lower plants have not as yet brought to light any plant possessing a three dimensional parenchymatous organisation that lacks plasmodesmata (Chapter 3). It is true that there are loose palmelloid aggregations and curved or flat two dimensional arrays of cells without connections, just as there are simple or branched filaments where the cells are self-sufficient. There are also organisations less complex than three dimensional parenchyma which *do* have plasmodesmata, and it seems quite reasonable to suggest that it is these which, in evolutionary terms, were able to exploit the architectural possibilities of the three dimensional multicellular habit, the ability to do so residing at least in part in their capacity to develop division of labour by using their cell to cell delivery system in nutrition, and possibly also in regulation of growth and development *via* hormonal or other gradients in their symplast.

It would be extremely interesting to have enough documentation to be able to say at what stage of tissue complexity the ability of plants to control the distribution and frequency of plasmodesmata (4.5.), and to develop specialisations such as pit fields and secondary plasmodesmata (4.3.), arose. The large laminarialean genera in the brown algae have all those abilities and use them in forming their long-distance transport systems (3.4.2.). There is as yet no report of lower plants forming secondary plasmodesmata in a non-division wall, but a likely place to look for this phenomenon would be among the more compact pseudo-parenchymatous organisations of branched filaments, as for example in the Desmarestiales.

What is known of the formation of plasmodesmata in division walls (i.e. walls that originate as cell plates at mitotic telophase (4.2.) suggests that the morphogenetic programme for plasmodesmatal formation is not easily suppressed. The absurdity of plasmodesmata traversing isolated and distorted fragments of wall floating around in cells whose division has been disordered by inhibitor treatments (4.2.1.) illustrates the point. There are, however, two categories of *normal* division process in which the daughters are destined to become symplastically isolated. One category is exemplified by guard cells of stomata, where the isolation (if it is achieved at all) occurs only after a considerable time, the plasmodesmata being eliminated after they have (presumably) played their part in nourishing the developing cells (4.5.). The other is a much more immediate isolation, such as occurs in the formation of reproductive cells. Even here, however, there very often seem to be transitory connections across the cell plates, connections which seemingly require the laying down of an equally transitory wall of callose to ensure that they are severed, leaving the egg, sperm, megaspore or microspore in the majestic isolation that characterises such cells before they embark on their unique developmental pathways (13.4.2.).

The consequences that isolation from the symplast has for cellular morphogenesis and expression of totipotency are very striking, and deserve further comment. There is in higher plants a spectrum of categories from complete isolation, to connection *via* plasmodesmata, to considerable loss of cellular identity in syncytia. Each of these states carries its own characteristic features, and the two extremes tell us something about the intermediate, that is, about the properties of plasmodesmata.

Complete isolation, whether of natural occurrence, or artifically imposed, brings about a process of de-differentiation, if not a tendency for the cell to behave as if it were a zygote or a spore (13.4.). Previous polarities may be lost, and an embryogenic process may begin. It follows that such effects are normally held in check - how, we do not know, but the existence of a hormonal or electrical 'message' that depends upon symplastic transport or membrane continuity *via* plasmodesmata is an obvious interpretation. A corollary to the hypothesis that the symplast somehow holds de-differentiation in abeyance is that a symplastic organisation represents a level of complexity that is capable of maintaining a differentiated state. Heslop-Harrison (1967) emphasises the point by contrasting the failure of *single cell* inocula to carry over origin-related behaviour patterns, with the maintenance of generalised properties of source tissues (in respect of flowering behaviour and growth differences between juvenile and adult tissues) seen when *massive* explants, i.e. whole tracts of symplast, are taken into *in vitro* culture. Hints that symplastic connections may be prerequisites for differentiation of various sorts in tissue cultures are reviewed in 13.4.2-3.

It is noteworthy that when all connections are broken during reproduction in higher plants, the isolation from the symplast creates the opportunity for full rein to be handed over to control systems housed in the apoplast. Modern work on the molecular basis of incompatibility mechanisms and the control of breeding systems has disclosed subtle processes of activation and suppression of cell development, mediated by *extracellular* recognition systems that operate at this time (13.2.). These systems are not, of course, developed during vegetative growth, as witness the fact that plants exhibiting sporophytic *incompatibility* at reproduction can also show vegetative *compatibility* through the successful establishment of grafts - this implying the establishment of symplastic continuity across the graft junction (13.3.3.3.). Clearly, one aspect of isolation from the symplast at reproduction is that it allows a fine control of reproductive compatibility to override what is a *relatively* wide-spectrum vegetative compatibility - wide enough to accommodate continuity of plasmalemma through plasmodesmata between host and parasite (4.3.2.; 13.9.), and across intergeneric chimaeral junctions (4.2.2.), to take but two examples.

The opposite extreme to complete isolation is seen in syncytia of various sorts, where the presence of large 'cytomictic' channels from cell to cell leads to a suppression of individual behavioural or developmental traits - quite the reverse of the stimulation of individuality that may be induced by isolation. Suppression of individuality in syncytia is evinced most strikingly by mitotic synchrony, as seen when the nuclei of interconnected meiocytes proceed through division much as do the nuclei in an algal coenocyte or a multinucleate cell, or when mitotic waves pass through non-cellular endosperm or along a non-articulated laticifer. It would seem at first sight a reasonable conclusion that channels of greater size than plasmodesmata are required for the synchronising stimuli (division proteins - Sachsenmaier *et al.*, 1972) to pass, in that cells in normal symplasts are normally asynchronous. Yet it may not be the size of the pores so much as their organisation that regulates their permeability to the stimuli. Thus cytoplasmic stimuli exert a synchronising influence when a nucleus at a given stage of mitosis is inserted into a cell in which the resident nucleus is at a different stage. This implies that the stimuli, whatever their nature, can pass between nucleoplasm and cytoplasm. A trans-membrane pathway at the nuclear envelope cannot be ruled out, but as for cell to cell junctions in plants, the more likely route is through pores, in this case nuclear envelope pores. Since it has been pointed out that the channel dimensions of nuclear envelope pores are quite similar to those of plasmodesmata (1.2.2.3.), we have the puzzle that the synchronising signals evidently pass through the one but not the other. The answer to this dilemma may be that at the nuclear boundary it is the cytoplasm that is continuous through the pore lumen, whereas in most higher plant plasmodesmata the cytoplasmic annulus is constricted (2.4.). An hypothesis that would be worth investigating thus is that the synchronising stimuli are cytoplasmic, and excluded from the endoplasmic reticulum - desmotubule system that (see later) is believed to traverse plasmodesmata.

Very much the same argument can be applied to informational molecules that direct processes of cell differentiation, many of which must be of nuclear origin, and must presumably pass out into the cytoplasm. Since cytomictically connected cells tend to be uniform in their developmental fate (e.g. meiocytes, sister guard cells in some grass stomata) it seems that the morphogenetic molecules, like the mitotic synchronisers, may be freely mobile within the syncytium.

Examination of the extremes thus enable us to surmise that plasmodesmata can transport signals that hold embryogenic, or de-differentiation, tendencies in check, but that they do not transport the molecules that bring about mitotic synchrony in syncytia and coenocytes. From the observation that neighbouring cells can differentiate in markedly different directions as well as sustain markedly different rates of mitosis, despite being symplastically connected (they may even *have* to be symplastically connected), we may also infer that the informational molecules or stimuli that govern such processes are likewise excluded (13.4.2.). Yet certain types of 'information' *are* able to move symplastically. Viruses can do so (Chapter 8), but consideration of dimensions makes it seem very likely that an ability to modify plasmodesmata from the shape and size that we are accustomed to seeing in electron micrographs is one part of the offence system of viruses. Setting aside this pathological type of information transport we are faced with a sad lack of evidence, and with the salutory fact that apart from what little we know about transmission of flowering hormones (13.3.3.3.), the best example of symplastic transfer of a morphogenetic influence was published in the 19th Century. Townsend's experiments, demonstrating that (presumed) plasmodesmata transmit whatever it is that a nucleus produces in order to allow cell wall regeneration on a neighbouring enucleate sub-protoplast, are long overdue for repetition and elaboration (13.3.1.). The recent discovery of plasmodesmata between the nucleated zoosporangium and the enucleate rhizoid of the chytrid *Entophlyctis* (3.6.) may have provided us with a most useful experimental material for such studies: however, in view of the knowledge that morphogenetic messengers persist and function for a considerable time in enucleated *Acetabularia* cells and pollen sub-protoplasts, it would be premature to assume that the *Entophlyctis* plasmodesmata are necessarily the means of providing the enucleate rhizoids with the information that they need for continued life after the wall separates them from the zoosporangium.

It should not be assumed that transport of free molecules is the only way in which morphogenetic influences can pass from cell to cell *via* plasmodesmata. There are at least two other possibilities. One is that the symplastically carried information could be biophysical rather than biochemical in nature. Conduction of electrical signals through tracts of symplast does occur (6.3.; 6.4.; 13.3.4.1.), and could be the basis of rapid transmission of a variety of morphogenetic messages, including, for instance, inter-organ phytochrome responses (Oelze-Karow and Mohr, 1973, 1974). The other possibility is that membrane-bound molecules could move from cell to cell by translational diffusion in or on the plasmalemma or desmotubule/endoplasmic reticulum, or could be moved as part of a membrane flow system (11.2.3.3.).

So much, then, for symplastic transport of morphogenetic information. What of symplastic transport of water, nutrients, and metabolic intermediates? Here at least both theoretical treatments (Chapter 5) and experimental observations are available, and there is a number of situations where not only can we say with some confidence that plasmodesmatal transport does occur, but we can also quantify the fluxes in question.

Discussion of the problems that arise when the ultrastructure of plasmodesmata is considered in relation to the pathway(s) of transport can be postponed if for the moment we look only at symplastic transport on the basis of a flux 'per plasmodesma'. Table 15.1. brings together data on water and solute fluxes for a number of situations considered in detail in earlier Chapters.

As mentioned frequently in this Volume, it is all too difficult to find situations where it can be said that the plasmodesmata 'must' be carrying the intercellular flux or flow. One of the arguments that has been used is that, where a flux has been found to be so high that it looks unacceptable as a trans-membrane flux, it therefore must be passing through plasmodesmata. How valid is this reasoning? MacRobbie (1971a) has suggested that the upper capacity of most higher plant cell membranes is probably the equivalent of 10^{-8}-10^{-7} mole m^{-2} s^{-1}. In, for example, the bundle sheath of C_4 plants, fluxes in excess of 100x greater than this are found (Table 12.1.) and hence are assumed to pass through plasmodesmata. But in other locations, in higher plants as well as in the algae, very high trans-membrane fluxes have been reported: loading of root xylem vessel elements from adjacent parenchyma cells (10.6.) looks as though it may sometimes require rates well in excess of 1×10^{-7} moles m^{-2} s^{-1}; and the very rapid movements of potassium and other ions across the cell membranes of stomatal guard cells - in the absence of large numbers of plasmodesmata (2.3.1.; 13.4.2.) - again seems to point to a greater than 'normal' solute permeability. The fluxes between guard cells and subsidiary cells are estimated to be in the range 1-1.5×10^{-7} mole m^{-2} s^{-1} (Fischer, 1972; data based on ^{42}K tracer experiments on *Vicia*) to 1.5-1.9×10^{-6} mole m^{-2} s^{-1} (Penny and Bowling, 1974; data based on the use of K^+ sensitive microelectrodes in *Commelina*). In such circumstances, it is at the moment unwise to infer plasmodesmatal function merely on the basis of intercellular fluxes marginally greater that those of most higher plant membranes.

Very high apparent flux rates are involved in glandular secretions both from phloem unloading (e.g. *Abutilon* nectary hairs - 11.4.), and

TABLE 15.1.

COMPARATIVE DATA ON ESTIMATED RATES OF TRANSPORT THROUGH PLASMODESMATA

Location and Reference	Plasmodesmatal frequency (μm^{-2})	Volume flow		Solute flux	
		Observed: m^3 plasmodesma^{-1} s^{-1}	Superiority Factor*	Observed: mole plasmodesma^{-1} s^{-1}	Superiority Factor**
Mesophyll, C_3 plant (11.2.4.1.)	3	-	-	$2\text{-}7 \times 10^{-19}$	6.5-210
Bundle sheath, C_4 (outer tangential wall) (12.4.; Table 12.1.)	15 (*Salsola*)	-	-	5×10^{-19} (*Amaranthus*) 10×10^{-19} (*Zea*)	164-1640 296-2960
Bundle sheath, C_3 (inner tangential wall) (11.2.4.1.)	7.8	-	-	2.9×10^{-19}	23-230
Sieve element - companion cell (11.2.4.1.)	6	-	-	8.2×10^{-19}	48-480
Chara node (9.5.; 9.6.)	4-5	-	-	9.5×10^{-19}	430-4300
Pisum root nodule (Table 10.9.)	?	-	-	-	34-340
Abutilon nectary (distal wall of stalk cell, 11.3.)	12.6	2.1×10^{-20}	2.3×10^{2} -1.4×10^{3} ***	-	-
Root endodermis (stage III) (inner tangential wall of barley, Tables 10.2.-7.)	1.05	2.4×10^{-20}	0.1 -41.8 ***	3.6×10^{-21} (PO_4)	0.04-0.1
ditto, *Cucurbita* (Table 10.8. and unpublished results)	6.2	1.5×10^{-21}	0.62 -2.47×10^{2} ***	2.2×10^{-20} (K^+)	1.4-14

For footnotes see opposite page.

xylem unloading (e.g. the salt secreting glands of *Aegialitis* (Atkinson *et al.*, 1967) where $5x10^{-5}$ mole Cl^- m^{-2} s^{-1} can be secreted over the total cross-section of the glands). However, in such cases it is most unlikely that the final symplast-apoplast movement occurs as a normal trans-membrane flux.

With this perspective in mind, the available data on solute fluxes passing across cell junctions that contain plasmodesmata can be examined. In the first six situations listed in Table 15.1. the flux per unit area of junction is known or can be estimated. The final column, labelled 'superiority factor' expresses the values by which the fluxes in question exceed MacRobbie's (1971a) upper limits for trans-membrane transport, i.e. $1\text{-}10x10^{-8}$ mole m^{-2} s^{-1}. The smallest of the 'superiority factors', at x6.5, could not in any sense be taken as good evidence that a pathway of lower resistance than the plasmalemma exists, but the majority of the factors, being greater than two orders of magnitude, lend credence to the idea that plasmodesmata, which are present in frequencies listed in the first column, do indeed function in cell to cell transport. More surprising than this is the uniformity that emerges when the observed or estimated flux per unit area is divided by the number of plasmodesmata per unit area. Despite five-fold variation in plasmodesmatal frequency, very great anatomical differences, and differences in the nature of the solutes being carried, the values for the flux per plasmodesma lie in a remarkably narrow range.

It is tempting to infer that somewhere in the range $2x10^{-19}\text{-}10^{-18}$ mole $plasmodesma^{-1}$ s^{-1} a saturating or limiting value is reached. This, however, is as yet unwarranted. We know that much higher fluxes *can* be carried, as in the bulk flow system of the *Abutilon* nectary. Even when the list is restricted to situations where transport is likely to be purely diffusive, there are still grounds for considering that coincidence enters into the uniformity of flux per plasmodesma. The *Chara* node plasmodesmata are much larger than the others (Table 9.1.); some of the calculations rest upon assumptions regarding rates of photosynthesis and translocation (11.2.; 4.1.); and above all, there is very good evidence that in large cells, transport is limited not by the plasmodesmata but by the cytoplasmic streaming that delivers the solutes to the cell to cell junctions. Our lack of knowledge of streaming rates and directions, save in easily observed cells such as giant algae, hairs and in epidermal strips, means that we cannot assess the relative contributions of plasmodesmata and cyclosis to transport in many physiologically important situations such as leaves and roots. Kamiya (1959), in reviewing the subject, reiterated earlier conclusions in stating that "*the role of rotation or circulation ... in the translocation of soluble substances is still uncertain*": his general conclusion still holds, though we do now know something, in theory (Chap-

---

◄ *Factor by which the plasmodesmatal volume flow per unit area of cell junction exceeds the flow that would occur across two successive plasmalemmas of unit area, each with hydraulic conductivity $2x10^{-8}$ m s^{-1} bar^{-1}

**Factor by which the plasmodesmatal flux per unit area of cell junction exceeds the flux that would occur if the solutes passed through two successive plasmalemmas of unit area, each capable of carrying a flux of $10^{-7}\text{-}10^{-8}$ mole m^{-2} s^{-1} (1-10 p mole cm^{-2} s^{-1}; see MacRobbie, 1971a)

***The two values listed are, for *Abutilon*, the desmotubular and cytoplasmic pathways respectively; and, for the roots, the extremes of the cases envisaged in Table 10.3.

ter 5) and in practice (Chapter 9), of how particular types and rates of cyclosis affect intercellular transport.

It is nevertheless desirable to seek additional data to see if other plasmodesmata carry amounts of solute that fall within the same range of values as in Table 15.1.: the bundle endodermis of legume nodules and the collecting cells of various glands such as salt glands are obvious candidates for investigation. If similar values are found there will indeed be some grounds for believing that plants can relate plasmodesmatal frequency to transport requirements - all we are doing is to add quantitation to what has long been claimed on qualitative grounds (2.3.1.). What we do not know, of course, is whether the plant predicts in advance how many plasmodesmata it should lay down in its cell plates, or whether continuing adjustments are made in the course of tissue development by secondary formation, branching, or occlusions in response to local concentration gradients or other stimuli (4.9.) - a question that is open to experimentation.

In the remaining three situations in Table 15.1. there is quite clearly a volume flow. Disparity of anatomy, function, and plasmodesmatal frequency notwithstanding, the flows per plasmodesma are similar in roots and nectary. Only in the *Abutilon* nectary stalk cell, however, does the plasmodesmatal frequency (together with the short length of the plasmodesmata) give a value for hydraulic conductivity that is convincingly 'superior' to that of the alternative membrane pathway. In the case of the root endodermis the low, or even sub-zero, 'superiority factors' for both volume flow of water, and potassium and phosphate fluxes imply that the plasmodesmata do little more than compensate for the presence of a (presumably) impervious cell wall.

It is commonly held that entry of water into the stele of a root is driven by an osmotic imbalance across the endodermis, the apoplast of which is 'sealed'. Yet it has been emphasised that the reflection coefficient of a plasmodesmatal pathway should, in theory, be zero (5.3.), and therefore there can be no question of osmotic water movement across a junction that is equipped with open plasmodesmata (5.8.). In the case of the root stele, the osmotic water movement occurs at the plasma membranes that separate the stele apoplast from the stele symplast, and is driven by an efflux of osmotically active solutes into that apoplast, the water inflow taking the path of least resistance, this including the plasmodesmatal pathway across the suberised inner tangential wall of the endodermis - in other words a Münch (1930) or Curran and MacIntosh (1962) model of bulk flow working in reverse. However, since the endodermal plasmodesmata are constricted, there is in addition a potential for a novel type of trans-symplastic osmosis, across the endoplasmic reticulum membrane at the beginning of the pathway, from cell to cell through interconnected cisternae, and across the endoplasmic reticulum membrane at the end of the pathway. This would, however, require osmotically "tight" seals at the neck constrictions of the plasmodesmata.

In roots as in other situations it is necessary to consider the possibility of bidirectional transport through plasmodesmata. Before doing so, however, ultrastructural details have to be examined. The following discussion mainly concerns higher plant plasmodesmata.

The reality of the plasmalemma is, of course, no longer in question, in the plasmodesmata or elsewhere (14.2.). We consider that the reality and widespread but not universal distribution of the desmotubule too should be accepted. It is seen in chemically-fixed material (with a variety of fixatives), after freeze-substitution, and most tellingly, in material that has been snap-frozen and cryo-sectioned without exposure to any fixative or dehydration solvent (2.4.2.). It may not always be present, but it is seen in the majority of cases where electron microscopists can describe the quality of fixation, somewhat subjectively, as 'good'. What it is and how it relates to other cell components is less clear. It is not a trapped microtubule, but is very likely to be a derivative of the endoplasmic reticulum, modified through processes that are acceptable according to modern concepts of the molecular architecture and dynamics of membranes (4.2.1.). In a few cases its continuity with the endoplasmic reticulum, as seen in fixed and sectioned material, is scarcely in doubt (Fig. 1.1.). The central rod that is so frequently observed in desmotubules is much more mysterious, and the possibility that it is an artifact, perhaps of staining, cannot be excluded: apart from this we know almost nothing of the interior of the desmotubule, and procedures which might be used to investigate the question of lumenal continuity with the endoplasmic reticulum - at present a matter for surmise - are urgently required.

Moving out from the axial region of the plasmodesma the next component is what has been referred to in this Volume as the cytoplasmic annulus. Since it is potentially a major channel of transport from cell to cell its structural details, as far as they are known, are important. Its lumen varies along the length of the plasmodesmatal canal, being in part open and in part constricted by plasmalemma and/or (where present) suberised lamella. It can in fact be constricted to dimensions that are too small for the electron microscopist to see and measure as open connections. Where it is 'open' there may be granular or fibrillar material ('spokes', Fig. 7.25.) in it, but whether this represents a regular structural component or accidental or artifactual inclusions is not known. Evidence for the reality of the constrictions seems strong. As with the desmotubule, they are seen following a variety of preparative procedures. They are retained when plasmodesmata are broken from the plasmalemma of the rest of the cell by severe plasmolysis. In at least one case they are found when the cell in question is immature, but are absent at maturity (11.3.) - pointing not only to their 'reality' (for the same preparation methods apply throughout the developmental sequence), but also to a dynamic quality which in turn implies the existence of specific functions, conceivably the regulation of the permeability and ultrafiltration characteristics of the cytoplasmic annulus.

We are thus led to the next component - the plasmalemma that bounds the plasmodesmatal canal. Once again, apart from its shape and approximate size, we know almost nothing about it. It is unfortunate that the behaviour of the fracture that yields freeze-fracture or freeze-etch images of cell membranes is not conducive to revealing details of plasmodesmata, for it seems very likely that this must be a remarkably specialised zone of the plasmalemma. Its constrictions must somehow be structurally supported; it seems to have specific properties in relation to cell wall metabolism (4.7.; 4.9.); the mode of formation of certain plasmodesmata implies that it carries recognition systems that allow it to fuse with the plasmalemma of the adjacent cell (13.3.). Since the total area of plasmalemma in the form of plasmodesmata is appreciable in comparison to the remaining plasmalemma we can anticipate that it will prove feasible to isolate it and examine some of its specialisations.

In listing the components of plasmodesmata we should include the sleeve of cell wall around the plasmodesmatal canal. There is evidence that this portion of wall is specialised, though in different ways in different tissues. Callose may be present; material resistant to wall-degrading enzymes is seen in aleurone (4.7.); virus-induced proliferation of the sleeve can occur (8.3.); differential staining has been observed (2.4.2.; 4.9.). And when protoplasts are being made from root tips there are intermediate stages when removal of wall allows plasmodesmata to expand and in effect pull neighbouring protoplasts together (13.8.), evidence that *in vivo* the wall itself may be a determinant of the dimensions of plasmodesmata.

We do not know how inter-related the above components of plasmodesmata are. The total structure is complicated considering its minute size, with two classes of membrane bounding two compartments in turn bounded by cell wall, but it can all be regarded as a macromolecular assemblage, the structure of which within limits transcends taxonomic boundaries throughout the plant kingdom. It is in every sense a "pore complex", though rather more elaborate than the structures in the nuclear envelope that are similarly designated. Nuclear pore complexes, unlike plasmodesmatal pore complexes, are very similar from one cell to another in an organism, but the two have in common dual modes of formation - either at inception of the structure they traverse (cell wall or nuclear envelope) - or secondarily by interpolation into the existing structure. In neither case are stages of development adequately described. Nuclear pore complexes have been isolated, and it is likely that plasmodesmatal pore complexes soon will be. It should be possible to prepare cell walls, and to isolate plasmodesmata from them: further details of ultrastructure might then be open to study using tracers such as lanthanum hydroxide, and using techniques of biochemical fractionation such as in comparable work on isolated animal gap junctions, which has revealed the presence of large amounts of a special protein (4.2.1.). There might also be some hope of investigating whether the 'constrictions' are based on some sort of agglutination reaction, pinching by a collar of cell wall, or compression by microfilament-membrane interaction.

We must now attempt to meld structure with function, bearing in mind that in terms of selective advantages, compartmentation of cell to cell transport and protection of solutes from the cytosol may have been as important a consideration as improving the total flux - so argue those who wish to pin a functional label on the desmotubule, and thereby provide a rationalisation for the existence of two concentric membranes in plasmodesmata. Certainly the ultrastructural data oblige us to consider more complex models than a simple hollow cylinder. The detailed biophysical treatments have in any case gone beyond such over-simplifications by inserting a solid rod along the axis of the cylinder (5.3.), on the assumption that the cytoplasmic annulus is the only operational pathway, and that it is open and of uniform dimensions. On this basis the fluxes so far found in symplastic short-distance transport systems emerge as feasible: the necessary driving forces, whether hydrostatic or derived from concentration differences, seem reasonably small. However, just as the reality of structural detail seen in electron micrographs is, rightly, the subject of challenge, so too the realism and relevance of models based on transport along hollow annuli has to be questioned. The ultrastructural evidence is that in most but not all cases the cytoplasmic annulus is constricted near its extremities. Mention of the controversy that clouds the issue of how 'tight' or leaky 'tight junctions' are in animal tissues (Diamond, 1974) serves to emphasise that no-one has as yet assessed what quantitative effects the constrictions might have. If, as suggested above, they *are* present *in vivo* as well as in ultra-thin sections, they must presumably increase ultrafiltration effects and diminish overall transport (1.2.1.).

Let us for the sake of argument take the view that the constrictions are in fact seals, and that the desmotubule is the only open pathway (as already pointed out we do not in fact know that it *is* open). It would be, of course, a transport pathway of lower cross-sectional area than the cytoplasmic annulus used as the basis for the models described in the preceding paragraph. Despite this, and the disadvantage of having a very small radius when bulk flow is proportional to the fourth power of the radius, it seems not unreasonable to suggest that it represents a feasible pathway. It could cope with the diffusive fluxes across the outer tangential wall of the bundle sheath of C_4 plants (12.4.). The pressure drops needed to drive bulk flow across the root endodermis (Table 10.4.) and the stalk cell wall of *Abutilon* nectary trichomes (11.3.) are, except where the most pessimistic assumptions are used, far below one bar. Whether the process of loading sieve elements from their companion cells is considered as a volume flow or as a diffusion process, the desmotubule would probably be adequate (11.2.4.). However, in all of the calculations that lead to these optimistic conclusions, there has been a total neglect of (a) the need to load and unload the compartment - endoplasmic reticulum (?) - that is continuous (?) with the desmotubules, and (b) the need to transport the solutes or solvent or both along the exceedingly tortuous cytoplasmic cisternae leading to and from the desmotubules: both of these requirements would increase the overall resistance of the postulated desmotubular pathway. Any attempt to calculate the extra resistance is complicated by the uncertain effects that cytoplasmic streaming would have on the geometry and temporal and spatial continuity of the system, as well as by lack of information about the hydraulic conductivity and solute permeability of the endoplasmic reticulum membrane (10.8.).

In assessing the functional potential of the desmotubule as above we are merely applying the conventional laws of transport through hollow cylinders, and are ignoring the possibility that desmotubular transport, if it occurs at all, might be exceedingly *un*conventional. It is known, for instance, that the time required for molecules to diffuse a set distance is reduced if the path that they follow is constrained from being random in three dimensions to a two dimensional plane, as in a cell membrane. The factor can be as large as 300 (see Edidin, 1974). One wonders, therefore, whether comparable surface phenomena might occur in or on the desmotubule, where the path between cells is so limited in its geometry that molecules in transit might have little opportunity for dissipation of their kinetic energy save by moving along the axis.

Measurements of electrical resistance of *Nitella* nodes and *Elodea* cell junctions (6.4.), of chloride transport across *Chara* nodes (9.5.), and of diffusion of uranin along *Tradescantia* stamen hairs (6.7.), all imply that plasmodesmata are less open than their overall geometry would indicate. Hence the outcome of calculations based on simple open models must be treated with caution. It is possible that some of the discrepancies may be traceable to ultrastructural features such as constrictions, or to the need to confine fluxes to desmotubules, but it is also probable that some may lie in the physical, as distinct from structural, properties of the pathways.

The hindrance to ions or molecules passing through a pore or cylindrical annulus is to a large extent determined by the viscosity of the substance filling that space. Reliable data for intracellular viscosity in plants are very restricted, and even more so for compartments within the cytoplasm. The viscosity of a vacuolar sap has been estimated at about 2×10^{-2} poise, and (e.g.) that of the cytoplasm of *Euglena* at 6×10^{-2} poise (10.5.1.2.). Other estimates based upon the rate of fall of statoliths, 0.24 or 8.0 poise (see Tyree, 1970), are almost certainly too high for present purposes in that the sedimentation would have been hindered by large-scale structures such as membranes within the cytoplasm. The possibility that the fluid within plasmodesmata is 'structured' and hence of high viscosity is, at the moment, inaccessible to direct experimental analysis, but is another problem that might become open to study if isolated preparations can be obtained. Perhaps the only valid generalisation that can be made at present is that in performing calculations, the viscosity value to be used for plasmodesmata that carry a bulk flow should be the viscosity of the moving fluid. That of pure water is about 1×10^{-2} poise, that of 15% sucrose is 1.6×10^{-2} poise. Tyree (1970), by contrast, calculated on the basis of the pore fluid being at viscosities of 0.5 to 2.0 poise (5.3.).

The value to be given to diffusion coefficients within plasmodesmatal pores is even less clear. The problem was discussed by Tyree (1970), who considered the magnitude of effects due to the colligative properties of cytoplasm and effects due to charges on the moving solutes. Factors which combine to reduce effective diffusion coefficients still further are described in Chapter 1. Until reliable estimates of viscosity and diffusion coefficients within plasmodesmata can be obtained there will be little hope of assessing accurately the functional potential of the structures described above.

Cytochemical approaches to studying the contents of plasmodesmata have as yet had only slight success. The ions of interest (for example sodium, potassium, chloride, phosphate) are all extremely diffusible, and the methods for their localisation, of necessity, rely on deposition of electron-opaque particles which are large enough to be resolved with the electron microscope. This inevitably means aggregation of ions, and some movement from the *in vivo* location. How much movement occurs depends on the technique used and in particular on the changes occurring as the precipitating agent itself diffuses through the tissue. The pyroantimonate technique appears to be anything but specific for sodium, according to the evidence of Chapter 7; and the precipitation of silver chloride to indicate the possible location of Cl^- in the living cell can probably only tell us that chloride was present within a particular compartment, without revealing anything about the amount - and even then the risk of artifacts is high. Other techniques involve analogues which, it is hoped, will follow a similar pathway to a naturally occurring ion: thus thallium has been used as a potassium substitute, and iodide to replace chloride (7.6.; 10.5.4.). Most commonly, and certainly in the case of thallium, the toxicity of the salts limits any comparison that can be made to the normal situation. It is nevertheless highly suggestive that the silver precipitation techniques (Stelzer *et al.*, 1975), and the thallium analogue technique (Van Iren and Van Der Spiegel, 1974), have given evidence for the presence of chloride and potassium in endoplasmic reticulum cisternae, encouraging the view that such cisternae can constitute a symplastic compartment, interconnected by the desmotubule.

Other technical possibilities are being explored: X-ray analysis to determine the elemental composition of either natural or introduced intracellular substances; and rapid freezing of cells, followed by appropriate methods, to minimize any chemical treatments or the possibility of leaching diffusible ions from the tissue. The difficulties are severe, particularly in relation to the present poor sensitivity of X-ray energy spectrometers to elements of low atomic number (the biologically 'interesting' ones!), but it can reasonably be hoped that it will soon be practicable to make useful advances, both in understanding the processes of chemical precipitation of diffusible ions as well as in carrying out X-ray microanalysis on naturally occurring elements in chemically untreated tissues.

There is little support from enzyme cytochemistry for the suggestion that plasmodesmata are sites where, by the utilisation of chemical energy, solute transport from cell to cell is facilitated by some form of carrier. The presence of phosphatase or ATPase within plasmodesmata is not open to any immediate or obvious explanation (7.3.3.; 7.4.). Both enzyme groups have also been found associated with the plasmalemma over the rest of the surface of the cell. In this aspect of the work two outstanding problems must be solved before further progress can be made: firstly, the Gomori-type precipitation reactions are notoriously prone to artifactual precipitation of heavy elements by purely physical effects, and therefore, extreme precautions must be taken to assure that the reaction product is both enzymatically produced, and is also precipitated at the site of formation; secondly, it is not easy to distinguish 'ATPase' from general 'phosphatase' activity in plant cells, and some citations of the former are probably open to question. In general, the study of enzyme activities associated with plasmodesmata requires much fuller study: so far, citations have arisen from experiments that have been mainly concerned with other structures, or undertaken for other reasons.

The early concept that plasmodesmata are centres of cell wall metabolism has, by contrast, been revived as a result of cytochemical (and other) work. It is known from indirect methods that plasmodesmata act as centres for the spread of some synthetic or hydrolytic enzymes into cell walls (4.7.), and it seems certain that plasmodesmata can serve as sources of, or distribution points for, various wall synthesizing or degrading enzymes. A regulatory role for the plasmodesmatal plasmalemma is suggested in such situations, following a controlled movement, perhaps *via* endoplasmic reticulum and desmotubule, of what are presumably cytoplasmically synthesized enzymes into the pores. There are other examples of alterations to the cell wall being accomplished through the intermediacy of specially located cisternae of endoplasmic reticulum (Gunning and Steer, 1975), and it is only the peculiarity of the desmotubular form of the endoplasmic reticulum that sets the present one apart. The problem of releasing the enzymes, first from the cavity of the endoplasmic reticulum, and secondly across the pore plasmalemma remains to be investigated.

Having considered the relevant background information, we can return to the subject of bidirectional transport through plasmodesmata - one that is likely to be as difficult as it is in the study of transport through sieve tubes. It is quite clear that, in some circumstances, plasmodesmata must carry fluxes or flows proceeding simultaneously in opposite directions. What is not known is whether it is necessary to postulate that a *single plasmodesma* can simultaneously carry opposing bulk flows. Let us consider some situations and some possibilities. Water and ions move inwards across the root endodermis, while sugars move outwards from the phloem to the cortical cells (10.5.5.; 11.3.); in the leaves of some grasses, there is transport of water and photosynthate in opposite directions across the mestome sheath (11.2.4.2.; 12.7.); in legume nodules sugars and products of nitrogen fixation move in opposite directions across the bundle endodermis (Table 10.9.). In such cases there is no absolute need to propose bidirectionality of individual plasmodesmata, in that *cells* in different positions (e.g. opposite xylem or phloem) might serve in unidirectional movement along one path *or* the other. At the moment there is no evidence to make an informed decision on the point.

However, there appear to be cases where plasmodesmata within a *single wall* probably combine to move different solutes in opposing directions. Such cases are across the walls of the stalk cell of *Abutilon* nectary hairs (11.3.; 11.4.); between bundle sheath and mesophyll cells in C_4 plants (Chapter 12); across the basal wall of the rhizoid of the *Polypodium* protonema (Fraser and Smith, 1974); as well as in other situations that have been cited or could be deduced.

Assuming, for the moment, that movement through plasmodesmata is necessary (e.g. on the basis that the fluxes are clearly greater than transmembrane fluxes, or there is some reason, such as a suberin lamella, suggesting that movement across the wall cannot occur), then what are the possibilities? Although the enquiry has narrowed to the level of the individual cell wall there are still many plasmodesmata, some of which could operate in one direction and some in the other; once again there are no data to prove or disprove the possibility. It is, however, unnecessary to make such a suggestion. Any individual open channel could serve for *diffusion* in two directions simultaneously: so long as the pore is not excessively long, and the concentration drop is not too small, bidirectional diffusive movement could occur without difficulty. If, however, a pressure flow were operating in one direction, what then would be the likelihood of an opposing movement in the same channel? Once again, diffusion would cope very effectively over relatively short distances. Examples may already be to hand in the pressure flow of sugar solution through the *Abutilon* hair with a diffusive flux of, say, potassium in the opposite direction (11.4.), and sugar moving outwards while water moves inwards across the root endodermis.

It seems, therefore, that the only circumstances that would require any special attributes of plasmodesmata would be those in which *pressure flows* took place *simultaneously* in opposite directions across the same wall. Even should this happen (and as yet there is no cited case of it) plasmodesmata appear to be equipped to accommodate the situation, as there are two potential conducting channels: the desmotubule; and the cytoplasmic annulus, which, in theory, could operate as envisaged in Fig. 1.2. However, it must be stressed that no situation has yet been reported where a plasmodesma with a single functional channel could not carry the bidirectional fluxes or flows measured. As a consequence, we are once again obliged to ask what lies behind the seemingly unnecessarily complicated ultrastructure of plasmodesmata. If a cytoplasmic route suffices, why have a desmotubule? On the other hand, if cytoplasmic continuity brings problems (see previous discussion on morphogenetic influences), a functional desmotubular pathway surrounded by a constricted cytoplasmic sleeve is at least a rational functional interpretation of the observed structures.

Any discussion of bidirectional transport would be incomplete if it did not mention the indications that the plant can exert a measure of control over directions and rates of transport through its existing plasmodesmata, that is, as distinct from control exerted through the insertion of new or occlusion of old plasmodesmata. Two situations where some form of valve action may be inferred are: the translocation of substances in a preferred direction despite the presence of plasmodesmata leading in other directions (11.3.); and the existence of turgor differences, sometimes very marked, across cell junctions where plasmodesmata occur (5.8.). Both phenomena can be seen in the phloem tissue of leaf veins (11.2.3.), and doubtless elsewhere too. They pose a most difficult problem for the microscopist who is searching for an explanation in ultrastructural terms, in that it is very probable that conventional methods of specimen preparation will introduce changes in conditions - e.g. abolish turgor gradients - so that any valve that exists might always, no matter what its *in vivo* state, be seen in a resting condition in the final specimen, and hence be unrecognisable as a valve. It is a sad commentary on our state of ignorance that, having discussed at some length the possibility of the desmotubule being a channel of transport, we have to admit that there are entirely different interpretations, one being that it can, with the cisternae to which it is attached, move back and forth as a sort of piston valve to effect a control over the entry into or exit from the cytoplasmic annulus of the plasmodesma. Alternatively, the neck constrictions that are so frequently seen to diminish the effective size of the cytoplasmic annulus could be the hypothetical valves, the role of the desmotubule merely being that of a solid object onto which the valves can close. Further speculation is unproductive at this stage.

Very nearly one hundred years ago, the first comprehensive paper dealing with protoplasmic connections was published. That work, written by Tangl, stimulated a spate of articles which, within a very short time, had countenanced most of the possible functions of plasmodesmata (14.1.). Both of the main early suggestions for plasmodesmatal function: the translocation of substances; and the transmission of stimuli do, in fact, occur.

Since Tangl, there have been periods of descriptive work on plasmodesmata and, more recently, an increasing interest in quantitative capabilities. Whatever the area of interest: solute movement; electrotonic coupling; transmission of stimuli; or the passage of viruses, there are two sets of parameters that must be studied in detail. These are: firstly the nature and amounts of the materials that can be shown to pass through the connections; and, secondly the physical data pertaining to the conducting pore itself. In the majority of all situations so far studied we remain far too ignorant about part, or all, of these vital data.

For the future, the main message must be to correlate structure and function by a multi-disciplinary attack on symplastic transport within the *same* cells and tissues. It is clear that variation in plasmodesmatal structure and frequency, as well as the markedly different permeabilities of different membranes to different ions, limits our ability to extrapolate from one situation to another. In both structure and function, and particularly so in higher plants, further progress will require the improvement of existing techniques and the development of new ones. We may then look forward to the next stage in the study of plasmodesmata, when we move beyond observations towards understanding.

BIBLIOGRAPHY

ABOU-MANDOUR, R., VOLK, O.H. and REINHARD, E. (1968). Über das Vorkommen eines cytokininartigen Faktors in *Cuscuta reflexa*. Planta 82, 153-163.

ALBERSHEIM, P. (1965). The substructure and function of the cell wall, in "Plant Biochemistry" (Eds. J. Bonner and J.E. Varner). Academic Press, London, 1965.

ALEXANDROV, W. (1925). Über ein neues Beispiel einer besonderen Art des Wassergewebes in den Blättern. Ber. dt. bot. Ges. 43, 418-426.

ALEXANDROV, W. and ABESSADZE, K.J. (1927). Über die Struktur der Seitenwände der Siebröhren. Planta 3, 77-89.

ALLAWAY, W.G. and SETTERFIELD, G. (1972). Ultrastructural observations on guard cells of *Vicia faba* and *Allium porrum*. Can. J. Bot. 50, 1405-1413.

ALLEN, R.D. and BOWEN, C.C. (1966). Fine structure of *Psilotum nudum* cells during division. Caryologia 19, 299-342.

ALLISON, A.V. and SHALLA, T.A. (1974). The ultrastructure of local lesions induced by potato virus X: A sequence of cytological events in the course of infection. Phytopathology 64, 784-793.

ANDERSON, W.P. (1975a). Long distance transport from roots, *in* "Ion Transport in Plant Cells and Tissues" (Eds. D.A. Baker and J.L. Hall), North Holland Publishing Co., Amsterdam.

ANDERSON, W.P. (1975b). Ion transport through roots, *in* "Development and Function of Plant Roots" (Eds. J.G. Torrey and D.T. Clarkson) p. 437-463, Academic Press, London.

ANDERSON, W.P. (1975c). Transport through roots, *in* "Encyclopaedia of Plant Physiology (Eds. A. Pirson and M. Zimmermann). Vol. 2. Transport in Plants, II. Springer-Verlag, Berlin.

ANDERSON, W.P., AIKMAN, D.P. and MEIRI, A. (1970). Excised root exudation - a standing-gradient osmotic flow. Proc. R. Soc. Lond. B. 174, 445-458.

ANDREWES, J.H. and SHALLA, T.A. (1974). The origin, development and conformation of amorphous inclusion body components in tobacco etch virus-infected cells. Phytopathology 64, 1234-1243.

ANGOLD, R.E. (1968). The formation of the generative cell in the pollen grain of *Endymion non-scriptus* (L). J. Cell. Sci. 3, 573-578.

APPLETON, T.C. (1974). A cryostat approach to ultrathin 'dry' frozen sections for electron microscopy: a morphological and X-ray analytical study. J. Microscopy 100, 49-74.

ARISZ, W.H. (1945). Contribution to a theory on the absorption of salts by the plant and their transport in parenchymatous tissue. Proc. K. ned. Akad. Wet. 48, 420-446.

ARISZ, W.H. (1948). Uptake and transport of chlorine by parenchymatic tissue of leaves of *Vallisneria spiralis*. III. Discussion of the transport and the uptake. Vacuole secretion theory. Proc. K. ned. Akad. Wet. Ser. C. 51, 25-33.

ARISZ, W.H. (1952). Transport of organic compounds. A. Rev. Pl. Physiol. 3, 109-130.

ARISZ, W.H. (1956). Significance of the symplasm theory for transport across the root. Protoplasma 46, 1-62.

ARISZ, W.H. (1958). Influence of inhibitors on the uptake and the transport of chloride ions in leaves of *Vallisneria spiralis*. Acta bot. neerl. 7, 1-32.

ARISZ, W.H. (1960). Symplasmatischer Salztransport in *Vallisneria*-Blättern. Protoplasma 52, 309-343.

ARISZ, W.H. (1961). Symplasm theory of salt uptake into and transport in parenchymatic tissue, *in* "Recent Advances in Botany". University of Toronto Press, 1125-1128.

ARISZ, W.H. (1963). Influx and Efflux by leaves of *Vallisneria spiralis*. I. Active uptake and permeability. Protoplasma, 57, 5-26.

ARISZ, W.H. (1969). Intercellular polar transport and the role of plasmodesmata in coleoptiles and *Vallisneria* leaves. Acta bot. neerl. 18, 14-38.

ARISZ, W.H. and SCHREUDER, M.J. (1956a). The path of salt transport in *Vallisneria* leaves. Proc. K. ned. Akad. Wet. Ser. C. 59, 454-460.

ARISZ, W.H. and SCHREUDER, M.J. (1956b). Influence of water withdrawal through transpiration on the salt transport in *Vallisneria* leaves. Proc. K. ned. Akad. Wet. Ser. C. 59, 461-470.

ARISZ, W.H. and WIERSEMA, E.P. (1966a). Symplastic long distance transport in *Vallisneria* plants investigated by means of autoradiograms. 1A. Proc. K. ned. Akad. Wet. Ser. C. 69, 223-232.

ARISZ, W.H. and WIERSEMA, E.P. (1966b). Symplastic long distance transport in *Vallisneria* leaves investigated by means of autoradiograms. 1B. Proc. K. ned. Akad. Wet. Ser. C. 69, 223-241.

ARMACOST, R.R. (1944). The structure and function of the border parenchyma and vein-ribs of certain dicotyledon leaves. Iowa Acad. Sci. 51, 157-169.

ARNAUDOW, N. (1925). Über Transplantation von Moosembryonen. Flora, Jena 118, 17-26.

ASHFORD, A.E. and JACOBSEN, J.V. (1974a). Cytochemical localisation of phosphatase in barley aleurone cells: The pathway of gibberellic-acid-induced enzyme release. Planta 120, 81-105.

ASHFORD, A.E. and JACOBSEN, J.V. (1974b). Cytochemical localisation of acid phosphatase in isolated aleurone layers treated with gibberellic acid and in germinating barley grains. Mechanisms of Regulation of Plant Growth (The Royal Society of New Zealand) Bulletin 12, 591-599.

AUDRAN, J.-C. (1974). Aspects ultrastructuraux de l'individualisation des microspores du *Ceratozamia mexicana* (Cycadées). C.r.hebd.Séanc. Acad. Sci., Paris. Sér. D. 278, 1023-1026.

ATKINSON, M.R., FINDLAY, G.P., HOPE, A.B., PITMAN, M.G., SADDLER, H.D.W. and WEST, K.R. (1967). Salt regulation in the mangroves *Rhizophora mucronata* Lam. and *Aegialitis annulata* R.BR. Aust. J. biol. Sci. 20, 589-599.

BACKS-HÜSEMANN, D. and REINERT, J. (1970). Embryobildung durch isolierte Einzelzellen aus Gewebekulturen von *Daucus carota*. Protoplasma 70, 49-60.

BADGER, M.R., ANDREWS, T.J. and OSMOND, C.B. (1974). Detection in C_3, C_4 and CAM plant leaves of a low K_m (CO_2) form of RuDP carboxylase, having high RuDP oxygenase at physiological pH, *in* "Proc. Third Int. Congress on Photosynthesis" (Ed. M. Avron), 1421-1429. Amsterdam, Elsevier.

BAJER, A. (1968a). Fine structure studies on phragmoplast and cell plate formation. Chromosoma 24, 383-417.

BAJER, A. (1968b). Behaviour and fine structure of spindle fibres during mitosis in endosperm. Chromosoma 25, 249-281.

BAKER, D.A. and HALL, J.L. (1973). Pinocytosis, ATP-ase and ion uptake by plants. New Phytol. 72, 1281-1291.

BAKER, J.R. (1948). The cell theory: a restatement, history and critique. Part 1. Q.Jl. Micros. Sci. 89, 103-125.

BALL, E. (1969). Histology of mixed callus cultures. Bull. Torrey bot. C Club. 96, 52-59.

BARBIER, C. (1972). Premières données ultrastructurales sur la différenciation de l'oosphère d'une Bryophyte, le *Mnium undulatum* (Mniacées). C.r.hebd. Séanc. Acad. Sci., Paris. Sér. D. 274, 3222-3224.

BARONDSE, S.H. (1970). *in* "The Neurosciences - second study programme" (Ed. F.O. Schmitt), 747-760. Rockefeller Uni. Press, N.Y.

BASSI, M., AUGUSTA FAVALI, M. and CONTI, G.G. (1974). Cell wall protrusions induced by cauliflower mosaic virus in Chinese cabbage leaves: a cytochemical and autoradiographic study. Virology 60, 353-358.

BATESON, W. (1921). Root cuttings and chimaeras. J. Genet. 11, 91-97.

BAUER, E. (1930). "Einführung in die experimentelle Vererbungslehre." Borntraeger, Berlin.

BAUM, H. (1948a). Über die postgenitale Verwachsung in Karpellen. Öst. bot. Z. 95, 86-94.

BAUM, H. (1948b). Postgenitale Verwachsung in und zwischen Karpell und Staubblattkreisen. Sber. Akad. Wiss. Wien. Math-nat Kl. Abt. 1. 157, 357-378.

BAWDEN, F.C., PIRIE, N.W., BERNAL, J.D. and FANKUCHEN, I. (1936). Liquid crystalline substances from virus-infected plants. Nature, Lond. 138, 1051-1052.

BECK, R.E. and SCHULTZ, J.S. (1970). Hindered diffusion in microporous membranes with known pore geometry. Science, N.Y., 170, 1002-1305.

BEHNKE, H.D. (1971). Über den Feinbau verdickter (Nacré) Wände und der Plastiden in den Siebröhren von *Annona* und *Myristica*. Protoplasma 72, 69-78.

BEHNKE, H.D. and PALIWAL, G.S. (1973). Ultrastructure of phloem and its development in *Gnetum gnemon*, with some observations on *Ephedra campylopoda*. Protoplasma 78, 305-319.

BELL, P.R. and MÜHLETHALER, K. (1962). The fine structure of the cells taking part in oogenesis in *Pteridium aquilinum* (L) Kuhn. J. Ultrastruct. Res. 7, 452-466.

BENBADIS, M.-C., LASSELAIN, M.-J. and DEYSSON, G. (1973). Imprégnation osmique de cellules méristématiques de la racine: Application à l'étude de l'appareil de Golgi normal et modifié par l'antipyrine. C.r.hebd.Séanc. Acad. Sci., Paris. Sér. D. 277, 2349-2352.

BENBADIS, M.-C., LASSELAIN, M.-J. and DEYSSON, G. (1974). Sur les modalités de la cytodiérèse dans les méristèmes radiculaire d'*Allium sativum* L. Étude ultrastructurale à l'aide de nouvelles techniques. C.r.hebd.Séanc. Acad. Sci., Paris. Sér. D. 278, 2523-2526.

BENBADIS, M.-C., LEVY, F. and DEYSSON, G. (1974). Interruption de la mitose en prophase avec persistance de la membrane nucléaire sous l'influence du cycloheximide: étude ultrastructurale. C.r.hebd. Séanc. Acad. Sci., Paris. Sér. D. 278, 1353-1355.

BENNETT, C.W. (1934). Plant tissue relations of the sugar-beet curly-top virus. J. agric. Res. 48, 665-701.

BENNETT, C.W. (1940). Relation of food translocation to movement of virus of tobacco mosaic. J. agric. Res. 60, 361-390.

BENNETT, C.W. (1944). Studies of dodder transmission of plant viruses. Phytopathology 34, 905-932.

BENNETT, M.V.L. (1973a). Permeability and structure of electrotonic junctions and intercellular movements of tracers. Ch. 8. *in* "Intracellular staining in Neurobiology" (Ed. S.B. Kater and C. Nicholson), Springer Verlag, Berlin.

BENNETT, M.V.L. (1973b). Function of electrotonic junctions in embryonic and adult tissues. Fedn. Proc. Fedn. Am. Socs. exp. Biol. 32, 65-75.

BERGANN, F. (1956). Untersuchungen an den Blüten und Früchten der Crataegomespili und ihrer Eltern. Flora, Jena 143, 219-268.

BERGANN, F. (1962). Über den Nachweis zwischenzelliger Genwirkungen (Partnerinduktion) bei der Pigmentbildung in den Brakteen der Periklinalchimaere *Euphorbia pulcherrima* Willd. "Eckes Rosa". Biol. Zbl. 81, 471-503.

BERKALOFF, C. (1963). Les cellules méristématiques d'*Himanthalia lorea* (L) S.F. Gray. Étude au microscope électronique. J. Microscopie 2, 213-228.

BERNAL, J.D. and FANKUCHEN, I. (1937). Structure types of protein 'crystals' from virus-infected plants. Nature, Lond. 139, 923-924.

BERNHARDI, J.J. (1805). "Beobachtungen über Pflanzengefässe und eine neue Art derselben". Erfurt, pp. vi+82.

BERRY, J.A., DOWNTON, W.J.S. and TREGUNNA, E.B. (1970). The photosynthetic carbon metabolism of *Zea mays* and *Gomphrena globosa*: The location of the CO_2 fixation and carboxyl transfer reactions. Can. J. Bot. 48, 777-786.

BIERBERG, W. (1909). Die Bedeutung der Protoplasmarotation für den Stofftransport in den Pflanzen. Flora, Jena 99, 52-80.

BINDING, H. (1966). Regeneration und Verschmelzung isolierter Laubmoosprotoplasten. Z. Pflphysiol. 55, 305-334.

BIRD, R.B., STEWART, W.E. and LIGHTFOOT, E.N. (1960). "Transport phenomena". Wiley, New York.

BISALPUTRA, T. (1966). Electron microscopic study of the protoplasmic continuity in certain brown algae. Can. J. Bot. 44, 89-93.

BISALPUTRA, T. and STEIN, J.R. (1966). The development of cytoplasmic bridges in *Volvox aureus*. Can. J. Bot. 44, 1697-1702.

BISALPUTRA, T., RUSANOWSKI, P.C. and WALKER, W.S. (1967). Surface activity, cell wall and fine structure of pit connections in the red alga *Laurencia spectabilis*. J. Ultrastruct. Res. 20, 277-289.

BJÖRKMAN, O., TROUGHTON, J.H. and NOBS, M. (1973). Photosynthesis in relation to leaf structure, *in* "Basic Mechanisms in Plant Morphogenesis", Brookhaven Symp. Biol. No. 25, 206-226. Upton, Brookhaven National Laboratory.

BLACK, C.C. (1973). Photosynthetic carbon fixation in relation to net CO_2 uptake. A. Rev. Pl. Physiol. 24, 253-286.

BLINN, D.W. and MORRISON, E. (1974). Intercellular cytoplasmic connections in *Ctenocladus circinnatus* Borzi Chlorophyceae with possible ecological significance. Phycologia 13, 95-97.

BLOCH, R. (1944). Developmental potency, differentiation and pattern in meristems of *Monstera deliciosa*. Am. J. Bot. 31, 71-77.

BLOCH, R. (1946). Differentiation and pattern in *Monstera deliciosa*. The idioblastic development of the trichosclereids in the air roots. Am. J. Bot. 33, 544-551.

BOCK, F. (1926). Experimentelle Untersuchungen an koloniebildenden Volvocaceen. Arch. Protistenk 56, 321-356.

BOEKE, J.H. (1971). Location of the postgenital fusion in the gynoecium of *Capsella bursa-pastoris* (L) Med. Acta bot. neerl. 20(6), 570-576.

BOEKE, J.H. (1973a). The postgenital fusion in the gynoecium of *Trifolium repens* L: Light and electron microscopical aspects. Acta bot. neerl. 22(5), 503-509.

BOEKE, J.H. (1973b). The use of light microscopy versus electron microscopy for the location of postgenital fusions in plants. Proc. K. ned. Akad. Wet. Ser. C. 76, 528-535.

BÖHMER, H. (1958). Untersuchungen über das Wachstum und den Feinbau der Zellwände in der *Avena*-Koleoptile. Planta 50, 461-497.

BONNETT, H.T. (1968). The root endodermis: fine structure and function. J. Cell Biol. 37, 199-205.

BONNETT, H.T. (1969). Cortical cell death during lateral root formation. J. Cell Biol. 40, 144-159.

BONNOT, E.J. (1967a). Cytologie végétale - L'infrastructure des plasmodesmes de deux Bryales. C.r.hebd.Séanc. Acad. Sci., Paris. Sér. D. 264, 2276-2279.

BONNOT, E.J. (1967b). Contributions à l'étude de la spermatogenèse muscinale. 1. Le plaste foliacé anthéridial (= linioplast) de *Pogonatum aloides* (Hedw.) P. Beauv. Bull. Soc. bot. Fr. 114, 138-144.

BONNOT, E.J. (1967c). Le plan d'organisation fondamental de la spermatide de *Bryum capillare* (L) Hedw. C.r.hebd. Séanc. Acad. Sci., Paris. Sér. D. 265, 958-961.

BORNMAN, C.H. (1967). Some ultrastructural aspects of abscission in *Coleus* and *Gossypium*. S. Afr. J. Sci. 63, 325-331.

BOSSE, A. WEBER-VAN (1890). Études sur les Algues de l'Archipel Malaisien. Annls. Jard. bot. Buitenz. 8, 165-188.

BOSTROM, T.E. and WALKER, N.A. (1975). Intercellular transport in plants. I. The rate of transport of chloride and the electric resistance. J. exp. Bot. (in press).

BOUCK, G.B. (1962). Chromatophore development, pits and other fine structure in the red alga *Lomentaria baileyana* (Harv.) Farlow. J. Cell Biol. 12, 553-569.

BOUCK, G.B. and GALSTON, A.W. (1967). Cell-wall deposition and the distribution of cytoplasmic elements after treatment of pea internodes with the auxin analog 2,3,5-Triiodobenzoic acid. Ann. N.Y. Acad. Sci. 144, 34-48.

BOURNE, U.K. and COLE, K. (1968). Some observations on the fine structure of the marine brown alga *Phaeostrophion irregulare*. Can. J. Bot. 46, 1369-1375.

BOWER, F.O. (1883). On plasmolysis and its bearing upon the relations between cell wall and protoplasm. Q. Jl. Micros. Sci. 23, 155-159.

BOWES, B.G. (1965). The ultrastructure of the shoot apex and young shoot of *Glechoma hederacea* L. Cellule 65, 351-356.

BOWES, G., OGREN, W.L. and HAGEMAN, R.H. (1971). Phosphoglycolate production catalysed by ribulose diphosphate carboxylase. Biochem. biophys. Res. Commun. 45, 716-722.

BOWLING, D.J.F. (1972). Measurement of profiles of potassium activity and electrical potential in the intact root. Planta 108, 147-151.

BRABEC, F. (1965). Propfung und Chimaeren unter besonderer Berücksichtigung der entwicklungsphysiologische Problematik, *in* "Encycl. Plant Physiol" 15 (Ed. A. Lang) 388-498, Springer Verlag, Berlin.

BRACKER, C.E. (1967). Ultrastructure of fungi. A. Rev. Phytopath. 5, 343-374.

BRACKER, C.E. and BUTLER, E.E. (1963). The ultrastructure and development of septa in hyphae of *Rhizoctonia solani*. Mycologia 55, 35-58.

BRAND, F. (1902). Zur näheren Kenntnis der Algengattung *Trentephoplia* Mart. Beih. bot. Zbl. 12, 200-225.

BRANTS, H.D. (1964). The susceptibility of tobacco and bean leaves to tobacco mosaic virus in relation to ectodesmata. Virology 23, 588-594.

BRANTS, H.D. (1965). Relation between ectodesmata and infection of leaves by C^{14}-labeled tobacco mosaic virus. Virology 26, 554-557.

BRÄUTIGAM, E. and MÜLLER, E. (1975a). Transportprozesse in *Vallisneria* - Blättern und die Wirkung von Kinetin und Kolchizin. I. Aufnahme von α-Aminobuttersäure in Gewebe von *Vallisneria* und die Wirking von Kinetin. Biochem. Physiol. Pflanzen 167, 1-15.

BRÄUTIGAM, E. and MÜLLER, E. (1975b). Transportprozesse in *Vallisneria* - Blättern und die Wirkung von Kinetin und Kolchizin. II. Symplastischer Transport von α-Aminobuttersäure in *Vallisneria* - Blättern und die Wirking von Kinetin. Biochem. Physiol. Pflanzen. 167, 17-28.

BRÄUTIGAM, E. and MÜLLER, E. (1975c). Transportprozesse in *Vallisneria* - Blättern und die Wirkung von Kinetin und Kolchizen. III. Induktion einer Senke durch Kinetin. Biochem. Physiol. Pflanzen 167, 29-39.

BRENNER, D.M. and CARROLL, G.C. (1968). Fine structural correlates of growth in the hyphae of *Ascodesmis sphaerospora*. J. Bact. 95, 658-671.

BRIGGS, G.E. (1957). Some aspects of free space in plant tissues. New Phytol. 56, 305-324.

BRIGGS, G.E. and ROBERTSON, R.N. (1957). Apparent free space. A. Rev. Pl. Physiol. 8, 11-30.

BRIGHIGNA, L. (1974). The ultrastructure of plasmodesmata in sucking scale in *Tillandsia*. Caryologia 27, 369-377.

BRINCKMANN, E. (1973). Zur Messung des Membranpotentials und dessen lichtabhängigen Änderungen an Blattzellen höherer Landpflanzen. Dissertation, Darmstadt.

BRINCKMANN, E. and LÜTTGE, U. (1974). Lichtabhängige Membranpotentialschwankungen und deren interzelluläre Weiterleitung bei panaschierten Photosynthese-Mutanten von *Oenothera*. Planta 119, 47-57.

BROUWER, R. (1965). Ion absorption and transport in plants. A. Rev. Pl. Physiol. 16, 241-266.

BROWN, W.V. (1975). Variations in anatomy, associations, and origins of Kranz tissue. Am. J. Bot. 62, 395-402.

BROWN, W.V. and JOHNSON, S.C. (1962). The fine structure of the grass guard cell. Am. J. Bot. 49, 110-115.

BROWNELL, P.F. (1965). Sodium as an essential micronutrient element for a higher plant (*Atriplex vesicaria*). Pl. Physiol. 40, 460-468.

BUCK, G.J. (1954). The histological development of the bud graft union in roses. Proc. Am. Soc. hort. Sci. 62, 497-502.

BUDER, J. (1911). Studien an *Laburnum Adami*. II. Z. Abstamm. lehre 5, 209-284.

BÜNNING, E. (1951). Über die Differenzierungsvorgänge in der Cruciferenwurze. Planta 39, 126-153.

BÜNNING, E. (1953). "Entwicklungs-und Bewegungsphysiologie der Pflanze". 3rd Ed. Springer Verlag, Berlin.

BÜNNING, E. (1955). Regeneration bei Pflanzen, *in* "Handbuch der Allg. Pathologie", 6 (Eds. F. Bücher, E. Letterer and F. Roulet), Springer Verlag, Berlin.

BÜNNING, E. and SAGROMSKY, H. (1948). Die Bildung des Spaltöffnungsmusters in der Blattepidermis. Z. Naturf. 3b, 203-216.

BURGESS, J. (1970). Cell shape and mitotic spindle formation in the generative cell of *Endymion non-scriptus*. Planta 95, 72-85.

BURGESS, J. (1971). Observations on structure and differentiation in plasmodesmata. Protoplasma 73, 83-95.

BURGESS, J. (1972). The occurrence of plasmodesmata-like structures in non-division wall. Protoplasma 74, 449-458.

BURGESS, J. and FLEMING, E.N. (1973). The structure and development of a genetic tumour of the pea. Protoplasma 76, 315-325.

BURGESS, J. and FLEMING, E.N. (1974). Ultrastructural studies of the aggregation and fusion of plant protoplasts. Planta 118, 183-193.

BURGESS, J. and NORTHCOTE, D.H. (1967). A function of the pre-prophase band of microtubules in *Phleum protense*. Planta 75, 319-326.

BURGESS, J., WATTS, J.W., FLEMING, E.N. and KING, J.M. (1973). Plasmalemma fine structure in isolated tobacco mesophyll protoplasts. Planta 110, 291-301.

BURR, F.A. and EVERT, R.F. (1973). Some aspects of sieve-element structure and development in *Selaginella kraussiana*. Protoplasma 78, 81-97.

BUTLER, R.D. and ALLSOPP, A. (1972). Ultrastructural investigation in the Stigonemataceae (Cyanophyta). Arch. Mikrobiol. 82, 283-299.

BUTTON, J., KOCHBA, J. and BORNMAN, C.H. (1974). Fine structure of and embryoid development from embryogenic ovular callus of 'Shamouti' orange (*Citrus sinensis* Osb.). J. exp. Bot. 25, 446-457.

BUTTROSE, M.S. (1963). Ultrastructure of the developing wheat endosperm. Aust. J. biol. Sci. 16, 305-317.

BUVAT, R. (1957). L'infrastructure des plasmodesmes et la continuité des cytoplasmes. C.r.hebd.Séanc. Acad. Sci., Paris. Sér. D. 245, 198-201.

BUVAT, R. (1960). L'infrastructure des plasmodesmes chez les cellules parenchymateuses des cordons conducteurs jeunes de *Cucurbita pepo* L. C.r.hebd.Séanc. Acad. Sci., Paris. Sér. D. 250, 170-172.

BUVAT, R. (1961). Le réticulum endoplasmique des cellules végétales. Ber. dt. bot. Ges. 74, 261-267.

BUVAT, R. (1963). Electron microscopy of plant protoplasm. Int. Rev. Cytol. 14, 41-155.

BUVAT, R. (1965). Les bases cytologiques de la différenciation et de la dedifférenciation chez les plantes, *in* "Encycl. Pl. Physiol." Vol. XV. 100-145 (Ed. A. Lang), Springer Verlag, Berlin.

BUVAT, R. (1969). "Plant cells". World University Library, London.

BUVAT, R. and PUISSANT, A. (1958). Observations sur la cytodiérèse et l'origine des plasmodesmes. C.r.hebd.Séanc. Acad. Sci., Paris. Sér. D. 247, 233-236.

CALDERINI, I. (1846). Essai d'experiences sur la greffe des Graminées. Annls. Sci. nat. (Bot.) Sér. III 6, 131-133.

CALDWELL, J. (1931). The physiology of virus diseases in plants. II. Further studies on the movement of mosaic in the tomato plant. Ann. appl. Biol. 18, 279-298.

CALDWELL, J. (1934). The physiology of virus diseases in plants. V. The movement of the virus agent in tobacco and tomato. Ann. appl. Biol. 21, 191-205.

CAMEFORT, H. (1962). L'organisation du cytoplasme dans l'oosphère et la cellule centrale du *Pinus laricio* Poir (var. *austriaca*). Annls. Sci. nat. (Bot.) Sér. 12, 3, 265-291.

CAMPBELL, N. and THOMSON, W.W. (1975). Chloride localization in the leaf of *Tamarix*. Protoplasma 83, 1-14.

CANNY, M.J. (1973). "Phloem Translocation", Cambridge University Press.

CARDALE, S. (1971). The Structure of the salt gland of *Aegiceras corniculatum*. Planta 99, 183-191.

CARDE, J.P. (1974). Le tissu de transfert (= cellules de Strasburger) dans les aiguilles du pin maritime (*Pinus pinaster* Ait.). II. Caractères cytochimiques et infrastructuraux de la paroi et des plasmodesmes. J. Microscopie 20, 51-72.

CAROLIN, R.C., JACOBS, S. and VESK, M. (1973). The structure of cells of mesophyll and parenchymatous sheath of the Gramineae. Bot. J. Linn. Soc. 66, 259-275.

CAROLIN, R.C., JACOBS, S. and VESK, M. (1975). Leaf structure in Chenopodiaceae. Bot. Jb. (in press).

CAROTHERS, Z.B. (1972). Studies of spermatogenesis in the Hepaticae. III. Continuity between plasma membrane and nuclear envelope in androgonial cells of *Blasia*. J. Cell Biol. 52, 273-282.

CAROTHERS, Z.B. (1973). Studies of spermatogenesis in the Hepaticae. IV. On the blepharoplast of *Blasia*. Am. J. Bot. 60, 819-828.

CARR, D.J. (1966). Metabolic and hormonal regulation of growth and development, *in* "Trends in plant morphogenesis" (Ed. Elizabeth G. Cutter), pp. 253-283, Longmans, London.

CARR, D.J. (1967). The relationship between florigen and the flower hormones. Ann. N.Y. Acad. Sci. 144, 305-312.

CARROLL, G. (1967). The fine structure of the ascus septum in *Ascodesmis sphaerospora* and *Saccobolus keverni*. Mycologia 59, 527-532.

CASS, D.D. (1973). An ultrastructural and Nomarski-interference study of the sperms of barley. Can. J. Bot. 51, 601-605.

CATALDO, D.A. (1974). Vein loading: the role of the symplast in intercellular transport of carbohydrate between the mesophyll and minor veins of tobacco leaves. Pl. Physiol. 53, 912-917.

CATALDO, D.A. and BERLYN, G.P. (1974). An evaluation of selected physical characteristics and metabolism of enzymatically separated mesophyll cells and minor veins of tobacco. Am. J. Bot. 61, 957-963.

CAVENEY, S. (1974). Intercellular communication in a positional field: movement of small ions between insect epidermal cells. Devl. Biol. 40, 311-322.

CHABOT, J.F. and CHABOT, B.F. (1975). Developmental and seasonal patterns of mesophyll ultrastructure in *Abies balsamea*. Can. J. Bot. 53, 295-304.

CHAILAKYAN, M.Kh. (1940a). Translocation of flowering hormones across various plant organs. I. Across the leaf. C.r.(Dokl.) Acad. Sci. SSSR 27, 160-163.

CHAILAKYAN, M.Kh. (1940b). Translocation of flowering hormones across various plant organs. II. Translocation across the stem. C.r.(Dokl.) Acad. Sci. SSSR 27, 255-258.

CHAILAKYAN, M.Kh. (1940c). Translocation of flowering hormones across various plant organs. III. Across the root. C.r.(Dokl.) Acad. Sci. SSSR 27, 373-376.

CHAILAKYAN, M.Kh. and BUTENKO, R.G. (1957). Movement of assimilates of leaves to shoots under differential photoperiodic conditions of leaves. Fiziol. Rast. 4, 450-462.

CHAMBERS, T.C. and FRANCKI, R.I.B. (1966). Localization and recovery of lettuce necrotic yellows virus from xylem tissues of *Nicotiana glutinosa*. Virology 29, 673-676.

CHAMBERS, T.C. and MERCER, F.V. (1964). Studies on the comparative physiology of *Chara australis*. II. The fine structure of the protoplast. Aust. J. biol. Sci. 17, 372-387.

CHAPMAN, E.A., BAIN, J.M. and GOVE, D.W. (1975). Mitochondria and chloroplast peripheral reticulum in the C_4 plants *Amaranthus edulis* and *Atriplex spongiosa*. Aust. J. Pl. Physiol. 2, 207-221.

CHARDARD, R. (1973). Observations aux microscopes électroniques à bas et haut voltages de coupes fines et épaisses de cellules méristématiques radiculaires de *Zea mays* L. imprégnées par des sels métalliques. C.r.hebd.Séanc. Acad. Sci., Paris. Sér. D. 276, 2155-2158.

CHAUVEAUD, G. (1891). Recherches embryogeniques sur l'appareil laticifère des Euphorbiacées, Urticacées, Apocyanacées et Asclepiadacées. Annls. Sci. nat. (Bot.) VII Sér. 14, 1-161.

CHEN, R.-F. and JONES, R.L. (1974). Studies on the release of barley aleurone cell proteins: autoradiography. Planta 119, 207-220.

CHEN, T.M., DITTRICH, P., CAMPBELL, W.H. and BLACK, C.C. (1974). Metabolism of epidermal tissues, mesophyll cells and bundle sheath strands resolved from mature nutsedge leaves. Archs. Biochem. Biophys. 163, 246-262.

CHESNOY, L. (1969). Sur l'origine du cytoplasme des embryons chez le *Biota orientalis* (Cupressacées). C.r.hebd. Séanc. Acad. Sci., Paris. Sér. D. 268, 1921-1924.

CHIEN, L.-C., CHING, Y.-H. and CHANG, P.-T. (1973). Cytological studies on cold resistance of plants. Ultrastructural changes of wheat cells in winter. Acta Bot. Sinica. 15, 22-36.

CHILD, C.M. (1910). Physiological isolation of parts and fission in *Planaria*. Arch. Entw. Mech. Org. 30, Pt. 2.

CHILD, C.M. (1915). "Individuality in Organisms". University of Chicago Press.

CHILD, C.M. (1917). Experimental alteration of the axial gradient in the alga, *Griffithsia bornetiana*. Biol. Bull. mar. biol. Lab., Woods Hole 32, 213-233.

CHILD, C.M. (1924). "Physiological foundations of behaviour" (Reprinted 1963), Holt Rinehart and Winston, N.Y.

CHRISTENSEN, B.N. (1973). Procion brown: an intracellular dye for light and electron microscopy. Science, N.Y. 182, 1255-1256.

CHUKHRII, M.G. (1971). Submicroscopic structure of plasmodesmata of succulent parenchyma. Iz. V. Akad. Nauk. Mold. SSR. Ser. Biol. Khim. Nauk. 4, 3-7.

CIOBANU, I.R. (1969). Electron microscope observations on the plasmodesmata of pollen mother cells of tomato (*Lycopersicum esculentum* Mill). Rev. Roum. Biol. Ser. Bot. 14(5), 269-273.

CITHAREL, J. (1972). Contribution à l' étude du métabolisme azoté des Algues marines. Utilization métabolique d' acide glutamique-^{14}C par *Ascophyllum nodosum* (Linné) Le Jolis et *Polysiphonia lanosa* (Linné) Tandy. Botanica mar. 15, 157-161.

CLARK, M.A. and ACKERMAN, G.A. (1971). A histochemical evaluation of the pyroantimonate-osmium reaction. J. Histochem. Cytochem. 19, 727-737.

CLARKE, A.E., KNOX, R.S. and JERMYN, M.A. (1975). (In press).

CLARKSON, D.T. and ROBARDS, A.W. (1975). The endodermis, its structural development and physiological role, *in* "Root Structure and Function (Eds. J. Torrey and D.T. Clarkson), Harvard Symposium, 1974. London: Academic Press.

CLARKSON, D.T., ROBARDS, A.W. and SANDERSON, J. (1971). The tertiary endodermis of barley roots: fine structure in relation to radial transport of ions and water. Planta 96, 292-305.

CLOWES, F.A.L. (1970). Nutrition and the quiescent centre of root meristems. Planta 90, 340-348.

CLOWES, F.A.L. (1972). Regulation of mitosis in roots by their caps. Nature (New Biol.) 235, 143-144.

CLOWES, F.A.L. and JUNIPER, B.E. (1964). The fine structure of the quiescent centre and neighbouring tissues in root meristems. J. exp. Bot. 15, 622-630.

CLOWES, F.A.L. and JUNIPER, B.E. (1968). "Plant cells", Blackwell Scientific Publications, Oxford.

COCHRAN, G.W. (1946). Effect of shading techniques on transmission of tobacco mosaic virus through dodder. Phytopathology 36, 396.

COCKING, E.C. (1960). A method for the isolation of plant protoplasts and vacuoles. Nature, Lond. 187, 927-929.

COHEN, W.D. and GOTTLEIB, T. (1971). C-microtubules in isolated mitotic spindles. J. Cell Sci. 9, 603-619.

COHEN, J. and LOEBENSTEIN, G. (1975). An electron microscopic study of starch lesions in cucumber cotyledons infected with tobacco mosaic virus. Phytopathology 65, 32-39.

COLE, K. (1969). The cytology of *Eudesme virescens* (Carm.) J. Ag. II Ultrastructure of cortical cells. Phycologia 8, 101-108.

COLE, K. (1970). Ultrastructural characteristics in some species of the order Scytosiphonales. Phycologia 9, 275-283.

COLE, K. and LIN, S.C. (1968). The cytology of *Leathesia difformis* l. Fine structure of the vegetative cells in field and cultured material. Syesis 1, 103-119.

COLE, K. and LIN, S.C. (1970). Plasmalemmasomes on sporelings of the brown alga *Petalonia debilis*. Can. J. Bot. 48, 265-268.

COLLANDER, R. (1957). Permeability of plant cells. A. Rev. Pl. Physiol. 8, 335-348.

CONTI, G.G., VEGETTI, G., BASSI, M. and FAVALI, M.A. (1972). Some ultrastructural and cytochemical observations on Chinese cabbage leaves infected with cauliflower mosaic virus. Virology 47, 694-700.

COOKE, R. and KUNTZ, I.D. (1974). The properties of water in biological systems. A. Rev. Biophys. Bioeng. 3, 95-126.

COOKSTON, R.K. and MOSS, D.N. (1973). A variation of leaf anatomy in *Arundinella hirta* (Gramineae). Pl. Physiol. 52, 397-402.

COOMBS, J. and BALDRY, C.W. (1972). C_4 pathway in *Pennisetum purpureum*. Nature, Lond. 238, 268-270.

CORMACK, R.G.H. (1947). A comparative study of developing epidermal cells in white mustard and tomato roots. Am. J. Bot. 34, 310-314.

CORRENS, C. (1909). Vererbungsversuche mit blass(gelb)grünen und buntblätterigen Sippen. Z. Vererblehre. 1, 291-329.

COSS, R.A. and PICKETT-HEAPS, J.D. (1974). Gametogenesis in the green alga *Oedog-*

onium cardiacum. II. Spermiogenesis. Protoplasma 81, 297-311.

COULOMB, P. and COULOMB, C. (1972). Localisation cytochimique ultrastructurale d' une adenosine triphosphatase - Mg^{++} dépendente dans des cellules de méristèmes radiculaires de la courge (*Cucurbita pepo* L. Cucurbitacée). C.r.hebd.Séanc. Acad. Sci.,Paris. Sér. D. 275, 1035-1038.

COX, G.C. (1971). "The structure and development of cells with thickened primary walls". D.Phil. Thesis, Oxford.

COX, R.P. (Ed.) (1974). "Cell communication". J. Wiley and Sons, N.Y.

COX, R.P., KRAUSS, M.R., BALIS, M.E. and DANCIS, J. (1974). Metabolic co-operation in cell culture. Ch. 4. in "Cell communication" (Ed. R.J. Cox). J. Wiley and Sons, N.Y.

CRAFTS, A.S. (1931). A technique for demonstrating plasmodesmata. Stain Technol. 6, 127-128.

CRAFTS, A.S. (1933). Sieve-tube structure and translocation in the potato. Pl. Physiol. 8, 81-104.

CRAMER, P.J.S. (1930). La greffe de l' *Hevea.* Revue de bot. appl. et d'Agriculture trop. 10, 3-10, 99-107, 156-174.

CRONSHAW, J. and BOUCK, G.B. (1965). The fine structure of differentiating xylem elements. J. Cell Biol. 24, 415-431.

CROWDY, S.H. and TANTON, T.W. (1970). Water pathways in higher plants 1. Free space in wheat leaves. J. exp. Bot. 21, 102-111.

CROWLEY, N.C., DAVISON, E.M., FRANCKI, R.I.B. and OWUSU, G.K. (1969). Infection of bean-root meristems by Tobacco Ringspot Virus. Virology 39, No. 2, 322-330.

CURRAN, P.F. and MACINTOSH, J.R. (1962). A model system for biological water transport. Nature 193, 347-348.

CUTTER, E.G. and HUNG, C.-Y. (1972). Symmetric and asymmetric mitosis and cytokinesis in the root tip of *Hydrocharis morsus-ranae* L. J. Cell Sci. 11, 723-737.

CZAJA, A.T. (1930). Zellphysiologische Untersuchungen an *Cladophora glomerata:* Isolierung, Regeneration, Polarität 11, 601-627.

CZANINSKI, Y. (1974). Formation des thylles dans le xylème de *Daucus carota* L. Étude ultrastructurale. C.r.hebd. Séanc. Acad. Sci., Paris. Sér. D. 278, 253-256.

DAINTY, J. (1962). Ion transport and electrical potentials in plant cells. A. Rev. Pl. Physiol. 13, 379-402.

DAINTY, J. and GINZBURG, B.Z. (1964). The measurement of hydraulic conductivity (osmotic permeability to water) of internodal Characean cells by means of transcellular osmosis. Biochim. biophys. Acta 79, 102-111.

DAINTY, J. and HOPE, A.B. (1959). Ionic relations of cells of *Chara australis.* I. Ion exchange in the cell wall. Aust. J. biol. Sci. 12, 395-411.

DANIEL, L. (1927). Variations de l' appareil sécréteur chez diverses plantes greffées. C.r.hebd.Séanc. Acad. Sci., Paris. Sér. D. 185, 1296-1298.

DASHEK, W.V., HARWOOD, H.I. and ROSEN, W.G. (1971). The significance of a wall-bound, hydroxyproline-containing glycopeptide in lily pollen tube elongation, *in* Pollen: Development and Physiology (Ed. J. Heslop-Harrison), Butterworths, London, 194-200.

DAVID-FERREIRA, J.F. and BORGES, M. de L.V. (1958). Virus na celula vegetal observacoes ao microscopio electronico 1. Virus Y da batateira. Bol. Soc. Broteriana 32, 329-332.

DAVIS, D.G. and RENKIN, E.M. (1974). Fluid flow through small pores. Biophys. J. 14, 514-515.

DAVISON, E.M. (1969). Cell to cell movement of tobacco ringspot virus. Virology 37, 694-696.

DAWES, C.J., SCOTT, F.M. and BOWLER, E. (1961). A light and electron microscopic survey of algal cell walls. I. Phaeophyta and Rhodophyta. Am. J. Bot. 48, 925-934.

DEASON, T.R., DARDEN, W.H. and ELY, S. (1969). The development of sperm packets of the M5 strain of *Volvox aureus.* J. Ultrastruct. Res. 26, 85-94.

DENISON, W.C. and CARROLL, G.C. (1966). The primitive Ascomycete: a new look at an old problem. Mycologia 58, 249-269.

DESHPANDE, B.P. (1974). Development of the sieve plate in *Saxifraga sarmentosa* L. Ann. Bot. 38, 151-158.

DEXHEIMER, J. (1970). Recherches cytophysiologiques sur les grains de pollen. Rev. Cytol. et Biol. vég. 33, 169-234.

DIAMOND, J.M. (1974). Tight and leaky junctions of epithelia. Fedn. Proc. Fedn. Am. Socs. exp. Biol. 33, 2220-2224.

DIBOLL, A.G. and LARSON, D.A. (1966). An electron microscope study of the mature megagametophyte in *Zea mays*. Am. J. Bot. 53, 391-402.

DICK, P.S. and ap REES, T. (1975). The pathway of sugar transport in roots of *Pisum sativum*. J. exp. Bot. 26, 305-314.

DICKENSON, P.B. (1964). The ultrastructure of the latex vessel of *Hevea brasiliensis*, *in* "Proceedings of the Natural Rubber Producers' Research Association Jubilee Conference" (Ed. L. Mullins). McLaren and Sons Ltd., London, 52-66.

DIERS, L. (1965a). Electronenmikroskopische Beobachtungen zur Archegoniumentwicklung des Lebermosses *Sphaerocarpus donnellii* Aust. Die Entwicklung des jungen Archegons bis zum Stadium der fertig ausgebildeten sekundären Zentralzelle. Planta 66, 165-190.

DIERS, L. (1965b). Electronenmikroskopische Untersuchungen über die Eizellbildung und Eizellreifung des Lebermooses, *Sphaerocarpus donnellii* Aust. Z. Naturf. 206, 795-800.

DIJKSTRA, J. (1962). On the early stages of infection by tobacco mosaic virus in *Nicotiana glutinosa* L. Virology 18, 142-143.

DIXON, P.S. (1963). The taxonomic implications of the "pit connexions" reported in the Bangiophycidae. Taxon. 12, 108-110.

DIXON, P.S. (1973). "Biology of the Rhodophyta". Oliver and Boyd, Edinburgh.

DOLZMANN, P. (1964). Sekundärer Verschluss von Plasmodesmen bei der Bildung von Intercellularräumen. Planta 63, 99-102.

DOLZMANN, P. (1965). Elektronenmikroskopische Untersuchungen an den Saughaaren von *Tillandsia usneoides*. II. Einige Beobachtungen zur Feinstruktur der Plasmodesmen. Planta 64, 76-80.

DOLZMANN, R. and DOLZMANN, P. (1964). Untersuchungen über die Feinstruktur und die Funktion der Plasmodesmen von *Volvox aureus*. Planta 61, 332-345.

DÖRING, H. (1932). Beiträge zur Frage der Hitzeresistenz pflanzlicher Zellen. Planta 18, 403-434.

DÖRR, I. (1967). Zum Feinbau der "Hyphen" von *Cuscuta odorata* und ihren Anschluss an den Siebröhren ihrer Wirtspflanzen. Naturwissenschaften 54, 474.

DÖRR, I. (1968a). Zur Lokalisierung von Zellkontakten zwischen *Cuscuta odorata* und verschieden höheren Wirtspflanzen. Protoplasma 65, 435-448.

DÖRR, I. (1968b). Plasmatische Verbindungen zwischen artfremden Zellen. Naturwissenschaften 55, 396.

DÖRR, I. (1968c). Feinbau der Kontakte zwischen *Cuscuta*-"Hyphen" und den Siebröhren ihrer Wirtspflanze. Ber. dt. bot. Ges. 2, 24-26.

DÖRR, I. (1969). Feinstruktur intrazellular wachsender *Cuscuta*-"Hyphen". Protoplasma 67, 123-137.

DÖRR, I. (1972). Der Anschluss der *Cuscuta*-"Hyphen" an die Siebröhren ihrer Wirtspflanzen. Protoplasma 75, 167-184.

DÖRR, I. and KOLLMAN, R. (1974). Strukturelle Grundlagen des Parasitismus bei *Orobanche*. I. Wachstum der Haustorialzellen im Wirtsgewebe. Protoplasma 80, 245-259.

DÖRR, I. and KOLLMAN, R. (1975). Strukturelle Grundlagen des Parasitismus bei *Orobanche*. II. Die Differenzierung der Assimilatleitungsbahn im Haustorialgewebe. Protoplasma 83, 185-199.

DRAWERT, H. and METZNER, I. (1956). Fluorescenz- und elektronenmikroskopische Beobachtungen an *Cylindrospermum* und einigen anderen Cyanophyceen. Ber. dt. bot. Ges. 69, 291-300.

DRAWERT, H. and METZNER-KUSTER, I. (1958). Fluorescenz- und elektronenmikroskopische Untersuchungen an *Beggiatoa alba* und *Thiothrix nivea*. Arch. Mikrobiol. 31, 422-434.

DREWS, G. (1959). Beiträge zur Kenntnis der phototaktischen Reaktionen der Cyanophyceen. Arch. Protistenk 104, 389-430.

DUCKETT, J.G. (1973). An ultrastructural study of the differentiation of the spermatozoid of *Equisetum*. J. Cell Sci. 12, 95-129.

DUCKETT, J.G., BUCHANAN, J.S., PEEL, M.C. and MARTIN, M.T. (1974). An ultrastructural study of pit connections and percurrent proliferations in the red alga *Nemalion helminthoides* (Vell. in With.) Batt. New Phytol. 73, 497-507.

DUNWELL, J.M. and SUNDERLAND, N. (1974a). Pollen ultrastructure in anther cultures of *Nicotiana tabacum*. I. Early stages of culture. J. exp. Bot. 25, 352-361.

DUNWELL, J.M. and SUNDERLAND, N. (1974b). Pollen ultrastructure in anther cultures of *Nicotiana tabacum*. II. Changes associated with embryogenesis. J. exp. Bot. 25, 363-372.

DUNWELL, J.M. and SUNDERLAND, N. (1975). Pollen ultrastructure in anther cultures of *Nicotiana tabacum*. III. The first sporophytic division. J. exp. Bot. 26, 240-252.

EDELSTEIN, T. (1972). *Halosacciocolax lundii*, sp. nov., a new red alga parasitic on *Rhodymenia palmata* (L.) Grev. Br. phycol. J. 7, 249-253.

EDIDIN, M. (1974). Two-dimensional diffusion in membranes. Symp. Soc. exp. Biol. 28, 1-14.

EDIDIN, M. and FAMBROUGH, D. (1973). Fluidity of the surface of cultured muscle fibres. Rapid lateral diffusion of marked surface antigens. J. Cell Biol. 57, 27-37.

EDWARDSON, J.R. (1966). Electron microscopy of cytoplasmic inclusions in cells infected with rod shaped viruses. Am. J. Bot. 53, 359-364.

EDWARDSON, J.R. (1974). Some properties of the potato virus Y-group. Florida Agr. Exp. Sta. Monogr. No. 4, 1-398.

ELFORD, W.J. (1931). A new series of graded Collodion membranes suitable for general bacteriological use, especially in filterable virus studies. J. Path. Bact. 34, 505-526.

EPSTEIN, E. (1955). Passive permeation and active transport of ions in plant roots. Pl. Physiol. 30, 529-535.

EPSTEIN, E. (1960). Spaces, barriers and ion carriers: ion absorption by plants. Am. J. Bot. 47, 393-399.

ERVIN, E.L. and EVERT, R.F. (1967). Aspects of sieve element ontogeny and structure in *Smilax rotundifolia*. Bot. Gaz. 128, 138-144.

ESAU, K. (1948a). Some anatomic aspects of plant virus disease problems. II. Bot. Rev. 14, 413-449.

ESAU, K. (1948b). Anatomic effects of Pierce's disease and phony peach. *Hilgardia* 18, 423-482.

ESAU, K. (1963). Ultrastructure of differentiated cells in higher plants. Am. J. Bot. 50, 495-506.

ESAU, K. (1965). "Plant anatomy". 2nd edn., Wiley, New York.

ESAU, K. (1967a). Anatomy of plant virus infections. Ann. Rev. Phytopath. 5, 45-76.

ESAU, K. (1967b). Minor veins in *Beta* leaves: Structure related to function. Proc. Am. phil. Soc. 111, 219-233.

ESAU, K. (1968). "Viruses in plant hosts: form, distribution and pathologic effects" (1968 John Charles Walker Lectures). University of Wisconsin Press, Madison, Milwaukee and London.

ESAU, K. (1969). The phloem. Encyclopaedia of Plant Anatomy 5, 1-496, Borntraeger, Berlin.

ESAU, K. (1972). Cytology of sieve elements in minor veins of sugar beet leaves. New Phytol. 71, 161-168.

ESAU, K. (1973). Comparative structure of companion cells and phloem parenchyma cells in *Mimosa pudica* L. Ann. Bot. 37, 625-632.

ESAU, K. and CHEADLE, VII. (1959). Size of pores and their contents in sieve elements of dicotyledons. Proc. natn. Acad. Sci. U.S.A. 45, 156-162.

ESAU, K. and CRONSHAW, J. (1968). Endoplasmic reticulum in the sieve element of *Cucurbita*. J. Ultrastruct. Res. 23, 1-14.

ESAU, K. and GILL, R.H. (1972). Nucleus and endoplasmic reticulum in differentiating root protophloem of *Nicotiana tabacum*. J. Ultrastruct. Res. 41, 160-175.

ESAU, K. and GILL, R.H. (1973). Correlations in differentiation of protophloem sieve elements of *Allium cepa* root. J. Ultrastruct. Res. 44, 310-328.

ESAU, K. and HOEFERT, L.L. (1971a). Cytology of beet yellows virus infection in *Tetragonia*. II. Vascular elements in infected leaf. Protoplasma 72, 255-273.

ESAU, K. and HOEFERT, L.L. (1971b). Composition and fine structure of minor veins in *Tetragonia* leaf. Protoplasma 72, 237-253.

ESAU, K. and HOEFERT, L.L. (1972a). Ultrastructure of sugar beet leaves infected with beet western yellows virus. J. Ultrastruct. Res. 40, 556-571.

ESAU, K. and HOEFERT, L.L. (1972b). Development of infection with beet western yellows virus in sugarbeet. Virology 48, 724-738.

ESAU, K. and HOEFERT, L.L. (1973). Particles and associated inclusions in sugarbeet infected with curly top virus. Virology 56, 454-464.

ESAU, K., CHEADLE, V.I. and RISLEY, E.B. (1962). Development of sieve-plate pores. Bot. Gaz. 123, 233-243.

ESAU, K., CRONSHAW, J. and HOEFERT, L.L. (1967). Relation of beet yellows virus to the phloem and to translocation in the sieve tube. J. Cell Biol. 32, 71-87.

ESCHRICH, W. (1953). Beiträge zur Kenntnis der Wundsiebrohrenentwicklung bei *Impatiens holstii*. Planta 43, 37-74.

ESCHRICH, W. and STEINER, M. (1968). Die submikroskopische Struktur der Assimilatleitbahnen von *Polytrichum commune*. Planta 82, 321-336.

EVANS, L.T. and WARDLAW, I.F. (1966). Independent translocation of ^{14}C-labelled assimilates and of the floral stimulus in *Lolium temulentum*. Planta 68, 310-326.

EVANS, L.V. and HOLLIGAN, M.S. (1972a). Correlated light and electron microscope studies on brown algae. I. Localization of alginic acid and sulphated polysaccharides in *Dictyota*. New Phytol. 71, 1161-1172.

EVANS, L.V. and HOLLIGAN, M.S. (1972b). Correlated light and electron microscope studies on brown algae. II. Physode production in *Dictyota*. New Phytol. 71, 1173-1180.

EVANS, L.V., CALLOW, J.A. and CALLOW, M.E. (1973). Structural and physiological studies on the parasitic red alga *Holmsella*. New Phytol. 72, 393-402.

EVENARI, M., KOLLER, D. and GUTTERMAN, Y. (1966). Effects of the environment of the mother plant on germination by control of seed-coat permeability to water in *Ononis sicula* Guss. Aust. J. biol. Sci. 19, 1007-1016.

EVERT, R.F. and DESHPANDE, B.F. (1970). An ultrastructural study of cell division in the cambium. Am. J. Bot. 57, 942-961.

EVERT, R.F. and EICHHORN, S.E. (1974). Sieve-element ultrastructure in *Platycerium bifurcatum* and some other polypodiaceous ferns: the nucleus. Planta 119, 301-318.

EVERT, R.F. and MURMANIS, L. (1965). Ultrastructure of the secondary phloem of *Tilia americana*. Am. J. Bot. 52, 95-106.

EVERT, R.F., ESCHRICH, W. and EICHHORN, S.E. (1971). Sieve-plate pores in leaf veins of *Hordeum vulgare*. Planta 100, 262-267.

EVERT, R.F., ESCHRICH, W. and EICHHORN, S.E. (1973). P-protein distribution in mature sieve elements of *Cucurbita maxima*. Planta 109, 193-210.

EVERT, R.F., BORNMAN, C.H., BUTLER, V. and GILLILAND, M.G. (1973). Structure and development of sieve areas in leaf veins of *Welwitschia*. Protoplasma 76, 23-34.

EVRARD, T.O. and CHAPPELL, W.E. (1967). "Translocation of growth regulators in *Chara vulgaris*". Water Resources Research Centre. Virginia Polytech. Inst., Blacksburg, Va.

EWART, A.J. (1903). "On the physics and physiology of protoplasmic streaming in plants". Oxford University Press, Oxford.

EYMÉ, J. and SUIRE, C. (1969). Ultrastructure des cellules sporogènes des Mousses: observations sur le plastidome et le chondriome. C.r.hebd.Séanc. Acad. Sci., Paris. Sér. D. 268, 290-293.

FABBRI, F. and PALANDRI, M. (1968). Indagini al microscopico ottico ed elettronico su apici vegetativi quiescenti in talee di *Psilotum nudum* (L) Beauv. Giorn. bot. Ital. 102, 33-53.

FAHN, A. (1967). "Plant anatomy". Pergamon Press, Oxford.

FALK, H. and SITTE, P. (1963). Zellfeinbau bei Plasmolyse. I. Der Feinbau der *Elodea*-Blattzellen. Protoplasma 57, 290-303.

FAWCETT, D.W. (1961). Intercellular bridges. Expl. Cell Res. Suppl. 8, 174-187.

FAY, P. (1973). The Heterocyst, *in* "The Biology of Blue-Green Algae" (Eds. N.G. Carr and B.A. Whitton), Botanical Monographs 9, Blackwell Scientific Publications, Oxford.

FELLOWS, R.J. and GEIGER, D.R. (1974). Structural and physiological changes in sugar beet leaves during sink to source conversion. Pl. Physiol. 54, 877-885.

FINDLAY, N. and MERCER, F.V. (1971a). Nectar production in *Abutilon*. I. Movement of nectar through the cuticle. Aust. J. biol. Sci. 24, 647-656.

FINDLAY, N. and MERCER, F.V. (1971b). Nectar production in *Abutilon*. II. Submicroscopic structure of the nectary. Aust. J. biol. Sci. 24, 657-664.

FINDLAY, N., REED, M.L. and MERCER, F.V. (1971). Nectar production in *Abutilon*. III. Sugar secretion. Aust. J. biol. Sci. 24, 665-675.

FINERAN, B.A. and LEE, M.S.L. (1975). Origin of quadrifid and bifid hairs in the trap of *Utricularia monanthos*. Protoplasma 84, 43-70.

FISCHER, R.A. (1972). Potassium accumulation by stomata of *Vicia faba*. Aust. J. biol. Sci. 25, 1107-1123.

FISCHER, R.A. and MacALISTER, T.J. (1975). A quantitative investigation of symplastic transport in *Chara corallina*. III. An evaluation of chemical and freeze substituting techniques in determining the *in situ* condition of the plasmodesmata. Can. J. Bot. (in press).

FISCHER, R.A., DAINTY, J. and TYREE, M.T. (1974). A quantitative investigation of symplasmic transport in *Chara corallina*. I. Ultrastructure of the nodal complex cell walls. Can. J. Bot. 52(6), 1209-1214.

FISHER, D.B. (1967). An unusual layer of cells in the mesophyll of the soybean leaf. Bot. Gaz. 128, 215-218.

FITTING, H. (1907). "Die Reizleitungsvorgänge bei den Pflanzen". Wiesbaden.

FLOYD, L., STEWART, K.D. and MATTOX, K.R. (1971). Cytokinesis and plasmodesmata in *Ulothrix*. J. Phycol. 7(4), 306-309.

FOGG, G. (1949). Growth and heterocyst production in *Anabaena cylindrica* in relation to carbon and nitrogen metabolism. Ann. Bot. 13, 241-259.

FÖRSTER, K. (1927). Die Wirkung äusserer Faktoren auf Entwicklung und Gestaltung bei *Marchantia polymorpha*. Planta 3, 325-390.

FOWKE, L.C. and PICKETT-HEAPS, J.D. (1969). Cell division in *Spirogyra*. II. Cytokinesis. J. Phycol. 5, 273-281.

FOWKE, L.C., BECH-HANSEN, C.W. and GAMBORG, O.L. (1974). Electron microscopic observations of cell regeneration from cultured protoplasts of *Ammi visnaga*. Protoplasma 79, 235-248.

FOWKE, L.C., BECH-HANSEN, C.W., CONSTABEL, F. and GAMBORG, O.L. (1974). A comparative study on the ultrastructure of cultured cells and protoplasts of soybean during cell division. Protoplasma 81, 189-203.

FOWKE, L.C., BECH-HANSEN, C.W., GAMBORG, O.L. and SHYLUK, J.P. (1973). Electron microscopic observations of cultured cells and protoplasts of *Ammi visnaga*. Am. J. Bot. 60, 304-312.

FRANK, H. von., LEFORT, M. and MARTIN, H.H. (1962). Elektronoptische und chemische Untersuchungen an Zellwänden der Blaualge *Phormidium uncinatum*. Z. Naturf. 17б, 262-268.

FRANKE, W. (1967). Mechanisms of foliar penetration of solutions. A. Rev. Pl. Physiol. 18, 281-300.

FRANKE, W. (1971). The entry of residues into plants *via* ectodesmata (ectocythodes). Residue Reviews 38, 81-115.

FRASER, T.W. and GUNNING, B.E.S. (1969). The ultrastructure of plasmodesmata in the filamentous green alga, *Bulbochaete hiloensis* (Nordst.) Tiffany. Planta 88, 244-254.

FRASER, T.W. and GUNNING, B.E.S. (1973). Ultrastructure of the hairs of the filamentous green alga *Bulbochaete hiloensis* (Nordst.) Tiffany: an apoplastidic plant cell with a well developed Golgi apparatus. Planta 113, 1-19.

FRASER, T.W. and SMITH, D.L. (1974). Young gametophytes of the fern *Polypodium vulgare* L. An Ultrastructural study. Protoplasma 82, 19-32.

FREDERICK, S.E., GRUBER, P.J. and TOLBERT, N.E. (1973). The occurrence of glycolate dehydrogenase and glycolate oxidase in green plants. An evolutionary survey. Pl. Physiol. 52, 318-323.

FREY-WYSSLING, A. and MÜHLETHALER, K. (1965). "Ultrastructural plant cytology". Elsevier, Amsterdam.

FREY-WYSSLING, A. and MULLER, H.R. (1957). Differentiation of plasmodesmata and sieve plates. J. Ultrastruct. Res. 1, 38-48.

FREY-WYSSLING, A., LÓPEZ-SÁEZ, J.F. and MÜHLETHALER, K. (1964). Formation and development of the cell plate. J. Ultrastruct. Res. 10, 422-432.

FRITSCH, F.E. (1935). Structure and reproduction of the algae. Vol. 1. Cambridge University Press.

FRITSCH, F.E. (1945). Structure and reproduction of the algae. Vol. II. Cambridge University Press.

FULCHER, R.G. and McCULLY, M.E. (1971). Histological studies on the genus *Fucus* V. An autoradiographic and electron microscopic study of the early stages of regeneration. Can. J. Bot. 49, 161-165.

FULCHER, R.G., O'BRIEN, T.P. and LEE, J.W. (1972). Studies on the aleurone layer. I. Conventional and fluorescence microscopy of the cell wall with emphasis on phenol-carbohydrate complexes in wheat. Aust. J. biol. Sci. 25, 23-34.

FULTON, R.W. (1941). The behaviour of certain viruses in plant roots. Phytopathology 31, 575-598.

FUNK, R. (1929). Untersuchungen heteroplastischer Transplantationen bei Solanaceen. Beitr. Biol. Pfl. 17, 404-468.

FUNK, G. (1955). Beiträge zur Kenntnis der Meeresalgen von Neapel. Pubbl. Staz. zool. Napoli. 25, Suppl.

FURSHPAN, E.J. and POTTER, D.D. (1968). Low-resistance junctions between cells in embryos and tissue culture, *in* "Current Topics in Developmental Biology" (Ed. A.A. Moscona). 3, 95-127, Academic Press, N.Y.

GAFF, D.F. and HALLAM, N.D. (1974). Resurrecting desiccated plants. Roy. Soc. N. Zealand Bull. 12, 389-393.

GALMICHE, J.M. (1973). Studies on the mechanism of glycerate 3-phosphate synthesis in tomato and maize leaves. Pl. Physiol. 51, 512-519.

GAMALEI, Yu.V. (1973). The role of plasmodesmata. Tsitologiya 15, 1427-1429.

GARDINER, W. (1882a). On the continuity of protoplasm in the motile organs of leaves. Proc. R. Soc. 34, 272-274.

GARDINER, W. (1882b). Note on open communications between cells in the pulvinus of *Mimosa pudica*. Q. Jl. Micros. Sci. 22, 365-366.

GARDINER, W. (1883). The continuity of the protoplasm through the walls of vegetable cells. Phil. Trans. R. Soc. Ser. B. 174, 817-863.

GARDINER, W. (1897). The histology of the cell wall with special reference to the mode of connexion of cells. Proc. R. Soc. B. 62, 100-112.

GARDINER, W. (1900). The genesis and development of the walls and connecting threads of the plant. Proc. R. Soc. B. 66, 186-188.

GARDINER, W. (1907). On the mode of formation of the initial cell wall, the genesis and neogenesis of the connecting threads, and the methods of connecting living tissues. Proc. Camb. Phil. Soc. biol. Sci. 14, 209-210.

GARDINER, W. and HILL, A.W. (1901). The histology of the cell wall with special reference to the mode of connection of cells. I. The distribution and character of "connecting threads" in the tissues of *Pinus sylvestris* and other allied species. Phil. Trans. R. Soc. Ser. B. 194, 83-125.

GARFIELD, R.E., HENDERSON, R.M. and DANIEL, E.E. (1972). Evaluation of the pyroantimonate technique for localisation of tissue sodium. Tissue and Cell 4, 575-589.

GARRETT, R.G. (1973). Non-persistent aphid-borne viruses, *in* "Viruses and invertebrates" (Ed. A.J. Gibbs), North Holland/Elsevier: Amsterdam and New York, p. 476-492.

GEIGER, D.R. (1976). Phloem loading, *in* "Encyclopaedia of Plant Physiology", Chapter 17 in Vol. 1 (Eds. U. Heber and C.R. Stocking), New Series, Heidelberg, Springer-Verlag (in press).

GEIGER, D.R. and CATALDO, D.A. (1969). Leaf structure and translocation in sugar beet. Pl. Physiol. 44, 45-54.

GEIGER, D.R., MALONE, J. and CATALDO, D.A. (1971). Structural evidence for a theory of vein loading of translocate. Am. J. Bot. 58, 672-675.

GEIGER, D.R., GIAQUINTA, R.T., SOVONICK, S.A. and FELLOWS, R.J. (1973). Solute distribution in sugar beet leaves in relation to phloem loading and translocation. Pl. Physiol. 52, 585-589.

GEIGER, D.R., SOVONICK, S.A., SHOCK, T.L. and FELLOWS, R.J. (1974). Role of free space in translocation in sugar beet. Pl. Physiol. 54, 892-898.

GEISY, R.M. and DAY, P.R. (1965). The septal pores of *Coprinus lagopus* in relation to nuclear migration. Am. J. Bot. 52, 287-293.

GEITLER, L. (1960). Schizophyzeen, *in* "Handbuch der Pflanzenanatomie" (Eds. W. Zimmermann and P. Ozenda), Borntraeger, Berlin.

GENEVÈS, L. (1966). Sur la structure et le comportement des plastes dans le tissu Sporogène au cours de sa proliferation, chez *Hypnum rusciforme*. C.r.hebd.Séanc. Acad. Sci., Paris. Sér. D. 262, 2215-2218.

GENEVÈS, L. (1967a). Activité des dictyosomes et formations extra cytoplasmiques globulaires et fibrillaires, dans les cellules Sporogènes d'*Hypnum rusciforne* (Hypnacées) avant et pendant la meiose. C.r.hebd.Séanc. Acad. Sci., Paris. Sér. D. 265, 17-20.

GENEVÈS, L. (1967b). Evolution comparée des ultrastructures nucleaires et des ribosomes cytoplasmiques au cours de la maturation des spermatozoides de *Polytrichum formosum* (Bryacées). C.r.hebd. Séanc. Acad. Sci., Paris. Sér D. 265, 602-605.

GENEVÈS, L. (1971). Sur la présence de polysaccharides dans diverses inclusions du cytoplasme et du noyau pendant la prophase réductionelle chez l'*Hypnum rusciforme* (Hypnacées). C.r.hebd.Séanc. Acad. Sci., Paris. Sér. D. 273, 2508-2511.

GENEVÈS, L. (1972a). Aspects ultrastructuraux de la formation de l'enveloppe des spores, chez l'*Hypnum rusciforme* (Hypnacées). C.r.hebd.Séanc. Acad. Sci., Paris. 275, 197-200.

GENEVÈS, L. (1972b). Étude cytochimique de la genèse des parois intersporales pendant la méiose, chez l'*Hypnum rusciforme*. C.r. hebd.Séanc. Acad. Sci., Paris. Sér. D. 274, 2875-2878.

GENKEL, P.A. (1950). The adaptive significance of dormancy in plants. Proc. 7th Int. Bot. Congr. Stockholm, 789-790.

GENKEL, P.A. and BAKANOVA, L.V. (1966). On plasmodesmata and Hechtian strands in some algae. Fiziol. Rast. 13, 867-870.

GENKEL, P.A. and KURKOVA, E.B. (1971). Plasmodesmata of common onion (*Allium cepa*, L.) studied by electron microscopy in the dormant state. Fiziol. Rast. 18, 777-780.

GENKEL, P.A. and OKNINA, E.Z. (1948). On the state of dormancy in plants. Dokl. Acad. Nauk SSSR 62, 409-412.

GENKEL, P.A. and PRONINA, D.N. (1968). Factors underlying dehydration resistance of poikiloxerophytes. Fiziol. Rast. 15, 68-74.

GENKEL, P.A. and PRONINA, D.N. (1969). Anabiosis with desiccation of the poikiloxerophytic flowering plant, *Mirothamnus flabellifolia*. Fiziol. Rast. 16, 745-749.

GERISCH, G. (1959). Die Zellendifferenzierung bei *Pleodorina californica* Shaw und die Organization der Phytomonadinen-kolonien. Arch. Protistenk 104, 292-358.

GEROLA, F.M. and BASSI, M. (1966). An electron microscopy study of leaf vein tumours from maize plants experimentally infected with maize rough dwarf virus. Caryologia 19, 13-40.

GEROLA, F.M., BASSI, M., LOVISOLO, O. and VIDANO, C. (1966). Virus-like particles in both maize plants infected with maize-rough dwarf virus and the vector *Laodelphax striatellus* Fallen. Phytopath. Z. 56, 97-99.

GIBBONS, G.S.B. and WILKINS, M.B. (1970). Growth inhibitor production by root caps in relation to geotropic responses. Nature, Lond. 226, 558-559.

GIBBS, A.J. (1969). Plant virus classification. Adv. Virus Res. 14, 263-328.

GIBBS, A.J. and SKEHEL, J.J. (1973). Viral RNA, *in* "The ribonucleic acids" (Eds. P.R. Stewart and D.S. Letham), p. 207-242, Springer-Verlag: Berlin, Heidelberg and New York.

GIBSON, R.A. and PALEG, L.G. (1972). Lysosomal nature of hormonally induced enzymes in wheat aleurone cells. Biochem. J. 128, 367-375.

GILDER, J. and CRONSHAW, J. (1973a). Adenosine triphosphatase in the phloem of *Cucurbita*. Planta 110, 189-204.

GILDER, J. and CRONSHAW, J. (1973b). The distribution of adenosine triphosphatase activity in differentiating and mature phloem cells of *Nicotiana tabacum* and its relationship to phloem transport. J. Ultrastruct. Res. 44, 388-404.

GILES, K.L. (1971). Dedifferentiation and regeneration in Bryophytes: a selective review. N.Z. Jl. Bot. 9, 689-694.

GILL, C.C. (1974). Inclusions and wall deposits in cells of plants infected with oat necrotic mottle virus. Can. J. Bot. 52, No. 3, 621-626.

GILULA, N.B. (1974). Junctions between cells. Chapter 1 *in* "Cell communication", (Ed. R.P. Cox), p. 1-29, J. Wiley and Sons, N.Y.

GILULA, N.B., REEVES, O.R. and STEINBACH, A. (1972). Metabolic coupling, ionic coupling and cell contacts. Nature, Lond. 235, 262-265.

GIMÉNEZ-MARTÍN, G., RISUEÑO, M.C. and SOGO, J.M. (1970). Development of the vegetative cell in the pollen grain. Cytologia 35, 77-90.

GINSBURG, H. (1971). Model for iso-osmotic water flow in plant roots. J. theor. Biol. 32, 147-158.

GINSBURG, H. (1972). Analysis of plant root electropotentials. J. theor. Biol. 37, 389-412.

GINSBURG, H. and GINZBURG, B.Z. (1970a). Radial water and solute flows in roots of *Zea mays*. I. Water flow. J. exp. Bot. 21, 580-592.

GINSBURG, H. and GINZBURG, B.Z. (1970b). Radial water and solute flows in roots of *Zea mays*. II. Ion fluxes across root cortex. J. exp. Bot. 21, 593-604.

GINSBURG, H. and GINZBURG, B.Z. (1971). Evidence for active water transport in a corn root preparation. J. Membrane Biol. 4, 29-41.

GINSBURG, H. and GINZBURG, B.Z. (1974). Radial water and solute flows in roots of *Zea mays*. IV. Electrical potential profiles across the root. J. exp. Bot. 25, 28-35.

GLAUERT, A.M. (1968). Electron microscopy of lipids and membranes. Jl. R. micros. Soc. 88, 49-70.

GODINEAU, J.-C. (1968). Ultrastructure des differents tissus de l'ovule du *Crepis tectorum* L. au moment de la prophase meiotique Données sur le cytoplasme de la cellule-mère des megaspores. C.r. hebd.Séanc. Acad. Sci., Paris. Sér. D. 266, 1008-1010.

GOEBEL, K. (1897). Über Jugendformen von Pflanzen und deren künstliche Wiederhervorrufung. Sber. Kgl. Bayer. Akad. Wiss. Math.-nat. Kl. 26, 447-497.

GOLDACRE, R.J. (1958). Morphogenesis and communication between cells, *in* "2nd International Congress on Cybernetics", p. 910-923, Ass. Int. Cybernétique.

GOLDACRE, R.J. and BEAN, A.D. (1960). A model for morphogenesis. Nature, Lond. 186, 294-295.

GOLDSMITH, M.H.M. (1968). The transport of auxin. A. Rev. Pl. Physiol. 19, 347-360.

GOLDSMITH, M.H.M. (1969). Transport of plant growth regulators, *in* "Physiology of plant growth and development" (Ed. M. B. Wilkins), McGraw-Hill, London, p. 127-162.

GOLDSMITH, M.H.M. and RAY, P.M. (1973). Intracellular localisation of the active process in polar transport of auxin. Planta 111, 297-314.

GOLDSMITH, M.H.M., FERNÁNDEZ, H.R. and GOLDSMITH, T.H. (1972). Electrical properties of parenchyma cell membranes in the oat coleoptile. Planta 102, 302-323.

GOLDSTEIN, M. (1967). Colony differentiation in *Eudorina*. Can. J. Bot. 45, 1591-1596.

GOODENOUGH, D.A. and STOEKENIUS, W. (1972). The isolation of mouse hepatocyte gap junctions. J. Cell Biol. 54, 646-656.

GORIN, N. (1969). The permeability of dead plant cells for some enzymes. Meded. LandbHoogesch. Wageningen 69(4), 1-93.

GÓRSKA-BRYLASS, A. (1967). Transitory callose envelope surrounding the generative cell in pollen grains. Acta Soc. Bot. Pol. 36, 419-422.

GÓRSKA-BRYLASS, A. (1968). Occurrence of callose in gametogenesis in *Pteridophyta*. Bull. Acad. pol. Sci. Cl. II Sér. Sci. biol. 16, 757-759.

GÓRSKA-BRYLASS, A. (1969). Callose in gametogenesis in liverworts. Bull. Acad. pol. Sci. Cl. II Sér. Sci. biol. 17, 549-554.

GÓRSKA-BRYLASS, A. (1970). The "callose stage" of the generative cells in pollen grains. Grana 10, 21-30.

GORTER, C.J. (1949). Action of 2,3,5-triiodobenzoic acid on growth of root hairs. Nature, Lond. 164, 800.

GRAHAM, J., CLARKSON, D.T. and SANDERSON, J. (1974). Water uptake by the roots of marrow and barley plants. Agricultural Research Council, Letcombe Laboratory, Annual Report, 1973. 9-12.

GREULACH, V.H. (1973). "Plant Function and Structure". New York, Macmillan.

GRIFFITHS, H.J. and AUDUS, L.J. (1964). Organelle distribution in the statocyte cells of the root tip of *Vicia faba* in relation to geotropic stimulation. New Phytol. 63, 319-333.

GROOT, S.R. De and MAZUR, P. (1962). "Nonequilibrium Thermodynamics", North Holland Publishing Co., Amsterdam.

GROVE, S.N., BRACKER, C.E. and MORRÉ, D.J. (1968). Cytomembrane differentiation in endoplasmic reticulum-golgi apparatus-vesicle complex. Science, N.Y. 161, 171-173.

GULLVÅG, B.M., SKAAR, H. and OPHUS, E.M. (1974). An ultrastructural study of lead accumulation within leaves of *Rhytidiadelphus squarrosus* (Hedw.) Warnst. A comparison between experimental and environmental poisoning. J. Bryol. 8, 117-122.

GUNNING, B.E.S. and PATE, J.S. (1969). "Transfer cells": plant cells with wall ingrowths, specialized in relation to short distance transport of solutes - their occurrence, structure, and development. Protoplasma 68, 107-133.

GUNNING, B.E.S., PATE, J.S. and BRIARTY, L.G. (1968). Specialized "transfer cells" in minor veins of leaves and their possible significance in phloem translocation. J. Cell Biol. 37, C7-C12.

GUNNING, B.E.S., PATE, J.S. and GREEN, L.W. (1970). Transfer cells in the vascular system of stems: taxonomy, association with nodes, and structure. Protoplasma 71, 147-171.

GUNNING, B.E.S., PATE, J.S., MINCHIN, F.R. and MARKS, I. (1974). Quantitative aspects of transfer cell structure in relation to vein loading in leaves and solute transport in legume nodules. Symp. Soc. exp. Biol. 28, 87-126.

GUNNING, B.E.S. and STEER, M.W. (1975). "Ultrastructure and the biology of plant cells". Edward Arnold, London.

GÜNTHER, E. (1956-7). Die Nachkommenschaft von Solanaceen-Chimaeren. Flora, Jena 144, 498-517.

GUTIERREZ, M., GRACEN, V.E. and EDWARDS, G.E. (1974). Biochemical and cytological relationships in C_4 plants. Planta 119, 279-300.

GUTKNECHT, J. and DAINTY, J. (1968). Ionic relations of marine algae. Oceanogr. Mar. Biol. Ann. Rev. 6, 163-200.

HAAS, D.L. and CAROTHERS, Z.B. (1975). Some ultrastructural observations on endodermal cell development in *Zea mays* roots. Am. J. Bot. 62, 336-348.

HABERLANDT, G. (1890). "Das reizleitende Gewebesystem der Sinnpflanze". Engelmann, Leipzig.

HABERLANDT, G. (1914). "Physiological plant anatomy". Trs. from 4th Germ. edn., Macmillan, London.

HABERLANDT, G. (1922). Über Zellteilungshormone und ihre Beziehung zur Wundheilung, Befruchtung und Adventivembryonie. Biol. Zbl. 42, 145-172.

HABESHAW, D. (1969). Translocation and the control of photosynthesis in sugar beet. J. exp. Bot. 20, 64-71.

HACCIUS, B. (1971). Zur derzeitigen Situation der Angiospermen - Embryologie. Bot. Jb. 91, 309-329.

HACCIUS, B. and ENGEL, I. (1968). Plasmodesmenartige Strukturen zwischen Zellen *in vitro*-kultivierter lockerer Kalli von *Cannabis sativa* L. Naturwissenschaften, 55, 45-46.

HACCIUS, B. and LAKSHMANAN, K.K. (1969). Adventiv-Embryonen-Embryoide-Adventiv-Knospen. Ein Beitrag zur Klärung der Begriffe. Öst. bot. Z. 116, 145-158.

HACKETT, C. (1968). A study of the root system of barley. I. Effects of nutrition on two varieties. New Phytol. 67, 287-299.

HAGEDORN, H. (1960). Elektronenmikroskopische Untersuchungen an Blaualgen, Naturwissenschaften 47, 430.

HAGEDORN, H. (1961). Untersuchungen über die Feinstruktur der Blaualgenzellen. Z. Naturf. 16b, 825-829.

HAGEN, F. (1918). Zur Physiologie des Spaltöffnungsapparates. Beitr. allg. Bot. 1, 260-291.

HALK, E.L. and McGUIRE, J.M. (1973). Translocation of tobacco ringspot virus in soybean. Phytopathology 63, 1291-1300.

HALL, J.L. (1969). Localization of cell surface adenosine triphosphatase activity in maize roots. Planta 85, 105-107.

HALL, J.L. and SEXTON, R. (1974). Fine structure and cytochemistry of the abscission zone cells of *Phaseolus* leaves. II. Localization of peroxidase and acid phosphatase in the separation zone cells. Ann. Bot. 38, 855-858.

HALL, T.A., ANDERSON, H. and APPLETON, T.C. (1973). The use of thin specimens for X-ray microanalysis in biology. J. Microscopy 99, 177-182.

HALLAM, N.D. and OSBORNE, D.J. (1974). Fine structure and histochemistry of ancient viable and non-viable seeds. Eighth Int. Congr. Electron Micr. Canberra, Vol. 2, 598-599.

HALPERIN, W. (1969). Morphogenesis in cell

cultures. A. Rev. Pl. Physiol. 20, 395-418.

HALPERIN, W. (1970). Embryos from somatic cells, *in* "Control mechanisms in the expression of cellular phenotypes". Symp. Int. Soc. Cell Biol. No. 9 (Ed. H.A. Padykula), Academic Press, N.Y.

HALPERIN, W. and JENSEN, W.A. (1967). Ultrastructural changes during growth and embryo development in carrot cell cultures. J. Ultrastruct. Res. 18, 428-443.

HAMNER, K.C. and BONNER, J. (1938). Photoperiodism in relation to hormones as factors in floral initiation and development. Bot. Gaz. 100, 388-411.

HAPPEL, J. and BRENNER, H. (1965). "Low Reynolds number hydrodynamics". Prentice-Hall, N.J.

HARRISON, B.D., FINCH, J.T., GIBBS, A.J., HOLLINGS, M., SHEPHERD, R.J., VALENTA, V. and WETTER, C. (1971). Sixteen groups of plant viruses. Virology 45, 356-363.

HARRISON-MURRAY, R.S. and CLARKSON, D.T. (1973). Relationships between structural development and the absorption of ions by the root system of *Cucurbita pepo*. Planta 114, 1-16.

HARTMANN, T. and ESCHRICH, W. (1969). Stofftransport in Rotalgen. Planta 85, 303-312.

HASHIMOTO, T., KISHI, T. and YOSHIDA, N. (1964). Demonstration of micropores in fungal crosswall. Nature, Lond. 202, 1353.

HASHIMOTO, T., MORGAN, J. and CONTI, S.F. (1973). Morphogenesis and ultrastructure of *Geotrichum candidum* septa. J. Bact. 116, 447-455.

HATCH, M.D. (1971a). The C_4-pathway of photosynthesis. Evidence for an intermediate pool of carbon dioxide and the identity of the donor C_4-dicarboxylic acid. Biochem. J. 125, 425-432.

HATCH, M.D. (1971b). Mechanism and function of the C_4-pathway of photosynthesis, *in* "Photosynthesis and Photorespiration" (Eds. M.D. Hatch, C.B. Osmond and R.O. Slatyer), p. 139-152. New York, Wiley-Interscience.

HATCH, M.D. (1975). Photosynthesis: the path of carbon, *in* "Plant Biochemistry" (Eds. J. Bonner and J. Varner). In press. New York, Academic Press.

HATCH, M.D. and OSMOND, C.B. (1976). Compartmentation and transport in C_4 photosynthesis, *in* "Encyclopedia of plant physiology" (Eds. U. Heber and C.R. Stocking), New Series. Heidelberg, Springer Verlag. (In press).

HATCH, M.D. and SLACK, C.R. (1970). The C_4-carboxylic acid pathway of photosynthesis, *in* "Progress in Phytochemistry" (Eds. L. Reinhold and Y. Liwschitz), p. 35-106. London, Wiley-Interscience.

HATCH, M.D., OSMOND, C.B. and SLATYER, R.O. (1971). "Photosynthesis and photorespiration". New York, Wiley-Interscience.

HATTA, T. and MATTHEWS, R.E.F. (1974). The sequence of early cytological changes in Chinese cabbage leaf cells following systemic infection with turnip yellow mosaic virus. Virology 59, 383-396.

HAVELANGE, A., BERNIER, G. and JACQMARD, A. (1974). Descriptive and quantitative study of ultrastructural changes in the apical meristem of mustard in transition to flowering. II. The Cytoplasm, mitochondria and proplastids. J. Cell Sci. 16, 421-432.

HAWKER, L.E. and GOODAY, M.A. (1967). Delimitation of the gametangia of *Rhizopus sexualis* (Smith) Callen. An electron microscope study of septum formation. J. gen. Microbiol. 49, 371-376.

HAWKER, L.E., GOODAY, M.A. and BRACKER, C.E. (1966). Plasmodesmata in fungal cell walls. Nature, Lond. 212, 635.

HAWKINS, E.K. (1972). Observations on the developmental morphology and fine structure of pit connections in red algae. Cytologia 37, 759-768.

HAX, W.M.A., VENROOIJ, G.E.P.M. van and VOSSENBERG, J.B.J. (1974). Cell communication: a cyclic-AMP mediated phenomenon. J. Membrane Biol. 19, 253-266.

HÉBANT, C. (1967). Sur la comparaison des tissus conducteurs des bryophytes et des plantes vasculaires. C.r.hebd. Séanc. Acad. Sci., Paris. Sér. D. 264, 901-903.

HÉBANT, C. (1969). Nouvelles observations sur le leptome de la tige feuillée des *Polytrichum*. C.r.hebd.Séanc. Acad. Sci., Paris. Sér. D. 269, 2530-2533.

HÉBANT, C. (1970a). Aspects infrastructuraux observés au cours de la différenciation du phloème (leptome) dans la tige feuillée de quelques mousses Polytrichales. C.r.hebd.Séanc. Acad. Sci., Paris, Sér. D. 271, 1361-1363.

HÉBANT, C. (1970b). A new look at the conducting tissues of mosses (Bryopsida): their structure, distribution, and significance. Phytomorphology 20, 390-410.

HÉBANT, C. (1972). Observations sur les traces foliaires des Mousses *s. str* (Bryopsida) II. Étude, chez quelques Polytrichales, des éléments à caractères "phloèmiens". Nova Hedwigia 23, 735-766.

HÉBANT, C. (1974). Polarized accumulations of endoplasmic reticulum and other ultrastructural features of leptoids in *Polytrichadelphus magellanicus* gametophytes. Protoplasma 81, 375-382.

HEITZMANN, J. (1873). Untersuchungen über das Protoplasma. Sber. Akad. Wiss. Wien. 67, see Wilson, E.B. The Cell in Development and Heredity, 3rd Ed. 1925. London, Macmillan.

HELDER, R.J. (1967). Translocation in *Vallisneria spiralis*, *in* "Handbook of Plant Physiology" (Ed. W. Ruhland), p. 30-43, Springer, Berlin 13.

HELDER, R.J. (1975). Polar potassium transport and electrical potential difference across the leaf of *Potamogeton lucens*, L. Proc. K. ned. Akad. Wet. Ser. C. 78, 189-197.

HELDER, R.J. and BOERMA, J. (1969). An electron microscopical study of the plasmodesmata in the roots of young barley seedlings. Acta bot. neerl. 18, 99-107.

HELLEBUST, J.A. and HAUG, A. (1972). Photosynthesis, translocation and alginic acid synthesis in *Laminaria digitata* and *Laminaria hyperborea*. Can. J. Bot. 50, 169-176.

HEPLER, P.K. and FOSKET, D.E. (1971). The role of microtubules in vessel member differentiation in *Coleus*. Protoplasma 72, 213-236.

HEPLER, P.K. and JACKSON, W.T. (1968). Microtubules and early stages of cell-plate formation in the endosperm of *Haemanthus katherinae* Baker. J. Cell Biol. 38, 437-446.

HEPLER, P.K. and NEWCOMB, E.H. (1967). Fine structure of cell plate formation in the apical meristem of *Phaseolus* roots. J. Ultrastruct. Res. 19, 498-513.

HEPLER, P.K. and PALEVITZ, B.A. (1974). Microtubules and microfilaments. A. Rev. Pl. Physiol. 25, 309-362.

HERSE, F. (1908). Beiträge zur Kenntnis der histologischen Erscheinungen bei der Veredelung der Obstbäume. Landw. Jbr. 37, 71-136.

HESLOP-HARRISON, J. (1964). Cell walls, cell membranes and cytoplasmic connections during meiosis and pollen development, *in* "Pollen physiology and fertilization" (Ed. H.F. Linskens). North Holland Pub. Co., Amsterdam.

HESLOP-HARRISON, J. (1966a). Cytoplasmic continuities during spore formation in flowering plants. Endeavour 25, 65-72.

HESLOP-HARRISON, J. (1966b). Cytoplasmic connections between angiosperm meiocytes. Ann. Bot. 30, 221-230.

HESLOP-HARRISON, J. (1967). Differentiation. A. Rev. Pl. Physiol. 18, 325-348.

HESLOP-HARRISON, J. (1968). Synchronous pollen mitosis and the formation of the generative cell in massulate orchids. J. Cell Sci. 3, 457-465.

HESLOP-HARRISON, J. (1971). Wall pattern formation in angiosperm microsporogenesis. Symp. Soc. exp. Biol. 25, 277-300.

HESLOP-HARRISON, J., KNOX, R.B. and HESLOP-HARRISON, Y. (1974). Pollen-wall proteins; exine-held fractions associated with the incompatibility response in Cruciferae. Theor. Appl. Genetics 44, 133-137.

HESS, D. (1972). Transformationen an höheren Organismen. Naturwissenschaften 59, 364-355.

HESS, W.M., HANSEN, D.J. and WEBER, D.J. (1975). Light and electron microscopy localization of chloride ions in cells of *Salicornia pacifica* var. *utahensis*. Can. J. Bot. 53, 1176-1187.

HICK, T. (1885). Protoplasmic continuity in the Fucaceae. J. Bot., Lond. 23, 97-102.

HICKS, G.S. and STEEVES, T.A. (1973). Plasmodesmata in the shoot apex of *Osmunda cinnamomea*. Cytologia 38, 449-453.

HIEBERT, E. and McDONALD, J.G. (1973). Characterization of some proteins associated with viruses in the potato Y group. Virology 56, 349-381.

HILL, A.E. and HILL, B.S. (1973). The *Limonium* salt gland: a biophysical and structural study. Int. Rev. Cytol. 35, 299-319.

HIROKI, C. and TU, J.C. (1972). Light and electron microscopy of potato virus M lesions and marginal tissue in red kidney bean. Phytopathology 62, 77-85.

HODSON, H.K. and HAMNER, K.C. (1970). Floral inducing extract from *Xanthium*. Science, N.Y. 167, 384-385.

HOEFERT, L.L. (1969a). Ultrastructure of *Beta* pollen. I. Cytoplasmic contents. Am. J. Bot. 56, 363-368.

HOEFERT, L.L. (1969b). Fine structure of sperm cells in pollen grains of *Beta*. Protoplasma 68, 237-240.

HÖFLER, K. (1933). Regenerationsvorgänge bei *Griffithsia schousboei*. Flora, Jena 127, 331-344.

HOFMEISTER, W. (1867). "Die Lehre von der Pflanzenzelle". Leipzig.

HOMÈS, J.L.A. (1958). Contribution a l' étude histologique de la greffe de *Gossypium hirsutum* L. C.r.Séanc. Soc. Biol. 152, 1205-1208.

HONDA, Y. and MATSUI, C. (1974). Electron microscopy of Cucumber mosaic virus-infected tobacco leaves showing mosaic symptoms. Phytopathology 64, 534-539.

HOOF, H.A. van (1958). An investigation of the biological transmission of a non-persistent virus. Doctoral thesis, Wageningen Agr. University, Van Putten and Oortmeijer, Alkmaar, The Netherlands.

HOPE, A.B. (1963). Ionic relations of cells of *Chara australis*. VI. Fluxes of potassium. Aust. J. biol. Sci. 16, 429-441.

HOPE, A.B. and WALKER, N.A. (1975). "The physiology of giant algal cells". Cambridge University Press, Cambridge.

HORINE, R.K. and RUESINK, A.W. (1972). Cell wall regeneration around protoplasts isolated from *Convolvulus* tissue culture. Pl. Physiol. 50, 438-445.

HORNER, H.T., LERSTEN, N.R. and BOWEN, C.C. (1966). Spore development in the liverwort *Riccardia pinguis*. Am. J. Bot. 53, 1048-1064.

HOUSE, C.R. (1974). "Water transport in cells and tissues". Ed. Arnold, London.

HOUWINK, A.L. (1935). The conduction of excitation in *Mimosa pudica*. Rec . Trav. bot. neerl. 32, 51-91.

HUBER, B. (1948). Zur Mikrotopographie der Saftströme im Transfusionsgewebe der Koniferennadel. Planta 35, 331-351.

HUGHES, J. and McCULLY, M.E. (1975). The use of optical brighteners in the study of plant structure. Stain Technol. (in press).

HUISINGA, B. and KNIJFF, A.M.W. (1974). On the functions of the Casparian strips in roots. Acta bot. neerl. 23, 171-175.

HUME, M. (1913). Connecting threads in graft hybrids. New Phytol. 12, 216-220.

HUMPHREYS, T.E. and GARRARD, L.A. (1971). Sucrose leakage from the maize scutellum: evidence for the participation of the phloem. Phytochemistry 10, 981-995.

IHLE, J.N. and DURE, L.S. III (1972). The temporal separation of transcription and translation and its control in cotton embryogenesis and germination, *in* "Plant Growth Substances 1970" (Ed. D.J. Carr), p. 216-221, Springer Verlag, Berlin.

IKUSHIMA, N. and MARUYAMA, S. (1968). The protoplasmic connection in *Volvox*. J. Protozool. 15, 136-140.

IMAMURA, S. and TAKIMOTO, A. (1957). Effect of ringing and incision given to the stem on the transmission of photoperiodic stimulus in *Pharbitis nil*. Bot. Mag., Tokyo 70, 13-22.

INAMDAR, J.A., PATEL, K.S. and PATEL, R.C. (1973). Studies on plasmodesmata in the trichomes and leaf epidermis of some Asclepiadaceae. Ann. Bot. 37, 657-660.

IREN, F. van and SPIEGEL, A. van der (1975). Subcellular localization of inorganic ions in plant cells by *in vivo* precipitation. Science, N.Y. 187, 1210-1211.

ISABURO-NAGAI (1914). Physiologische Untersuchungen über Farnprothallien. Flora, Jena 6, 281-330.

ISRAEL, H.W. and SAGAWA, Y. (1964a). Post pollination ovule development in *Dendrobium* orchids. I. Introduction. Caryologia 17, 53-64.

ISRAEL, H.W. and SAGAWA, Y. (1964b). Post pollination ovule development in *Dendrobium* orchids. II. Fine structure of the nucellar and archesporial stages. Caryologia 17, 301-316.

ISRAELACHVILI, J.N. (1973). Theoretical considerations on the asymmetric distribution of charged phospholipid molecules on the inner and outer layers of curved bilayer membranes. Biochim. biophys. Acta 323, 659-663.

ISRAELACHVILI, J.N. and MITCHELL, D.J. (1975). A model for the packing of lipids in bilayer membranes. Biochim. biophys. Acta 389, 13-19.

ITO, M. (1962). Studies on the differentiation of fern gametophytes. I. Regeneration of single cells isolated from cordate gametophytes of *Pteris vittata*. Bot. Mag., Tokyo 75, 19-28.

ITO, S., SATO, E. and LOEWENSTEIN, W.R. (1974). Studies on the formation of a permeable cell membrane junction. I. Coupling under various conditions of membrane contact. Effects of Colchicine, Cytochalasin B, Dinitrophenol. J. Membrane Biol. 19, 305-337.

JACOB, F. and NEUMANN, S. (1968). Studien an *Cuscuta reflexa* Roxb. I. Zur Funktion der Haustorien bei der Aufnahme von Saccharose. Flora Abt. A. Physiol. Biochem. Pflanze 159, 191-203.

JARVIS, P. and HOUSE, C.R. (1970). Evidence for symplasmic ion transport in maize roots. J. exp. Bot. 21, 83-90.

JENNINGS, D.H., THORNTON, J.D., GALPIN, W.M.F. and COGGINS, C.R. (1974). Translocation in Fungi. Symp. Soc. exp. Biol. 28, 139-156.

JENSEN, T.E. and VALDOVINOS, J.G. (1967). Fine structure of abscission zones. I. Abscission zones of the pedicels of tomato and tobacco flowers at anthesis. Planta 77, 298-318.

JENSEN, T.E. and VALDOVINOS, J.G. (1968). Fine structure of abscission zones. III. Cytoplasmic changes in abscising pedicels of tobacco and sunflower flowers. Planta 83, 303-313.

JENSEN, W.A., FISHER, D.B. and ASHTON, M.E. (1968). Cotton embryogenesis: the pollen cytoplasm. Planta 81, 206-228.

JERMYN, M.A. (1974). A class of lectins widespread in the seeds of higher plants. Proc. Aust. Biochem. Soc. 7, 32.

JONES, M.G.K. and DROPKIN, V.H. (1975). Cellular alterations induced in soybean roots by three endoparasitic nematodes. Physiol. Pl. Path. 5, 119-124.

JONES, M.G.K. and DROPKIN, V.H. (in press). Scanning electron microscopy of nematode-induced giant transfer cells.

JONES, M.G.K. and NORTHCOTE, D.H. (1972). Multinucleate transfer cells induced in *Coleus* roots by the root-knot nematode, *Meloidogyne arenaria*. Protoplasma 75, 381-395.

JONES, M.G.K., NOVACKY, A. and DROPKIN, V.H. (in press). Transmembrane potentials of parenchyma cells and nematode-induced transfer cells. Protoplasma.

JONES, R.L. (1972). Fractionation of enzymes of the barley aleurone layer. Evidence for a soluble mode of enzyme release. Planta 103, 95-109.

JUNGERS, V. (1930). Recherches sur les plasmodesmes chez les végétaux, I. Cellule 40, 7-82.

JUNGERS, V. (1933). Recherches sur les plasmodesmes chez les végétaux, II. Les synapses des algues rouges. Cellule 42, 7-28.

JUNIPER, B.E. (1963). Origin of plasmodesmata between sister cells of the root tips of barley and maize. Jl. R. microsc. Soc. 82, 123-126.

JUNIPER, B. (1972). Mechanisms of perception and patterns of organization in root caps, *in* "The dynamics of meristem cell populations" (Ed. M.W. Miller and C.C. Kuehnert), Plenum Press, N.Y.

JUNIPER, B.E. (in press). Junctions between plant cells. Chapter 3.3., *in* "Textbook of developmental biology" (Ed. C.F. Graham and P.F. Wareing), Blackwell, Oxford.

JUNIPER, B.E. and BARLOW, P.W. (1969). The distribution of plasmodesmata in the root tip of maize. Planta 89, 352-360.

JUNIPER, B.E. and FRENCH, A. (1970). The fine structure of the cells that perceive gravity in the root tip of maize. Planta 95, 314-329.

JUNIPER, B.E., GROVES, S., LANDAU-SCHACHAR, B. and AUDUS, L.J. (1966). Root cap and the perception of gravity. Nature, Lond. 20, 93-94.

KAMIYA, N. (1959). Protoplasmic streaming. Protoplasmatologia. VIII (Physiologie des Protoplasmas) 3a, 1-199.

KANAMURA, S. (1973). Optimal postfixation washing time for ultrastructural demonstration of glucose-6-phosphatase activity. J. Histochem. Cytochem. 21, 1086-1089.

KAO, K.N. and MICHALYUK, M.R. (1974). A method for high frequency intergeneric fusion of plant protoplasts. Planta 115, 355-367.

KARAS, I. and McCULLY, M.E. (1973). Further studies of the histology of lateral root development in *Zea mays*. Protoplasma 77, 243-269.

KARLING, J.S. (1929). The lactiferous system of *Achras zapota* L.1. A preliminary account of the origin, structure, and distribution of the latex vessels in the apical meristem. Am. J. Bot. 16, 803-824.

KARZEL, R. (1926). Uber die Nachwirkungen der Plasmolyse. Jb. wiss. Bot. 65, 551-591.

KASSANIS, B. (1967). Plant tissue culture, *in* "Methods in virology 1." (Eds. K. Maramorosch and H. Koprowski), p. 537-566, Academic Press, N.Y.

KASSANIS, B., TINSLEY, T.W. and QUAK, F. (1958). The inoculation of tobacco callus tissue with tobacco mosaic virus. Ann. appl. Biol. 46, 11-19.

KATER, S.B. and NICHOLSON, C. (1973). "Intracellular staining techniques in neurobiology". Springer, Berlin.

KATO, Y. (1964). Consequences of ultraviolet radiation on the differentiation and growth of fern gametophytes. New Phytol. 63, 21-27.

KAUFMAN, P.B., PETERING, L.B., YOCUM, C.S. and BAIC, D. (1970a). Ultrastructural studies on stomata development in internodes of *Avena sativa*. Am. J. Bot. 57, 33-49.

KAUFMANN, P.B., PETERING, L.B. and SMITH, J.G. (1970b). Ultrastructural development of cork-silica cell pairs in *Avena* internodal epidermis. Bot. Gaz. 131, 173-185.

KEDEM, O. and KATCHALSKY, A. (1958). Thermodynamic analysis of the permeability of biological membranes to non-electrolytes. Biochim. biophys. Acta 27, 229-246.

KEENAN, D.W. and MORRÉ, D.J. (1970). Phospholipid class and fatty acid composition of golgi apparatus isolated from rat liver and comparison with other cell fractions. Biochemistry, N.Y. 9, 19-24.

KIENITZ-GERLOFF, F. (1891). Die Protoplasmaverbindungen zwischen benachbarten Gewebeselementen in der Pflanze. Bot. Ztg. 49, 1-10, 17-26, 32-46, 48-60, 64-74.

KIENITZ-GERLOFF, F. (1902). Neue Studien über Plasmodesmen. Ber. dt. bot. Ges. 20, 93-117.

KIM, K.S. and FULTON, J.P. (1971). Tubules with viruslike particles in leaf cells infected with bean pod mottle virus. Virology 43, 329-337.

KIM, K.S. and FULTON, J.P. (1973). Plant virus-induced cell wall overgrowth and associated membrane elaboration. J. Ultrastruct. Res. 45, 328-342.

KIM, K.S. and FULTON, J.P. (1975). An association of plant cell microtubules and virus particles. Virology 64, 560-565.

KIM, S.H., QUIGLEY, G., SUDDATH, F.L., McPHERSON, A., SNEDEN, D., KIM, J.J., WEINZIERL, J., BLATTMANN, P. and RICH, A. (1972). The three dimensional structure of yeast phenylalanine transfer RNA: shape of the molecule at 5.5Å resolution. Proc. natn. Acad. Sci. U.S.A. 69, 3746-3750.

KIRK, B.T. and SINCLAIR, J.B. (1966). Plasmodesmata between hyphal cells of *Geotrichum candidum*. Science, N.Y. 153, 1646.

KIRK, J.T.O. and TILNEY-BASSETT, R.A.E. (1967). "The Plastids". Freeman and Co., London.

KITAJIMA, E.W. and LAURITIS, J.A. (1969). Plant virions in plasmodesmata. Virology 37, 681-684.

KITAJIMA, E.W., LAURITIS, J.A. and SWIFT, H. (1969). Fine structure of *Zinnia* leaf tissues infected with *Dahlia* mosaic virus. Virology 39, 240-249.

KITE, G.L. (1915). Studies on the permeability of the internal cytoplasm of animal and plant cells. Am. J. Physiol. 37, 282-299.

KLASOVA, A. (1975). Intercellular communication in maize root endodermis. Biologia Plantarum 17, 79-85.

KLEBS, G. (1884). Über die neueren Forschungen betreffs der Protoplasmaverbindungen benachbarter Zellen. Bot. Ztg. 42, 443-448.

KLEBS, G. (1887). Beiträge zur Physiologie der Pflanzenzelle. Ber. dt. bot. Ges. 5, 181-188.

KLEBS, G. (1888). Beiträge zur Physiologie der Pflanzenzelle. Unters. Bot. Inst. Tübingen 2, 489-568.

KOHL, F.G. (1897). Die Protoplasmaverbindungen der Spaltöffnungsschliesszellen und der Moosblätter . Bot. Cbl. 72, 257-265.

KOK, A.C.A. (1933). Über den Transport körperfremder Stoffe durch parenchymatisches Gewebe. Recl. Trav. bot. neerl. 30, 23-139.

KOLLMANN, R. and DÖRR, I. (1969). Strukturelle Grundlagen des zwischenzelligen Stoffaustausches. Ber. deutsch bot. Ges. 82, 415-425.

KOLLMAN, R. and SCHUMACHER, W. (1959). Über die Feinstruktur des Phloems von *Metasequoia glyptostroboides* und seine jahreszeitlichen Veränderungen. I. Das Ruhephloem. Planta 57, 583-607.

KOLLMANN, R. and SCHUMACHER, W. (1962). Über die Feinstruktur des Phloems von *Metasequoia glyptostroboides* und seine jahreszeitlichen Veränderungen. II. Mitt. Vergleichende Untersuchungen der Plasma - tischenVerbindungsbrücken in Phloemparenchymzellen und Siebzellen. Planta 58, 366-386.

KOLLMANN, R. and SCHUMACHER, W. (1963). Über die Feinstruktur des Phloems von *Metasequoia glyptostroboides* und seine jahreszeitlichen Veränderungen. IV. Mitt. Weitere Beobachtungen zum Feinbau der Plasmabrücken in den Siebzellen. Planta 60, 360-389.

KOLODNY, G.M. (1974). Transfer of macromolecules between cells in contact. Ch. 5, *in* "Cell communication" (Ed. R.P. Cox), p. 97-111, J. Wiley and Sons, N.Y.

KOLOMIEC, P.T. (1955). Convex plasmolysis as a criterion of the state of dormancy and growth in red clover. Fiziol. Rast. 2, 141-147.

KOMNICK, H. (1962). Elektronenmikrosko - pische Lokalisationvon Na^+ und Cl^- in Zellen und Geweben. Protoplasma 55, 414-418.

KOMNICK, H. and ABEL, J.H. (1971). Location and fine structure of the chloride cells and their porous plates in *Callibaetis* spec. (Ephemeroptera, Baetidae). Cytobiologie 4, 467-479.

KOMNICK, H. and BIERTHER, M. (1969). Zur histochemischen Ionenlokalisation mit Hilfe des Elektronenmikroskopie unter besonderer Berücksichtigung der Chloridreaktion. Histochemie 18, 337-362.

KOMNICK, H. and STOCKEM, W. (1973). The porous plates of coniform chloride cells in mayfly larvae. High resolution analysis and demonstration of solute pathways. J. Cell Sci. 12, 665-681.

KOMNICK, H. and WOHLFARTH-BOTTERMANN, K.E. (1964). Morphologie des Cytoplasmas. Fortschr. Zool. 17, 1-154.

KOMNICK, H., RHEES, R.W. and ABEL, J.H. (1972). The function of ephemerid chloride cells. Histochemical autoradiographic and physiological studies with radioactive chloride on *Callibaetis*. Cytobiologie 5, 65-82.

KONAR, R.N., THOMAS, E. and STREET, H.E. (1972). Origin and structure of embryoids arising from epidermal cells of the stem of *Ranunculus sceleratus* L. J. Cell Sci. 11, 77-93.

KOPECKÁ, M., GABRIEL, M. and NEČAS, O. (1974). A method of isolating anucleated yeast protoplasts unable to synthesize the glucan fibrillar component of the wall. J. gen. Microbiol. 81, 111-120.

KORN, R.W. (1969). A stochastic approach to the development of *Coleochaete*. J. theor. Biol. 24, 147-168.

KREGER-VAN RIJ, N.J.W. and VEENHUIS, M. (1969a). Septal pores in *Endomycopsis platypodis* and *Endomycopsis monospora*. J. gen. Microbiol. 57, 91-96.

KREGER-VAN RIJ, N.J.W. and VEENHUIS, M. (1969b). A study of vegetative reproduction in *Endomycopsis platypodis* by electron microscopy. J. gen. Microbiol. 58, 341-346.

KREGER-VAN RIJ, N.J.W. and VEENHUIS, M. (1971). A comparative study of the cell wall structure of basidiomycetous and related yeasts. J. gen Microbiol. 68, 87-95.

KREGER-VAN RIJ, N.J.W. and VEENHUIS, M. (1972). Some features of vegetative and sexual reproduction in *Endomyces* species. Can. J. Bot. 50, 1691-1695.

KRENKE, N.P. (1933). "Wundkompensation, Transplantationen und Chimaeren bei Pflanzen". Springer Verlag, Berlin.

KRUATRACHUE, M. and EVERT, R.F. (1974). Structure and development of sieve elements in the leaf of *Isoetes muricata*. Am. J. Bot. 61, 253-266.

KRULL, R. (1960). Untersuchungen über den Bau und die Entwicklung der Plasmodesmen im Rindenparenchym von *Viscum album*. Planta 55, 598-629.

KUGRENS, P. and WEST, J.A. (1972a). Ultrastructure of spermatial development in the parasitic red alga *Levringiella gardneri* and *Erythrocystis saccata*. J. Phycol. 8, 331-343.

KUGRENS, P. and WEST, J.A. (1972b). Ultrastructure of tetrasporogenesis in the parasitic red alga *Levringiella gardneri* (Setchell) Kylin. J. Phycol. 8, 370-383.

KUHLA, F. (1900). Die Plasmaverbindungen bei *Viscum album*. Bot. Ztg. 58, 29-58.

KUO, J. and O'BRIEN, T.P. (1974a). Lignified sieve elements in the wheat leaf. Planta 117, 349-353.

KUO, J. and O'BRIEN, T.P. (1974b). Development of the suberized lamella in the mestome sheath of wheat leaves. Eighth Int. Congr. Electron Micr. Vol. 2, 604-605.

KUO, J., O'BRIEN, T.P. and CANNY, M.J. (1974). Pit field distribution, plasmodesmatal frequency and assimilate flux in the mestome sheath cells of wheat leaves. Planta 121, 97-118.

KUO, J., O'BRIEN, T.P. and ZEE, S.-Y. (1972). The transverse veins of the wheat leaf. Aust. J. biol. Sci. 25, 721-737.

KURKOVA, E.B., VAKHMISTROV, D.B. and SOLOVYEV, V.A. (1974). Ultrastructure of some cells in the barley root as related to the transport of substances, *in* "Structure and Function of Primary Root Tissues (Ed. J. Kolek), p. 75-86, Bratislava, Slovak Acad. Sci.

KURSANOV, A.L. and BROVCHENKO, M.I. (1970). Sugars in the free space of leaves: their origin and possible involvement in transport. Can. J. Bot. 48, 1243-1250.

KÜSTER, E. (1916). "Pathologische Pflanzenanatomie". 2nd Edition, G. Fischer, Jena.

KÜSTER, E. (1933). Die Plasmodesmen von *Codium*. Protoplasma 19, 335-349.

KÜSTER, E. (1939). Über Membranbildung an kontrahierte Protoplasten höherer Pflanzen. Z. wiss. Mikrosk. 56, 63-65.

KÜSTER, E. (1956). "Die Pflanzenzelle". 3rd Edition, G. Fischer, Jena.

LAETSCH, W.M. (1971). Chloroplast structural relationships in leaves of C_4 plants, *in* "Photosynthesis and Photorespiration" (Eds. M.D. Hatch, C.B. Osmond and R.O. Slatyer), p. 323-349, New York, Wiley-Interscience.

LAETSCH, W.M. (1974). The C_4 syndrome: a structural analysis. A. Rev. Pl. Physiol. 25, 27-52.

LAMBERTZ, P. (1954). Untersuchungen über das Vorkommen von Plasmodesmen in den Epidermis - aussenwänden. Planta 44, 147-190.

LAMONT, H.C. (1969). Sacrificial cell death and trichome breakage in an Oscillatoriacean blue-green alga: The role of murein. Arch. Mikrobiol. 69, 237-259.

LAMPORT, D.T.A. (1970). Cell wall metabolism. A. Rev. Pl. Physiol. 21, 235-270.

LAMPORT, D.T.A. (1974). The role of hydroxyproline-rich proteins in the extracellular matrix of plants, *in* "Macromolecules Regulating Growth and Development" (Eds. E.D. Hay, T.J. King and J. Papaconstantinou), 13th Symp. Soc. for Developmental Biol., Academic Press, N.Y.

LANDRÉ, P. (1972). Origine et développement des epidermes cotyledonaires et foliaires de la moutarde (*Sinapis alba* L.). Différentiation ultrastructurale des stomates. Annls. Sci. nat. (Bot.) 12 Sér. 13, 247-322.

LANG, N.J. (1968). The fine structure of blue-green algae. A. Rev. Microbiol. 22, 15-46.

LANG, N.J. and FAY, P. (1971). The heterocysts of blue-green algae. II. Details of Ultrastructure. Proc. R. Soc. Lond. B. 178, 193-203.

LATIES, G.G. (1959). Active transport of salt into plant tissues. A. Rev. Pl. Physiol. 10, 87-112.

LÄUCHLI, A. (1972). Electron probe analysis, *in* "Microautoradiography and electron probe analysis. Their application to plant physiology" (Ed. U. Lüttge), Springer Verlag, Berlin, 1972.

LÄUCHLI, A. (1973). Investigation of ion transport in plants by electron probe analysis. Principles and perspectives, *in* "Ion transport in plants" (Ed. W.P. Anderson), Academic Press, London, 1973.

LÄUCHLI, A., SPURR, A.R. and WITTKOPP, R.W. (1970). Electron probe analysis of freeze-substituted, epoxy resin embedded tissue for ion transport studies in plants. Planta 95, 341-350.

LÄUCHLI, A., STELZER, R., GUGGENHEIM, R. and HENNING, L. (1974a). Precipitation techniques as a means for intracellular ion localization by use of electron probe analysis, *in* "Microprobe analysis as applied to cells and tissues" (Eds. T.A. Hall, P. Echlin and R. Kaufmann), Academic Press, London, 1974.

LÄUCHLI, A., KRAMER, D. and STELZER, R. (1974b). Ultrastructure and ion localisation in xylem parenchyma cells of roots, *in* "Membrane Transport in Plants" (Eds. U. Zimmermann and J. Dainty), p. 363-371, Springer Verlag, 1974.

LÄUCHLI, A., KRAMER, D., PITMAN, M.G. and LÜTTGE, U. (1974c). Ultrastructure of xylem parenchyma cells of barley roots in relation to ion transport to the xylem. Planta 119, 85-99.

LAWRENCE, P.A. (1975). The structure and properties of a compartment border: the intersegmental boundary in *Oncopeltus*, *in* "Cell Patterning", p. 3-23, Ciba Fdn. Symp. 29.

LAWSON, R.H. and HEARON, S.S. (1971). The association of pinwheel inclusions with plasmodesmata. Virology 44, 454-456.

LAWSON, R.H. and HEARON, S.S. (1973). Ultrastructure of carnation etched ring virus-infected *Saponaria vaccaria* and *Dianthus caryophyllus*. J. Ultrastruct. Res. 48, 201-205.

LAWSON, R.H., HEARON, S.S. and SMITH, F.F. (1971). Development of pinwheel inclusions associated with sweet potato russet crack virus. Virology 46, 453-463.

LEDBETTER, M.C. and PORTER, K.R. (1970). "Introduction to the fine structure of plant cells". Springer-Verlag, Berlin.

LEDOUX, L. (Ed.) (1971). "Informative molecules in biological systems". North Holland Publ. Co., Amsterdam.

LEE, A.G., BIRDSALL, N.J.M. and METCALFE, J.C. (1973). Measurement of fast lateral diffusion of lipids in vesicles and in biological membranes by nuclear magnetic resonance. Biochemistry 12, 1650-1659.

LEE, P.E. (1967). Morphology of wheat striate mosaic virus and its localisation in infected cells. Virology 33, 84-94.

LEE, R.E. (1971). The pit connections of some lower red algae: ultrastructure and phylogenetic significance. Br. phycol. J. 6, 29-38.

LEFORT, M. (1963). Infrastructure plastidiale du *Pandorina morum* (Müll). C. r.hebd.Séanc. Acad. Sci., Paris. Sér. D. 256, 4717-4720.

LESHEM, B. (1973). The fine structure of intercalary meristem cells of the bulb internode in *Hordeum bulbosum* L. Z. Pflphysiol. Bd. 69, 293-298.

LEVITT, J. (1960). In defence of the plasma membrane-theory of cell permeability. Protoplasma 52, 161-163.

LEWIS, I.F. (1909). The life history of *Griffithsia bornetiana*. Ann. Bot. 23, 639-690.

LIBERMAN-MAXE, M. (1971). Étude cytologique de la différenciation des cellules criblées de *Polypodium vulgare* (Polypodiacée). J. Microscopie 12, 271-288.

LIESKE, R. (1921). Propfversuche IV. Untersuchung über die Reizleitung der Mimosen. Ber. dt. bot. Ges. 39, 353-361.

LILIEN, J.E. (1969). Toward a molecular explanation for specific cell adhesion, Chapter 6, *in* "Current Topics in Developmental Biology" Vol. 4 (Eds. A.A. Moscona and A. Monroy), Academic Press, N.Y.

LINSBAUER, K. (1926). Über Regeneration von Farnprothallien und die Frage der "Teilungsstoffe". Biol. Zbl. 46, 80-96.

LINTILHAC, P.M. (1974a). Differentiation, organogenesis, and the tectonics of cell wall orientation. II. Separation of stresses in a two-dimensional Model. Am. J. Bot. 61, 135-140.

LINTILHAC, P.M. (1974b). Differentiation, organogenesis, and the tectonics of cell wall orientation. III. Theoretical considerations of cell wall mechanisms. Am. J. Bot. 61, 230-237.

LINTILHAC, P.M. and JENSEN, W.A. (1974). Differentiation, organogenesis, and the tectonics of cell wall orientation. I. Preliminary observations on the development of the ovule in cotton. Am. J. Bot. 61, 129-134.

LIS, H. and SHARON, N. (1973). The biochemistry of plant lectins (Phytohaemagglutinins). A. Rev. Biochem. 42, 541-574.

LITTLEFIELD, L. and FORSBERG, C. (1965). Absorption and translocation of phosphorous - 32 in *Chara globularis* Thuill. Physiologia Pl. 18, 291-296.

LITZ, R.E. and KIMMINS, W.C. (1968). Plasmodesmata between guard cells and accessory cells. Can. J. Bot. 46, 1603-1605.

LIVINGSTON, L.G. (1935). The nature and distribution of plasmodesmata in the tobacco plant. Am. J. Bot. 22, 75-87.

LIVINGSTON, L.G. (1964). The nature of plasmodesmata in normal (living) plant tissue. Am. J. Bot. 51, 950-957.

LOEWENSTEIN, W.R. (1968). Communication through cell junctions. Implications in growth control and differentiation. Devl. Biol. Suppl. 2, 151-183.

LOEWENSTEIN, W.R. (1973). Membrane junctions in growth and differentiation. Fedn. Proc. Fedn. Am. Socs. exp. Biol. 32, 60-64.

LOMBARDO, G. and GEROLA, F.M. (1968). Cytoplasmic inheritance and ultrastructure of the male generative cell of higher plants. Planta 82, 105-110.

LOPEZ-ABELLA, D. and BRADLEY, R.H.E. (1969). Aphids may not acquire and transmit stylet-borne viruses while probing intercellularly. Virology 39, 338-342.

LÓPEZ-SÁEZ, J.F., GIMÉNEZ-MARTÍN, G. and RISUEÑO, M.C. (1966a). Fine structure of plasmodesm. Protoplasma 61, 81-84.

LÓPEZ-SÁEZ, J.F., RISUEÑO, M.C. and GIMÉNEZ-MARTÍN, G. (1966b). Inhibition of cytokinesis in plant cells. J. Ultrastruct. Res. 14, 85-94.

LOU, C.H. (1955). Protoplasmic continuity in plants. Acta bot. Sin. 4, 183-222.

LOU, C.H., WU, S.-H., CHANG, W.-C. and SHAO, L.-M. (1957). Intercellular movements of protoplasm as a means of translocation of organic material in garlic. Scientia Sinica 6, 139-157.

LUCY, J.A. (1964). Globular lipid micelles and cell membranes. J. theor. Biol. 7, 360-373.

LUGARDON, B. (1968). Sur l'existence de liaisons protoplasmiques entre les cellules-mères des microspores de Ptéridophytes au cours de la prophase hétérotypique. C.r.hebd.Séanc. Acad. Sci., Paris. Sér. D. 267, 593-596.

LUNDEGÅRDH, H. (1922). Zelle und Zytoplasma, *in* "Handbuch der Pflanzenanatomie" Vol. 1. Abt. 1. 1st. Teil. Zytologie (Ed. K. Linsbauer), Borntraeger, Leipzig.

LUNDEGÅRDH, H. (1941). Untersuchungen über das chemische-physikalische Verhalten der Oberfläche lebender Pflanzenzellen. Protoplasma 35, 548-587.

LUNDEGÅRDH, H. (1950). Translocation of salts and water through wheat roots. Physiologia Pl. 3, 103-151.

LUNDEGÅRDH, H. (1955). Mechanisms of absorption, transport, accumulation and secretion of ions. A. Rev. Pl. Physiol. 6, 1-24.

LÜNING, K., SCHMITZ, K. and WILLENBRINK, J. (1971). Translocation of ^{14}C-labelled assimilates in two *Laminaria* species, p. 420-425 in Proc. 7th Int. Seaweed Symp., Sapporo, Japan.

LÜTTGE, U. (1971). Structure and function of plant glands. A. Rev. Pl. Physiol. 22, 23-44.

LÜTTGE, U. (1974). Co-operation of organs in intact higher plants: a review, *in* "Membrane transport in plants" (Eds. U. Zimmermann and J. Dainty), p. 353-362, Springer-Verlag, Berlin.

LÜTTGE, U., LÄUCHLI, A., BALL, E. and PITMAN, M.G. (1974). Cycloheximide: a specific inhibitor of protein synthesis and intercellular ion transport in plant roots. Experientia 30, 470-471.

LUTZ, R.W. and SJOLUND, R.D. (1973). Development of the generative cell wall in *Monotropa uniflora* L. pollen. Pl. Physiol. 52, 498-500.

McBRIDE, G.E. (1970). Cytokinesis and ultrastructure in *Fritschiella tuberosa* Iyengar. Arch. Protistenk 112, 365-375.

MacCALLUM, A.B. (1905). On the nature of the silver reaction in animal and vegetable tissues. Proc. R. Soc. B 76, 217-229.

McCULLY, M.E. (1968). Histological studies on the genus *Fucus*. III. Fine structure and possible functions of the epidermal cells of the vegetative thallus. J. Cell Sci. 3, 1-16.

McDONALD, J.G. and HIEBERT, E. (1975). Characterization of the capsid and cylindrical inclusion proteins of three strains of turnip mosaic virus. Virology 63, 295-303.

McLEAN, R.J. and BOSMANN, H.B. (1975). Cell-cell interactions: Enhancement of glucosyl transferase ectoenzyme systems during *Chlamydomonas* gamete contact. Proc. natn. Acad. Sci. (Washington) 72, 310-313.

McNEIL, M., ALBERSHEIM, P., TAIZ, L. and JONES, R.L. (1975). The structure of plant cell walls. VII. Barley aleurone cells. Pl. Physiol. 55, 64-68.

McNUTT, N.S. and WEINSTEIN, R.S. (1973). Membrane ultrastructure at mammalian intercellular junctions. Prog. Biophys. molec. Biol. 26, 45-101.

MacROBBIE, E.A.C. (1964). Factors affecting the fluxes of potassium and chloride ions in *Nitella translucens*. J. gen. Physiol. 47, 859-877.

MacROBBIE, E.A.C. (1969). Ion fluxes to the vacuole of *Nitella translucens*. J. exp. Bot. 20, 236-256.

MacROBBIE, E.A.C. (1971a). Phloem translocation, facts and mechanisms: a comparative survey. Biol. Rev. 46, 429-481.

MacROBBIE, E.A.C. (1971b). Fluxes and compartmentation in plant cells. A. Rev. Pl. Physiol. 22, 75-96.

MAHLBERG, P.G. (1963). Development of non-articulated laticifers in seedling axis of *Nerium oleander*. Bot. Gaz. 124, 224-231.

MAHLBERG, P.G. and SABHARWAL, P.S. (1967). Mitosis in the non-articulated laticifer of *Euphorbia marginata*. Am. J. Bot. 54, 465-472.

MANGENOT, G. (1924). Sur les communications protoplasmiques dans l'appareil sporogène de quelques Floridées. Revue algol. 1, 376-421.

MARCHALONIS, J.J., CLARK, A.E. and KNOX, R.B. (1975). Isolation and partial characterization of pollen-stigma recognition factors in *Gladiolus*. Proc. Aust. Biochem. Soc. 8, 93.

MARCHANT, H.J. (1974). Mitosis, cytokinesis and colony formation in the green alga *Sorastrum*. J. Phycol. 10, 107-120.

MARCHANT, H.J. and PICKETT-HEAPS, J.D. (1973). Mitosis and cytokinesis in *Coleochaete scutata*. J. Phycol. 9, 461-471.

MARCHANT, H.J., PICKETT-HEAPS, J.D. and JACOBS, K. (1973). An ultrastructural study of zoosporogenesis and the mature zoospore of *Klebsormidium flaccidum*. Cytobios 8, 95-107.

MARCHANT, R. and ROBARDS, A.W. (1968). Membrane systems associated with the plasmalemma of plant cells. Ann. Bot. 32, 457-471.

MARCHASE, R.B., BARBERA, A.J. and ROTH, S. (1975). A molecular approach to retinotectal specificity, *in* "Cell Patterning", p. 315-341, Ciba Fdn. Symp. 29.

MARES, D.J., NORSTOG, K. and STONE, B.A. (1975). Early stages in the development of wheat endosperm. 1. The change from free nuclear to cellular endosperm. Aust. J. Bot. 23, 311-326.

MARKHAM, R., SMITH, K.M. and LEA, D.E. (1942). The sizes of viruses and the methods employed in their estimation. Parasitology 34, 315-352.

MARKS, I. (1973). "Ultrastructural studies of minor veins". Thesis, Queen's University of Belfast.

MARTENS, P. (1966). Du megasporange cryptogamique a l'ovule gymnospermique, *in* "Trends in plant morphogenesis" (Eds. E.G. Cutter, A. Allsopp, F. Cusick and I.M. Sussex), p. 155-169, Longmans Green and Co., London.

MARTIN, L.F. and McKINNEY, H.H. (1938). Tobacco mosaic virus concentrated in cytoplasm. Science, N.Y. 88, 458-459.

MAST, S.O. (1924). Structures and locomotion in *Amoeba proteus*. Anat. Rec. 29, 88.

MATTHEWS, R.E.F. (1970). "Plant Virology", p. 778, Academic Press: New York and London.

MATTSON, O., KNOX, R.B., HESLOP-HARRISON, J. and HESLOP-HARRISON, Y. (1974). Protein pellicle of stigmatic papillae as a probable recognition site in incompatibility reactions. Nature, Lond. 247, 298-300.

MAZIA, D., PETZELT, C., WILLIAMS, R.O. and MEZA, I. (1972). A Ca-activated ATPase in the mitotic apparatus of the sea urchin egg (isolated by a new method). Expl. Cell Res. 70, 325-332.

MEEUSE, A.D.J. (1941a). A study of intercellular relationships among vegetable cells with special reference to "sliding growth" and to "cell shape". Rec. trav. bot. neerl. 38, 18-40.

MEEUSE, A.D.J. (1941b). Plasmodesmata. Bot. Rev. 7, 249-262.

MEEUSE, A.D.J. (1957). Plasmodesmata (Vegetable kingdom). Protoplasmatologia IIA1c, 1-43.

MELCHERS, G. (1937). Die Wirkung von Genen, tiefen Temperaturen und blühenden Propfpartnern auf die Blühreife von *Hyoscyamus niger*. Biol. Zbl. 57, 568-614.

MELCHERS, G. and LANG, A. (1948). Die Physiologie der Blütenbildung. Biol. Zbl. 67, 105-174.

MENKE, W. and FRICKE, B. (1964). Beobachtungen über die Entwicklung der Archegonien von *Dryopteris filix-mas*. Z. Naturf. 19b, 520-524.

MERCER, F.V. (1956). Cytology and the

electron microscope. Proc. Linn. Soc. N.S.W. 81, 4-19.

MERCER, F.V. (1960). The submicroscopic structure of the cell. A. Rev. Pl. Physiol. 11, 1-24.

MERCER, F.V. and RATHGEBER, N. (1962). Nectar secretion and cell membrane. Fifth Int. Congr. for Electron Microscopy, Vol. 2, WW-11.

MERKENS, W.S.W., ZOETEN, G.A. de and GAARD, G. (1972). Observations on ectodesmata and the virus infection process. J. Ultrastruct. Res. 41, 397-405.

MESQUITA, J.F. (1967). Sur la perturbation de la cytodiérèse dans les cellules méristématiques des racines d'*Allium cepa* traitées par l'acénaphtène. C.r.hebd.Séanc. Acad. Sci., Paris. Sér. D. 265, 322-325.

MESQUITA, J.F. (1970). Ultrastructura do meristema radicular de *Allium cepa* L. e suas alterações induzidas por agentes mitoclásicos e radiomiméticos. Rev. Fac. Ciênc. Coimbra 44, 1-201.

MESQUITA, J.F. and MANGENOT, S. (1967). Action de la vincaleucoblastine sur les cellules méristématiques des racines d' *Allium cepa* L. C.r.hebd.Séanc. Acad. Sci., Paris. Sér. D. 265, 1917-1919.

METZNER, I. (1955). Zur Chemie und zum submikroskopischen Aufbau der Zellwände, Scheiden und Gallerten von Cyanophyceen. Arch. Mikrobiol. 22, 45-77.

MEYER, A. (1896a). Die Plasmaverbindungen und die Membran in *Volvox globator*, *aureus* and *tertius*, mit Rücksicht auf die tierische Zelle. Bot. Ztg. 54, 187-217.

MEYER, A. (1896b). Das Irrtümliche der Angaben über das Vorkommen dicker Plasmaverbindungen zwischen den Parenchymzellen einiger Filicineen und Angiospermen. Ber. dt. bot. Ges. 14, 154-158.

MEYER, A. (1896c). Das Vorkommen von Plasmaverbindungen bei den Pilzen. Ber. dt. bot. Ges. 14, 280-281.

MEYER, A. (1897). Über die Methoden zur Nachweisung der Plasmaverbindungen. Ber. dt. bot. Ges. 15, 166-172.

MEYER, A. (1914). Notiz über die Bedeutung der Plasmaverbindungen fur die Propfbastarde. Ber. dt. bot. Ges. 32, 447-456.

MEYER, A. (1920). "Morphologische und physiologische Analyse der Zelle". Jena.

MEYER, A. and SCHMIDT, E. (1910). Über die gegenseitige Beeinflussung der Symbionten heteroplastischen Transplantationen, mit besonderer Berücksichtigung der Wanderung der Alkaloide durch die Propfstellen. Flora, Jena 100, 317-397.

MEYER, D.E. (1953). Über das Verhalten einzelner isolierter Prothalliumzellen und dessen Bedeutung für Korrelation und Regeneration. Planta 41, 642-645.

MEYER, F.J. (1962). Das trophische Parenchym. A. Assimilationsgewebe.Encyclopedia of Plant Anatomy 4(7A), 1-188.

MIEHE, A. (1901). Über die Wanderungen des pflanzlichen Zellkernes. Flora, Jena 88, 105-142.

MIEHE, H. (1905). Wachstum, Regeneration und Polarität isolierter Pflanzenzellen. Ber. dt. bot. Ges. 23, 257-264.

MILLER, J.H. (1968). Fern gametophytes as experimental material. Bot. Rev. 34, 361-440.

MILLER, R.A., GAMBORG, O.L., KELLER, W.A. and KAO, K.N. (1971). Fusion and division of nuclei in multinucleated soybean protoplasts. Can. J. Genet. Cytol. 13, 347-353.

MILNE, R.G. (1966). Electron microscopy of tobacco mosaic virus in leaves of *Nicotiana glutinosa*. Virology 28, 527-532.

MINCHIN, F.R. and BAKER, D.A. (1970). A mathematical analysis of water and solute transport across the root of *Ricinus communis*. Planta 94, 16-26.

MINCHIN, F.R. and PATE, J.S. (1973). The carbon balance of a legume and the functional economy of its root nodules. J. exp. Bot. 24, 259-271.

MINCHIN, F.R. and PATE, J.S. (1974). Diurnal functioning of the legume root nodule. J. exp. Bot. 25, 295-308.

MIYAMOTO, Y. (1972). Correlation of ectodesmata and plasmodesmata numbers with susceptibility of *Nicotiana glutinosa* leaves to initial tobacco mosaic virus infection. Ann. Phytopathol. Soc. Jap. 38(1), 86-87.

MOLINE, H.E. (1973). Ultrastructure of *Datura stramonium* leaves infected with the *Physalis* mottle strain of belladonna mottle virus. Virology 56, 123-133.

MOLISCH, H. (1915). Über einige Beobachtungen an *Mimosa pudica*. Sber. Akad. Wiss. Wien. 124, Abt. 1, 507-528.

MOORE, R.T. and McALEAR, J.H. (1962). Fine structure of mycota. 7. Observations on septa of Ascomycetes and Basidiomycetes. Am. J. Bot. 49, 86-94.

MORETON, R.B., ECHLIN, P., GUPTA, G.L., HALL, T.A. and WEIS-FOGH, T. (1974). Preparation of frozen hydrated tissue sections for X-ray microanalysis in the scanning electron microscope. Nature, Lond. 247, 113-115.

MORRÉ, D.J. (1968). Cell wall dissolution and enzyme secretion during leaf abscission. Pl. Physiol. 43, 1545-1559.

MOSES, H.L. and ROSENTHAL, A.S. (1968). Pitfalls in the use of lead ion for histochemical localisation of nucleoside phosphatases. J. Histochem. Cytochem. 16, 530-539.

MOSHKOV, B.S. (1935). Transfer of photoperiodic treatments from leaves to growing points. Dokl. Akad. Nauk. SSSR 24, 489-491.

MOTHES, K. and ROMEIKE, A. (1955). Nicotin als Ursache der Unverträglichkeit von Propfungen. Flora, Jena 142, 109-131.

MOTHES, K. and ROMEIKE, A. (1958). Die Alkaloide, *in* "Encyclop. Plant Physiol." Vol. 8, p. 989-1049 (Ed. K. Mothes), Springer-Verlag, Berlin.

MUELLER, D.M.J. (1972). Observations on the ultrastructure of *Buxbaumia* protonema: plasmodesmata in the cross-walls. Bryologist 75(1), 63-68.

MUELLER, D.M.J. (1974). Spore wall formation and chloroplast development during sporogenesis in the moss, *Fissidens limbatus*. Am. J. Bot. 61, 525-534.

MÜHLDORF, A. (1937). Das plasmatische Wesen der pflanzlichen Zellbrücken. Beih. bot. Zbl. 56(A), 171-364.

MÜHLETHALER, K. (1967). Ultrastructure and formation of plant cell walls. A. Rev. Pl. Physiol. 18, 1-24.

MÜLLER, E. and BRÄUTIGAM, E. (1973). Symplastic translocation of α-aminoisobutyric acid in *Vallisneria* leaves and the action of kinetin and colchicine, *in* "Ion Transport in Plants" (Ed. W.P. Anderson), p. 555-562, Academic Press, London.

MÜLLER-STOLL, W.R. (1965). Regeneration bei niederen Pflanzen (in physiologischen Betrachtung), *in* "Encyclop. Plant Physiol." Vol. 15, p. 92-155 (Ed. A. Lang), Springer-Verlag, Berlin.

MÜNCH, E. (1930). "Die Stoffbewegung in der Pflanze". Jena, Gustav Fischer.

MURANT, A.F. and ROBERTS, I.M. (1971). Cylindrical inclusions in coriander leaf cells infected with parsnip mosaic virus. J. gen. Virol. 10, 65-70.

MURANT, A.F., ROBERTS, I.M. and GOOLD, R.A. (1973). Cytopathological changes and extractable infectivity in *Nicotiana clevelandii* leaves infected with carrot mottle virus. J. gen. Virol. 21, 269-283.

MURANT, A.F., ROBERTS, I.M. and HUTCHESON, A.M. (1975). Effects of parsnip yellow fleck virus on plant cells. J. gen. Virol. 26, 277-285.

MURMANIS, L. and EVERT, R.F. (1967). Parenchyma cells of secondary phloem in *Pinus strobus*. Planta 73, 301-318.

MURMANIS, L. and SACHS, I.B. (1969). Seasonal development of secondary xylem in *Pinus strobus*, L. Wood Sci. Technol. 3, 177-193.

MUZIK, T.J. (1958). Role of parenchyma cells in graft union in vanilla orchid. Science, N.Y. 127, 82.

MUZIK, T.J. and LaRUE, C.D. (1952). The grafting of large monocotyledonous plants. Science, N.Y. 116, 589-591.

MUZIK, T.J. and LaRUE, C.D. (1954). Further studies on the grafting of monocotyledons. Am. J. Bot. 41, 448-455.

NÄF, U. (1961). Mode of action of antheridium-inducing substances in ferns. Nature, Lond. 189, 900-903.

NÄGELI, C. and CRAMER, L. (1855). "Pflanzenphysiologische Untersuchungen No. 1". Schultheiss, Zurich.

NAKAZAWA, S. (1963). Role of the protoplasmic connections in the morphogenesis of fern gametophytes. Sci. Rep. Tôhoku Univ. Ser. IV (Biol.) 29, 247-255.

NAPP-ZINN, K. (1961). Über Ektodesmen und verwandte Erscheinungen. Ber. dt. bot. Ges. 74, 61-65.

NEMEČ, B. (1924). Methoden zum Studien der Regeneration der Pflanzen, *in* "Handbuch der biologischen Arbeitsmethoden" (Ed. E. Abderhalden), Abt XI, Teil 2.

NEUSHUL, M. and LIDDLE, L. (1968). A light- and electron-microscopic study of primary heterogeneity in the eggs of two brown algae. Am. J. Bot. 55, 1068-1073.

NEWCOMB, E.H. (1969). Plant microtubules. A. Rev. Pl. Physiol. 20, 253-288.

NEWCOMB, W. (1973). The development of the embryo sac of sunflower *Helianthus annuus* after fertilization. Can. J. Bot. 51, 879-890.

NEWCOMB, W. and FOWKE, L.C. (1973). The fine structure of the change from the free-nuclear to cellular condition in the endosperm of chickweed *Stellaria media*. Bot. Gaz. 134, 236-241.

NEWMAN, E.I. (in press). Discussion on water relations in plants. Proc. R. Soc. B.

NICHOLSON, N.L. and BRIGGS, W.R. (1972). Translocation of the photosynthate in the brown alga *Nereocystis*. Am. J. Bot. 59, 97-106.

NIELSEN, A.E. (1964). "Kinetics of precipitation". Pergamon Press, Oxford.

NIMS, R.C., HALLIWELL, R.S. and ROSBERG, D.W. (1967). Disease development in cultured cells of *Nicotiana tabacum* L. var. Samsun NN injected with tobacco mosaic virus. Cytologia 32, 224-235.

NIR, I., KLEIN, S. and POLJAKOFF-MAYBER, A. (1969). Effect of moisture stress on submicroscopic structure of maize roots. Aust. J. biol. Sci. 22, 17-33.

NIXON, H.L. (1956). An estimate of the number of tobacco mosaic virus particles in a single hair cell. Virology 2, 126-128.

NOAILLES, M.-C. (1974). Comparaison de l'ultrastructure du parenchyme des tiges et feuilles d'une mousse normalement hydratée et en cours de dessiccation (*Pleurozium schreberi* [Willd.] Mitt.). C.r.hebd.Séanc. Acad. Sci., Paris. Sér. D. 278, 2759-2762.

NORSTOG, K. (1967). Fine structure of the spermatozoid of *Zamia* with special reference to the flagellar apparatus. Am. J. Bot. 54, 831-840.

NORSTOG, K. (1972). Early development of the barley embryo: Fine structure. Am. J. Bot. 59, 123-132.

NORTHCOTE, D.H. and WOODING, F.B.P. (1966). Development of sieve tubes in *Acer pseudoplatanus*. Proc. R. Soc. B. 163, 524-537.

NORTHCOTE, D.H. and WOODING, F.B.P. (1968). The structure and function of phloem tissue. Sci. Prog., Oxf. 56, 35-58.

O'BRIEN, T.P. (1970). Further observations on hydrolysis of the cell wall in the xylem. Protoplasma 69, 1-14.

O'BRIEN, T.P. and CARR, D.J. (1970). A suberized layer in the cell walls of the bundle sheath of grasses. Aust. J. biol. Sci. 23, 275-287.

O'BRIEN, T.P. and THIMANN, K.V. (1967a). Observations on the fine structure of the oat coleoptile. II. The parenchyma cells of the apex. Protoplasma 63, 417-442.

O'BRIEN, T.P. and THIMANN, K.V. (1967b). Observations on the fine structure of the oat coleoptile. III. Correlated light and electron microscopy of the vascular tissues. Protoplasma 63, 445-477.

OELZE-KAROW, H. and MOHR, H. (1973). Quantitative correlation between spectrophotometric phytochrome assay and physiological response. Photochem. Photobiol. 18, 319-330.

OELZE-KAROW, H. and MOHR, H. (1974). Interorgan correlation in a phytochrome-mediated response in the mustard seedling. Photochem. Photobiol. 20, 127-131.

OKNINA, E.Z. (1948). On the plasmodesmata of cells in the resting condition. Dokl. Acad. Nauk. SSSR. 62, 705-708.

OLAH, L.V. and HANZELY, L. (1973). Effect of digitonin on cellular division. V. The distribution of microtubules. Cytologia 38, 55-72.

OLESEN, P. (1975). Plasmodesmata between mesophyll and bundle sheath cells in relation to the exchange of C_4-Acids. Planta 123, 199-202.

OLIVEIRA, L. and BISALPUTRA, T. (1973). Studies in the brown alga *Ectocarpus* in culture. I. General ultrastructure of the sporophytic vegetative cells. J. Submicr. Cytol. 5, 107-120.

OLIVER, J.W. (1887). Über Fortleitung des Reizes bei reizbaren Narben. Ber. dt. bot. Ges. 5, 162-169.

OPHUS, E.M. and GULLVÅG, B.M. (1974). Localization of lead within leaves of *Rhytidiadelphus squarrosus* (Hedw.) Warnst. by means of transmission electron microscopy and X-ray microanalysis. Cytobios 10, 45-58.

OSMOND, C.B. (1971) Metabolite transport in C_4 photosynthesis. Aust. J. biol. Sci. 24, 159-163.

OSMOND, C.B. (1975). Ion absorption and carbon metabolism in cells of higher plants, *in* "Encyclopedia of plant physiology" (Eds. U. Lüttge and M.G. Pitman),

New Series. Heidelberg, Springer-Verlag (in press).

OSMOND, C.B. and BJÖRKMAN, O. (1972). Simultaneous measurements of oxygen effects on net photosynthesis and glycolate metabolism in C_3 and C_4 species of *Atriplex*. Carnegie Inst. Wash. Yearbook 71, 141-148.

OSMOND, C.B. and HARRIS, B. (1971). Photorespiration during C_4-photosynthesis. Biochim. biophys. Acta 234, 270-282.

PAINE, P.L. and SCHERR, P. (1975). Drag coefficients for the movement of rigid spheres through liquid-filled cylindrical pores. Biophys. J. 15, 1087-1091.

PAINE, P.L., MOORE, L.C. and HOROWITZ, S.B. (1975). Nuclear envelope permeability. Nature, Lond. 254, 109-114.

PALEVITZ, B.A. and HEPLER, P.K. (1974). The control of the plane of division during stomatal differentiation in *Allium*. II. Drug studies. Chromosoma 46, 327-341.

PALLAGHY, C.K. (1973). Electronprobe micro-analyses of potassium and chloride in freeze-substituted leaf sections of *Zea mays*. Aust. J. biol. Sci. 26, 1015-1034.

PALLAGHY, C.K., LUTTGE, U. and WILLERT, K. von (1970). Cytoplasmic compartmentation and parallel pathways of ion uptake in plant root cells. Z. PflPhysiol. 62, 51-57.

PALLAS, J.E. Jr. and MOLLENHAUER, H.H. (1972a). Physiological implications of *Vicia faba* and *Nicotiana tabacum* guard cell ultrastructure. Am. J. Bot. 59, 504-514.

PALLAS, J.E. JR. and MOLLENHAUER, H.H. (1972b). Electron microscopic evidence for plasmodesmata in dicotyledonous guard cells. Science, N.Y. 175, 1275-1276.

PANKRATZ, H.S. and BOWEN, C.C. (1963). Cytology of blue-green algae. I. The cells of *Symploca muscarum*. Am. J. Bot. 50, 387-399.

PAOLILLO, D.J. Jr. (1969). The plastids of *Polytrichum*. II. The sporogenous cells. Cytologia 34, 133-144.

PAPPENHEINER, J.R. (1953). Passage of molecules through capillary walls. Physiol. Rev. 33, 384-423.

PARKER, B.C. (1963). Translocation in the giant kelp *Macrocystis*. Science, N.Y. 140, 891-892.

PARKER, B.C. (1964). Chemical nature of sieve tube callose in *Macrocystis*. Phycologia 4, 27-42.

PARKER, B.C. (1965). Translocation in the giant kelp *Macrocystis*. I. Rates, direction, quantity of ^{14}C-labelled products and fluorescein. J. Phycol. 1, 41-46.

PARKER, B.C. (1966). Translocation in *Macrocystis*. III. Composition of sieve tube exudate and identification of the major ^{14}C-labelled products. J. Phycol. 2, 38-41.

PARKER, B.C. and FU, M. (1965). The internal structure of the egg kelp (*Pelagophycus* species). Can. J. Bot. 43, 1293-1305.

PARKER, B.C. and HUBER, J. (1965). Translocation in *Macrocystis*. II. Fine structure of the sieve tubes. J. Phycol. 1, 172-179.

PARKER, J. (1964). Further notes on the sieve plates of *Macrocystis pyrifera*. Protoplasma 58, 681-684.

PARKER, J. and PHILPOTT, D.E. (1961). The ultrastructure of sieve plates of *Macrocystis pyrifera*. Bull. Torrey bot. Club 88, 85-90.

PASZEWSKI, A. and ZAWADZKI, T. (1974). Action potentials in *Lupinus angustifolius* L. shoots. II. Determination of the strength-duration relation and the all or nothing law. J. exp. Bot. 25, 1097-1103.

PATE, J.S. and GUNNING, B.E.S. (1969). Vascular transfer cells in angiosperm leaves. A taxonomic and morphological survey. Protoplasma 68, 135-156.

PATE, J.S. and GUNNING, B.E.S. (1972). Transfer Cells. A. Rev. Pl. Physiol. 23, 173-196.

PATE, J.S., GUNNING, B.E.S. and BRIARTY, L.G. (1969). Ultrastructure and functioning of the transport system of the leguminous root nodule. Planta 85, 11-34.

PAUL, D.C. and GOFF, C.W. (1973). Comparative effects of caffeine, its analogues and calcium deficiency on cytokinesis. Exp. Cell Res. 78, 399-413.

PAULSON, R.E. and SRIVASTAVA, L.M. (1968). The fine structure of the embryo of *Lactuca sativa*. I. Dry embryo. Can. J. Bot. 46, 1437-1445.

PEAT, A. and WHITTON, B.A. (1968). Vegetative cell structure in *Anabaenopsis* sp. Arch. Mikrobiol. 63, 170-176.

PENNY, M.G. and BOWLING, D.J.F. (1974).

A study of potassium gradients in the epidermis of intact leaves of *Commelina communis*, L. in relation to stomatal opening. Planta 119, 17-25.

PERNER, E. (1965). Elektronenmikroskopische Untersuchungen an Zellen von Embryonen im Zustand völliger Samenruhe. I. Die zelluläre Strukturordnung in der Radicula lufttrockener Samen von *Pisum sativum*. Planta 65, 334-357.

PFEFFER, W. (1877). "Osmotische Untersuchungen". Leipzig.

PFEFFER, W. (1885). Zur Kenntnis der Kontaktreize. Unters. Bot. Inst. Tübingen Vol. 1, 483-535.

PFEFFER, W. (1897). "Pflanzenphysiologie". 2nd Edition. Engelmann, Leipzig.

PFEFFER, W. (1900-1906). "The Physiology of Plants". 2nd Edition. Translated by A.J. Ewart, London: Oxford.

PFEIFFER-WELLHEIM, F.(1924). Über ein Silberimprägnierungsverfahren zur Darstellung der Plasmodesmen in einigen Endospermgeweben und bei Moosblättchen. Z. wiss. Mikrosk. 41, 324-334.

PHILLIPS, H.L. and TORREY, J.G. (1974a). The ultrastructure of the quiescent centre in the apex of cultured roots of *Convolvulus arvensis* L. Am. J. Bot. 61, 871-878.

PHILLIPS, H.L. and TORREY, J.G. (1974b). The ultrastructure of the root cap in cultured roots of *Convolvulus arvensis* L. Am. J. Bot. 61, 879-887.

PHILLIS, E. and MASON, T.G. (1933). Studies on the transport of carbohydrate in the cotton plant. III. The polar distribution of sugar in the foliage leaf. Ann. Bot. 47, 585-634.

PICKARD, B. (1973). Action potentials in higher plants. Bot. Rev. 39, 172-201.

PICKETT-HEAPS, J.D. (1967). Ultrastructure and differentiation in *Chara* sp. I. Vegetative cells. Aust. J. biol. Sci. 20, 539-551.

PICKETT-HEAPS, J.D. (1968). Ultrastructure and differentiation in *Chara* (*fibrosa*). IV. Spermatogenesis. Aust. J. Biol. Sci. 21, 655-690.

PICKETT-HEAPS, J.D. (1969). Preprophase microtubules and stomatal differentiation in *Commelina cyanea*. Aust. J. Biol. Sci. 22, 375-391.

PICKETT-HEAPS, J.D. (1970). Some ultrastructural features of *Volvox*, with particular reference to the phenomenon of inversion. Planta 90, 174-190.

PICKETT-HEAPS, J.D. (1971). Reproduction by zoospores in *Oedogonium*. I. Zoosporogenesis. Protoplasma 72, 275-314.

PICKETT-HEAPS, J.D. (1972). Variation in mitosis and cytokinesis in plant cells: its significance in the phylogeny and evolution of ultrastructural systems. Cytobios 5, 59-77.

PICKETT-HEAPS, J.D. and MARCHANT, H.J. (1972). The phylogeny of the green algae: a new proposal. Cytobios 6, 255-264.

PICKETT-HEAPS, J.D. and NORTHCOTE, D.H. (1966a). Organisation of microtubules and endoplasmic reticulum during mitosis and cytokinesis in wheat meristems. J. Cell Sci. 1, 109-120.

PICKETT-HEAPS, J.D. and NORTHCOTE, D.H. (1966b). Cell division in the formation of the stomatal complex of the young leaves of wheat. J. Cell Sci. 1, 121-128.

PIERCE, G.J. (1893). On the structure of the haustoria of some Phanerogamic parasites. Ann. Bot. 7, 291-324.

PIERRE, N. (1970). Étude des infrastructures des prothalles femelles de *Pinus nigra* et *Pinus silvestris* aux derniers stades cénocytiques, en cours de cellularisation et aux premiers stades cellulaires. C.r.hebd.Séanc. Acad. Sci., Paris. Sér. D. 270, 3207-3209.

PILET, P.-E. (1971). Root cap and georeaction. Nature, Lond. 233, 115-116.

PILET, P.-E. (1972). Root cap and root growth. Planta 106, 169-171.

PILET, P.-E. (1974). Control by the root cap on growth and georeaction of roots, *in* "Plant growth substances, 1973", p. 1104-1110. Hirokawa, Tokyo.

PISKERNIK, A. (1914). Die Plasmaverbindungen bei Moosen. Öst. bot. Z. 64, 107-120.

PITMAN, M.G. (1965a). The location of the Donnan free space in disks of beetroot tissue. Aust. J. Biol. Sci. 18, 547-553.

PITMAN, M.G. (1965b). Sodium and potassium uptake by seedlings of *Hordeum vulgare*. Aust. J. biol. Sci. 18, 10-24.

PITMAN, M.G., MERTZ, S.M., GRAVES, J.S.,

PIERCE, W.S. and HIGINBOTHAM, N. (1970). Electrical potential differences in cells of barley roots and their relation to ion uptake. Pl. Physiol. 47, 76-80.

PITTS, T.D. (1971). *In* "Direct Interaction between Cells". 3rd Lepetit Colloquium North Holland, Amsterdam.

PLOWE, J.Q. (1931a). Membranes in the plant cell. I. Morphological membranes at protoplasmic surfaces. Protoplasma 12, 196-220.

PLOWE, J.Q. (1931b). Membranes in the plant cell. II. Localization of differential permeability in the plant protoplast. Protoplasma 12, 221-240.

POCOCK, M.A. (1933). *Volvox* in South Africa. Ann. S. African Mus. 16, 523-646.

POIRAULT, G. (1893). Recherches anatomiques sur les cryptogames vasculaires. Annls. Sci. nat. (Bot.) Sér. VII. 18, 113-256.

POJNAR, E., WILLISON, J.H.M. and COCKING, E.C. (1967). Cell wall regeneration by isolated tomato-fruit protoplasts. Protoplasma 64, 460-480.

POL, P.A. van der (1972). Floral induction, floral hormones and flowering. Meded. LandbHoogesch. Wageningen 72(9), 1-89.

PORTER, K.R. and BONNEVILLE, M.A. (1968). "Fine structure of cells and tissues". Lea and Febiger, Philadelphia.

PORTER, K.R. and MACHADO, R.D. (1960a). Studies on the endoplasmic reticulum. IV. Its form and distribution during mitosis in cells of onion root tip. J. biophys. biochem. Cytol. 7, 167-180.

PORTER, K.R. and MACHADO, R.D. (1960b). The endoplasmic reticulum and the formation of plant cell walls. Proc. Eur. Reg. Conf. Electron Mic., Delft. Vol. II, 754-758.

POSTE, G. and ALLISON, A.C. (1971). Membrane fusion reaction: a theory. J. theor. Biol. 32, 165-184.

POSTE, G. and ALLISON, A.C. (1973). Membrane fusion. Biochim. biophys. Acta 300, 421-465.

POTTER, D.D., FURSHPAN, E.J. and LENNOX, E.S. (1966). Connections between cells of the developing squid as revealed by electrophysiological methods. Proc. natn. Acad. Sci. (Washington) 55, 328-336.

POUX, N. (1973). Observation en microscopie électronique de cellules végétales imprégnées par l'osmium. C.r. hebd.Séanc. Acad. Sci., Paris. Sér. D. 276, 2163-2166.

POWELL, M.J. (1974). Fine structure of plasmodesmata in a chytrid. Mycologia 66(4), 606-614.

POWER, J.B., CUMMINGS, S.E. and COCKING, E.C. (1970). Fusion of isolated plant protoplasts. Nature, Lond. 225, 1016-1018.

POWER, J.B., FREARSON, E.M. and COCKING, E.C. (1971). The preparation and culture of spontaneously fused tobacco leaf spongy-mesophyll protoplasts. Biochem. J. 123, 29P-30P.

PRAT, R. (1972). Contribution a l'étude des protoplasts végétaux. 1. Effet du traitement d'isolement sur la structure cellulaire. J. Microscopie 14, 85-114.

PRATT, L.H. and COLEMAN, R.A. (1974). Phytochrome distribution in etiolated grass seedlings as assayed by an indirect antibody-labelling method. Am. J. Bot. 61, 195-202.

PRESTON, R.D. (1974). "The physical biology of plant cell walls". Chapman and Hall, London.

PRICE, W.C. (1938). Studies on the virus of tobacco necrosis. Am. J. Bot. 25, 603-612.

PURCIFULL, D.E., HIEBERT, E. and McDONALD, J.G. (1973). Immunochemical specificity of cytoplasmic inclusions induced by viruses of the potato Y group. Virology 55, 276-279.

RAMUS, J. (1969a). Pit connection formation in the red alga *Pseudogloiophloea*. J. Phycol. 5, 57-63.

RAMUS, J. (1969b). Dimorphic pit connections in the red alga *Pseudogloiophloea*. J. Cell Biol. 41, 340-345.

RAMUS, J. (1971). Properties of septal plugs from the red alga *Griffithsia pacifica*. Phycologia 10, 99-103.

RANDLES, J.W. and FRANCKI, R.I.B. (1972). Infectious nucleocapsid particles of lettuce necrotic yellows virus with RNA dependent RNA polymerase activity. Virology 50, 297-300.

RAPPAPORT, I. and WILDMAN, S.G. (1957). A kinetic study of local lesion growth on *Nicotiana glutinosa* resulting from tobacco mosaic virus infection. Virology 4, 265-274.

RASHID, A. and STREET, H.E. (1974). Segmentations in microspores of *Nicotiana sylvestris* and *Nicotiana tabacum* which lead to embryoid formation in anther cultures. Protoplasma 80, 323-334.

RAVEN, J.A. and SMITH, F.A. (1974). Significance of hydrogen ion transport in plant cells. Can. J. Bot. 52, 1035-1048.

RAWLENCE, D.J. and TAYLOR, A.R.A. (1970). The rhizoids of *Polysiphonia lanosa*. Can. J. Bot. 48, 607-611.

REDDY, G.M. and COE, E.H. JR. (1962). Inter-tissue complementation: a simple technique for direct analysis of gene-action sequence. Science, N.Y. 138, 149-150.

RÉDEI, G.P. (1969). Genetic estimate of cellular autarky. Experientia 23, 584-586.

REED, M.L., FINDLAY, N. and MERCER, F.V. (1971). Nectar production in *Abutilon*. IV. Water and solute relations. Aust. J. biol. Sci. 24, 677-688.

REICHLE, R.E. and ALEXANDER, J.V. (1965). Multiperforate septations, Woronin bodies and septal plugs in *Fusarium*. J. Cell Biol. 24, 489-496.

REICHLE, R.E. and LICHTWARDT, R.W. (1972). Fine structure of the Trichomycete *Harpella melusinae*, from black-fly guts. Arch. Mikrobiol. 81, 103-125.

REINKE, J. (1876). Beiträge zur Kenntnis der Tange. Jb. wiss. Bot. 10, 317-382.

RENKIN, E.M. (1955). Filtration, diffusion, and molecular sieving through porous cellulose membranes. J. gen. Physiol. 38, 225-243.

RENKIN, E.M. and GILMORE, J.P. (1973). Glomerular filtration. Chapter 9, *in* "Handbook of Physiology", Section 8, p. 185-248. Amer. Physiol. Soc., Washington, D.C.

RETALLACK, B. and BUTLER, R.D. (1970). The development and structure of the zoospore vesicle in *Bulbochaete hiloensis*. Arch. Mikrobiol. 72, 223-237.

RETALLACK, B. and BUTLER, R.D. (1973). Reproduction in *Bulbochaete hiloensis* (Nordst.) Tiffany II. Sexual reproduction. Arch. Mikrobiol. 90, 343-364.

REVEL, J.P. and KARNOVSKY, M.J. (1967). Hexagonal array of subunits in intercellular junctions of the mouse heart and liver. J. Cell Biol. 33, C7-C12.

RHOADES, M.M. (1952). The effect of the bronze locus on anthocyanin formation in maize. Am. Nat. 86, 105-108.

RISUEÑO, M.C., GIMÉNEZ-MARTÍN, G. and LÓPEZ-SÁEZ, J.F. (1968). Experimental analysis of plant cytokinesis. Expl. Cell Res. 49, 136-147.

ROBARDS, A.W. (1968a). Desmotubule - a plasmodesmatal substructure. Nature, Lond. 218, 784.

ROBARDS, A.W. (1968b). A new interpretation of plasmodesmatal ultrastructure. Planta 82, 200-210.

ROBARDS, A.W. (1971). The ultrastructure of plasmodesmata. Protoplasma 72, 315-323.

ROBARDS, A.W. (1975). Plasmodesmata. A Rev. Pl. Physiol. 26, 13-29.

ROBARDS, A.W. (in press). Differentiation of plant root cells relevant to intercellular transport. Proc. Int. Symp. Plant Cell Differentiation, Lisbon, August 1974.

ROBARDS, A.W. and JACKSON, S.M. (1975). Root structure and function - an integrated approach. Proc. 50th Anniv. Meeting, Society for Experimental Biology, Cambridge, July 1974 (in press).

ROBARDS, A.W. and ROBB, M.E. (1972). Uptake and binding of uranyl ions by barley roots. Science, N.Y. 178, 980-982.

ROBARDS, A.W. and ROBB, M.E. (1974). The entry of ions and molecules into roots: an investigation using electron-opaque tracers. Planta 120, 1-12.

ROBARDS, A.W., PAYNE, H. and GUNNING, B.E.S. (in press). Isolation of the endodermis using wall-degrading enzymes. (Submitted to Cytobiologie).

ROBARDS, A.W., JACKSON, S.M., CLARKSON, D.T. and SANDERSON, J. (1973). The structure of barley roots in relation to the transport of ions into the stele. Protoplasma 77, 291-311.

ROBARDS, A.W. and KIDWAI, P. (1969). Cytochemical localization of phosphatase in differentiating secondary vascular cells. Planta 87, 227-238.

ROBERT, D. (1969). Evolution de quelques organites cytoplasmiques au cours de la maturation de l'oosphère de *Selaginella kraussiana* A. Br. C.r.hebd.Séanc. Acad. Sci.,Paris. Sér. D. 268, 2775-2778.

ROBERTS, I.M. and HARRISON, B.D. (1970). Inclusion bodies and tubular structures in *Chenopodium amaranticolor* plants infected with strawberry latent ringspot virus. J. gen. Virol. 7, 47-54.

ROBERTS, K. and NORTHCOTE, D.H. (1970). The structure of sycamore callus cells during division in a partially synchronised suspension culture. J. Cell Sci. 6, 299-321.

ROBERTSON, J.D. (1964). Unit membranes: a review with recent new studies of experimental alterations and a new subunit structure in synaptic membranes, *in* "Cellular membranes in development" (Ed. M. Locke), p. 1-81, Academic Press, New York and London.

ROBERTSON, R.N. (1950). The last haunts of demons: a comparative study of secretion and accumulation (Presidential Address). Proc. Linn. Soc. N.S.W. 75, iv-xx.

RODKIEWICZ, B. (1970). Callose in cell walls during megasporogenesis in Angiosperms. Planta 93, 39-47.

ROECKL, B. (1949). Nachweis eines Konzentrationshubs zwischen Palisadenzellen und Siebröhren. Planta 36, 530-550.

ROELOFSEN, P.A. (1959). The plant cell-wall. Encyclopedia of plant anatomy 3(4), 1-336.

ROELOFSEN, P.A. (1965). Ultrastructure of the wall in growing cells and its relation to the direction of growth. Adv. Bot. Res. 2, 69-149.

ROLAND, J.C. (1973). The relationship between the plasmalemma and the plant cell wall. Int. Rev. Cytol. 36, 45-92.

ROLAND, J.C. and VIAN, B. (1971). Réactivité du plasmalemme végétal: étude cytochimique. Protoplasma 73, 121-137.

ROSS. A.F. and ISRAEL, H.W. (1970). Use of heat treatments in the study of acquired resistance to tobacco mosaic virus in hypersensitive tobacco. Phytopathology 60, 755-769.

ROSS, M.K., THORPE, T.A. and COSTERTON, J.W. (1973). Ultrastructural aspects of shoot initiation in tobacco callus cultures. Am. J. Bot. 60, 788-795.

RUBERY, P.H. and SHELDRAKE, A.R. (1974). Carrier-mediated auxin transport. Planta 118, 101-121.

RUSSOW, E. (1883). Über die Perforation der Zellwand und den Zusammenhang der Protoplasmakörper benachbarter Zellen. Sber. Ges. naturf. Dorpat. 6, 562.

RUTTER, W.J., PICTET, R.L. and MORRIS, P.W. (1973). Toward molecular mechanisms of developmental processes. A. Rev. Biochem. 42, 601-646.

SABATINI, D.D., BENSCH, K. and BARRNETT, R.J. (1963). The preservation of cellular ultrastructure and enzymatic activity by aldehyde fixation. J. Cell Biol. 17, 19-58.

SABNIS, D.D., GORDON, M. and GALSTON, A.W. (1970). Localization of adenosine triphosphatase activity on the chloroplast envelope in tendrils of *Pisum sativum*. Pl. Physiol. 45, 25-32.

SACHS, J. (1882). "Vorlesungen über Pflanzenphysiologie". Leipzig.

SACHS, T. (1972). The pattern of plasmolysis as a criterion for intercellular relations. Israel J. Bot. 21, 90-98.

SACHSENMAIER, W., REMY, U. and PLATTNER-SCHOBEL, R. (1972). Initiation of synchronous mitosis in *Physarum polycephalum*. Expl. Cell Res. 73, 41-48.

SAMUEL, G. (1931). Some experiments on inoculating methods with plant viruses and on local lesions. Ann. appl. Biol. 18, 494-506.

SAMUEL, G. (1934). The movement of tobacco mosaic virus within the plant. Ann. appl. Biol. 21, 90-111.

SANDER, K. (1975). Pattern specification in the insect embryo, *in* "Cell Patterning". Ciba Fdn. Symp. 29, 241-263.

SANGER, J.M. and JACKSON, W.T. (1971). Fine structure study of pollen development in *Haemanthus katherinae* Baker. I. Formation of vegetative and generative cells. J. Cell Sci. 8, 289-301.

SASSEN, M.M.A. (1964). Fine structure of *Petunia* pollen grain and pollen tube. Acta bot. neerl. 13, 175-181.

SAUTER, J.J. (1974). Structure and physiology of Strasburger cells. Ber. dt. bot. Ges. 87, 327-336.

SAVILLE, D.B.O. (1955). A phylogeny of the Basidiomycetes. Can. J. Bot. 33, 60-104.

SCHAFFSTEIN, G. (1932). Untersuchungen an gegliederten Milchröhren. Beih. bot. Zbl. 49, 197-220.

SCHEER, C. van der and GROENEWEGEN, J. (1971). Structure in cells of *Vigna unguiculata* infected with cowpea mosaic virus. Virology 46, 493-497.

SCHINDLER, A.M. and IBERALL, A.S. (1973).

The need for a kinetics for biological transport. Biophys. J. 13, 804-806.

SCHMALHAUSEN, J. (1877). Beiträge zur Kenntnis der Milchsaftbehälter der Pflanzen. Mem. Acad. Imp. St. Petersbourg. Sér. 7. 24, 1-27.

SCHMITZ, K. and SRIVASTAVA, L.M. (1974a). Fine structure and development of sieve tubes in *Laminaria groenlandica*, Rosenv. Cytobiologie 10, 66-87.

SCHMITZ, K. and SRIVASTAVA, L.M. (1974b). The enzymatic incorporation of ^{32}P into ATP and other organic compounds by sieve-tube sap of *Macrocystis integrifolia*. Planta 116, 85-89.

SCHMITZ, K. and SRIVASTAVA, L.M. (1975). On the fine structure of sieve tubes and the physiology of assimilate transport in *Alaria marginata*. Can. J. Bot. 53, 861-876.

SCHMITZ, K., LÜNING, K. and WILLENBRINK, J. (1972). CO_2-fixierung und Stofftransport in benthischen marinen Algen. II. Ferntransport ^{14}C-markierter Assimilate bei *Laminaria hyperborea* und *Laminaria saccharina*. Z. Pflphysiol. 67, 418-429.

SCHNEIDER, I.R. (1965). Introduction, translocation and distribution of viruses in plants. Adv. Virus Res. 11, 163-221.

SCHNEPF, E. (1974). Septum of *Geotrichum candidum* or valve of a centric diatom? J. Bact. 119, 330-331.

SCHOOLAR, A.I. and EDELMAN, J. (1971). The site and active nature of sucrose secretion from sugar-cane leaf tissue. J. exp. Bot. 22, 809-817.

SCHOSER, G. (1956). Über die Regeneration bei den Cladophoraceen. Protoplasma 47, 103-134.

SCHUBERT, O. (1913). Bedingungen zur Stecklingsbildung und Propfung von Monokotylen. Cbl. Bakt. Abt. 2. 38, 390-443.

SCHULZ, P. and JENSEN, W.A. (1968a). *Capsella* embryogenesis: The early embryo. J. Ultrastruct. Res. 22, 376-392.

SCHULZ, P. and JENSEN, W.A. (1968b). *Capsella* embryogenesis: The egg, zygote and young embryo. Am. J. Bot. 55, 807-819.

SCHULZ, P. and JENSEN, W.A. (1969). *Capsella* embryogenesis: The suspensor and the basal cell. Protoplasma 67, 139-163.

SCHULZ, P. and JENSEN, W.A. (1971). *Capsella* embryogenesis: The chalazal proliferating tissue. J. Cell Sci. 8, 201-227.

SCHUMACHER, W. (1934). Die Absorptionsorgane von *Cuscuta odorata* und der Stoffübertritt aus den Siebröhren der Wirtspflanze. Jb. wiss. Bot. 80, 74-91.

SCHUMACHER, W. (1942). Über plasmodesmenartige Strukturen in Epidermisaussenwänden. Jb. wiss. Bot. 90, 530-545.

SCHUMACHER, W. and HALBSGUTH, W. (1939). Über den Anschluss einiger höherer Parasiten an die Siebröhren der Wirtspflanzen. Ein Beitrag zum Plasmodesmenproblem. Jb. wiss. Bot. 87, 324-355.

SCHUMACHER, W. and LAMBERTZ, P. (1956). Über die Beziehung zwischen der Stoffaufnahme durch Blattepidermen und der Zahl der Plasmodesmen in den Aussenwänden. Planta 47, 47-52.

SCHWEIDLER, J.H. (1910). Über traumatogene Zellsaft - und Kernüberschritte bei *Moricandia arvensis* D.C. Jb. wiss. Bot. 48, 540-590.

SCHWEMMLE, J. (1969). Der Einfluss des Standortes auf die Samenkeimung. Biol. Zbl. 87, 37-46.

SCOTT, F.M. (1949). Plasmodesmata in xylem vessels. Bot. Gaz. 110, 492-495.

SCOTT, F.M., SCHROEDER, M.S. and TURRELL, F.M. (1948). Development, cell shape, suberization of internal surface and abscission in the leaf of the Valencia Orange, *Citrus sinensis*. Bot. Gaz. 109, 381-411.

SCOTT, F.M., HAMNER, K.C., BAKER, E. and BOWLER, E. (1956). Electron microscope studies of cell wall growth in the onion root. Am. J. Bot. 43, 313-324.

SCOTT, T.K. and WILKINS, M.B. (1969). Auxin transport in roots. IV. Effects of light on IAA movement and geotropic responsiveness in *Zea* roots. Planta 87, 249-258.

SCOTT-BLAIR, G.W. and SPANNER, D.C. (1974). "An introduction to biorheology". Elsevier, Amsterdam.

SELIM, H.H.A. (1947). Translocation of the floral stimulus in two-branched plants of *Perilla*. Proc. K. ned. Akad. Wet. Amsterdam. Ser. C. 60, 67-75.

SEMENOVA, G.A. and TAGEEVA, S.V. (1972). Ultrastructural organization of intercellular bridges (plasmodesmata) of plant cells. Dokl. Akad. Nauk. SSSR. Ser. Biol. 202(6), 1427-1428.

SERVAITES, J.C. and GEIGER, D.R. (1974). Effects of light intensity and oxygen on photosynthesis and translocation in sugar beet. Pl. Physiol. 54, 575-578.

SETTERFIELD, G. and BAYLEY, S.T. (1957). Studies on the mechanism of deposition and extension of primary cell walls. Can. J. Bot. 35, 435-444.

SETTERFIELD, G. and BAYLEY, S.T. (1958). Arrangement of cellulose microfibrils in walls of elongating parenchyma cells. J. biophys. biochem. Cytol. 4, 377-382.

SEVERIN, H.H.P. (1924). Curly leaf transmission experiments. Phytopathology 14, 80-93.

SEXTON, R. and HALL, J.H. (1974). Fine structure and cytochemistry of the abscission zone cells of *Phaseolus* leaves. 1. Ultrastructural changes occurring during abscission. Ann. Bot. 38, 849-854.

SHALLA, T.A. (1959). Relations of tobacco mosaic virus and barley stripe mosaic virus to their host cells as revealed by ultrathin tissue-sectioning for the electron microscope. Virology 7, 193-219.

SHAW, S. and WILKINS, M.B. (1973). The source and lateral transport of growth inhibitors in geotropically stimulated roots of *Zea mays* and *Pisum sativum*. Planta 109, 11-26.

SHEETZ, M.P. and SINGER, S.J. (1974). Biological membranes as bilayer couples. A molecular mechanism of drug - erythrocyte interactions. Proc. natn. Acad. Sci. U.S.A. 71, 4457-4461.

SHEFFIELD, F.M.L. (1936). The role of plasmodesms in the translocation of virus. Ann. appl. Biol. 23, 506-508.

SHERIDAN, J.D. (1974). Electrical coupling of cells and cell communication, *in* "Cell Communication" (Ed. R.P. Cox), p. 31-42. J. Wiley and Sons, N.Y.

SHERIFF, D.W. and MEIDNER, H. (1974). Water pathways in leaves of *Hedera helix* L. and *Tradescantia virginiana*. J. exp. Bot. 25, 1147-1156.

SHIH, C.Y. and CURRIER, H.B. (1969). Fine structure of phloem cells in relation to translocation in the cotton seedling. Am. J. Bot. 56, 464-472.

SHIKATA, E. (1966). Electron microscope studies on plant viruses. J. Fac. Agric. Hokkaido (imp.) Univ. 55, 1-110.

SHIKATA, E. and MARAMOROSCH, K. (1966). An electron microscope study of plant neoplasia induced by wound tumour virus. J. natn. Cancer Inst. 36, 97-116.

SIBAOKA, T. (1962). Excitable cells in *Mimosa*. Science, N.Y. 137, 226.

SIBAOKA, T. (1966). Action potentials in plant organs. Symp. Soc. exp. Biol. 20, 49-73.

SIEGEL, A., ZAITLIN, M. and SEHGAL, O.P. (1962). The isolation of defective tobacco mosaic virus strains. Proc. natn. Acad. Sci. U.S.A. 48, 1845-1851.

SILVERSTEIN, S.C., ASTELL, C., LEVEN, D.H., SCHONBERG, M. and ACS, G. (1972). The mechanisms of Reovirus uncoating and gene activation *in vivo*. Virology 47, 797-806.

SIMON, S.V. (1930). Transplantationsversuche zwischen *Solanum melongena* und *Iresine Lindeni*. Jb. wiss. Bot. 72, 137-160.

SIMONCIOLI, C. (1974). Ultrastructural characteristics of *Diplotaxis erucoides* (L.) D.C. suspensor. Giorn. bot. Ital. 108, 175-189.

SIMONIESCU, N., SIMONIESCU, M. and PALADE, G.E. (1975). Permeability of muscle capillaries to small hemepeptides: evidence for the existence of patent transendothelial channels. J. Cell Biol. 64, 586-607.

SINGER, S.J. (1974). The molecular organization of membranes. Ann. Rev. Biochem. 43, 805-833.

SINGH, A.P. and MOGENSEN, H.L. (1975). Fine structure of the zygote and early embryo in *Quercus gambelii*. Am. J. Bot. 62, 105-115.

SINGH, A.P. and SRIVASTAVA, L.M. (1972). The fine structure of corn phloem. Can. J. Bot. 50, 839-846.

SINGH, A.P. and SRIVASTAVA, L.M. (1973). The fine structure of pea stomata. Protoplasma 76, 61-82.

SINGH, M. and HILDEBRANDT, A.C. (1966). Movements of TMV inclusion bodies within tobacco callus cells. Virology 30, 314-342.

SINNOTT, E.W. (1960). "Plant Morphogenesis". McGraw Hill, N.Y.

SINYUKHIN, A.M. and GORCHAKOV, V.V. (1968). Role of the vascular bundles of the stem in long distance transmission of stimulation by means of bioelectric impulses. Fiziol Rast. 15, 400-407.

SKIERCZYŃSKA, J. (1968). Some of the electrical characteristics of the cell membrane of *Chara australis*. J. exp. Bot. 19, 389-406.

SLACK, C.R., HATCH, M.D. and GOODCHILD, D.J. (1969). Distribution of enzymes in mesophyll and parenchyma sheath chloroplasts of maize in relation to the C_4 dicarboxylic acid pathway of photosynthesis. Biochem. J. 114, 489-498.

SMITH, A.I. (1939). The comparative histology of some of the Laminariales. Am. J. Bot. 26, 571-585.

SMITH, D.C. (1974). Transport from symbiotic algae and symbiotic chloroplasts to host cells. Symp. Soc. exp. Biol. 28, 485-520.

SMITH, D.L. (1972). Staining and osmotic properties of young gametophytes of *Polypodium vulgare* L. and their bearing on rhizoid formation. Protoplasma 74, 465-479.

SMITH, F.A. (1966). Active phosphate uptake by *Nitella translucens*. Biochim. biophys. Acta 126, 94-99.

SMITH, F.A. (1971). Transport of solutes during C_4 photosynthesis: assessment, *in* "Photosynthesis and Photorespiration" (Eds. M.D. Hatch, C.B. Osmond and R.O. Slatyer), p. 502-306. New York, Wiley-Interscience.

SMITH, G.M. (1944). A comparative study of the species of *Volvox*. Trans. Am. microsc. Soc. 63, 265-310.

SMITH, K.M. and MacCLEMENT (1941). Further studies on the ultrafiltration of plant viruses. Parasitology 33, 320-330.

SOVONICK, S.A., GEIGER, D.R. and FELLOWS, R.J. (1974). Evidence for active phloem loading in the minor veins of sugar beet. Pl. Physiol. 54, 886-891.

SPANSWICK, R.M. (1972). Electrical coupling between cells of higher plants: A direct demonstration of intercellular communication. Planta 102, 215-227.

SPANSWICK, R.M. (1974). Symplasmic transport in plants. Symp. Soc. exp. Biol. 28, 127-137.

SPANSWICK, R.M. (1975). Symplasmic transport in tissues, *in* "Encyclopaedia of Plant Physiology", New Series, Vol. 2. Transport in Plants (Eds. U. Lüttge and M.G. Pitman). Springer-Verlag, Berlin.

SPANSWICK, R. and COSTERTON, J. (1967). Plasmodesmata in *Nitella translucens*: Structure and electrical resistance. J. Cell Sci. 2, 451-464.

SPEARING, J.K. (1937). Cytological studies on the Myxophyceae. Arch. Protistenk. 89, 209-278.

SPENCER, D.F. and KIMMINS, W.C. (1969). Presence of plasmodesmata in callus cultures of tobacco and carrot. Can. J. Bot. 47, 2049-2050.

SPENCER, D.F. and KIMMINS, W.C. (1971). Ultrastructure of tobacco mosaic virus lesions and surrounding tissue in *Phaseolus vulgaris* Pinto. Can. J. Bot. 49, 417-421.

SPITZER, N.C. (1970). Low resistance connexions between cells in the developing anther of the lily. J. Cell Biol. 45, 565-575.

SPRENT, I. (1972). The effects of water stress on nitrogen-fixing nodules: II. Effects of the fine structure of detached soya bean nodules. New Phytol. 71(3), 443-450.

SRIVASTAVA, L.M. and SINGH, A.P. (1972a). Certain aspects of xylem differentiation in corn. Can. J. Bot. 50, 1795-1804.

SRIVASTAVA, L.M. and SINGH, A.P. (1972b). Stomatal structure in corn leaves. J. Ultrastruct. Res. 39, 345-363.

STADELMANN, E. (1956). Plasmolyse und Deplasmolyse, *in* "Encyclopedia of Plant Physiology" Vol. 2, 71-115 (Eds. H.J. Bogen and H. Ullrich), Springer-Verlag, Berlin.

STAMBOLTSYAN, E. Yu (1972). Symplastic connections of the endodermis in the germ root of tomato and wheat. Bot. Zh. 57, 607-614.

STANGE, Luise (1957). Untersuchungen über Umstimmungs - und Differenzierungsvorgänge in regenerierende Zellen des Lebermooses, *Riella*. Z. Bot. 45, 197-244.

STARR, R.C. (1970). Control of differentiation in *Volvox*. Devl. Biol. Supplement 4, 59-100.

STEELE, S.D. and FRASER, T.W. (1973). The ultrastructure of *Geotrichum candidum* hyphae. Can. J. Bot. 19, 1507-1512.

STEIN, J.R. (1965). On cytoplasmic strands in *Gonium pectorale* (Volvocales). J. Phycol. 1, 1-5.

STEINBISS, H.H. and SCHMITZ, K. (1973). CO_2 Fixierung und Stofftransport in benthischen marinen Algen. V. Zur autoradiographischen Lokalisation der Assimilattransportbahnen im Thallus von *Laminaria hyperborea*. Planta 112, 253-263.

STELZER, R., LÄUCHLI, A. and KRAMER, D. (1975). Pathways of intercellular chloride transport in roots of intact barley seedlings. Cytobiologie 10, 449-457.

STEVENINCK, R.F.M. van (1975). Cell differentiation, aging and ion transport, *in* "Encyclopedia of Plant Physiology", New Series (Eds. U. Lüttge and M.G. Pitman). Springer-Verlag (in press).

STEVENINCK, R.F.M. van, ARMSTRONG, W.D., PETERS, P.D. and HALL, T.A. (in press b). Ultrastructural localisation of ions. III. Distribution of chloride in mesophyll cells of mangrove (*Aegiceras corniculatum*, Blanco). Aust. J. Pl. Physiol.

STEVENINCK, R.F.M. van, BALLMENT, B., PETERS, P.D. and HALL, T.A. (in press a). Ultrastructural localisation of ions. II. X-ray analytical verification of silver precipitation products and distribution of chloride in mesophyll cells of barley seedlings. Aust. J. Pl. Physiol.

STEVENINCK, R.F.M. van and CHENOWETH, A.R.F. (1972). Ultrastructural localisation of ions. I. Effect of high external sodium chloride concentration on the apparent distribution of chloride in leaf parenchyma cells of barley seedlings. Aust. J. biol. Sci. 25, 499-516.

STEVENINCK, R.F.M. van, STEVENINCK, M.E. van, HALL, T.A. and PETERS, P.D. (1974a). X-ray microanalysis and distribution of halides in *Nitella translucens*, *in* "Electron microscopy 1974" (Eds. J.V. Sanders and D.J. Goodchild). The Australian Academy of Science, Canberra 1974.

STEVENINCK, R.F.M. van, STEVENINCK, M.E. van, HALL, T.A. and PETERS, P.D. (1974b). A chlorine free medium for use in X-ray analytical electron microscope localisation of chloride in biological tissues. Histochemie 38, 173-180.

STEVENINCK, M.E. van, STEVENINCK, R.F.M. van, MITTELHEUSER, C.J., PETERS, P.D. and HALL, T.A. (in preparation b). Ultrastructural localisation of ions. V. An electron microscope investigation of the cobaltinitrite method for potassium localisation.

STEVENINCK, M.E. van, STEVENINCK, R.F.M. van, PETERS, P.D. and HALL, T.A. (in preparation a). Ultrastructural localisation of ions. VI. Detection of iron and phosphate in the vacuoles of barley root tip cells by means of the antimony precipitation method.

STEVENSON, D.W. (1974). Ultrastructure of the nacreous leptoids (sieve elements) in the polytrichaceous moss *Atrichum undulatum*. Am. J. Bot. 61, 414-421.

STEWARD, F.C., MAPES, M.O., KENT, A.E. and HOLSTEN, R.D. (1964). Growth and development of cultured plant cells. Science, N.Y. 143, 20-27.

STEWART, K.D., MATTOX, K.R. and CHANDLER, C.D. (1974). Mitosis and cytokinesis in *Platymonas subcardiformis*; a scaly green monad. J. Phycol. 10, 65-79.

STEWART, K.D., MATTOX, K.R. and FLOYD, G.L. (1973). Mitosis, cytokinesis, the distribution of plasmodesmata, and other cytological characteristics in the Ulotrichales, Ulvales and Chaetophorales: Phylogenetic and taxonomic considerations. J. Phycol. 9, 128-141.

STEWART, W.D.P., HAYSTEAD, A. and PEARSON, H.W. (1969). Nitrogenase activity in heterocysts of blue-green algae. Nature, Lond. 224, 226-228.

STIGTER, H.C.M. de (1966). Parallelism between the transport of ^{14}C-photosynthates and the flowering response in grafted *Silene armeria* L. Z. Pflphysiol. 55, 11-19.

STOREY, H.H. (1928). Transmission studies of maize streak disease. Ann. appl. Biol. 15, 1-25.

STRASBURGER, E. (1882). "Über den Bau und das Wachstum der Zellhäute". Jena.

STRASBURGER, E. (1901). Ueber Plasmaverbindungen pflanzlicher Zellen. Jb. wiss. Bot. 36, 493-610.

STRASBURGER, E. (1923). Das botanische Praktikum (bearbeitet von M. Koernicke). Jena.

STREET, H.E. (Ed.) (1973). "Plant tissue and cell culture". Bot. Monographs No. 11, Blackwells, Oxford.

STRETTON, A.O.W. and KRAVITZ, E.A. (1968). Neuronal geometry: determination with a technique of intracellular dye injection. Science, N.Y. 162, 132-134.

STRUGGER, S. (1957a). Der elektronenmikroskopische Nachweis von Plasmodesmen mit Hilfe der Uranylimprägnierung an Wurzelmeristemen. Protoplasma 48, 231-236.

STRUGGER, S. (1957b). Elektronenmikroskopische Beobachtungen an den Plasmodesmen des Urmeristems der Wurzelspitze von *Allium cepa*. Protoplasma 48, 365-367.

STUBBE, W. (1958). Dreifarbenpanaschierung bei Oenotheren. II. Z. Vererb-Lehre 89, 189-203.

STURCH, H.H. (1924). On the life history of *Harveyella pachyderma* and *H. mirabilis*. Ann. Bot. 38, 27-42.

SUBAK-SHARPE, H. (1969). Metabolic cooperation between cells, *in* "Homeostatic Regulation", p. 276-288, Ciba Fdn. Symp.

SUBAK-SHARPE, H., BUCK, P. and PITTS, T.D. (1969). Metabolic co-operation between biochemically marked cells. J. Cell Sci. 4, 353-368.

SUKHORUKOV, K.T. and PLOTNIKOVA, Y.M. (1965). O strukture i funktsiyakh plazmodesm i ektodesm. Usp. Sovrem. Biol. 60, 299-315.

SUNDERLAND, N. and WICKS, F.M. (1971). Embryoid formation in pollen grains of *Nicotiana tabacum*. J. exp. Bot. 22, 213-226.

SUSSEX, I.M. and CLUTTER, M.E. (1968). Differentiation in tissues, free cells and reaggregated plant cells. In Vitro 3, 3-12.

SUTCLIFFE, J.F. (1962). "Mineral salts absorption in plants". Pergamon, Oxford.

TAINTER, F.H. (1971). The ultrastructure of *Arceuthobium pusillum*. Can. J. Bot. 49, 1615-1622.

TAIZ, L. and JONES, R.L. (1970). Gibberellic acid, β-1, 3-glucanase and the cell walls of barley aleurone layers. Planta 92, 73-84.

TAIZ, L. and JONES, R.L. (1973). Plasmodesmata and an associated cell wall component in barley aleurone tissue. Am. J. Bot. 60, 67-75.

TAKADA, H., YAGI, T. and HIRAOKA, J. (1965). Elektronenoptische Untersuchungen "*Endomycopsis fibuliger*" auf festen Nährboden. Protoplasma 59, 494-505.

TAKEBE, I., LABIB, G. and MELCHERS, G. (1971). Regeneration of whole plants from isolated mesophyll protoplasts of tobacco. Naturwissenschaften 58, 318-320.

TANGL, E. (1879). Ueber offene Communicationen zwischen den Zellen des Endosperms einiger Samen. Jb. wiss. Bot. 12, 170-190.

TANTON, T.W. and CROWDY, S.H. (1970). Water pathways in higher plants. I. Free space in wheat leaves. J. exp. Bot. 21, 102-111.

TANTON, T.W. and CROWDY, S.H. (1972a). Water pathways in higher plants. II. Water pathways in roots. J. exp. Bot. 23, 600-618.

TANTON, T.W. and CROWDY, S.H. (1972b). Water pathways in higher plants. III. The transpiration stream within leaves. J. exp. Bot. 23, 619-625.

TAYLOR, G. (1953). Dispersion of soluble matter in solvent flowing slowly through a tube. Proc. R. Soc. A. 219, 186-203.

THAINE, R. (1965). The preservation of plant cells, particularly sieve tubes, by vacuum freeze-drying. J. exp. Bot. 16, 192-196.

THODAY, M.G. (1911). On the histological relation between *Cuscuta* and its host. Ann. Bot. 25, 655-681.

THOMAS, E., KONAR, R.N. and STREET, H.E. (1972). The fine structure of the embryogenic callus of *Ranunculus sceleratus* L. J. Cell Sci. 11, 95-109.

THOMAS, P.E. and FULTON, R.W. (1968). Correlation of ectodesmata number with non specific resistance to initial virus infection. Virology 34, 459-469.

THOMPSON, E. and COLVIN, J.R. (1970). Electron cytochemical localisation of cystine in plant cell walls. J. Microscopy 91, 87-98.

THOMSON, W.W. and JOURNETT, R. de (1970). Studies on the ultrastructure of the guard cells of *Opuntia*. Am. J. Bot. 57, 309-316.

THOMSON, W.W. and LIU, L.L. (1967). Ultrastructural features of the salt gland of *Tamarix aphylla*. Planta 73, 201-220.

THORNE, J.H. and KOLLER, H.R. (1974). Influence of assimilate demand on photosynthesis, diffusive resistances, translocation, and carbohydrate levels of soybean leaves. Pl. Physiol. 54, 201-207.

TOBLER, F. (1902). Zerfall und Reproduktionsvermögen des Thallus einer Rhodomelacee. Ber. dt. bot. Ges. 20, 351-365.

TOBLER, F. (1903). Über Eigenwachstum der Zelle und Pflanzenform. Jb. wiss. Bot. 39, 527-580.

TOBLER, F. (1906). Über Regeneration und Polarität. Jb. wiss. Bot. 42, 461-502.

TOWNSEND, C.O. (1897). Der Einfluss des Zellkernes auf die Bildung der Zellhaut. Jb. wiss. Bot. 30, 484-510.

TRAPP, G. (1933). A study of the foliar endodermis in the Plantaginaceae. Trans. Roy. Soc. Edinb. 57(II), 523-546.

TRINCI, A.P.J. and COLLINGE, A.J. (1973). Structure and plugging of septa of wild type and spreading colonial mutants of *Neurospora*. Arch. Mikrobiol. 91, 355-364.

TRIPODI, G. (1971). Some observations on the ultrastructure of the red alga *Pterocladia capillacea* (Gmel.) Bon.et Thur. J. Submicr. Cytol. 3, 63-70.

TRÖNDLE, A. (1913). Eine neue Methode zur Darstellung der Plasmodesmen. Verh. Schweiz. Naturf. Ges. 44, 213-217.

TUCKER, S. (1974). Dedifferentiated guard cells in Magnoliaceous leaves. Science, N.Y. 185, 445-447.

TURGEON, R., WEBB, J.A. and EVERT, R.F. (1975). Ultrastructure of minor veins in *Cucurbito pepo* leaves. Protoplasma 83, 217-232.

TURRELL, F.M. (1936). The area of the internal exposed surface of dicotyledon leaves. Am. J. Bot. 23, 255-264.

TYREE, M.T. (1970). The symplast concept. A general theory of symplastic transport according to the thermodynamics of irreversible processes. J. theor. Biol. 26, 181-214.

TYREE, M.T. and TAMMES, P.M.L. (1975). Translocation of uranin in the symplasm of staminal hairs of *Tradescantia*. Can. J. Bot. (in press).

TYREE, M.T., FISCHER, R.A. and DAINTY, J. (1974). A quantitative investigation of symplasmic transport in *Chara corallina*. II. The symplasmic transport of chloride. Can. J. Bot. 52, 1325-1334.

UMRATH, K. (1959). Der Erregungsvorgang, *in* "Encyclopedia of Plant Physiology", Vol. 17, p. 24-110 (Ed. E. Bünning). Springer-Verlag, Berlin.

UNDERBRINK, A.G. and OLAH, L.V. (1968). Effect of digitonin on cellular division. Part III. Fine structural aspects of early phragmoplast development in the absence of an organised mitotic spindle. Cytologia 33, 155-164.

UPPAL, B.N. (1934). The movement of tobacco mosaic virus in leaves of *Nicotiana sylvestris*. Indian J. agric. Sci. 4, 865-873.

VAKHMISTROV, D.B. (1971). Possible ways and mechanisms of radial ion transport in plant roots. Agrokhimiya 9, 138-152.

VAKHMISTROV, D.B., KURKOVA, E.B. and SOLOVYEV, V.A. (1972). Characteristics of plasmodesmata and lomasome-like formations in barley roots in connection with transport of substances. Fiziol. Rast. 19, 951-960.

VALDOVINOS, J.G. and JENSEN, T.E. (1968). Fine structure of abscission zones. II. Cell wall changes in abscising pedicels of tobacco and tomato flowers. Planta 83 295-302.

VALDOVINOS, J.G., JENSEN, T.E. and SICKO, L.M. (1972). Fine structure of abscission zones. IV. Effect of ethylene on the ultrastructure of abscission of tobacco flower pedicels. Planta 102, 324-333.

VALDOVINOS, J.G., JENSEN, T.E. and SICKO, L.M. (1974). Abscission: cellular changes at the ultrastructural level, *in* "Plant growth substances, 1973", p. 1034-1041. Hirokawa, Tokyo.

VAN'T HOF, J. (1973). The regulation of cell division in higher plants. Brookhaven Symp. Biol. 25, 152-165.

VASIL, I.K. and ALDRICH, H.C. (1970). A histochemical and ultrastructural study of the ontogeny and differentiation of pollen in *Podocarpus macrophyllus* D. Don. Protoplasma 71, 1-37.

VASIL'YEV, I.M. (1971). "Wintering of plants" (Eng. transl. of Zymovka Rastenii, Moscow, 1956). Am. Inst. Biol. Sci., Washington.

VAZART, B. (1963). Différenciation des cellules sexuelles et fécondation chez les cryptogames. Protoplasmatologia VII 36, 1-363, Springer-Verlag, Wien.

VAZART, B. (1969). Structure et évolution de la cellule génératrice du Lin, *Linum usitatissimum* L. au cours des premiers stades de la maturation du pollen. Rev. Cytol. et Biol. vég 32, 101-114.

VAZART, J. (1970). Aspects infrastructuraux de la reproduction sexuée chez le Lin. Derniers stades de la différenciation du pollen. Structure inframicroscopique de la cellule génératrice et des gamètes. Rev. Cytol. et Biol. vég. 33, 289-310.

VAZART, M.B. (1971). Infrastructure de microspores de *Nicotiana tabacum* L. susceptibles de se developper en embryoides après excision et mise en culture des anthères. C.r.hebd.Séanc. Acad. Sci., Paris. Sér. D. 27, 549-552.

VEEN, B.W. (1970). Control of plant cell shape by cell wall structure. Proc. K. ned. Akad. Wet. 73, 118-121.

VERNIORI, A., DU BOIS, R., DEGOODT, P., GASSEE, J.P. and LAMBERT, P.P. (1973). Measurement of the permeability of biological membranes. J. gen. Physiol. 62, 489-507.

VIAN, B. (1970). Observations sur l'évolution des substances intercellulaires pendant la spermatogenèse chez un hépatique, *Fossombronia angulosa*. C.r. hebd.Séanc. Acad. Sci., Paris.Sér. D. 270, 2140-1243.

VIAN, B. and ROLAND, J.-C. (1972). Différenciation des cytomembranes et renouvellement du plasmalemme dans les phénomènes de sécrétions végétales. J. Microscopie 13, 119-136.

VIAN, B. and ROUGIER, M. (1974). Ultrastructure des plasmodesmes après cryoultramicrotomie. J. Microscopie 20, 307-312.

VIGIL, E. and RUDDAT, M. (1973). Effect of gibberellic acid and actinomycin D on the formation and dsitribution of the rough endoplasmic reticulum in barley aleurone cells. Pl. Physiol. 51, 549-558.

VÖCHTING, H. (1892). "Über Transplantation am Pflanzenkörper". Tübingen.

VRIES, H. de (1885). Plasmolytische Untersuchungen über die Wand der Vakuolen. Abhandl. sächs. Akad. Wiss., Math-nat. Kl. 16, 185.

WAALAND, S.D. and CLELAND, R.E. (1974). Cell repair through cell fusion in the red alga, *Griffithsia pacifica*. Protoplasma 79, 185-196.

WALKER, N.A. (1957). Ion permeability of the plasmalemma of the plant cell. Nature, Lond. 180, 94-95.

WALKER, N.A. (1974). Chloride transport to the charophyte vacuole, *in* "Membrane transport in plants" (Eds. U. Zimmerman and J. Dainty), p. 173-179, Springer-Verlag, Berlin.

WALKER, N.A. and BOSTROM, T.E. (1973). Intercellular movement of chloride in *Chara* - a test of models for chloride influx, *in* "Ion Transport in Plants" (Ed. W.P. Anderson), p. 447-458, Academic Press, London.

WALKER, N.A. and SMITH, F.A. (1975). Intracellular pH in *Chara corallina* measured by DMO distribution. Pl. Sci. Letters 4, 125-132.

WALKEY, D.G.A. and WEBB, M.J.W. (1968). Virus in plant apical meristems. J. gen. Virol. 3, 311-313.

WALLES, B. (1967). Use of biochemical mutants in analyses of chloroplast morphogenesis, *in* "Biochemistry of Chloroplasts" (Ed. T.W. Goodwin), p. 633-653, Vol. 2, Academic Press, London.

WALTON, A.G. (1967). "The formation and properties of precipitates". Interscience Publishers, John Wiley and Sons, New York.

WARDLAW, I.F. (1974). Phloem transport: Physical, chemical or impossible. A. Rev. Pl. Physiol. 25, 515-539.

WARDROP, A.B. (1955). The mechanism of surface growth in parenchyma of *Avena* coleoptiles. Aust. J. Bot. 3, 137-148.

WARDROP, A.B. (1956). The nature of surface growth in plant cells. Aust. J. Bot. 4, 193-199.

WARDDROP, A.B. (1965). Cellular differentiation in xylem, *in* "Cellular ultrastructure of woody plants (Ed. W.A. Côté), p. 61-97, Syracuse University Press, Syracuse.

WARK, M.C. (1965). Fine structure of the phloem of *Pisum sativum*. 2. The companion cell and phloem parenchyma. Aust. J. Bot. 13, 185-193.

WARMBRODT, R.D. and EVERT, R.F. (1974a). Structure and development of the sieve element in the stem of *Lycopodium lucidulum*. Am. J. Bot. 61, 267-277.

WARMBRODT, R.D. and EVERT, R.F. (1974b). Structure of the vascular parenchyma in the stem of *Lycopodium lucidulum*. Am. J. Bot. 61, 437-443.

WEAST, R.C. (Ed.) (1963). "Handbook of chemistry and physics". 49th Edition. Cleveland, Chemical Rubber Company 1963.

WEBER, F. (1925). Plasmolyseform und Kernform funktionierende Schliesszellen. Jb. wiss. Bot. 64, 687.

WEBER, F., KENDA, G. and THALER, I. (1953). Schliesszellen-Chloroplasten vergilben nicht. Protoplasma 42, 246-249.

WEBSTER, B.D. (1968). Anatomical aspects of abscission. Pl. Physiol. 43, 1512-1544.

WEBSTER, B.D. (1973). Ultrastructural studies of abscission in *Phaseolus*. Ethylene effects on cell walls. Am. J. Bot. 60, 436-447.

WEBSTER, D.H. and CURRIER, H.B. (1968). Heat-induced callose and lateral movement of assimilates from phloem. Can. J. Bot. 46, 1215-1220.

WEIDE, A. (1938). Uber die Regenerationsleistungen der *Callithamnion*. Arch. Protistenk. 91, 209-221.

WEILING, F. (1965). Zur Feinstruktur der Plasmodesmen und Plasmakanäle bei Pollenmutterzellen. Planta 64, 97-118.

WEINTRAUB, M. and RAGETLI, H.W.J. (1961). Cell wall composition of leaves with localized virus infection. Phytopathology 51, 215-219.

WEINTRAUB, M. and RAGETLI, H.W.J. (1964). An electron microscope study of tobacco mosaic virus lesions in *Nicotiana glutinosa*. J. Cell Biol. 23, 499-509.

WEINTRAUB, M., RAGETLI, H.W.J. and LO, E. (1974). Potato virus Y particles in plasmodesmata of tobacco leaf cells. J. Ultrastruct. Res. 46, 131-148.

WENT, J.L. van and TAMMES, P.M.L. (1972). Experimental fluid flow through plasmodesmata of *Laminaria digitata*. Acta bot. neerl. 21(4), 321-326.

WENT, J.L. van and TAMMES, P.M.L. (1973). Trumpet filaments in *Laminaria digitata* as an artefact. Acta bot. neerl. 22, 112-119.

WENT, J.L. van, AELST, A.C. van and TAMMES, P.M.L. (1973a). Transverse connections between cortex and translocating medulla in *Laminaria digitata*. Acta bot. neerl. 22, 77-78.

WENT, J.L. van, AELST, A.C. van and TAMMES, P.M.L. (1973b). Open plasmodesmata in sieve plates of *Laminaria digitata*. Acta bot. neerl. 22, 120-123.

WENT, J.L. van, AELST, A.C. van and TAMMES, P.M.L. (1975). Anatomy of staminal hairs from *Tradescantia* as a background for translocation studies. Acta bot. neerl. 24, 1-6.

WERKER, E. and VAUGHAN, J.G. (1974). Anatomical and ultrastructural changes in aleurone and myrosin cells of *Sinapis alba* during germination. Planta 116, 243-255.

WESTON, G.D. and CASS, D.D. (1973). Observations on the development of the paraveinal mesophyll of soybean leaves. Bot. Gaz. 134, 232-235.

WETMORE, C.M. (1973). Multiperforate septa in lichens. New Phytol. 72, 535-538.

WHALEY, W.G. and MOLLENHAUER, H.H. (1963). The Golgi-apparatus and cell plate formation. J. Cell Biol. 17, 216-221.

WHALEY, W.G., MOLLENHAUER, H.H. and LEECH, J.H. (1960). The ultrastructure of the meristematic cell. Am. J. Bot. 47, 401-449.

WHEELER, H. and HANCHEY, P. (1968). Permeability phenomena in plant disease. A. Rev. Phytopath. 6, 331-350.

WHELAN, E.D.P. (1974). Discontinuities in the callose wall, intermeiocyte connections, and cytomixis in angiosperm meiocytes. Can. J. Bot. 52, 1219-1224.

WHELAN, E.D.P., HAGGIS, G.H. and FORD, E.J. (1974). Scanning electron microscopy of the callose wall and intermeiocyte connection in angiosperms. Can. J. Bot. 52, 1215-1218.

WICHARD, W. and KOMNICK, H. (1971). Electronmicroscopical and histochemical evidence of chloride cells in tracheal gills of mayfly larvae. Cytobiologie 3, 215-228.

WILBUR, F.H. and RIOPEL, J.L. (1971a). The role of cell interaction in the growth and differentiation of *Pelargonium hortorum* cells *in vitro*. I. Cell interaction and growth. Bot. Gaz. 132 (3), 183-193.

WILBUR, F.H. and RIOPEL, J.L. (1971b). The role of cell interaction in the growth and differentiation of *Pelargonium hortorum* cells *in vitro*. II. Cell interaction and differentiation. Bot. Gaz. 132(3), 193-202.

WILCOX, M., MITCHISON, G.J. and SMITH, R.J. (1973). Pattern formation in the blue-green alga, *Anabaena*. J. Cell Sci. 12, 707-723.

WILDON, D.C. and MERCER, F.V. (1963a). The ultrastructure of the vegetative cell of blue-green algae. Aust. J. biol. Sci. 16, 585-596.

WILDON, D.C. and MERCER, F.V. (1963b). The ultrastructure of the heterocyst and akinete of the blue-green algae. Arch. Mikrobiol. 47, 19-31.

WILDY, P. (1971). "Classification and nomenclature of Viruses" Monographs in Virology (Ed. J.L. Melnick), Vol. 5, p. 81. Basel: Karger.

WILKINS, H., LARQUÉ-SAAVEDRA, A. and WAIN, R.L. (1974). Studies on factors influencing root growth, *in* "Plant growth substances, 1973", p. 1231-1234. Hirokawa, Tokyo.

WILKINS, H. and WAIN, R.L. (1975). The role of the root cap in the response of the primary roots of *Zea mays* L. seedlings to white light and gravity. Planta 123, 217-222.

WILLIAMS, M.E. and MOZINGO, H.N. (1971). The fine structure of the trigger hair in Venus's flytrap. Am. J. Bot. 58, 532-539.

WILLIAMS, S.E. and PICKARD, B.G. (1972a). Receptor potentials and action potentials in *Drosera* tentacles. Planta 103, 193-221.

WILLIAMS, S.E. and PICKARD, B.G. (1972b). Properties of action potentials in *Drosera* tentacles. Planta 103, 222-240.

WILLIAMS, S.E. and PICKARD, B.G. (1974). Connections and barriers between cells of *Drosera* tentacles in relation to their electrophysiology. Planta 116, 1-16.

WILLIAMS, S.E. and SPANSWICK, R.M. (1972). Intracellular recordings of the action potentials which mediate the thigmonastic movements of *Drosera*. Pl. Phys. Suppl. 50, 64.

WILLIAMSON, R.E. (1975). Cytoplasmic streaming in *Chara*: A cell model activated by ATP and inhibited by cytochalasin B. J. Cell Sci. 17, 655-668.

WILSENACH, R. and KESSEL, M. (1965). Micropores in the cross-wall of *Geotrichum candidum*. Nature, Lond. 207, 545-546.

WILSON, E.B. (1928). "The Cell in Development and Heredity". Macmillan and Co., N.Y.

WILSON, H.J., ISRAEL, H.W. and STEWARD, F.C. (1974). Morphogenesis and fine structure of cultured carrot cells. J. Cell Sci. 15, 57-73.

WINKLER, H. (1910). Über die Nachkommenschaft der *Solanum* - Propfbastarde und die Chromosomenzahlen ihrer Keimzellen. Z. Bot. 2, 1-38.

WITHERS, L.A. and COCKING, E.C. (1972). Fine-structural studies on spontaneous and induced fusion of higher plant protoplasts. J. Cell Sci. 11, 59-75.

WITHROW, Alice P. and WITHROW, R.B. (1943). Translocation of the floral stimulus in *Xanthium*. Bot. Gaz. 104, 409-416.

WOESLER, A. (1934). Beitrag zur Kenntnis der vegetative Vermehrung von *Sphagnum cymbifolium* Ehrh. Beitr. Biol. Pfl. 22, 13-24.

WOLK, C.P. (1968). Movement of carbon from vegetative cells to heterocysts in *Anabaena cylindrica*. J. Bact. 96, 2138-2143.

WOLK, C.P. (1973). Physiology and cytological chemistry of blue-green algae. Bact. Rev. 37, 32-101.

WOLK, C.P. and WOJCIUCH, E. (1971). Photoreduction of acetylene by heterocysts. Planta 97, 126-134.

WOLPERT, L. (1969). Positional information and the spatial pattern of cellular differentiation. J. theor. Biol. 25, 1-47.

WOLPERT, L. (1971). Positional information and pattern formation. Current topics in Developmental Biology 6, 183-224.

WOLSWINKEL, P. (1974a). Complete inhibition of setting and growth of fruits of *Vicia faba* L. resulting from the draining of the phloem system by *Cuscuta* species. Acta. bot. neerl. 23, 48-60.

WOLSWINKEL, P. (1974b). Enhanced rate of ^{14}C-solute release to the free space by the phloem of *Vicia faba* parasitised by *Cuscuta*. Acta bot. neerl. 23, 177-188.

WOODCOCK, C.L.F. and BELL, P.R. (1968). Features of the ultrastructure of the female gametophyte of *Myosurus minimus*. J. Ultrastruct. Res. 22, 546-563.

WOODING, F.B.P. (1968). Fine structure of callus phloem in *Pinus pinea*. Planta 83, 99-110.

WOODING, F.B.P. (1974). Development and fine structure of angiosperm and gymnosperm sieve tubes. Symp. Soc. exp. Biol. 28, 27-41.

WOODING, F.B.P. and NORTHCOTE, D.H. (1965). The fine structure and development of the companion cell of the phloem of *Acer pseudoplatanus*. J. Cell Biol. 24, 117-128.

WORLEY, J.F. (1968). Rotational streaming in fibre cells and its role in translocation. Pl. Physiol. 43(10), 1648-1655.

WOZNY, A. and MLODZIANOWSKI, F. (1973). Views of the ultrastructure of plasmodesmata. Wiad. Bot. 17, 93-99.

WU, J.H. and DIMITMAN, J.E. (1970). Leaf structure and callose formation as determinants of TMV movement in bean leaves as revealed by UV irradiation studies. Virology 40, 820-827.

WYLIE, R.B. (1939). Relations between tissue organization and vein distribution in dicotyledon leaves. Am. J. Bot. 26, 219-225.

WYLIE, R.B. (1943). The role of the epidermis in foliar organization and its relations to the minor venation. Am. J. Bot. 30, 273-280.

WYLIE, R.B. (1952). The bundle sheath extension in leaves of dicotyledons. Am. J. Bot. 39, 645-651.

YOSHIDA, Y. (1961). Role of nucleus in cytoplasmic activities with special reference to the formation of surface membrane in *Elodea* leaf. Pl. Cell Physiol. 2, 139-150.

YOUNG, T.W.K. (1969). Ultrastructure of aerial hyphae in *Linderina pennispora*. Ann. Bot. 33, 211-216.

ZEE, S.-Y. (1969). The fine structure of differentiating sieve elements of *Vicia faba*. Aust. J. Bot. 17, 441-456.

ZEE, S.-Y. and CHAMBERS, T.L. (1968). Fine structure of the primary root phloem of *Pisum*. Aust. J. Bot. 16, 37-47.

ZEE, S.-Y. and O'BRIEN, T.P. (1970). Studies on the ontogeny of the pigment strand in the caryopsis of wheat. Aust. J. biol. Sci. 23, 1153-1171.

ZEEVAART, J.A.D. (1958). Flower formation as studied by grafting. Med. Landbouwhogeschool Wageningen 58(3), 1-88.

ZEEVAART, J.A.D. (1966). Reduction of the gibberellin content of *Pharbitis nil* by CCC and after-effects in the progeny. Pl. Physiol. 41, 856-862.

ZEPF, E. (1952). Über die Differenzierung des Sphagnumblattes. Z. Bot. 40, 87-118.

ZIEGLER, H. (1963). Untersuchungen über die Feinstruktur des Phloems. II. Die Siebplatten bei der Braunalge *Macrocystis pyrifera* (L) Ag. Protoplasma 57, 786-799.

ZIEGLER, H. (1964). Storage, mobilization and distribution of reserve material in trees, *in* "The formation of wood in forest trees" (Ed. M.H. Zimmerman), p. 303-320. Academic Press, London and New York.

ZIEGLER, H. (1968). La sécrétion du nectar, *in* "Traité de biologie de l'abeille" Vol. 3 (Ed. R. Chauvin), Masson et Cie, Paris.

ZIEGLER, H. (1974). What do we know about the function of plasmodesmata in transcellular transport?, *in* "Membrane transport in plants" (Eds. U. Zimmermann and J. Dainty), p. 372-380. Springer-Verlag.

ZIEGLER, H. and LÜTTGE, U. (1966). Die Salzdrüsen von *Limonium vulgare*. I. Mitt. Die Feinstruktur. Planta 70, 193-206.

ZIEGLER, H. and LÜTTGE, U. (1967). Die Salzdrüsen von *Limonium vulgare*. II. Mitt. Die Lokalisierung des Chlorids. Planta 74, 1-17.

ZIEGLER, H. and RUCK, I. (1967). Untersuchungen über die Feinstruktur des Phloems. III. Die 'Trompetenzellen' von *Laminaria*-Arten. Planta 73, 62-73.

ZIEGLER, H., SHMUELI, E. and LANGE, G. (1974). Structure and function of the stomata of *Zea mays*. 1. Development. Cytobiologie 9, 162-168.

ZINSMEISTER, D.D. and CAROTHERS, Z.B. (1974). The fine structure of oogenesis in *Marchantia polymorpha*. Am. J. Bot. 61, 499-512.

ZOETEN, G.A. de and GAARD, G. (1969). Possibilities for inter- and intracellular translocation of some icosahedral plant viruses. J. Cell Biol. 40, 814-823.

Author Index

Subject Index

Page numbers in small type refer to brief references in the text; page numbers in large type refer to the beginnings of extended text sections on the particular topic; page numbers in italics refer to Figures or Tables on the page in question.

Related Titles

Encyclopedia of Plant Physiology, New Series
Editors: A. Pirson, M. H. Zimmermann

Vol. 1: **Transport in Plants I**

Phloem Transport
Edited by M. H. Zimmermann, J. A. Milburn
With contributions by numerous experts

Vol 2: (in 2 parts)

Transport in Plants II
Edited by U. Lüttge, M. G. Pitman
With a Foreword by R. A. Robertson
With contributions by numerous experts
Part A: **Cells**
Part B: **Tissues and Organs**

Membrane Transport in Plants
Edited by U. Zimmermann, J. Dainty

D. Hess

Plant Physiology
Molecular, Biochemical, and Biophysical Fundamentals of Metabolism and Development
Springer Study Edition

Structure and Function of Chloroplasts
Edited by M. Gibbs

Planta

An International Journal of Plant Biology

PLANTA publishes original articles on all branches of botany with the exception of taxonomy and floristics. Papers on cytology, genetics, and related fields are included providing they shed light on general botanical problems.

Oecologia

In cooperation with the International Association for Ecology (Intecol)

OECOLOGIA reflects the dynamically growing interest in ecology. Emphasis is placed on the functional interrelationship of organisms and environment rather than on morphological adaptation. The journal publishes original articles, short comunications, and symposium reports on all aspects of modern ecology, with particular reference to physiological and experimental ecology, population ecology, organic production, and mathematical models.

Sample copies as well as subscription and back-volume information available upon request

Please address:
Springer-Verlag
Werbeabteilung 4021
D 1000 Berlin 33
Heidelberger Platz 3
or
Springer-Verlag New York Inc.
New York Inc.
Promotion Department
175 Fifth Avenue
New York, NY 10010